Advanced Signal Analysis with Applications

Advanced Signal Analysis with Applications

HARISH PARTHASARATHY

Division of Electronics and Communication
Netaji Subhash Institute of Technology
New Delhi

I.K. International Publishing House Pvt. Ltd.

NEW DELHI • BANGALORE

Published by

I.K. International Publishing House Pvt. Ltd.
S-25, Green Park Extension
Uphaar Cinema Market
New Delhi – 110 016 (India)
E-mail : info@ikinternational.com
Website: www.ikbooks.com

ISBN 978-93-81141-09-0

10 9 8 7 6 5 4 3 2 1

Published by Krishan Makhijani for I.K. International Publishing House Pvt. Ltd., S-25, Green Park Extension, Uphaar Cinema Market, New Delhi – 110 016 and Printed by Rekha Printers Pvt. Ltd., Okhla Industrial Area, Phase II, New Delhi – 110 020.

The author is grateful to
Prof. Raj Senani for his constant
encouragement and support and providing
the lab facilities. He is also grateful to
Tarun Kumar Rawat for helping with the
typesetting work

Preface

The mathematical tools used in signal analysis involve differential and difference equations, integral equations, matrix algebra and calculus, complex analysis and probability theory and random processes. This book applies these tools to problems in various branches of physics like fluid dynamics, electromagnetism and quantum theory. The book will be of use to research workers in signal processing as well as to research workers in physics and applied mathematics. Partial differential equations have been introduced here as an additional tool in signal analysis since they are used to describe quantum, electromagnetic and fluid dynamical phenomena not to forget Einstein's equations of gravitation. The book will be of use to research students of applied mathematics and engineering who are in search for a new research problem or who wish to develop new techniques for solving an existing research problem.

Harish Parthasarathy

Contents

Chapter 1

Signal Theory

Introduction: This chapter is a collection of problems and projects related to diverse branches of signal theory including basic probability theory, stochastic processes and Brownian motion, linear and nonlinear systems, differential equations and wavelets in signal processing and parameter estimation.

[1] Explain how you would cast a second order Volterra system in data matrix form so that the least squares method can be applied. Assume that the Volterra system has the form

$$y(n) = \sum_{m=0}^{p} h(m)x(n-m) + \sum_{m,k=0}^{p} g(m,k)x(n-m)x(n-k)$$

The problem is to estimate the kernels $h(.), g(.,.)$ from input-output data. Let $e(m)$ denote the $p \times 1$ vector having a one in its mth position and zeroes at all the other positions. Then define the input data vectors and filter coefficient vectors

$$\mathbf{x}_{n,1} = \sum_{m=0}^{p} x(n-m)\mathbf{e}(m), \mathbf{h} = \sum_{m=0}^{p} h(m)\mathbf{e}(m),$$

$$\mathbf{g} = \sum_{m,k=0}^{p} g(m,k)\mathbf{e}(m) \otimes \mathbf{e}(k)$$

$$\mathbf{x}_{n,2} = \sum_{m,k=0}^{p} x(n-m)x(n-k)\mathbf{e}(m) \otimes \mathbf{e}(k)$$

Then we easily see that

$$\mathbf{x}_{n,1}^{T}\mathbf{h} = \sum_{m} h(m)x(n-m), \mathbf{x}_{n,2}^{T}\mathbf{g} = \sum_{m,k} g(m,k)x(n-m)x(n-k)$$

Note that $\mathbf{x}_{n,1}^{T}$ is a $1 \times p+1$ row vector while $\mathbf{x}_{n,2}$ is a $1 \times p^2$ row vector. Define a $N \times p+1$ input data matrix $\mathbf{X}_1$ whose n^{th} row is $\mathbf{x}_{1,n}$ where $n = p, p+1, ..., p+N-1$ and define another $N \times (p+1)^2$ input data matrix whose n^{th} row is $\mathbf{x}_{2,n}$ where $n = p, p+1, ..., p+N-1$. Then let $\mathbf{y}$ be the $N \times 1$ output data vector whose n^{th} entry is $y(n)$. Then we have

$$\mathbf{y} = \mathbf{X}_1 \mathbf{h} + \mathbf{X}_2 \mathbf{g}$$

In the presence of noise, these equations are approximate and we may estimate $\mathbf{h}, \mathbf{g}$ by minimizing

$$\| \mathbf{y} - \mathbf{X}_1 \mathbf{h} - \mathbf{X}_2 \mathbf{g} \|^2$$

over $\mathbf{h}, \mathbf{g}$. See Ref.[7].

[2]. Explain how you would estimate the parameters of continuous time signals in white Gaussian noise.

hint: The measured process is given by

$$x(t) = s(t, \theta) + w(t), t \in [0, T]$$

where w is white Gaussian noise. We choose an orthonormal basis $\phi_i(t), i = 1, 2, ...$ for $L^2[0, T]$ and represent the measured process as

$$x_i = s_i(\theta) + w_i, x_i = \int_0^T x(t)\phi_i(t)dt, w_i = \int_0^T w(t)\phi_i(t)dt$$

Then $\{w_i\}$ is a white Gaussian sequence:

$$\mathbb{E}(w_i w_j) = \sigma^2 \delta_{ij},$$

where

$$\sigma^2 \delta(t - s) = \mathbb{E}(w(t)w(s))$$

Then, the probability density of $\{x_i\}$ given the parameters θ is given by

$$p(\{x_i\}|\theta) = C.exp(-\frac{1}{2\sigma^2} \sum_{i=1}^{\infty} (x_i - s_i(\theta))^2)$$

$$= C.exp(-\frac{1}{2} \int_0^T (x(t) - s(t,\theta))^2 dt)$$

and hence the parameters θ may be estimated by minimizing $\int_0^T (x(t) - s(t,\theta))^2 dt$ with respect to θ.

[3]. Explain how you would use a least squares method for estimating both the time constants of a circuit having two time constants.

hint: Suppose τ_1, τ_2 are the two time constants of the circuit. Then the impulse response of the circuit has the form

$$x(t) = a_1.exp(-t/\tau_1) + a_2.exp(-t/\tau_2), t \geq 0$$

We sample this signal at discrete time instants $n\Delta$ and get

$$x[n] = a_1.exp(-\alpha_1 n) + a_2.exp(-\alpha_2 n), n = 0, 1, 2, ...$$

where

$$\alpha_1 = \Delta/\tau_1, \alpha_2 = \Delta/\tau_2$$

Define the vector

$$\mathbf{e}(\alpha) = [1, exp(-\alpha), exp(-2\alpha), ..., exp(-(N-1)\alpha)]^T$$

Also, define the data vector

$$\mathbf{x} = [x[0], x[1], ..., x[N-1]]^T$$

Then

$$\mathbf{x} = a_1 \mathbf{e}(\alpha_1) + a_2 \mathbf{e}(\alpha_2)$$

and we can estimate $a_i, \alpha_i, i = 1, 2$, by minimizing

$$E(a_1, a_2, \alpha_1, \alpha_2) = \parallel \mathbf{x} - a_1 \mathbf{e}(\alpha_1) - a_2 \mathbf{e}(\alpha_2) \parallel^2$$

with respect to $a_i, \alpha_i, i = 1, 2$. We define the $N \times 2$ matrix

$$\mathbf{E}(\alpha_1, \alpha_2) = [\mathbf{e}(\alpha_1), \mathbf{e}(\alpha_2)]$$

and the amplitude vector

$$\mathbf{a} = [a_1, a_2]^T$$

Then,

$$\mathbf{x} = \mathbf{E}(\alpha_1, \alpha_2)\mathbf{a}$$

In the presence of noise, this equation holds only approximately and the parameters $\alpha_i, i = 1, 2, \mathbf{a}$ are estimated by minimizing $\parallel \mathbf{x} - \mathbf{E}(\alpha_1, \alpha_2)\mathbf{a} \parallel^2$. First minimizing this over $\mathbf{a}$ gives

$$\hat{\mathbf{a}} = (\mathbf{E}^*\mathbf{E})^{-1}\mathbf{E}^*\mathbf{x}$$

Then α_i' can be estimated by maximizing

$$\mathbf{x}^*\mathbf{P}\mathbf{x}$$

where

$$\mathbf{P} = \mathbf{E}(\mathbf{E}^*\mathbf{E})^{-1}\mathbf{E}$$

the orthogonal projection operator on the range of $\mathbf{E}$. This equation it may be noted also arises in the estimation of multiple frequencies in noisy sinusoids.

[4]. This project forms a summary of a piece of research in wavelet based signal processing carried out by one of the author's colleagues Ref.[8] related to estimation of parameters of linear and nonlinear systems using wavelets. The first phase of this work deals with estimating the parameters of linear differential equations using wavelets. We consider a second order linear differential equation driven by noise and work on estimating the coefficients of this differential equation. We observe that such an estimation could on the one hand be carried out using the least squares method. This method would

however involve all the samples of the output signal over the given observation range. Thus a lot of data storage would be needed here. On the other hand, we could choose an orthonormal basis for the signal space and consider estimating the coefficients of this differential equation from knowledge of only the inner product of the output signal with this basis. If the basis is chosen properly, then a great deal of information about the output signal would be contained in these inner products. Such a good basis can be chosen using the theory of wavelets for the wavelet coefficients contain information about the resolution as well as the signal duration. If the correct choice of the wavelets basis is made, then almost all the information about the chief characteristics of the signal would be compressed in a few wavelet coefficients. Thus if we base our system parameter estimation on these few wavelet coefficients, then with lesser data storage, we could get improved estimates of the parameters. A paper containing this work is currently in review. The second part of her work dealt with constructing an approximate Volterra model for a transistor amplifier based on the Ebers-Moll model of the transistor. The Kirchchoff voltage and current relations for the amplifier are written down and then the nonlinear functions in the Ebers-Moll model are decomposed into a strong linear component and a weak nonlinear component. Using standard methods of perturbation theory for differential equations, we then obtain a second order Volterra model for the transistor. The computations leading to the determination of the first and second order Volterra kernels are very heavy and the highlight of this work is the exact expression for the second order Volterra kernel in terms of the transistor parameters and the circuit elements. A paper containing these expressions is currently in review. The final part of her work deals with using the Volterra model for the transistor amplifier combined with wavelets to estimate the transistor parameters. The idea is applicable to any Volterra system whose kernels depend on parameters that we wish to estimate. The idea is to expand the Volterra kernels in terms of the wavelet basis and then plug these into the input output equation for the Volterra model and then to form the wavelet transform of the output. What results is a quadratic relation connecting the kernel wavelet coefficients and the output wavelet coefficients. The unknown parameters are now hidden inside the kernel wavelet coefficients and these may be estimated using a gradient search algorithm. This idea can be generalized to higher order Volterra kernels. The general philosophy of this method is contained in the following argument: Consider a Volterra system

$$y(t) = \sum_{i=1}^{p} \int h_i(u_1, ..., u_i, \theta) x(t - u_1)...x(t - u_i) du_1...du_i$$

θ is an unknown parameter vector to be estimated from knowledge of input-output data $\{x(t), y(t)\}$. Expand the Volterra kernels h_i in terms of a wavelet basis truncated to a few terms:

$$h_i(u_1, ..., u_i, \theta) = \sum_{n_1,...,n_i,k_1,...,k_i} c(n_1, ..., n_i, k_1, ..., k_i, \theta) \psi_{n_1,k_1}(u_1)...\psi_{n_i,k_i}(u_i)$$

Plugging this into the Volterra integral equation gives us

$$y(t) = \sum_{i=1}^{p} \sum_{n_1,...,n_i,k_1,...,k_i} c(n_1, ..., n_i, k_1, ..., k_i, \theta) \int x(t - u_1)...x(t - u_i) \psi_{n_1,k_1}(u_1)...\psi_{n_i,k_i}(u_i) du_1...du_i$$

Now take the wavelet transform on both sides to get

$$y_{a,b} = \int y(t) \psi_{a,b}(t) dt$$

$$= \sum_i \sum_{n_1,...,n_i,k_1,...,k_i} c(n_1,...,n_i,k_1,...,k_i,\theta) \int x(t-u_1)...x(t-u_i)\psi_{a,b}(t)\psi_{n_1,k_1}(u_1)...\psi_{n_i,k_i}(u_i)dtdu_1...du_i$$

We define the input data coefficients

$$f(a,b,n_1,...,n_i,k_1,...,k_i) = \int x(t-u_1)...x(t-u_i)\psi_{a,b}(t)\psi_{n_1,k_1}(u_1)...\psi_{n_i,k_i}(u_i)dtdu_1...du_i$$

Then we can write

$$y_{a,b} = \sum_{n_1,...,n_i,k_1,...,k_i} f(a,b,n_1,...,n_i,k_1,...,k_i)c(n_1,...,n_i,k_1,...,k_i,\theta)$$

[5]. Heat equation

$$(k^2/2)T_{,xx}(t,x) = T_{,t}(t,x)$$

Assume that T has period 1 in time. Then

$$T(t,x) = \sum_{n=-\infty}^{\infty} a_n(x)exp(2\pi int)$$

Plug this into the heat equation to get

$$k^2 a_n''/2 = 2\pi in a_n$$

or

$$a_n'' = (4\pi in/k^2)a_n$$

The solution is

$$a_n(x) = c_n.exp((1+i)\sqrt{2\pi n/k^2}x) + d_n exp(-(1+i)\sqrt{2\pi n/k^2}x)$$

for $n \geq 0$ and

$$a_n(x) = c_n.exp((i-1)\sqrt{2\pi |n|/k^2}x) + d_n.exp((1-i)\sqrt{2\pi |n|/k^2}x)$$

for $n < 0$. Now we assume that

$$T(t,x) \to A, x \to \infty$$

where A is a finite constant. Then it follows that for $n > 0$, $c_n = 0$, $c_0 = A$ and for $n < 0$, $d_n = 0$. Thus, the solution can be expressed as

$$T(t,x) = A + \sum_{n>0}(d_n.exp(-(1+i)\sqrt{2\pi n/k^2}x)exp(2\pi int) + c_{-n}exp((i-1)\sqrt{2\pi n/k^2}x)exp(-2\pi int)$$

We also require that $T(t,0) = B(t)$, some prescribed function of time. Thus,

$$B(t) = A + \sum_{n>0} d_n.exp(2\pi int) + c_{-n}exp(-2\pi int)$$

It follows that

$$A = \int_0^1 B(t)dt, d_n = \int_0^1 B(t).exp(-2\pi int)dt, c_{-n} = \int_0^1 B(t).exp(2\pi int)dt$$

and that completes the derivation.

Ref:Problem posed in "Signals and Systems" by Oppenheim, Willsky and Nawab" and conveyed to the author by Dr.Shambunath Sharma.

[6]. Justification of the workstations and related equipment for the DSP and SSP laboratories. The workstations will enable us to perform heavy computations at a very high speed. The computations are very expensive and a dedicated workstation will really be useful. The major applications are to the simulation of (a) galactic evolution and (b) tot he solution of the problem of determining the energy band structure in semiconductor crystals using quantum mechanics. For simulating galactic evolution, we first of all need to provide input to the workstation in the form of the initial positions $\mathbf{r}_i(0), i = 1, 2, ..., N$ and initial velocities $\mathbf{v}_i(0), i = 1, 2, ..., N$ of each individual star in the galaxy. The number N is of the order 10^{10}. The workstation will compute the forces between each pair of stars $\mathbf{F}_{ij} = \frac{GM^2(\mathbf{r}_i - \mathbf{r}_j)}{|\mathbf{r}_i - \mathbf{r}_j|^3}$ using Newton's inverse square law of gravitation. These forces will then be input into another programme in the same workstation for updating the positions and velocities in accordance with Newton's equations of motion:

$$\mathbf{r}_i''(t) = \sum_{j:j\neq i} \frac{GM^2(\mathbf{r}_j(t) - \mathbf{r}_i(t))}{|\mathbf{r}_j(t) - \mathbf{r}_i(t)|^3}$$

which in discretized for reads

$$\mathbf{r}_i(t + \Delta) = \mathbf{r}_i(t) + \Delta.\mathbf{v}_i(t)$$

$$\mathbf{v}_i(t + \Delta) = \mathbf{v}_i(t) + \Delta. \sum_{j:j\neq i} \frac{GM^2(\mathbf{r}_j(t) - \mathbf{r}_i(t))}{|\mathbf{r}_j(t) - \mathbf{r}_i(t)|^3}$$

In other words, a system of around 10^{10} ordinary differential equations will be solved using this programme and the ensemble of trajectories of the stars will be traced on a graph. Globular clusters as well as spiral galaxies can thus be simulated using this workstation. This procedure may also be used for simulating molecular dynamics. The other application is to the solution of the problem of determining the energy eigenvalues and eigenfunctions of a crystal. The system of equations to be solved is the stationary state Schrodinger equation:

$$(T_e + T_n + V_e + V_n + V_{en})\psi = E\psi$$

where T_e is the kinetic energy operator of the electrons, T_n is the kinetic energy operator of the nuclei, V_e is the potential energy of the electrons, V_n is the potential energy of the nuclei and V_{en} is the interaction potential energy between the electrons and nuclei. This is a complex system of partial differential equations and is solved iteratively using the Born- Oppenheimer method wherein first the energy levels of the electrons with the nuclear positions fixed are determined. These energy levels are functions of the nuclear positions. Then the nuclear wave functions are obtained by using the electronic energy levels as a potential field in which the nuclei are moving. The computation is very complex and a dedicated workstation of the above kind is needed. There are other applications involving heavy computations for which the workstations will be needed. These include simulation complex mechanical systems, fluid dynamical systems in which the entire field of the fluid is partitioned into pixels and a velocity is assigned to each pixel. The system to be solved is the Navier-Stokes differential equation which is quadratically nonlinear in the fluid velocity. The other equation involving heavy computational load is the Boltzmann kinetic transport equation for the particle distribution function. The solution of this nonlinear integro differential equation will enable us to study kinetic phenomena in semiconductors. The equation has the following form:

$$f_{,t}(t,\mathbf{r},\mathbf{v}) + (\mathbf{v},\nabla_r f(t,\mathbf{r},\mathbf{v})) + (\mathbf{F}(t,\mathbf{r},\mathbf{v}),\nabla_v f(t,\mathbf{r},\mathbf{v})) =$$
$$= I(f(t,\mathbf{r},\mathbf{v}))$$

where I is a quadratically nonlinear integral operator called the Boltzmann collision integral.

[7]. Let V be a vector space of dimension n. Let $f,g \in V^*$ be nonzero and let $f \neq g$. Then prove that if $W = \{x \in V : f(x) = g(x) = 0\}$, then $dimW = n - 2$ and W^c is connected in V. On the other hand if $W_1 = \{x \in V : f(x) = 0\}$, then $dimW_1 = n - 1$ and W_1^c is not connected in V.

hint: For the latter part, consider the sets

$$W_{1+} = \{x \in V : f(x) > 0\}, W_{1-} = \{x \in V : f(x) < 0\}$$

Then, W_{1+} and W_{1-} are both open in W_1^c, both are disjoint and their union equals W_1^c. This proves that W_1^c is disconnected in V. On the other hand, consider the first case here. Let $x,y \in W^c$. Then either $f(x) \neq 0$ or $g(x) \neq 0$ and $f(y) \neq 0$ or $g(y) \neq 0$. The we must show that a curve in W^c can be constructed connecting x and y. This picture is obvious if $V = \mathbb{R}^3$ so that W is a line. Then given any two points not falling on the line, we can construct a curve joining these two points such that the curve never touches the line.

[8]. What do you understand by the discrete topology ?

ans:A topology in which every single point set is open.

43.If a topological space is discrete and compact, then it must be finite. For suppose X is compact and discret. Then since

$$X = \bigcup_{x \in X} \{x\}$$

and $\{x\}$ is open in X for every $x \in X$, it follows that $\{\{x\} : x \in X\}$ is an open covering of X and hence it has a finite subcovering $\{\{x_i\} : i = 1, 2, ..., N\}$. This proves that

$$X = \{x_1, x_2, ..., x_N\}$$

and hence X is finite.

[9]. Distinguish between the Frobenius norm and the spectral norm for linear operators in a finite dimensional normed linear space V over the field of complex numbers.

ans:

Let A be an $n \times n$ complex matrix and let $e_i, i = 1, 2, ..., n$ be an orthonormal basis for $\mathbb{C}^n$. Then, the Frobeinus norm is defined by

$$\| A \|_{fr}^2 = \sum_{i=1}^{n} \| Ae_i \|^2 = \sum_{i=1}^{n} e_i^* A^* Ae_i$$

$$= Tr(A^*A) = \sum_{i,j} |a_{ij}|^2$$

On the other hand the spectral norm is defined by

$$\| A \|_{sp} = sup(\| Ax \|: \| x \| \le 1)$$

and it is easy to see that $\| A \|_{sp}$ equals the maximum eigenvalue of $(A^*A)^{1/2}$. On the other hand, it easily follows from above that $\| A \|_{fr}^2$ equals the sum of all the eigenvalues of A^*A. Hence, we have in particular, that

$$\| A \|_{fr} \ge \| A \|_{sp}$$

[10]. Complex lowpass representation of bandpass signals

Let $x(t)$ be a real bandpass signal with spectrum $X(\omega)$ nonzero only for $|\omega \pm \omega_c| \le \omega_B$ where $\omega_c > \omega_B$. Define

$$X_+(\omega) = X(\omega)U(\omega)$$

and set

$$S(\omega) = X_+(\omega + \omega_c)$$

Then,

$$S(\omega - \omega_c) = X_+(\omega) = X(\omega)U(\omega)$$

and

$$\bar{S}(-\omega - \omega_c) = \bar{X}_+(-\omega) = \bar{X}(-\omega)U(-\omega) = X(\omega)U(-\omega)$$

since $x(t)$ is real. Thus,

$$S(\omega - \omega_c) + \bar{S}(-\omega - \omega_c) = X(\omega)(U(\omega) + U(-\omega)) = X(\omega)$$

It follows on taking the inverse Fourier transform of this equation that

$$x(t) = Re(s(t)exp(j\omega_c t))$$

[11]

Integrating factor for a system of first order differential equations(Problem posed by Dr.Shambunath Sharma)

$$\frac{d\mathbf{x}(t)}{dt} = \mathbf{A}(t)\mathbf{x}(t) + \mathbf{w}(t)$$

The state transition matrix $\Phi(t,s)$ satisfies the differential equation

$$\frac{\partial \Phi(t,s)}{\partial t} = \mathbf{A}(t)\Phi(t,s), t \ge s$$

with the initial condition $\Phi(s,s) = \mathbf{I}$ Then define

$$\psi(t,s) = \Phi(t,s)^{-1}$$

Then

$$\frac{\partial \psi(t,s)}{\partial t} = -\Phi(t,s)^{-1}\frac{\partial \Phi(t,s)}{\partial t}\Phi(t,s)^{-1}$$

$$= -\Phi(t,s)^{-1}\mathbf{A}(t)\Phi(t,s)\Phi(t,s)^{-1} = -\psi(t,s)\mathbf{A}(t)$$

It follows that

$$\frac{\partial}{\partial t}(\psi(t,s)\mathbf{x}(t)) = \psi(t,s)\mathbf{x}'(t) - \psi(t,s)\mathbf{A}(t)\mathbf{x}(t)$$

$$= \psi(t,s)(\mathbf{x}'(t) - \mathbf{A}(t)\mathbf{x}(t)) = \psi(t,s)\mathbf{w}(t)$$

and we get on integrating this equation from $t = 0$ to t,

$$\psi(t,s)\mathbf{x}(t) - \mathbf{x}(s) = \int_s^t \psi(u,s)\mathbf{w}(u)du$$

or

$$\mathbf{x}(t) - \Phi(t,s)\mathbf{x}(s) = \int_s^t \Phi(t,s)\Phi(u,s)^{-1}\mathbf{w}(u)du$$

Now observe the state transition property:

$$\Phi(t,s) = \Phi(t,u)\Phi(u,s)$$

and hence

$$\Phi(t,s)\Phi(u,s)^{-1} = \Phi(t,u)$$

It follows that

$$\mathbf{x}(t) = \Phi(t,s)\mathbf{x}(s) + \int_s^t \Phi(t,u)\mathbf{w}(u)du$$

[12]. Problem posed by Dr.Satish Chand: (a) Suppose V is a signal subspace of $L^2(\mathbb{R})$ and $\phi_n(t), n = 1, 2, \ldots$ is a basis for this space. Suppose $x(t)$ belongs to this signal space and has zero mean, ie,

$$\mu_x = \int_{-\infty}^{\infty} x(t)dt = 0$$

We expand x in terms of ϕ_n and truncate to retain only a finite number of terms. Suppose $\phi_n's$ have non-zero mean, ie,

$$\mu_n = \int \phi_n(t)dt \neq 0$$

Then the approximation

$$\hat{x}(t) = \sum_{n=1}^{N} c_n\phi_n(t)$$

will in general have non-zero mean. In fact,

$$\int_{-\infty}^{\infty} \hat{x}(t)dt = \sum_{n=1}^{N} c_n \mu_n$$

How to remedy this biasing problem ?

Ans: Let

$$\psi_n(t) = \phi_n(t) - \mu_n.exp(-t^2/2)/\sqrt{2\pi}$$

Then the basis ψ_n is unbiased, ie,

$$\int \psi_n(t)dt = 0$$

However if $\{\phi_n\}$ is orthonormal, $\{\psi_n\}$ need not be orthonormal. So using the unbiased basis, the coefficients are harder to compute.

(b) Bi-orthogonal wavelet basis: Suppose $\phi_n, n = 1, 2, ...$ is a basis for an inner product space V and let $\psi_n, n = 1, 2, ...$ be a dual basis in the sense that $< \phi_n, \psi_m >= \delta_{nm}$. Suppose we are given a vector $x \in V$ and we expand x in terms of the ϕ_n, ie,

$$x = \sum_n c_n \phi_n$$

where the $c_n's$ are complex scalars. Then

$$< x, \psi_m >= \sum_n c_n < \phi_n, \psi_m >= c_m$$

and hence

$$x = \sum_n < x, \psi_n > \phi_n$$

This is equivalent to the operator equation

$$\mathbf{I} = \sum_n |\psi_n >< \phi_n|$$

[13]. Nonlinear systems relative to orthonormal bases: Let V be an inner product space and $V \otimes V$ be the tensor product of V with itself. Consider the input-output equation

$$y = Ax + B(x \otimes x)$$

where $A : V \to V$ and $B : V \otimes V \to V$ are linear operators. Let $e_n, n = 1, 2, ...$ be an orthonormal basis for V. We expand the signal x as a linear combination of the $e_n's$ so that

$$x = \sum_n < x, e_n > e_n = \sum_n x_n e_n, \; x_n =< x, e_n >

Then,

$$y_n = < y, e_n > = < Ax, e_n > + < B(x \otimes x), e_n >$$

$$= \sum_m < Ae_m, e_n > x_m + \sum_{p,q} < B(e_p \otimes e_q), e_n > x_p x_q$$

Define

$$h(n, m) = < Ae_m, e_n >, \; g(n, p, q) = < B(e_p \otimes e_q), e_n >$$

so that the above equation can be expressed as

$$y_n = \sum_m h(n, m) x_m + \sum_{p,q} g(n, p, q) x_p x_q$$

and this is clearly a second order Volterra system. Thus a quadratic nonlinear system on a vector space, when expressed in terms of a basis acquires the form of a Volterra type of nonlinear system.

[14]. This argument can be generalized to higher order Volterra systems as follows: Let V be a vector space and $\otimes$ a tensor product on V. Let $V^{\otimes n}$ denote the n-fold tensor product of V with itself. Consider a system on V having input-output equation

$$y = \sum_{n=1}^{N} B_n(x^{\otimes n})$$

where $B_n : V^{\otimes n} \to V$ are linear transformations. Then let $e_n, n = 1, 2, ...,$ be an orthonormal basis for V. We write

$$x = \sum_n x_n e_n, \; x_n = < x, e_n >$$

Then,

$$y = \sum_{n=1}^{N} B_n \sum_{k_1,...,k_n} x_{k_1}...x_{k_n} e_{k_1} \otimes ... \otimes e_{k_n}$$

so that

$$y_m = < y, e_m > = \sum_{n=1}^{N} h_n(m, k_1, ..., k_n) x_{k_1}...x_{k_n}$$

where

$$h_n(m, k_1, ..., k_n) = < B_n(e_{k_1} \otimes ... \otimes e_{k_n}), e_m >$$

This shows that any polynomial type of nonlinear system on a vector space can be represented using a Volterra type of system relative to a basis.

[15]. Explain how using creation and annihilation operators in momentum space you would construct a quantum field $\phi(t, x)$ such that if the associated Hamiltonian $H(\phi, \pi)$ is quadratic, then the equal time commutation relation

$$[\phi(t, x), \pi(t, y)] = i\delta^3(x - y)$$

holds.

[16]. Consider the diffusion equation

$$\frac{\partial N(t,x)}{\partial t} = D\frac{\partial^2 N(t,x)}{\partial x^2}$$

Suppose $N(0,x)$ is a random Gaussian field with mean zero and covariance

$$< N(0,x).N(0,y) >= R_N(x-y)$$

Then determine the correlation at time t, ie,

$$R_N(t,x-y) =< N(t,x).N(t,y) >$$

Hint:Express the solution as

$$N(t,x) = \int K(t,u)N(0,x-u)du$$

so that

$$R_N(t,x-y) =< N(t,x).N(t,y) >= \int K(t,u)K(t,v) < N(0,x-u).N(0,y-v) > dudv$$

$$= \int K(t,u)K(t,v)R_N(x-y-u+v)dudv$$

Now take the Fourier transform on both sides with respect to $x-y$ and show that if $\hat{K}(t,\omega)$ is the Fourier transform of $K(t,x)$, then the Fourier transform of $R_N(t,x)$ is given by

$$\hat{R}_N(t,\omega) = |\hat{K}(t,\omega)|^2 \hat{R}_N(0,\omega)$$

Evaluate the kernel $\hat{K}(t,\omega)$ explicitly.

[17]. Syllabus for a course on nonlinear circuits and systems.

1.Definition of a linear system as a linear transformation from one vector space to another.

$$T:V \to V, T\{c_1 x_1 + c_2 x_2\} = c_1 T\{x_1\} + c_2 T\{x_2\}$$

2.Nonlinear systems as transformations on a vector space that violate linearity.

$$T\{x\} = \sum_i a_i x_i + \sum_{i,j} b_{ij} x_i x_j$$

3.Examples of nonlinear systems in terms of transformations on a vector space:

4.Volterra type nonlinear systems in continuous and discrete time.

$$y(n) = \sum_k g_1(n,k)x(k) + \sum_{k,m} g_2(n,k,m)x(k)x(m) + \sum_{k,m,p} g_3(n,k,m,p)x(k)x(m)x(p) + ...$$

$$y(t) = \sum_{n=1}^{\infty} \int_{\mathbb{R}^n} h_i(t, u_1, ..., u_n) x(u_1) x(u_2)...x(u_n) du_1 du_2...du_n$$

4.Nonlinear systems in discrete time defined via difference equations. Non-recursive

$$y(n) = f(n, x(n), x(n-1), ..., x(n-p))$$

and recursive

$$y(n) = f(n, y(n-1), y(n-2), ..., y(n-p)) + g(n, x(n), x(n-1), ..., x(n-q))$$

5. Nonlinear systems in continuous time described via differential equations.

$$\frac{d^n y(t)}{dt^n} = f\left(\frac{d^{n-1}y(t)}{dt^{n-1}}, ..., \frac{dy(t)}{dt}\right), y(t)) + g\left(\frac{d^m x(t)}{dt^m}, \frac{d^{m-1}x(t)}{dt^{m-1}}, ..., \frac{dx(t)}{dt}, x(t)\right)$$

6.Systems invariant under a group of transformations. For example, consider the vector system

$$\frac{d\mathbf{x}}{dt} = \mathbf{A}\mathbf{x}(t)$$

where $\mathbf{x}(t) \in \mathbb{R}^n$ and $\mathbf{A} \in \mathbb{R}^{n \times n}$. Let T be a non-singular matrix. and set $y(t) = T(x(t))$. Then,

$$dy(t)/dt + T(dx(t)/dt) = TAx(t) = TAT^{-1}y(t)$$

so that if $TAT^{-1} = A$, then $y(t) = x(t)$ for all t. Now, observe that the set of all nonsingular matrices T for which $TAT^{-1} = A$ is a group. Hence, if we denote this group by G_A, it follows that the above dynamical system is invariant under the group G of transformations of the state space. More generally, suppose $\mathbf{f} : \mathbb{R}^n \to \mathbb{R}^n$ is a nonlinear transformation. Consider the dynamical system

$$d\mathbf{x}(t)/dt = \mathbf{f}(\mathbf{x}(t))$$

Let $g : R^n \to \mathbb{R}^n$ be a nonsingular linear transformation such that $gfg^{-1} = \mathbf{f}$, or equivalently, such that $gof = fog$. Then,

$$dg\mathbf{x}(t)/dt = gd\mathbf{x}(t)/dt = g\mathbf{f}(\mathbf{x}(t)) = \mathbf{f}(g\mathbf{x}(t))$$

and it follows from the uniqueness theorem for solutions to differential equations with same initial conditions that if $gx(0) = x(0)$, then $gx(t) = x(t)$ for all t. As an example, suppose $\mathbf{f} : \mathbb{R}^3 \to \mathbb{R}^3$ has the form $\mathbf{f}(\mathbf{x}) = \phi(|\mathbf{x}|)\mathbf{x}$. Then $Rf(\mathbf{x}) = \mathbf{f}(\mathbf{R}\mathbf{x})$ for any rotation $\mathbf{R}$. This is in view of the identity $|\mathbf{R}\mathbf{x}| = |\mathbf{x}|$. It follows that the dynamical system in $\mathbb{R}^3$ defined by the differential equation

$$d\mathbf{x}(t)/dt = \phi(|\mathbf{x}(t)|)\mathbf{x}(t)$$

is invariant under rotations. Consider now the motion of a particle under a central potential $V(|x|)$ in $\mathbb{R}^3$. The equations of motion have the form

$$x'(t) = v(t), v'(t) = \phi(|x(t)|)x(t)$$

Here, $x(t), v(t) \in \mathbb{R}^3$ are respectively the position and velocity vectors. Let R be a rotation of $\mathbb{R}^3$. Then

$$(Rx(t))' = Rx'(t) = Rv(t), (Rv(t))' = Rv'(t) = \phi(|Rx(t)|)Rx(t)$$

since $|Rx(t)| = x(t)$. It follows that $(Rx(t), Rv(t))$ also satisfies the equation of motion and hence, the motion is invariant under spatial rotations. Note that the motion $(x(t), v(t))$ is not invariant under $SO(6)$. It is invariant, as shown under the subgroup of $SO(6)$ consisting of elements of the form $diag[R, R]$ with $R \in SO(3)$.

6. Examples of nonlinear systems from circuit theory. Current dependent resistors and inductors and voltage dependent capacitors. Suppose that the resistance R, the inductance L and the capacitance C are functions of the current i flowing through the circuit. Assume that these three components are connected in series to a voltage source $v_i(t)$. The differential equation for the current is then given by

$$\frac{d(L(i)i)}{dt} + R(i)i + \frac{1}{C(i)} \int_0^t i\,dt = 0$$

This can be put in the form

$$(C(i)(L(i)i)')' + (C(i)R(i)i)' + i = 0$$

This equation can be solved only approximately using the perturbation theory for differential equations. We shall not go into the details here but just state that if we write this nonlinear differential equation as the sum of a linear component and a nonlinear component, and introduce a small parameter into the nonlinear component and then expand the current as a power series in the parameter, then we get a sequence of linear differential equations for each term in the perturbation series for the current.

7. Quantum mechanics as a bilinear theory: Let the potential $V(t, \mathbf{x})$ be the "input signal" and the wave function $\psi(t, \mathbf{x})$ the "output signal" of a quantum mechanical system. The relationship between the input and the output is defined by the Schrodinger wave equation:

$$-\frac{1}{2m}\nabla^2 \psi(t, \mathbf{x}) + V(t, \mathbf{x})\psi(t, \mathbf{x}) = i\frac{\partial \psi(t, \mathbf{x})}{\partial t}$$

We can write this equation symbolically as

$$id\psi_t/dt = (H_0 + V_t)\psi_t$$

where $H_0 = -\frac{1}{2m}\nabla^2$ is the kinetic energy operator of the system and V_t is the perturbing potential. The solution to this equation can be expressed as a Volterra series in the input signal V_t as follows. Write

$$\psi_t = exp(-itH_0)\phi_t$$

and plug this into the wave equation to get

$$exp(-itH_0)(id\phi_t/dt) + H_0 exp(itH_0)\phi_t = H_0 exp(-itH_0)\phi_t + V_t exp(-itH_0)\phi_t$$

or

$$id\phi_t/dt = exp(itH_0).V_t.exp(-itH_0)\phi_t$$

the solution to which is a Volterra series:

$$\phi_t = \phi_0 + (\sum_{n=1}^{\infty}((-i)^n/n!)\int_{0<t_1<t_2<...<t_n<t}$$

$$exp(it_nH_0)V_{t_n}.exp(-i(t_n-t_{n-1}H_0)V_{t_{n-1}}...exp(-i(t_2-t_1)H_0)V_{t_1}exp(-t_1H_0)dt_1dt_2...dt_{n-1}dt_n)\phi_0$$

This is a Volterra series for the output (the wave function) in terms of the input (the potential). Suppose V_t is a random Gaussian potential. Then, to calculate the mean of the output, we need all the moments of the potential. The computation is based on the formula that if $X_1,...,X_n$ is a set of zero mean jointly Gaussian random variables, then

$$< X_1X_2...X_n >= \sum < X_{i_1}X_{j_1} >< X_{i_2}X_{j_2} > ... < X_{i_k}X_{j_k} >$$

provided that n is even, $k = n/2$ and the sum is over all the possible pairings $(i_1,j_1),...,(i_k,j_k)$ of $1,2,...,n$. How many pairs are there ? To prove this, we assume without loss of generality that $X_1,...,X_n$ are zero mean Gaussian random variables with correlation

$$R(a,b) = \mathbb{E}(X_aX_b)$$

Then, the joint moment generating function of $X_1,...,X_n$ is given by

$$\mathbb{E}(exp(t_1X_1+...+t_nX_n)) = exp(\frac{1}{2}\sum_{a,b=1}^{n}R(a,b)t_at_b)$$

The coefficient of the product $t_1t_2...t_n$ on the left side is given by

$$\mathbb{E}(X_1X_2...X_n)$$

while the coefficient of the same on the right side is

$$\sum R(a_1,b_1).R(a_2,b_2)...R(a_k,b_k)$$

where $k = n/2$ and the sum is over all pairings. Note that the only term in the expansion of

$$exp(\frac{1}{2}\sum_{a,b}R(a,b)t_at_b)$$

that contains the product $t_1t_2...t_n$ is given by

$$\frac{1}{2^kk!}(\sum_{a,b}R(a,b)t_at_b)^k$$

and the coefficient of $t_1...t_n$ in the expansion of this is the required expression.

[18]. Development of a music laboratory:

1.Instruments needed: (1) Veena, (2) Flute (both eight and six holed), (3) Sitar, (4) Tanpura, (5) Mridangam, (6) Tabla, (7) Piano, (8) Harmonium.

Aim of the lab: To analyze the sounds produced by various instruments corresponding to various raagas using standard signal processing tools like the short time Fourier transform and wavelet transform. Note that if $h(t)$ is a window function, then the corresponding short time Fourier transform of a signal $x(t)$ is given by

$$S_{hx}(\tau, \omega) = \int_{-\infty}^{\infty} h(t - \tau)x(t)exp(-i\omega t)dt$$

dynamics of flute: The point where a hole is closed is a node while the point where a hole is open is an antinode. In this way, we can make out the wave pattern prevalant inside the flute for a given combination of open and closed holes. A more detailed analysis of flute dynamics would involve studying the air fluid within the flute using the Navier-Stokes equation. According to [1], the dynamics can be described effectively by regarding the air inside to consist of different kinds of particles. The ith kind of particle has a distribution function $f_i(\mathbf{r}, t)$ and it propagates with a speed $\mathbf{c}_i$ satisfying a Boltzmann equation of the kind

$$f_i(\mathbf{r} + \mathbf{c}_i, t + 1) = f_i(\mathbf{r}, t) + \Omega_i(f)(\mathbf{r}, t)$$

where Ω_i is the Boltzmann collision term. Note that if $\mathbf{J}(\mathbf{r}, t)$ is the total particle current and $\rho(\mathbf{r}, t)$ is the total particle density, then

$$\rho(\mathbf{r}, t) = \sum_i f_i(\mathbf{r}, t), \rho(\mathbf{r}, t)\mathbf{J}(\mathbf{r}, t) = \sum_i \mathbf{c}_i f_i(\mathbf{r}, t)$$

and the mass conservation equation can be applied to this pair to furnish one of the equations for the air fluid inside the flute. If u_α denote the components of the average velocity and p the pressure then we have

$$u_\alpha = \sum_i c_{i\alpha} f_i / \sum_i f_i$$

Example:Analysis of Bhairavi raaga:

Find out the frequencies of sa, ri(big) ga(small) ma(small) pa da(big) ni(small) Sa. While coming down from Sa to sa, big da is replaced by small da. We shall henceforth denote big ri by Ri, big ga by Ga and so on. The swarams at the higher scale shall be denoted by the same letters with a bar above them.

Song:Viriboni, raagam Bhairavi

sa, ri, ni,da, ni sa ri ga sa ri, ga ri ga ga rii, ga ga ri ga ma pa da ma pa ga ri sa ni da ma ga ri sa, ni sa ri ga sa ri ni ni da ni sa ga ri ni sa ni da pa pa da pa da, ni ni da pa da ni sa ri, pa da ni sa ri ga ma pa ma ga ri ga ma ni da pa ma ga ri ni sa ri ga ma ga ri ga ga ri, $\bar{s}a$ ni da pa ma ga ri ga ma ni da pa ma ga ri pa ma ga ri sa, pa da ni sa ri ga, ni sa ri ga ma pa, ni ni da pa, da ma pa ga ri ga ma pa da ni, $\bar{s}a$.

First thing is to represent this song by a time domain signal. For a given swaram x, let $\omega(x)$ denote the associated frequency. The duration of the signal representing a given swaram is T so that if the swaram is sung during the time interval $[kT, (k + 1)T]$, then the associated signal is $cos(\omega(x)(t - kT))p(t - kT)$

where $p(t) = u(t) - u(t - T)$ or equivalently, $p(t) = 1$ for $kT \leq t \leq (k+1)T$ and $p(t) = 0$ for $t < kT$ and $t > (k+1)T$. Given a swaram sequence of the above kind, say $\{x_1 x_2 ... x_N\}$, the corresponding time domain signal is given by

$$x(t) = \sum_k A_k cos(\omega(x_k)(t - kT))p(t - kT)$$

where A_k is the amplitude of the kth swaram.

Ref: Helmut Kuhnelt, Simulating the mechanism of sound generation in flutes using the Lattice Boltzmann method, Proceedings of the Stockholm MUSIC Acoustics Conference, August 6-9, 2003, Stockholm, Sweden.

[19]. Syllabus for Circuits and Systems

1.Signals and their classification

2.Operations on signals

3.Definition of a linear system

4.Properties of systems:Linearity, time invariance, causality, stability, memoryless property.

5.Systems described by differential and difference equations, block diagrammatic representation using delay, differential and integral operators.

6.Convolution representation of LTI systems.

7.Fourier series representation of periodic signals

8.Fourier transform for aperiodic signals and stable LTI system analysis.

9.Laplace transform for stable and unstable systems.

10.Solution of differential equations using the Laplace transform

11.The Z-transform and its properties

12.Solution of difference equations using the Z-transform.

13.Discretization of continuous time systems described by differential equations

14.Pole-Zero plots

15.Kirchchoff's laws for a circuit

16.Linear circuit elements:Resistor, capacitor and inductor

17.Analogy of electric circuits with mechanical systems(LCR and spring-mass with damping)

18.Mutual inductance and coupled circuits

19.Analysis of linear circuits using KCL and KVL equations

20.Graph theoretic analysis of networks (Incidence matrix, cutset matrix and fundamental loop matrix)

21.Laplace transform based network analysis

22.Two port networks

23.Sinusoidal steady state and phasor analysis of linear networks.

[20]. Syllabus for Digital Signal Processing

1.Definition and properties of the Z-transform

(a) Definition, ROC, Linearity, shift, modulation and convolution property, initial and final value theorems.

(b).Solving difference equations using the Z-transform.

(c) Block diagrammatic representation of LTI systems based on the Z-transform.

2. Discrete time Fourier transform

Definition and analysis of stable LTI systems using the DTFT, Inverse DTFT.

3. Discrete and Fast Fourier transform for arbitrary radix.

Decimation in time and frequency.

4. Discrete Hilbert transform

Definition in the frequency domain and time domain version, Hilbert transform relations between the real and imaginary parts of the Fourier transform of a causal signal, representation of bandpass signals using complex lowpass signals.

5. Homomorphic signal processing

For defining linear systems, we need the notion of signal addition and scalar multiplication. If the linear system preserves these operations, we say that the system is linear. If it preserves other kinds of operations, we get the notion of homomorphic systems. Specifically, let $(x, y) \rightarrow F(x, y)$ be the generalized sum of input signals and $(c, x) \rightarrow G(c, x)$ the generalized scalar multiplication of input signals. Let $(u, v) \rightarrow \tilde{F}(u, v)$ be the generalized sum of output signals and $(c, u) \rightarrow \tilde{G}(c, u)$ the generalized scalar multiplication of output signals. Then a map T from the input signal space to the output signal space is said to be homomorphic (generalized linear) if for all input signals x_1, x_2 and scalars c_1, c_2, we have

$$T(F(G(c_1, x_1), G(c_2, x_2))) = \tilde{F}(\tilde{G}(c_1, T(x_1)), \tilde{G}(c_2, T(x_2)))$$

In the special case when $F, \tilde{F}$ are signal additions in the input and output vector spaces respectively and $G, \tilde{G}$ are scalar multiplications in the respective spaces, T becomes a linear system. Note that for addition and scalar multiplication to be defined, we require that the input and output signal spaces be respectively vector spaces over a common field.

6.Image processing fundamentals based on the two dimensional Z-transform.

7. Introduction to one and two dimensional Volterra systems for speech and image processing. In two dimensions, the second order Volterra filter has the form

$$y(n, m) = \sum_{k,l} h(k, l)x(n - k, m - l) + \sum_{k,l,p,q} g(k, l, p, q)x(n - k, m - l)x(n - p, m - q)$$

This equation admits a matrix-vector representation defined as follows. Let $\mathbf{e}_k$ be the $N \times 1$ vector having a one in the kth position and zeros at all the other positions. Define the vectors

$$\mathbf{h} = \sum h(k,l)\mathbf{e}_k \otimes \mathbf{e}_l, \quad \mathbf{g} = \sum g(k,l,p,q)\mathbf{e}_k \otimes \mathbf{e}_l \otimes \mathbf{e}_p \otimes \mathbf{e}_q$$

$$\mathbf{y} = \sum_{n,m} y(n,m)\mathbf{e}_n \otimes \mathbf{e}_m$$

Also define the input data matrices

$$\mathbf{X}_1 = \sum_{n,m,k,l} x(n-k,m-l)(\mathbf{e}_n \otimes \mathbf{e}_m)(\mathbf{e}_k \otimes \mathbf{e}_l)^T$$

$$\mathbf{X}_2 = \sum_{n,m,k,l,p,q} x(n-k,m-l)x(n-p,m-q)(\mathbf{e}_n \otimes \mathbf{e}_m)(\mathbf{e}_k \otimes \mathbf{e}_l \otimes \mathbf{e}_p \otimes \mathbf{e}_q)^T$$

Then the above Volterra system can be represented as

$$\mathbf{y} = \mathbf{X}_1\mathbf{h} + \mathbf{X}_2\mathbf{g}$$

In case, there is noise in the system, the above equations holds only approximately. In this case, the first and second order Volterra filter coefficient vectors $\mathbf{h}$ and $\mathbf{g}$ can be estimated given input and output data by minimizing

$$E(\mathbf{h},\mathbf{g}) = \parallel \mathbf{y} - \mathbf{X}_1\mathbf{h} + \mathbf{X}_2\mathbf{g} \parallel^2$$

Setting the gradient of E with respect to the coefficient vectors $\mathbf{h}, \mathbf{g}$ results in the optimal equations

$$\mathbf{X}_1^T(\mathbf{y} - \mathbf{X}_1\mathbf{h} - \mathbf{X}_2\mathbf{g}) = \mathbf{0},$$

$$\mathbf{X}_2^T(\mathbf{y} - \mathbf{X}_1\mathbf{h} - \mathbf{X}_2\mathbf{g}) = \mathbf{0}$$

[21]. Syllabus for an elective in signal processing:Group theory and image processing

1.Definition of a group

2.Subgroups of a group

3.Examples of groups, Abelian and non-Abelian

4.The rotation and translation groups in the plane

5.The Euclidean motion group in space

6.Direct product of groups

7.Basic theory of differential manifolds:Definition, curves, tangent space, tangent bundle, diffeomorphisms, vector fields, one forms, differential forms, push forward and pull back maps.

8.Definition of a Lie group

9.Lie algebra of a Lie group, Jacobi identity, structure constants.

10.Solvable Lie algebras

11.Radical of a Lie algebra

12.Nilpotent Lie algebras

13.Adjoint representation of a Lie algebra.

14.Cartan-Killing form and Cartan's criterion for solvability and semisimplicity of a Lie algebra: Let $\mathfrak{g}$ be a Lie algebra and for $X \in \mathfrak{g}$, define $ad(X) \in \mathcal{L}(\mathfrak{g})$ so that $ad(X)(Y) = [X,Y]$. Then consider $B(X,Y) = Tr(ad(X).ad(Y))$ for $X,Y \in \mathfrak{g}$. If $\mathfrak{g}$ is semisimple then the bilinear form B on $\mathfrak{g}$ is non-degenerate and the converse is also true.

15. Cartan subalgebras of a Lie algebra

16.The theory of roots based on simultaneous diagonability of commuting semisimple matrices.

17.Compact real forms of a Lie algebra and a Lie group

18.The decomposition

$$\} = \approx + \sqrt{}$$

19.Conjugations with respect to real forms and compact real forms

20.Compact and non-compact roots.

21.The differential of a representation of a Lie group as a representation of the corresponding Lie algebra.

22.Orbital integrals:

23.Universal enveloping algebra of a Lie algebra.

24.The Weyl group associated with a Cartan subalgebra: If $\langle$ is a Cartan subalgebra of a Lie algebra $\}$ and A is the corresponding Lie group, then the normalizer of A in G is denoted by A'. It consists of all elements $g \in G$ for which $gAg^{-1} \subset A$. Then $W = A'/A$ is a finite group. If the Lie algebra is complex and semisimple, then upto conjugacy, there is just one Cartan subalgebra while if it is real, there may be more than one.

25.Symmetric algebra on a Lie algebra and HarishChandra's λ-map:Let $\mathfrak{g}$ be a Lie algebra and $S(\mathfrak{g})$ the symmetric algebra over $\mathfrak{g}$. Let $\mathcal{B}$ denote the universal enveloping algebra of $\mathfrak{g}$. This is defined as follows: Let τ denote the tensor algebra on $\}$ and for $X,Y \in \mathfrak{g}$, define $u_{X,Y} \in \tau$ so that

$$u_{X,Y} = X \otimes Y - Y \otimes X - [X,Y]$$

Let $\mathcal{L}$ denote the two sided ideal in τ defined by the span of all elements of the form $t \otimes u_{X,Y} \otimes t'$ where $t,t' \in \tau$ and $X,Y \in \}$. Then $\mathcal{B} = \tau/\mathcal{L}$. We note that for $X,Y \in \}$,

$$X \otimes Y - Y \otimes X + \mathcal{L} = [X,Y] + \mathcal{L}$$

Thus, we define $\gamma : \tau \to \tau/\mathcal{L}$ so that

$$\gamma(t) = t + \mathcal{L}, t \in \tau$$

Then,

$$\gamma(X)\gamma(Y) - \gamma(Y)\gamma(X) = \gamma([X,Y]), X, Y \in \}$$

For $X_1, ..., X_n \in \}$, consider the element $X_1...X_n \in S(\})$ defined by

$$X_1...X_n = \sum_{\sigma \in S_n} X_{\sigma 1} \otimes ... \otimes X_{\sigma n}$$

Then define

$$\lambda : S(\}) \to \mathcal{B}$$

so that

$$\lambda(X_1...X_n) = \frac{1}{n!} \sum_{\sigma \in S_n} \gamma(X_{\sigma 1})...\gamma(X_{\sigma n})$$

and extend λ linearly to the entire $S(\})$. We claim that λ is one one and onto. For example

$$\gamma(X)\gamma(Y) = \gamma(Y)\gamma(X) + \gamma([X,Y])$$

so that

$$\gamma(X)\gamma(Y) = \frac{1}{2}(\gamma(X)\gamma(Y) + \gamma(Y)\gamma(X)) + \frac{1}{2}\gamma([X,Y])$$

$$= \lambda(XY) + \frac{1}{2}\lambda([X,Y])$$

In general, it is easily shown that

$$\gamma(X_1...X_n) - \gamma(X_{\sigma 1}...X_{\sigma n}) \in \mathcal{B}_{n-1}$$

for all permutations σ. Here, $\mathcal{B}_{n-1}$ is the subspace of $\mathcal{B}$ spanned by elements having degree $\leq n-1$. It follows that

$$\gamma(X_1...X_n) - \lambda(X_1...X_n) \in \mathcal{B}_{n-1}$$

and hence using induction on n and linearity of λ, it is readily seen that $\gamma(X_1...X_n)$ is the image under λ of and element in $S_n(\})$, where $S_n(\})$ is the subspace of $S(\})$ spanned by elements having degree $\leq n$.

[22]. Syllabus for an elective Stochastic processes in engineering systems.

1.Definition of a probability space.

2.Properties of the probability measure

3.The limit supremum and limit infimum of a sequence of sets

4.The Borel-Cantelli lemmas

5.Definition of simple random variables

6.The notion of measurability of a map between two measurable spaces

7.Integration of measurable functions

8.The expectation of a random variable

9.Examples of random variables.

10.Central limit theorem for sums of independent random variables.

11.Various notions of convergence of a sequence of random variables.

Convergence in probability, convergence almost surely, convergence in mean square, convergence in distribution.

Proof that if $X_n, n = 0, 1, 2, \ldots$ is a sequence of random variables and for each $\epsilon > 0$, $\sum_n P(|X_n| > \epsilon) < \infty$, then X_n converges almost surely to zero.

Proof that if a sequence of random variables converges in probability, then a subsequence can be extracted which converges almost surely.

hint:Let X_n converge in probability to zero. Then for each $k = 1, 2, \ldots$, there exists an integer n_k such that $P(|X_{n_k}| > 1/k) < 1/2^k$. Further, it is clear that the $n'_k s$ can be chosen to be increasing. Then let k_0 be any fixed positive integer. We have

$$\sum_{k > k_0} P(|X_{n_k}| > 1/k_0) \leq \sum_{k > k_0} 2^{-k} < \infty$$

proving that the sequence X_{n_k} converges almost surely to zero.

11.Stochastic processes: Definition in terms of joint probability distribution functions, notion of a stationary process, Definition of a Markov process and examples.

$$P(X_t \in B | \mathcal{F}_s) = P(X_t \in B | X_s) a.s.P, s < t$$

where $\mathcal{F}_t = \sigma(X_s : s \leq t)$. This is the definition of a Markov process. It follows that the Markov process statistics is completely determined by the initial distribution μ (distribution of X_0) and the transition probability density functions

$$p(s, t, x, y) dy = P(X_t \in dy | X_s = x), s < t$$

We have that

$$P(X_{t_i} \in dx_i, i = 1, 2, \ldots, n) = P(t_1, dx_1) P(t_1, x_1; t_2, dx_2) \ldots P(t_{n-1}, x_{n-1}, t_n, dx_n)$$

so that

$$P(X_{t_i} \in E_i, i = 1, 2, \ldots, n) = \int_{E_1 \times E_2 \times \ldots \times E_n} P(t_1, dx_1) P(t_1, x_1, t_2, dx_2) \ldots P(t_{n-1}, x_{n-1}, t_n, dx_n)$$

Note that

$$P(t_1, E) = \int_E P(t_1, dx_1) = \int \mu(dx_0) P(0, x_0, t_1, E)$$

Problem:Verify the consistency of the family of finite dimensional probability distributions obtained from an initial distribution and a transition probability function. Hence, show that given the initial distribution and a transition probability distribution, a Markov process can be constructed (hint:Use Kolmogorov's consistency theorem).

12.Let $X_t, t \geq 0$ be a stochastic process with autocorrelation function $R(t,s) = \mathbb{E}(X_t X_s)$. Then determine the differential equations for the correlations $\mathbb{E}(Y_t X_s)$ and $\mathbb{E}(Y_t Y_s)$ where Y satisfies the differential equation

$$D^n Y_t + a_1 D^{n-1} Y_t + \ldots + a_{n-1} D Y_t + a_n Y_t = X_t$$

Further, express these correlations in terms of the impulse response of the system.

Ans: for $t > s$,

$$D^n R_{YX}(t,s) + a_1 D^{n-1} R_{YX}(t,s) + \ldots + a_{n-1} D R_{YX}(t,s) + a_n R_{YX}(t,s) = R_{XX}(t,s)$$

where $D = \frac{\partial}{\partial t}$. If h_t is the impulse response, then

$$R_{YX}(t,s) = \int h(u) R_{XX}(t-u,s) du$$

12. Proof of the Kolmogorov existence theorem for stochastic processes.

13. Almost sure non-differentiability of the Brownian sample paths.

11.Brownian motion:Definition, continuity of sample paths, the space $C[0,\infty)$ and its σ field $\mathcal{B}(C[0,\infty))$, Brownian martingales, examples.

12.Modifications of a stochastic process, notion of indistinguishability of two processes and equality of two processes in distribution.

13.Filtration, completion of a filtration, augmentation of a filtration. If $\mathcal{F}_t$ is a filtration, let $\mathcal{N}_t$ denote the set of all subsets G of Ω for which there exists an $N \in \mathcal{F}_t$ with $P(N) = 0$ and $G \subset N_t$. Then we define $\bar{\mathcal{F}}_t = \sigma(\mathcal{F}_t \cup N_t)$. We define $\mathcal{F}_t^0 = \sigma(\mathcal{F}_t \cup N_\infty)$. The first is the augmented filtration and the second is the completed filtration.

14.Given a filtration $\{\mathcal{F}_t : t \geq 0\}$, the notion of $\mathcal{F}_{t+} = \bigcap_{s>t} \mathcal{F}_s$ and its properties.

15.Optional times and stopping times relative to a filtration. If $\{\mathcal{F}_t\}$ is a filtration and T a stop time, then the definition of $\mathcal{F}_T$ as the collection of all events E for which $E \cap \{T \leq t\} \in \mathcal{F}_t, t \geq 0$. Proof that $\mathcal{F}_T$ is a σ-field.

16.Markov semigroups:

$$\mathbb{E}(f(X_t)|X_s = x) = U_{t-s} f(x), t > s$$

Then from the Markov property, for $t > s > u$,

$$U_{t-u} f(x) = \mathbb{E}(f(X_t)|X_u = x) = \mathbb{E}(\mathbb{E}(f(X_t)|X_s)|X_u = x) = U_{s-u} U_{t-s} f(x)$$

so that for $u, v \geq 0$,

$$U_{u+v} = U_u U_v$$

17.Construction of the Brownian motion process using the Haar functions and sequences of independent Gaussian random variables. Definition of the Haar functions: $H_{n,k}(t)$ equals $-2^{-n/2}$ for $(2k-1)/2^n \leq t \leq 2k/2^n$, $2^{-n/2}$ for $2k/2^n \leq t \leq (2k+1)/2^n$ and zero at all the other points. We are assuming that n takes values $2r, r = 0, 1, 2, ...$ and $k = 1, 2, ..., 2^{n/2} - 1$. Then $\{H_0\} \cup \{H_{n,k}\}$ forms an orthonormal basis for $L^2[0, 1]$. Here $H_0(t) = 1$ for $0 < t < 1$. We define

$$S_{n,k}(t) = \int_0^t H_{n,k}(s)ds$$

$S'_{n,k}s$ are called the Schauder functions. We want to show that $\sum_{n,k} S_{n,k}(t)S_{n,k}(s) = t \wedge s$. For suppose $t, s \in [0, 1]$ and $s \le t$. We have

$$S_{n,k}(t) = \int_0^1 H_{n,k}(u)\chi_{[0,t]}(u)du$$

so that since $H'_{n,k}s$ form an onb for $L^2[0, 1]$, it follows that

$$\sum_{n,k} H_{n,k}(v)S_{n,k}(t) = \chi_{[0,t]}(v)$$

Integrate both sides with respect to v from $v = 0$ upto $v = s$ to get

$$\sum_{n,k} S_{n,k}(s)S_{n,k}(t) = s$$

and this proves the claim. Let ξ_{nk} be iid $N(0, 1)$ random variables and define the process

$$B(t) = \sum_{n,k} \xi_{nk} S_{n,k}(t)$$

Then we can show that $B(.)$ has almost surely continuous sample paths and that its covariance equals

$$\mathbb{E}(B(t).B(s)) = \sum_{n,k} S_{n,k}(t).S_{n,k}(s) = t \wedge s$$

To prove the continuity of the sample paths of $B(.)$, we must make use of the Borel-Cantelli lemma combined with uniform convergence.

18.Progressively measurable processes: For each $T > 0$, the mapping $(\omega, t) \rightarrow X_t(\omega)$, from $\Omega \times [0, T]$ into $\mathbb{R}$ should be $\mathcal{F}_T \otimes \mathcal{B}[0, T]$ measurable.

19.Definition of an adapted process and the filtration generated by a stochastic process.

20.Computing probabilities related to stoptimes of the Brownian motion using the optional stopping theorem for martingales. Let T_a be the first hitting time of the process at the level a. Then from the optional stopping theorem,

$$\mathbb{E}exp(\lambda B(T_a) - \lambda^2 T_a/2) = 1$$

or

$$\mathbb{E}exp(-\lambda^2 T_a/2) = exp(-\lambda a)$$

and this gives the moment generating function of T_a. Let T_b be the first hitting time at level $b > 0$ and T_{-a} the first hitting time at level $-a < 0$. Let $T_{a,b} = T_{-a} \wedge T_b$. Then, from the optional stopping theorem,

$$\mathbb{E}B(T_{a,b}) = 0$$

or

$$aP(T_{-a} < T_b) + bP(T_{-a} > T_b) = 0$$

and this determines the probability $P(T_{-a} < T_b)$, the probability that the process will hit the level $-a$ before it hits b. Again from the optional stopping theorem,

$$\mathbb{E}exp(\lambda B(T_{a,b}))exp(-\lambda^2 T_{a,b}/2) = 1$$

It follows that

$$\mathbb{E}exp(\lambda(B(T_{a,b}) - (b-a)/2))exp(-\lambda^2 T_{a,b}/2) = exp(-\lambda(b-a)/2)$$

Changing λ to $-\lambda$ gives

$$\mathbb{E}exp(-\lambda(B(T_{a,b}) - (b-a)/2))exp(-\lambda^2 T_{a,b}/2) = exp(\lambda(B-a)/2)$$

Adding the two equations gives us

$$\mathbb{E}cosh(\lambda(B(T_{a,b}) - (b-a)/2))exp(-\lambda^2 T_{a,b}/2) = cosh(\lambda(b-a)/2)$$

Now $B(T_{a,b})$ assumes values $-a, b$ only and hence $B(T_{a,b}) - (b-a)/2$ assumes values $\pm(a+b)/2$ from which it follows that

$$\mathbb{E}exp(-\lambda^2 T_{a,b}/2) = \frac{cosh(\lambda(b-a)/2)}{cosh(\lambda(b+a)/2)}$$

which yields the desired formula for the moment generating function of $T_{a,b}$.

The exponential martingale: If $B(t), t \geq 0$ is standard Brownian motion and x a parameter, then

$$M_x(t) = exp(x.B(t) - x^2 t/2)$$

is a Martingale as can be verified by using the formula for the conditional probability densities of jointly normal random variables. This can also be readily verified using Ito's formula for Brownian motion:

$$dM_x(t) = M_x(t)dB(t)$$

so that

$$M_x(t) - M_x(s) = \int_s^t M_x(u)dB(u)$$

21.Doob's maximal inequality for martingales as a generalization of Kolmogorov's maximal inequality for sums of independent random variables.

Suppose $X_n, n = 0, 1, 2, ...$ is a martingale. Define for $\epsilon > 0$, the stop time

$$\tau = min(n : |X_n| > \epsilon)$$

Let $\mathcal{F}_n = \sigma(X_k : k \leq n)$, the filtration generated by the process X. Then,

$$P(max_{0 \leq k \leq n}|X_k| > \epsilon) = P(\tau \leq n)$$

On the other hand,

$$\mathbb{E}|X_n| = \sum_{k=0}^{n} \mathbb{E}|X_n|\chi_{\tau=k} \geq \sum_{k=0}^{n} \mathbb{E}|X_k|\chi_{\tau=k}$$

$$\geq \epsilon \sum_{k=0}^{n} P(\tau = k) = \epsilon.P(\tau \leq n)$$

and this is the desired maximal inequality of Doob.

Suppose X_t is a continuous semimartingale. The Doob-Meyer decomposition gives

$$X_t = A_t + M_t$$

where A is a process of bounded variation and M_t is a martingale. The Ito differential rule gives

$$df(X_t) = f'(X_t)(dA_t + dM_t) + \frac{1}{2}f''(X_t)d < M, M >_t$$

which in integral form, translates to

$$f(X_t) - f(X_0) = \int_0^t f'(X_s)dA_s + \int_0^t f'(X_s)dM_s + \frac{1}{2}\int_0^t f''(X_s)d < M, M >_s$$

The integral with respect to M is a stochastic integral.

Let M be a martingale and $c_1, c_2, \dots$ real constants and $0 = t_0 < t_1 < \dots < t_n = T$. Consider the random variable

$$X = \sum_{k=0}^{n-1} c_k(M(t_{k+1}) - M(t_k))$$

We have for $k < m$,

$$\mathbb{E}(M(t_{k+1}) - M(t_k))(M(t_{m+1}) - M(t_m)) = BbbE((M(t_{k+1}) - M(t_k))\mathbb{E}(M(t_{m+1}) - M(t_m)|\mathcal{F}_{t_m}) = 0$$

and hence,

$$\mathbb{E}(X^2) = \sum_{k=0}^{n-1} c_k^2 \mathbb{E}(M(t_{k+1}) - M(t_k))^2$$

Further, note that

$$\mathbb{E}M(t_{k+1})M(t_k) = \mathbb{E}M(t_k)^2$$

and hence

$$\mathbb{E}(M(t_{k+1}) - M(t_k))^2 = \mathbb{E}M(t_{k+1})^2 - \mathbb{E}M(t_k)^2$$

and we get

$$\mathbb{E}X^2 = \sum_{k=0}^{n-1} c_k^2(\mathbb{E}M(t_{k+1})^2 - \mathbb{E}M(t_k)^2)$$

More generally, we can assume that c_k is an $\mathcal{F}_{t_k}$-measurable random variable. Consider the process

$$\xi(t) = \sum_{k=0}^{n-1} c_k \chi_{[t_k,t_{k+1})}(t)$$

This is an adapted process. Indeed let $t \in [t_k, t_{k+1})$. Then, $\xi(t) = c_k$ which is $\mathcal{F}_{t_k}$-measurable and hence $\mathcal{F}_t$-measurable. The stochastic integral of ξ with respect to M is defined by

$$= I_M(\xi) = \int_0^T \xi(t) dM(t) = \sum_{k=0}^{n-1} c_k(M(t_{k+1}) - M(t_k))$$

and hence,

$$\mathbb{E} I_M(\xi) = 0, \mathbb{E} I_M(\xi)^2 = \sum_k \mathbb{E} c_k^2 \mathbb{E}((M(t_{k+1}) - M(t_k))^2 | \mathcal{F}_{t_k})$$

$$= \sum_k \mathbb{E}[c_k^2 [\mathbb{E}((M(t_{k+1})^2 | \mathcal{F}_{t_k}) - M(t_k)^2]]$$

Consider $s < t$. The Ito rule gives

$$d(M^2) = 2M.dM + d < M, M >$$

so that

$$M_t^2 - M_s^2 = 2 \int_s^t M dM + < M, M >_t - < M, M >_s$$

and hence

$$\mathbb{E}(M_t^2 | \mathcal{F}_s) - M_s^2 = \mathbb{E}(< M, M >_t | \mathcal{F}_s) - < M, M >_s$$

This formula can also be proved directly without the use of Ito's formula. In fact, one normally uses this formula to define the Ito integral and derive its properties. We start by noting that if $s = t_0 < t_1 < ... < t_n = t$, then

$$(M_t - M_s)^2 = (\sum_{k=0}^{n-1} (M(t_{k+1}) - M(t_k)))^2 = \sum_k (M(t_{k+1}) - M(t_k))^2$$

$$+ 2 \sum_{k<m} (M(t_{k+1}) - M(t_k))(M(t_{m+1}) - M(t_m))$$

so that

$$\mathbb{E}((M_t - M_s)^2 | \mathcal{F}_s) = \sum_k \mathbb{E}((M(t_{k+1}) - M(t_k))^2 | \mathcal{F}_s)$$

which on letting the partition become finer and finer results in

$$\mathbb{E}((M_t - M_s)^2 | \mathcal{F}_s) = \mathbb{E}(< M, M >_t - < M, M >_s | \mathcal{F}_s) = \mathbb{E}(< M, M >_t | \mathcal{F}_s) - < M, M >_s$$

From this, we readily conclude that

$$\mathbb{E} I_M(\xi)^2 = \mathbb{E}(\int_0^T \xi(t)^2 d < M, M >_t)$$

Reference:Karatzas and Shreve, Brownian motion and stochastic calculus.

22.Existence and uniqueness theorems for solutions to stochastic differential equations, proof based on martingale inequalities.

23.Derivation of the Fokker-Planck equation from the Chapman-Kolmogorov equation and from the stochastic differential equation using the Ito calculus.

24.Applications of stochastic differential equations to problems of economics.

A company has $d+1$ assets. At time t, the prices of these assets are $P_0(t), P_1(t), ..., P_d(t)$ where $P_0(t)$ is the bond and $P_1, ..., P_d$ are the other assets. The bond evolves in time in accord with a deterministic rate function $r(t)$ while the other assets evolve with randomly fluctuating rates:

$$dP_0(t) = r(t)P_0(t)dt, dP_i(t) = P_i(t)(b_i(t)dt + \sum_j \sigma_{ij}(t)dB_j(t)), i = 1, 2, ..., d$$

A person owns $N_i(t)$ assets of type i at time t. He consumes an amount $C(t)dt$ in time $[t, t+dt]$. Let $X(t)$ denote the amount he has at time t. Then it follows that

$$dX(t) = \sum_{i=0}^{d} N_i(t)dP_i(t) - C(t)dt$$

This is the fundamental stochastic differential equation that determines the evolution of the amount owned by the person with time. Note that

$$\sum_{i=0}^{d} N_i(t)dP_i(t) = r(t)N_0(t)P_0(t)dt + \sum_{i=1}^{d} N_i(t)(b_i(t)dt + \sum_j \sigma_{ij}(t)dB_j(t))P_i(t)$$

$$= r(N_0P_0 + \sum_{i=1}^{d} N_iP_i)dt + \sum_{i=1}^{d} N_i(b_i - r + \sum_j \sigma_{ij}dB_j)P_i$$

$$= r(t)X(t)dt + \sum_{i=1}^{d}(b_i(t) - r(t) + \sum_j \sigma_{ij}(t)dB_j(t))\pi_i(t)$$

where

$$\pi_i(t) = N_i(t)P_i(t)$$

25.Brownian local time. If B_t is Brownian motion then we define for $x \in \mathbb{R}$ and $t \geq 0$,

$$L(t, x, \omega) = lim_{\epsilon \to 0} \frac{1}{4\epsilon} meas\{s : 0 \leq s \leq t, |B_s - x| \leq \epsilon\}$$

It follows that for any Borel subset B of $\mathbb{R}$,

$$2\int_B L(t, x)dx = \int_0^t \chi_B(B_s)ds$$

and hence

$$\int_0^t P(B_s \in B)ds = 2 \int_B \mathbb{E}L(t,x)dx$$

Proof that if Brownian local time exists, then the sample paths of Brownian motion are almost surely non-differentiable.

23.Estimation of Volterra kernels in image processing(Vireshwar's M.Tech thesis work)

$$y(n,m) = \sum_{k,l} h(k,l)x(n-k,m-l) + \sum_{k,l,p,q} g(k,l,p,q)x(n-k,m-l)x(n-p,m-q)$$

24.The heat equation and related partial differential equations associated with the Brownian motion process

Suppose $B(t)$ is Brownian motion and f a twice differentiable function on $\mathbb{R}$. Then, $u(t,x) = \mathbb{E}f(x+B(t))$ satisfies the Heat equation. Conversely suppose $v(t,x)$ satisfies the Backward equation, ie,

$$v_{,t} = \frac{1}{2}v_{,xx}$$

Then $M(t) = v(t,B(t))$ is a Brownian martingale as can be verified immediately using the Ito formula:

$$dM(t) = v_{,t}(t,B(t)) + v_{,x}(t,B(t))dB(t) + \frac{1}{2}v_{,xx}(t,B(t))dt$$

$$= (v_{,t} + \frac{1}{2}v_{,xx})(t,B(t))dt + v_{,x}(t,B(t))dB(t)$$

$$= v_{,x}(t,B(t))dB(t)$$

and hence

$$M(t) - M(s) = \int_s^t v_{,x}(u,B(u))dB(u)$$

proving that $M(.)$ is a Martingale. Suppose F is a nondecreasing function on $[0,\infty)$ and consider

$$v(t,x) = \int_0^\infty exp(xy - ty^2/2)dF(y)$$

Then, for every t, $v(t,x)$ is nonnegative an increasing in x and

$$v_{,t} = -\int (y^2/2)exp(xy - ty^2/2)dF(y), v_{,xx} = \int y^2 exp(xy - ty^2/2)dF(y)$$

so that

$$v_{,t} + \frac{1}{2}v_{,xx} = 0$$

ie v satisfies the backward heat equation. For every $b > 0$, we can find an $A(t,b)$ such that $v(t,A(t,b)) = b$ since v is non-decreasing in the second variable. Now $v(t,B(t))$ is a martingale. Let $T_b = inf(t \geq 0 : B(t) > A(t,b))$. Then

$$T_b = inf(t \geq 0 : v(t,B(t)) > b)$$

and we have

$$P(s < T_b < \infty | \mathcal{F}_s) = P(\sup_{t>s} v(t, B(t)) > b | \mathcal{F}_s^B) = v(s, B(s))/b$$

from a well known result in martingales.

Suppose $Z(t)$ is a non-negative martingale. Let $T = inf(t > 0 : Z(t) = b)$. Let $A \in \mathcal{F}_s$. Let $s < t$. Then,

$$\int_{A \cap \{T > s\}} Z(t \wedge T) dP = \int_{A \cap \{T > s\}} Z(s \wedge T) dP = \int_{A \cap \{T > s\}} Z_s dP$$

The left side equals

$$\int_{A \cap \{T > s\}} Z(t) \chi_{T=\infty} dP + \int_{A \cap \{T > s\}} Z(T) \chi_{\{T < \infty\}} dP$$

Assuming that $Z(t) \to 0$ as $t \to \infty$, the limit of the above as $t \to \infty$ equals

$$bP(A \cap \{s < T < \infty\}$$

since $Z(T) \chi_{T < \infty} = b \chi_{T < \infty}$. We thus get

$$b.P(\{\infty > T > s | \mathcal{F}_s) = Z(s)$$

on the event $\{T > s\}$. This proves the claim. We have thus obtained the probabilities that the Brownian motion will cross a large class of curves $A(t, b)$.

Reference:Karatzas and Shreve, Brownian motion and stochastic calculus.

[23]. Reflected Brownian motion:Let $B_t, t \geq 0$ be standard Brownian motion. Then $|B_t|$ is the reflected Brownian motion, ie, it is obtained by reflecting B at the origin. Claim:Reflected Brownian motion is a Markov process. Indeed, let $s < t$. Then,

$$P(|B_t| \in dy | \mathcal{F}_s^B) = P(|B_t| \in dy | B_s) = (p_t(B_s, y) + p_t(B_s, -y)) dy$$

where

$$p_t(x, y) dy = P(B_t \in dy | B_0 = x)$$

is the transition probability density of B. Now, $p_t(x, -y) = p_t(-x, y)$ and hence

$$P(|B_t| \in dy | \mathcal{F}_s^B) = (p_t(B_s, y) + p_t(-B_s, y)) dy$$

It is clear that this function depends on B_s only via $|B_s|$ and therefore

$$P(|B_t| \in dy | \mathcal{F}_s^{|B|}) = (p_t(B_s, y) + p_t(-B_s, y)) dy$$

Thus, $|B_t|$ is a Markov process with transition probability density

$$p_+(t, x, y) = p_t(x, y) + p_t(-x, y)$$

Reference:Karatzas and Shreve, Brownian motion and stochastic calculus.

[24]. Brownian motion reflected between two walls. Let $x = 0$ be the first wall and $x = a$ the other. Define $\phi(2na) = 0, \phi((2n+1)a) = a$ and ϕ is linear in the intervals $[2na, (2n+1)a]$ and $[(2n+1)a, (2n+2)a]$, ie, $\phi(x) = x - 2na$ for $x \in [2na, (2n+1)a]$ and $\phi(x) = (2n+2)a - x$ for $x \in [(2n+1)a, (2n+2)a]$. Then, $\phi(B_t)$ is obtained by reflecting B_t between the two walls $x = 0$ and $x = a$. We have for $y \in [0, a]$,

$$P^x(\phi(B_t) \in dy) = \sum_n P^x(B_t \in [2na, (2n+1)a], B_t - 2na \in dy)$$

$$+ P^x(B_t \in [(2n+1)a, (2n+2)a], (2n+2)a - B_t \in dy)$$

$$= (\sum_n (p_t(2na + y - x) + p_t((2n+2)a - y - x))dy = \sum_n p_t(2na + y - x) + p_t(2na - y - x))dy$$

$$= \sum_n p_+(t, x, 2na + y)$$

where

$$p_+(t, x, y) = p(t, x, y) + p(t, x, -y)$$

is the transition probability density for absorbed Brownian motion.

Reference:Karatzas and Shreve, Brownian motion and stochastic calculus.

[25]. Absorbed Brownian motion: Let T_0 be the first time at which Brownian motion B_t starting at some point $x > 0$ hits the origin, ie,

$$T_0 = min(t \geq 0 : B_t = 0)$$

Then the process $X(t) = B(t \wedge T_0)$ is absorbed Brownian motion. We have for $y > 0$,

$$P^x(X(t) \in dy) = P^x(T_0 > t, B(t) \in dy)$$

$$= P^x(min_{0<s<t}B(s) > 0, B(t) \in dy) = P^x(min_{0<s<t}B(s) - x > -x, B(t) - x \in dy - x)$$

$$= P^0(min_{0<s<t}B(s) > -x, B(t) \in dy - x) = P^0(max_{0<s<t} - B(s) < x, -B(t) \in -dy + x)$$

$$= P^0(M(t) < x, B(t) \in -dy + x) = P(B(t) \in -dy + x) - P(M(t) > x, B(t) \in -dy + x)$$

$$= (p_t(y - x) - p_t(y + x))dy = p_-(t, x, y)dy$$

where

$$p_-(t, x, y) = p_t(y - x) - p_t(y + x)$$

Here,

$$M(t) = max(B(s) : s \leq t)$$

Reference:Karatzas and Shreve, Brownian motion and stochastic calculus.

Note that the probability $P(M(t) > x, B(t) < x - y)$ for $x > y > 0$ can be calculated using the reflection principle.

[26] Strong and weak solutions to stochastic differential equations

[27] Martingales associated with a diffusion process. Suppose X_t is a diffusion process that satisfies the stochastic differential equation

$$dX_t = \mu(X_t)dt + \sigma(X_t)dB_t$$

where B is a standard Brownian motion process. Let f be a twice continuously differentiable function on $\mathbb{R}$. Then,

$$df(X_t) = f'(X_t)dX_t + \frac{1}{2}f''(X_t)\sigma^2(X_t)dt$$

$$= f'(X_t)(\mu(X_t)dt + \sigma(X_t)dB_t) + \frac{1}{2}f''(X_t)\sigma^2(X_t)dt$$

$$= A_t f(X_t)dt + \sigma(X_t)f'(X_t)dB_t$$

where A_t is the generator of the diffusion process, ie,

$$A_t f(x) = \mu(x)f'(x) + \frac{1}{2}\sigma^2(x)f''(x)$$

It follows that

$$M_t^f = f(X_t) - f(X_0) - \int_0^t A_s f(X_s)ds$$

is a martingale. The basic theorem of Stroock and Varadhan states that if for a large class of functions f, M_t^f is a martingale, then (X, W) is a weak solution to the stochastic differential equation.

Girsanov's theorem: Let X_t be and adapted process and B_t a Brownian motion process. X is adapted with respect to the filtration generated by B. Define the process

$$Z_t = exp(\int_0^t X_s dB_s - \frac{1}{2}\int_0^T X_s^2 ds)$$

Then

$$dZ_t = X_t dB_t$$

and hence, it follows that Z is a Brownian martingale. Let $T > 0$ and define a probability measure Q so that $dQ = Z_T dP$. We want to show that the process

$$Y_t = B_t - \int_0^t X_s ds$$

is a Brownian motion process relative to the probability measure Q over the interval $[0, T]$. To do so, we observe that the moment generating functional of the process Y over this interval is given by

$$\Phi_Y(f) = \mathbb{E}_Q(exp(\int_0^T f(t)dY_t) = \mathbb{E}(Z_T.exp(\int_0^T f(t)dY_t))$$

Now,

$$\int_0^T f(t)dY_t = \int_0^T f(t)dB_t - \int_0^T f(t)X_t dt$$

so that

$$\Phi_Y(f) = \mathbb{E}(exp(\int_0^T (f(t) + X_t)dB_t - \int_0^T (f(t)X_t + X_t^2/2)dt))$$

We note that

$$\mathbb{E}((exp((f(t) + X_t)dB_t - (f(t)X_t + X_t^2/2)dt)|\mathcal{F}_t) = exp((f(t) + X_t)^2 dt/2 - (f(t)X_t + X_t^2/2)dt)$$

$$= exp(f^2(t)dt/2)$$

and hence

$$\Phi_Y(f) = exp(\frac{1}{2}\int_0^T f^2(t)dt)$$

proving that Y is a Brownian motion relative to the measure Q.

Example: Let B_t be standard Brownian motion consider the sde

$$dX_t = sgn(X_t)dB_t$$

Define

$$W_t = \int_0^t sgn(X_s)dX_s$$

It is easy to see that (X, W) is a weak solution to the above sde. Indeed $dW_t = sgn(X_t)dX_t$ and hence

$$\int_0^t sgn(X_s)dW_s = \int_0^t sgn(X_s)^2 dX_s = \int_0^t dX_s = X_t$$

We also have that $(-X, W)$ is also a solution to the same sde and hence pathwise uniqueness does not hold for the above sde.

Suppose W is a process such that for every twice continuously differentiable function f,

$$f(W_t) - \frac{1}{2}\int_0^t \Delta f(W_s)ds$$

is a martingale. Then W is a standard Brownian motion process. Indeed take $f(x) = x$ to conclude that W_t is a martingale. Now take $f(x) = x^2$ and conclude that $W_t^2 - t$ is a martingale. But from the stochastic calculus for martingales, it follows that

$$d(W_t^2) = 2W_t dW_t + d < W, W >_t$$

and hence the Doob-Meyer decomposition of the submartingale W_t^2 is given by

$$W_t^2 = < W, W >_t + M_t$$

where M is a martingale. From the uniqueness of the Doob-Meyer decomposition, it follows that $< W, W >_t = t$ and $M_t = W_t^2 - t$. The former implies due to a theorem of Levy that W is a Brownian motion process.

Let $\sigma_{ij}(t,x), 1 \leq i \leq d, 1 \leq j \leq r$ be smooth functions on $\mathbb{R}_+ \times \mathbb{R}^d$ and let $b_i(t,x), 1 \leq i \leq d$ also be smooth functions on the same space. Consider the sde

$$dX_{it} = b_i(t, X_t)dt + \sum_j \sigma_{ij}(t, X_t)dB_j$$

where $(B_1, ..., B_r)$ is a standard r-dimensional Brownian motion process. Let $f : \mathbb{R}^d \to \mathbb{R}$ be a smooth function. By the Ito differential rule,

$$df(X_t) = \sum_i \frac{\partial f(X_t)}{\partial x_i}(b_i(t, X_t)dt + \sum_j \sigma_{ij}(t, X_t)dB_j(t)) + (\frac{\partial f(t, X_t)}{\partial t} + \frac{1}{2}\sum_{i,j} a_{ij}(t, X_t)\frac{\partial^2 f(t, X_t)}{\partial x_i \partial x_j})dt$$

where

$$a_{ij}(t, x) = \sum_k \sigma_{ik}(t, x)\sigma_{jk}(t, x)$$

Define A_t, the generator of the Markov process X so that:

$$A_t f(x) = \sum_i b_i(t, x)\frac{\partial f(t, x)}{\partial x_i} + \frac{1}{2}\sum_{i,j} a_{ij}(t, x)\frac{\partial^2 f(t, x)}{\partial x_i \partial x_j}$$

Then, the above formula can be expressed as

$$df(X_t) = (\frac{\partial f(t, X_t)}{\partial t} + A_t f(X_t))dt + dM_t^t$$

where

$$M_t^f = \int_0^t \sum_{i,j} \frac{\partial f(X_s)}{\partial x_i}\sigma_{ij}(s, X_s)dB_j(s)$$

is a martingale. Thus,

$$M_t^f = f(t, X_t) - f(0, X_0) - \int_0^t (\frac{\partial f(s, X_s)}{\partial s} + A_s f(X_s))ds$$

Conversely, the Stroock-Varadhan theorem asserts that if for all well behaved f, M_t^f as defined above is a martingale, then there exists a Brownian motion W such that (X, W) is a weak solution to the above sde. Let us work out the details for the scalar case. The extension to the vector case is straightforward.

$$M_t^f = f(X_t) - \int_0^t A_s f(X_s)ds$$

is a martingale where

$$A_t f(x) = \underline{(t, x)}f'(x) + \frac{1}{2}\sigma^2(t, x)f''(x)$$

Take $f(x) = x$ to get the result that

$$M_t = X_t - \int_0^t b(s, X_s)ds$$

is a martingale. Take $f(x) = x^2$ to get the result that

$$N_t = X_t^2 - \int_0^t (2X_s b(s, X_s) + \sigma^2(s, X_s))ds$$

is a martingale. Now,

$$N_t = (M_t + \int_0^t b(s, X_s)ds)^2 - \int_0^t (2X_s b(s, X_s) + \sigma^2(s, X_s))ds$$

$$= M_t^2 + (\int_0^t bds)^2 + 2M_t \int_0^t bds - \int_0^t (Xb + \sigma^2)ds = M_t^2 + Z_t$$

say. We have

$$dZ_t = 2b_t \int_0^t b_s ds + 2(\int_0^t b_s ds)dM_t + 2M_t b_t dt - (2X_t b_t + \sigma_t^2)dt$$

so that the martingale component of Z is the integral of $2(\int_0^t b_s ds)dM_t$ and the bounded variation component of the same is the integral of

$$2b_t \int_0^t b_s ds + 2M_t b_t db_t - 2(M_t + \int_0^t b_s ds)b_t dt - \sigma_t^2 dt = -\sigma_t^2 dt$$

Denoting the martingale part of Z by V, we get

$$N_t = M_t^2 + V_t - \int_0^t \sigma_s^2 ds$$

and hence

$$< M, M >_t = \int_0^t \sigma_s^2 ds$$

It follows from the basic martingale representation theorem that there exists a Brownian motion W such that

$$M_t = \int_0^t \sigma^2(s, X_s)dW_s$$

and this shows that (X, W) is a weak solution to the sde

$$dX_t = b(t, X_t)dt + \sigma(t, X_t)dW_t$$

Reference:Karatzas and Shreve, Brownian motion and stochastic calculus.

[28]. Ito's formula using local time. Formally we have

$$L(t, x) = \int_0^t \delta(x - B(s))ds$$

so that

$$\int_A L(t, x)dx = \int_0^t \chi_A(B(s))ds$$

Then,

$$f''(B(t)) = \int_{\mathbb{R}} f''(x)\delta(x - B(t))dx = \int f''(x)dL(t,x)dx/dt$$

or

$$\int_0^t f''(B(s))ds = \int f''(x)L(t,x)dx$$

and the Ito formula can be cast as

$$f(B(t)) - f(B(0)) = \int_0^t f'(B(s))dB(s) + \frac{1}{2}\int_{\mathbb{R}} f''(x)L(t,x)dx$$

This is also called the Ito-Meyer formula.

[29]. Poisson measure: Let a stochastic process field $N(t, E)$ be defined for $t \geq 0$ and Borel subsets E of $\mathbb{R}^n$ such that if $E_1, ..., E_k$ are disjoint Borel sets, the processes $N(., E_i), i = 1, 2, .., k$ are independent and for each Borel set E, $N(., E)$ is a Poisson process with mean $\lambda.\mu(E)$ where μ is a measure on the Borel subsets of $\mathbb{R}^n$. Consider the random variable

$$X = \int_{[0,T]\times E} f(t,x)N(dt, dx)$$

Compute its moment generating function. Suppose first that f is simple. Then, the above integral can be evaluated by noting that if $F_i, i = 1, 2, ..., n$ are disjoint subsets of $[0, T] \times \mathbb{R}^n$, then

$$\mathbb{E}exp(\sum_i c_i N(F_i)) = \Pi_{i=1}^n \mathbb{E}exp(c_i N(F_i)) = \Pi_{i=1}^n exp(\int_{F_i} lambdadtdF(x)(exp(c_i) - 1))$$

or equivalently, if μ denotes the Lebesgue measure on $\mathbb{R}$, then, the above equals

$$exp(\sum_i (exp(c_i) - 1)\mu \times F(F_i))$$

In the general case, this becomes

$$exp(\int_{[0,T]\times E} \lambda(exp(f(t,x)) - 1))dtdF(x))$$

[30]. Integration with respect to semimartingales. The Doleans-Dade-Meyer formula.

Let $X(t) = X_d(t) + X_c(t)$ where X is a semimartingale and X_c is the continuous component of X while X_d is the discontinuous component of X. Let $\Delta X(t) = X(t) - X(t-)$ denote the jump of X at time t. We have

$$dX(t) = X(t + dt) - X(t-) = \Delta X(t) + dX_c(t)$$

It follows that

$$df(X(t)) = f(X(t-) + dX(t)) - f(X(t-)) = f(X(t-) + \Delta X(t) + dX_c(t)) - f(X(t-))$$

$$= f(X(t) + \Delta X(t)) - f(X(t-)) + f'(X(t-) + \Delta X(t))dX_c(t) + \frac{1}{2}f''(X(t-) + \Delta X(t))d[X,X]_c(t)$$

$$= \Delta f(X(t)) + f'(X(t-))dX_c(t) + \frac{1}{2}f''(X(t-))d[X,X]_c(t)$$

Note that this formula can also be expressed as

$$df(X(t)) = \Delta f(X(t)) - f'(X(t-))\Delta X(t) + f'(X(t-))dX(t) + \frac{1}{2}f''(X(t-))d[X,X]_c(t)$$

This formula can be expressed in integral form as

$$f(X(t)) - f(X(s)) = \sum_{s \leq u \leq t} (\Delta f(X(u)) - f'(X(u-))\Delta X(u)) + \int_s^t f'(X(u-))dX(u)$$

$$+ \frac{1}{2} \int_s^t f''(X(u-))d[X,X]_c(u)$$

[31]. Definition of the variation and p^{th}-order variation of a process. Suppose $X(t)$ is a process defined over the interval $[0, T]$. Its pth-order variation over the interval $[0, t]$ is defined by

$$V_p(X)_t = lim_{n \to \infty} \sum_{i=0}^{n-1} |X(t_{n,i+1}) - X(t_{n,i})|^p$$

where $0 = t_{n,0} < t_{n,1} < ... < t_{n,n} = t$ and $max(|t_{n,i+1} - t_{n,i}| : 0 \leq i \leq n - 1) \to 0$ as $n \to \infty$.

Problem:Show that if the pth order variation is finite, then the qth order variation is zero for all $q > p$. Indeed, we have

$$V_q(X) == lim_n \sum_i |X(t_{i+1}) - X(t_i)|^q \leq lim_n (sup_i |X(t_{i+1}) - X(t_i)|^{q-p}) \sum_i |X(t_{i+1}) - X(t_i)|^p$$

which converges to zero since the first term converges to zero by continuity and the second term converges to the finite number $V_p(X)_t$.

[32]. Ito's formula for a Levy process. Assume that $N(t, dx)$ is a Poisson random measure and let $f(t, x)$ be a function of two variables, a time variable $t \in \mathbb{R}$ and a Borel subset B of $\mathbb{R}^n$:

$$N(t, B) = \int_B N(t, dx)$$

Then define the process

$$X(t) = \int_{[0,t] \times E} f(s, x)N(ds, dx)$$

We have $dX(t) = f(t, x)$ if $N(dt, dx) = 1$ and zero other wise. Thus,

$$d\psi(X(t)) = \int (\psi(x + f(t, x)) - \psi(x))N(dt, dx)$$

This is Ito's formula for the Levy process X.

Note:Suppose E, F are two disjoint Borel subsets of $\mathbb{R}^n$. Then

$$P(N(dt, E) = N(dt, F) = 1) = o(dt)$$

which follows from the facts that the random variables $N(dt, E)$ and $N(dt, F)$ are independent and $P(N(dt, E) = 1) = \lambda.F(E)dt$.

[33]. Notion of a stochastic differential equation driven by a Levy process:

$$Z(t) - Z(0) = \int_{[0,t]\times E} \psi(s, x, Z(s))N(ds, dx) + \int_0^t \phi(s, Z(s))dB(s) + \int_0^t \eta(s, Z(s))ds$$

[34]. Martingale representation theorem.

Let M_t be a continuous Martingale relative to the filtration $\{\mathcal{F}_t\}$. Let $\tau(t) = min(s > 0 :< M, M >_s = t\}$. Let

$$B_t = M_{\tau(t)}$$

Then, B is a continuous Martingale relative ot the filtration $\{\mathcal{F}_{\tau(t)}\}$ and

$$< B, B >=< M, M >_{\tau(t)} = t$$

Thus, by Levy's theorem, B is a Brownian motion process. We have

$$M_t = B_{<M,M>_t}$$

[35] Let X, Y, Z be jointly Gaussian random variables. Let $p(x, y|z)$ denote the joint density of X, Y given Z. Determine the value of z at which $p(x, y|z)$ attains its maximum. Assume x, y, z have zero mean.

hint: Let Q denote the inverse of the correlation matrix of the vector (X, Y, Z). Then,

$$p(x, y|z) = p(x, y, z)/p(z)$$

$$= K.exp(-\frac{1}{2}(Q(11)x^2 + Q(22)y^2 + Q(33)z^2 + 2Q(12)xy + 2Q(23)yz + 2Q(13)xz))/exp(-\frac{1}{2\sigma^2}z^2)$$

where $\sigma^2 = \mathbb{E}z^2$. The optimal value of z at which this attains its maximum is obtained by minimizing

$$Q(33)z^2 + 2Q(23)yz + 2Q(13)xz - z^2/\sigma^2$$

The minimum is obtained by setting the partial derivative of the above with respect to z equal to zero. This gives

$$Q(33)z + Q(23)y + Q(13)x - z/\sigma^2 = 0$$

or

$$z = \frac{(Q(23)y + Q(13)x)}{(1/\sigma^2 - Q(33))}$$

[36] Problem: Let B be standard Brownian motion and let $u < s < t$. Determine the value of z so that $P(B(u) \in dx, B(t) \in dy | B(s) = z)$ is maximized. Specialize to the case when $u = (s+t)/2$.

hint: We first need the correlation matrix of the Gaussian random vector $(B(s), B(t), B(u))$. It is given by

$$R(11) = s, R(12) = R(21) = s, R(13) = R(31) = s, R(22) = t, R(23) = R(32) = u, R(33) = u$$

or in matrix notation

$$\mathbf{R} = \begin{pmatrix} s & s & s \\ s & t & u \\ s & u & u \end{pmatrix}$$

To compute the inverse of this matrix, we must solve the linear equations

$$R(\xi, \eta, \zeta)^T = (x, y, z)^T$$

ie,

$$s\xi + s\eta + s\zeta = x, s\xi + t\eta + u\zeta = y, s\xi + u\eta + u\zeta = z$$

Solving, we find that

$$\eta = (y - z)/(t - u), \xi = (ux - sz)/(s(u - s)), \zeta = z/s - (ux - sz)/(s(u - s)) - (y - z)/(t - u)$$

so that

$$Q(13) = 1/(s - u), Q(23) = 1/(u - t), Q(33) = 1/s + 1/(u - s) + 1/(t - u)$$

Further,

$$\sigma^2 = u$$

and the optimum value of z' that maximizes $p(x', y' | z')$ can be written down. Specifically,

$$z' = (Q(23)y + Q(13)x)/(\sigma^{-2} - Q(33)) = (y/(u - t) + x/(s - u))/(1/u - 1/s - 1/(u - s) - 1/(t - u))$$

Now observe that

$$1/u - 1/s - 1/(u-s) - 1/(t-u) = (s(u-s)(t-u) - u(u-s)(t-u) - us(t-u) - us(u-s))/us(u-s)(t-u)$$

[37] Martingale representation theorem:

1.Suppose $X(t, \omega)$ is a stochastic process on a probability space $(\Omega, \mathcal{F}, P)$ and $Y(t)$ is a stochastic process on another probability space $(\Omega', \mathcal{F}, P')$. Define the processes $\tilde{X}_t$ and $\tilde{Y}_t$ on the probability space $(\Omega \times \Omega', \mathcal{F} \otimes \mathcal{F}', P \times P')$ so that

$$\tilde{X}_t(\omega, \omega') = X_t(\omega), \tilde{Y}_t(\omega, \omega') = Y_t(\omega')$$

Then the processes $\tilde{X}, \tilde{Y}$ are independent and the former has the same law as X while the latter has the same law as Y. This is a useful device in probability theory and we shall apply it to obtain the representation theorem for continuous martingales.

Let $M_t^i, i = 1, 2, ..., n$ be components of a vector valued martinagle with respect to a filtration. Assume that M^i are almost surely continuous. Define the matrix $Q(t) = (q_{ij}(t))$ with

$$\int_0^t q_{ij}(s)ds = <M^i, M^j>_t = \lim \sum_n (M^i(t_{n+1}) - M^i(t_n))(M^j(t_{n+1}) - M^j(t_n))$$

where $0 = t_0 < t_1 < ... < t_n = t$ and the as n increases, the size of the partition converges to zero. Then $Q(t)$ is continuous and positive semidefinite for each t. It has the spectral factorization

$$Q(t) = U(t)\Lambda(t)U^T(t)$$

where $U(t)$ is a real orthogonal matrix and $\Lambda(t) = diag(\lambda_i(t), i = 1, 2, ..., n)$ with $\lambda_i(t) \geq 0$ for all t. Define the vector valued process

$$N(t) = \int_0^t U^T(s)dM(s)$$

or equivalently, in component form,

$$N_i(t) = \sum_j \int_0^t u_{ji}(s)dM_j(s)$$

Clearly, N_i is a martingale for every i. Define

$$Z_i(t) = \int_0^t \chi_{\lambda_i(s)>0}dN_i(s)/\sqrt{\lambda_i(s)} + \int \chi_{\lambda_i(s)=0}dB_i(s)$$

where (B^i) is a vector valued Brownian motion independent of N. Then,

$$dZ_i(t)dZ_j(t) = dt.\delta_{ij}$$

In fact, we have

$$<N_i, N_j>_t = \int_0^t \sum_{m,k} u_{mi}(s)u_{kj}(s)d<M_m, M_k>_s = \int_0^t \sum_{m,k} u_{mi}(s)u_{kj}(s)q_{mk}(s)ds = (\int_0^t \lambda_i(s)ds)\delta_{ij}$$

It follows that

$$d<N_i, N_j>_t = \lambda_i(t)dt.\delta_{ij}$$

Thus,

$$d<Z_i, Z_j>_t = \chi_{\lambda_i(t)>0}d<N_i, N_j>_t /\lambda_i(t) + \chi_{\lambda_i(t)=0}\delta_{ij}dt$$

$$= \delta_{ij}(\chi_{\lambda_i(t)>0} + \chi_{\lambda_i(t)=0})dt = \delta_{ij}dt$$

This proves the claim and hence $Z = (Z_i)$ is a vector valued Brownian motion process.

Reference:Karatzas and Shreve, Brownian motion and stochastic calculus.

[38] Solution of the general bilinear sde:

$$dX(t) = (aX(t) + b)dt + (cX(t) + d)dB(t)$$

Suppose that $B(.)$ were a function of bounded variation. Then let us derive the general solution. We can express the above sde as

$$dX(t) = (a + cdB(t))X(t) + bdt + d.dB(t)$$

the solution to which is

$$X(t) = exp(a + cB(t))X(0) + \int_0^t exp(a(t - s) + c(B(t) - B(s))))(b.ds + d.dB(s))$$

We therefore try a solution of the form

$$X(t) = \int_0^t exp(\alpha(t - s) + \beta(B(t) - B(s)))(\gamma.ds + \delta.dB(s))$$

We compute its differential:

$$dX(t) = \gamma.dt + \delta.dB(t) + \alpha.X(t)dt + \beta.X(t)dB(t) + (\beta^2/2)X(t)dt$$

$$= (\gamma + (\alpha + \beta^2/2).X(t))dt + (\delta + \beta X(t))dB(t)$$

In order that this satisfy the given sde, it is sufficient that

$$\gamma = b, \alpha + \beta^2/2 = a, \delta = d, \beta = c$$

so that

$$\alpha = a - c^2/2$$

We have thus obtained a particular solution to the given bilinear sde.

[39] Skorohod equation: Let $y(t)$ be a continuous process with $y(0) = 0$. Then there exists a unique nondecreasing process $k(t)$ such that if

$$x(t) = y(t) + k(t)$$

then $x(t) \geq 0$ for all t and $k(t)$ is flat off $\{x(t) = 0\}$, ie, $\int_0^T \chi_{x(t)>0} dk(t) = 0$ for all $T > 0$. This process is given by

$$k(t) = max(0, max_{s \leq t}(-y(s)))$$

From this formula, we verify that

$$k(t) + y(t) = max(y(t), max_{s \leq t}(y(t) - y(s))) \geq 0$$

since $y(t) - y(s)|_{s=t} = 0$. We need verify that $k(t)$ is flat off the set $\{t : x(t) = 0\}$. Suppose $x(t) > 0$ for some t. Then since $k(t) = max(0, max_{s \leq t} - y(s))$ we have

$$x(t) = k(t) + y(t) = max(y(t), max_{s \leq t}(y(t) - y(s))) > 0$$

Two cases are therefore possible. One $y(s) > 0$ for all $s \in [0, t]$ in which case, $x(t) = y(t)$. In this case, $k(t) = 0$. Moreover since y is continous and $y(t) > 0$, it follows that there exists an $h > 0$ such that $y(u) > 0$ for all $u \in [t, t+h]$. Hence, $k(u) = 0$ for all $u \in [t, t+h]$ which means that k is flat over $[t, t+h]$ proving the claim. In the second case $y(s) \leq 0$ for some $s \in [0, t]$. Then there exists an $s_0 \in [0, t]$ such that $y(s_0) = min_{s \leq t} y(s) \leq 0$. In this case $k(t) = -y(s_0)$ and $x(t) = y(t) - y(s_0)$. Since $x(t) > 0$, it follows that $x(t) = y(t) - y(s_0) > 0$ and hence, there exists an $h > 0$ such that $x(u) = y(u) - y(s_0) > 0$ for all $u \in [t, t+h]$. It is also clear that $y(u) > y(s_0)$ for all $u \in [t, t+h]$ and hence $-y(u) < -y(s_0)$ for all $u \in [t, t+h]$. This implies that $k(u) = -y(s_0)$ for all $u \in [t, t+h]$ and hence k is flat off $x > 0$.

Now let $W(t)$ be standard Brownian motion. The Ito rule gives

$$d|W(t)| = sgn(W(t))dW(t) + dL_t(0)$$

where

$$L_t(0) = \int_0^t \delta(W(s))$$

is the Brownian local time at 0. Equivalently,

$$|W(t)| = \int_0^t sgn(W(s))dW(s) + L_t(0)$$

it is known that $L_t(0)$ is nondecreasing, continuous and flat off $|W(t)| > 0$. Hence from the Skorohod equation, it follows that $L_t(0) = max_{-s \leq t}(-B(s))$ where $B(t) = \int_0^t sgn(W(s))dW(s)$ is again a Brownian motion process. It follows that $L_t(0)$ and M_t have the same law and $M_t - B_t$ has the same law as $|W(t)|$. The Skorohod equation thus provides a remarkable method of deducing equivalence in law of important stochastic processes associated with the Brownian motion process.

It remains to prove the uniqueness of the process $k(t)$. Suppose that $k(t), x(t)$ satisfy all the given conditions. Consider first an interval $[a, b]$ in which $x(u) = 0$. Then $k(u) + y(u) = 0$ for all $u \in [a, b]$. It follows that $k(u) = -y(u)$ for all $u \in [a, b]$. Now suppose $x(u) > 0$ for $u \in (b, c)$. Then, since k is flat off $x > 0$, it follows that $k(c) = k(b) = -y(b)$. Suppose

[40]. Conditional expectation as an orthogonal projection.

[41]. Quantum Ito's formula and quantum stochastic integration.

1.Symmetric and antisymmetric tensor products of a Hilbert space.

2.Boson and Fermion Fock space.

3.Exponential vectors in the Boson Fock space.

4.Weyl operator

5.Creation, annihilation and conservation fields in Boson Fock space

6.Notion of a quantum martingale process

6.The creation, annihilation and conservation processes in Boson Fock space.

7.The quantum stochastic integral, definition.

$\mathcal{H}$ is a Hilbert space and $u \in \mathcal{H}$. $\Gamma_s(\mathcal{H})$ is the Boson Fock space. Specifically, let $\otimes_s$ denote the symmetric tensor product in $\mathcal{H}$. Then,

$$\Gamma_s(\mathcal{H}) = \mathbb{C} \oplus \bigoplus_{n \geq 1} \mathcal{H}^{\otimes_s n}$$

For $u \in \mathcal{H}$, define $e(u) \in \Gamma_s(\mathcal{H})$ by the equation

$$e(u) = 1 \oplus \bigoplus_{n \geq 1} u^{\otimes n}/\sqrt{n!}$$

Then

$$< e(u), e(v) >= exp(< u, v >), u, v \in \mathcal{H}$$

The closure of the span of the vectors $\{e(u) : u \in \mathcal{H}\}$ is the Boson Fock space $\Gamma_s(\mathcal{H})$.

[42] Let $\mathcal{U}(\mathcal{H})$ denote the algebra of unitary operators in $\mathcal{H}$. For $u \in \mathcal{H}$ and $U \in \mathcal{U}(\mathcal{H})$, define the operator $W(u, U)$ in $\Gamma_s(\mathcal{H})$ so that

$$W(u, U)e(v) = exp(- \parallel u \parallel^2 /2- < u, Uv >)e(Uv + u)$$

We verify that $W(u, U)$ is a unitary operator.

$$< W(u, U)e(w), W(u, U)e(v) >=< e(u), e(v) >$$

Let $\xi(.)$ be a nonatomic spectral measure on $\mathbb{R}_+ = [0, \infty)$ with values in $P(\mathcal{H})$. Define $\mathcal{H}_{t]} = \xi[0, t]\mathcal{H}$. Then for $0 < t_1 < t_2 < ... < t_n$, we define

$$\mathcal{H}_{[t_i, t_{i+1}]} = \xi[t_i, t_{i+1}]\mathcal{H}$$

We then have the direct sum decomposition

$$\mathcal{H} = \bigoplus_{i=0}^{n} \mathcal{H}_{[t_i, t_{i+1}]}$$

where

$$\mathcal{H}_{[t_0, t_1]} = \mathcal{H}_{t_1]} = \xi[0, t_1]\mathcal{H}$$

and

$$\mathcal{H}_{[t_n, t_{n+1}]} = \mathcal{H}_{[t_n, \infty)} = \xi[t_n, \infty)\mathcal{H}$$

Now we use an important result: Suppose $\mathcal{H} = \mathcal{H}_1 \otimes \mathcal{H}_2$. Then $\Gamma_s(\mathcal{H})$ is isomorphic to $\Gamma_s(\mathcal{H}_1) \otimes \Gamma_s(\mathcal{H}_2)$. This isomorphism can be defined by $e(u + v) \rightarrow e(u) \otimes e(v)$ for $u \in \mathcal{H}_1$ and $v \in \mathcal{H}_2$. The unitary isomoprhism property follows from the equations

$$< e(u_1 + v_1), e(u_2 + v_2) >= exp(< u_1 + v_1, u_2 + v_2 >)$$

$$= exp(<u_1, u_2> + <v_1, v_2>) = <e(u_1), e(u_2)><e(v_1), e(v_2)>$$

$$= <e(u_1) \otimes e(v_1), e(u_2) \otimes e(v_2)>, u_1, u_2 \in \mathcal{H}_1, v_1, v_2 \in \mathcal{H}_2$$

It follows that $\Gamma_s(\mathcal{H})$ is unitarily isomorphic to $\otimes_i \Gamma_s(\mathcal{H}_{[t_i, t_{i+1}]})$ Let H be a unitary operator in $\mathcal{H}$ and define

$$\lambda(H) = -i\frac{d}{dt}W(0, exp(itH))|_{t=0}$$

Then since $t \to W(0, exp(itH))$ is a one parameter group, it follows that

$$W(0, exp(itH)) = exp(it\lambda(H))$$

We have

$$W(0, exp(itH))e(u) = e(exp(itH)u)$$

and hence,

$$<e(v), W(0, exp(itH))e(u)> = <e(v), e(exp(itH)u)> = exp(<v, exp(itH)u>)$$

and hence,

$$<e(v), \lambda(H)e(u)> = <v, Hu><e(v), e(u)>$$

[43] Let L_i be an operator in $\Gamma_s(\mathcal{H})$ such that $L_i e(u) = (L_i e(u_{t_i}])).e(u_{[t_i})$ where $u_{t]} = \xi[0, t]u$ and $u_{[t} = \xi[t, \infty)u$. Then define the process

$$L(t) = \sum_i L_i \chi_{[t_i, t_{i+1}]}(t)$$

$L(t)$ is then a simple adapted process in the quantum sense. (See An introduction to quantum stochastic calculus, by K.R.Parthasarathy) Define the processes $A(t), A^*(t), \Lambda_H(t)$ by the equations

$$A(t) = a(m_t), A^*(t) = a^*(m_t), \Lambda_H(t) = \lambda(H_t)$$

where m_t is a ξ-martingale, ie,

$$\xi[0, s]m_t = m_s, t \geq s$$

We have

$$A(t)e(u) = a(m_t)e(u) = <m_t, u> e(u) = <m, \xi[0, t]u>$$

$$= <<m, u>>[0, t]e(u) = <m_t, u_{t]}> e(u_{[t}) = (a(m_t)e(u_{t]})).e(u_{[t})$$

where $m = m_\infty$. Define the integral

$$\int L(t)dM(t) = \sum_i L_i(M(t_{i+1}) - M(t_i))$$

By this we mean that

$$\left(\int L(t)dM(t)\right)e(u) = \sum_i (L_i e(u_{t_i}]))(M(t_{i+1}) - M(t_i))e(u_{[t_i})$$

Note that

$$< e(v), a^*(m)e(u) >=< a(m)e(v), e(u) >=<< m, v > e(v), e(u) >=< v, m >< e(v), e(u) >$$

On the other hand,

$$\frac{d}{dt} < e(v), e(u+tm) > |_{t=0} = \frac{d}{dt} exp(< v, u+tm >)|_{t=0} =< v, m >< e(v), e(u) >$$

and therefore

$$a^*(m) = \frac{d}{dt}e(u+tm)|_{t=0}$$

We have

$$A(t)e(u) = a(m_t)e(u) = (a(m_t)e(u_{t]}))e(u_{[t})$$

Define the process

$$\Lambda_H(t) = \lambda(H_t) = \lambda(H\xi[0,t])$$

We are assuming that H commutes with the spectral measure ξ. We have

$$< e(v), \Lambda_H(t)e(u) >=< e(v), \lambda(H_t)e(u) >< e(v), e(u) >=< v, H_t u >=< v, Hu > [0,t] < e(v), e(u) >$$

We have

$$< e(u), W(exp(iH_t))e(v) >=< e(u), e(exp(iH_t)v) >$$

$$= exp(< u, exp(iH_t)v >) = exp(< u, v > + << u, (exp(iH) - 1)v >> [0,t])$$

which implies that

$$d < e(u), W(exp(iH_t))e(v) >=< e(u), W(exp(iH_t))e(v) > d << u, (exp(iH) - 1)v >> (t)$$

or

$$< e(u), W(exp(iH_t))e(v) >= \int_0^t < e(u), W(exp(iH_s))e(v) > d << u, (exp(iH) - 1)v >> (s)$$

This means that the process $X(t) = W(exp(iH_t))$ satisfies the quantum stochastic differential equation

$$dX(t) = X(t).d\Lambda_{exp(iH)-1}(t), X(0) = I$$

Now consider the process

$$Y(t) = W(m_{t]}, I)$$

We have

$$Y(t)e(v) = exp(-\frac{1}{2} >< m, m >> [0,t]- << m, v >> [0,t])e(v + m_{t]})$$

and therefore

$$< e(u), Y(t)e(v) >= exp(-\frac{1}{2} << m, m >> [0,t]- << m, v >> [0,t]+ < u, v > + << u, m >> [0,t])$$

It follows that

$$d < e(u), Y(t)e(v) >=< e(u), Y(t)e(v) > (d << m, m >> (t) - d << m, v >> (t) + << u, m >> (t))$$

from which it follows that $Y(t)$ satisfies the quantum stochastic differential equation

$$dY(t) = Y(t)(d << m, m >> (t) - dA^*(t) + dA(t))$$

We now come back to the definition of the quantum stochastic integral above for adapted processes:

$$\left(\int L dM \right)e(u) = \sum (L_i e(u_{t_i]}))(M(t_{i+1}) - M(t_i))e(u_{t_i})$$

Take $M(t) = A(t) = a(m_{t]})$. Then

$$(M(t_{i+1}) - M(t_i))e(u_{t_i]}) = << m, u >> [t_i, t_{i+1}]e(u_{[t_i})$$

[44] Tarun's thesis

Conclusions and scope for further work: This work, broadly speaking, deals with the applications of Ito's stochastic calculus and stochastic differential equations to a class of engineering problems covering estimation of the trajectories of objects subject to random forces and the problem of determining the stochastic behaviour of weights in the least mean square algorithm applied to system identification. Using Ito's stochastic calculus, Kushner had formulated and solved the problem of constructing an evolution equation for the conditional probability density of the state of a system at a given time based on noisy observations. The state of the system in this formulation is assumed to obey a vector stochastic differential equation. This evolution equation, however, is highly nonlinear and intractable as far as implementation issues are concerned. Several engineers working on this problem therefore suggested many kinds of approximations to the Kushner equation. Such approximations are based on Taylor series expansions of the conditional moments of observables. When the noisy observations are discrete then the problem becomes more useful and approximations can be derived for this too. This formulation has been applied in this thesis to the trajectory estimation problem. Secondly, the least mean square algorithm tells us how to adapt the weights of a system to get good approximations to the true system. This adaptation is based on matching input to output data using an instantaneous error minimization algorithm. This discrete time algorithm can well be approximated by a stochastic differential equation when the noise is white and one can derive from this approximation a the Fokker-Planck equation for the conditional probability density of the weight. This enables us to determine the convergence properties of the weight and one of the goals of this thesis is to demonstrate this by different kinds of examples. It is worth looking at higher order Taylor approximations to the trajectory estimation problem. This thesis uses a truncated second order Taylor approximation. Another point worth looking at is the quantization of the filtering algorithm with applications to quantum phenomena. A quantum stochastic calculus has been developed by Hudson and Parthasarathy [] and using this, one can formulate noisy versions of the Schrodinger and Heisenberg dynamics. The noise is introduced in the form of creation, annihilation and conservation operator fields along with a quantum Ito formula for these processes. Using this one can develop a nonlinear filtering theory. The concept of measurement is difficult to understand in quantum mechanics. For example two noncommuting observables cannnot be measured simultaneously and hence in general developing a filtering theory for noncommutative measurements becomes impossible. Applications of quantum filtering theory include estimating the trajectories of atoms and elementary particles moving inside particle accelerators. Not much success has been achieved in this direction and it would be interesting to look at this. Suppose that we are given a signal $x(t)$, which may be real or complex and suppose that this signal is the output of a nonlinear system. We wish to choose the coefficients of the approximating system so that a given set of characteristics of the output data is matched to the same set of characteristics of a desired signal. Then an LMS like algorithm can be developed for that purpose. For example, we can have the least mean phase algorithm in which the phase of the complex output signal is matched to those of a given signal. More generally, we express the given characteristics of the output as a vector valued functional of the output at a given time and those of the desired signal in a similar manner. Then the squared error between these two vectors is adaptively minimized using the LMS algorithm. In this thesis, we have focussed our attention on stochastic differential equations driven by Brownian motion. More general kinds of noise can also be considered, for example, Poisson shot noise and more generally, a white nonGaussian noise. There is a general calculus of stochastic differential equations driven by Levy processes (which are independent increment processes) and one can consider developing a filtering theory for such equations.

[45] Problems in signal processing.

[1] Given a time series $\{x[n] : n \in \mathbb{Z}\}$, explain how you would obtain the coefficients of the first order linear predictor based on past p data samples using the least squares method.

hint:

$$\hat{x}[n] = -\sum_{k=1}^{p} a[klx[n-k]$$

where $a[k], k = 1, 2, ..., p$ are obtained by minimizing the error energy

$$\sum_{n=p}^{N-1} (x[n] - \hat{x}[n])^2$$

3.Explain the construction of the causal Wiener filter obtained by solving the Wiener-Hopf convolutional integral equation. Explain Kolmogorov's innovations approach to the construction of the causal Wiener filter.

4.Consider a data stream $x[n], n = 0, 1, ..., N-1$. Estimate the autocorrelation sequence using the scheme

$$\hat{R}[\tau] = \frac{1}{N-\tau} \sum_{n=0}^{N-\tau-1} x[n]x[n+\tau], \tau \geq 0$$

and put $R[-\tau] = R[\tau]$ for $\tau < 0$. Construct the $p \times p$ autocorrelation matrix

$$\mathbf{R} = ((R[i-j]))_{0 \leq i,j \leq p-1}$$

Prove that this matrix is Toeplitz but not necessarily positive semidefinite.

5.Develop a prediction theory for Volterra systems. Specifically, suppose that $x(n)$ is to be predicted based on $x(n-k), k = 1, 2, ..., p$ using the model

$$\hat{x}(n) = \sum_{k=1}^{p} h(k)x(n-k) + \sum_{k,m=1}^{p} g(k,m)x(n-k)x(n-m)$$

Then formulate the optimal equations satisfied by $\{h(k)\}$ and $\{g(k,m)\}$ in matrix form and obtain the solution.

6.Consider a function $T(t,x)$ and explain how you would fit a model of the form

$$T_{,t}(t,x) = DT_{,xx}(t,x)$$

where D is an adjustable parameter. Derive the optimal equations satisfied by D based on data collected over the set $[0, N] \times [a, b]$.

7.Write down the discretized form of the Klein-Gordon equation

$$\phi_{,tt}(t,x) - \phi_{,xx}(t,x) + m^2 \phi(t,x) = 0$$

and obtain the general solution of this discretized equation.

8.Derive the array signal model

$$\mathbf{x} = \mathbf{As} + \mathbf{w}|$$

for the signal collected by an array of sensors from plane wave sources in the sky in p different directions. Obtain the expressions for the array signal matrix $\mathbf{A}$ in terms of the directions of the sources and the position of the sensor elements.

9.Suppose $x(t)$ is a signal. Let $x_e(t)$ denote its even part and $x_o(t)$ its odd part. Show that

$$\int_{-\infty}^{\infty} x_e(t)x_o(t)dt = 0$$

and hence,

$$\int_{-\infty}^{\infty} x^2(t)dt = \int_{-\infty}^{\infty} x_e(t)^2 dt + \int_{-\infty}^{\infty} x_o(t)^2 dt$$

10.Suppose $x[n]$ is a discrete time signal with period N. Define the signal $y[n] = \sum_k x[k]\delta[pk - n]$ where p is a positive integer. Show that $y[n] = x[n/p]$ if p divides n and $y[n] = 0$ if p does not divide n. Deduce that $y[n + pN] = y[n]$ for all n. Show that if $Y(e^{j\omega})$ and $X(e^{j\omega})$ are respectively the DTFT's of y and x, then

$$Y(e^{j\omega}) = X(e^{jp\omega})$$

11.Suppose $x[n]$ is a discrete time signal and p a positive integer. Define the signal $y[n] = x[pn]$. Compute the DTFT of $y[n]$ in terms of the DTFT of $x[n]$.

hint:

$$X(\omega + 2\pi k/p) = \sum_{n} x[n]exp(-j\omega n).exp(-j2\pi kn/p)$$

so that

$$\sum_{k=0}^{p-1} X(\omega + 2\pi k/p) = p \sum_{n,r} x[n]exp(-j\omega n)\delta[n-rp] = p\sum_{r} x[rp]exp(-j\omega rp) = p.Y(p\omega)$$

12. Suppose $x(u,v)$ is a two dimensional signal with Fourier transform

$$X(\omega_1, \omega_2) = \int_{\mathbb{R}^2} x(u,v)exp(-j(\omega_1 u + \omega_2 v))dudv = F(x)(\omega_1, \omega_2)$$

Then prove the following statements:

(a) The Fourier transform F is a linear operator, ie, if

$$F(c_1 x + c_2 y) = c_1 F(x) + c_2 F(y)$$

(b) Time shifting property: If $x_1(u,v) = x(u-a, v-b)$, then

$$F(x_1)(\omega_1, \omega_2) = exp(-j(a\omega_1 + b\omega_2))F(x)(\omega_1, \omega_2)$$

(c) Modulation property: If $x_1(u,v) = exp(j(t_1 u + t_2 v))x(u,v)$, then

$$F(x_1)(\omega_1, \omega_2) = X(\omega_1 - t_1, \omega_2 - t_2) = F(x)(\omega_1 - t_1, \omega_2 - t_2)$$

13.Suppose $f(x,y)$ is a function defined on $\mathbb{R}^2$. The projection of this function on the line $px + qy - 1 = 0$ is defined by the formula

$$F(p,q) = \int f(x,y)\delta(px + qy - 1)dxdy$$

We have

$$\delta(px + qy - 1) = \frac{1}{2\pi}\int_{-\infty}^{\infty} exp(it(px + qy - 1))dt$$

and hence,

$$F(p,q) = \int f(x,y)exp(it(px+qy-1))dtdxdy = \frac{1}{2\pi}\int exp(-it)dt \int f(x,y)exp(i(tpx + tqy))dxdy$$

$$= \int exp(-it)\hat{f}(tp, tq)dt$$

where $\hat{f}$ is the two dimensional Fourier transform of f.

14.Suppose V is a vector space over $\mathbb{R}$ and $B : V \times V \to \mathbb{R}$ is a nondegenerate bilinear form. Given any $X \in V$, we can associate an $f_X \in V^*$ so that $f_X(Y) = B(X,Y), Y \in V$. A function $p : V \to \mathbb{R}$ is called a polynomial if it has the form

$$p(X) = \sum_{(r,i)\in D} f_{1,i}(X)f_{2,i}(X)...f_{r,i}(X)$$

where $f_{j,i} \in V^*$ and D is a finite set. We can associate with each polynomial p of the above form the element ξ_p of the symmetric algebra $S(V)$ given by

$$\xi_p = \sum_{r,i} \frac{1}{r!} \sum_{\sigma \in S_r} X_{\sigma 1,i} \otimes ... \otimes X_{\sigma r,i}$$

where

$$f_{j,i} = f_{X_{j,i}}$$

For $X_1, ..., X_m \in V$, it is usual to denote $\frac{1}{m!} \sum_{\sigma \in S_m} X_{\sigma 1} \otimes ... \otimes X_{\sigma m}$ by $X_1 X_2 ... X_m$. Then we have

$$\xi_p = \sum_{(r,i) \in D} X_{1,i} X_{2,i} ... X_{r,i}$$

The correspondence $p \to \xi_p$ is an isomorphism between the algebra of polynomials in V and the symmetric algebra $S(V)$.

15.Calculate $\mathbb{E}(y(n_1)...y(n_k))$ in terms of $\mathbb{E}(x(m_1)...x(m_k))$ when

$$y(n) = \sum_k h(n,k) x(k)$$

[46] Abstract:This paper deals with the problem of estimating the parameters of a transistor using wavelets. An amplifier is constructed using this transistor and using the Ebers-Moll model of the transistor, the basic Kirchchoff voltage and current equations for the amplifier are written down. This is a system of nonlinear ordinary differential equations, the nonlinearity arising due to the transistor characteristics. The voltages at the different nodes are expanded in powers of a small parameter that defines the nonlinearity and using this perturbation expansion, a second order Volterra model for the amplifier is set up. The parameter in the first order and second order Volterra kernel to be estimated is V_T, the cutin voltage of the transistor. The Volterra kernels are then expanded using a wavelet basis in which the coefficients now become functions of V_T. The output is also expanded in the wavelet basis and hence a linear model that relates the wavelet coefficients of the first and second order Volterra kernels to those of the output is derived. Using this model combined with a gradient search method, V_T is estimated. The advantage of using the wavelet representation of the amplifier model in contrast to estimating V_T directly from the sampled time domain model is that the latter after sampling may not contain all the significant characteristics of the signal unless the sampling time interval is very small. On the other hand, the wavelet method will preserve all the important characteristics of the signal even if only a few coefficients are used because the wavelet method is adapted to time-scale resolution. Specifically, if the signal has different frequencies over different time intervals, sampling at a uniform duration will not be effective. If the sampling interval is too small, then too many samples will be needed while if it is too large, then signal characteristics in the high frequency region will be lost. On the other hand, we can capture the essential characteristics of the signal by choosing the time delay index ranges appropriately at different resolutions. We first choose a large resolution index and then adjust the time delay index

differently at different resolutions. The wavelet based method is particularly suitable for signals where the frequency changes with time.

[47]. Prove that if $p > 1$ and $x, y \in \mathbb{C}^N$, then

$$|\sum_i x_i \bar{y}_i| \leq (\sum_i |x_i|^p)^{1/p} (\sum_i |y_i|^q)^{1/q}$$

where $q = p/p - 1$, or equivalently, $1/p + 1/q = 1$. Deduce that if $p \geq 1$, then

$$(\sum_i |x_i + y_i|^p)^{1/p} \leq (\sum_i |x_i|^p)^{1/p} + (\sum_i |y_i|^p)^{1/p}$$

The first is called Holder's inequality and the second Minkowski's inequality. The second is deduced from the first as follows:

$$\sum_i |x_i + y_i|^p = \sum_i |x_i + y_i|^{p-1} |x_i + y_i| \leq \sum_i |x_i + y_i|^{p-1} |x_i| + \sum_i |x_i + y_i|^{p-1} |y_i|$$

and

$$\sum_i |x_i + y_i|^{p-1} |x_i| \leq (\sum_i |x_i + y_i|^p)^{1-1/p} (\sum |x_i|^p)^{1/p}$$

an d likewise,

$$\sum_i |x_i + y_i|^{p-1} |y_i| \leq (\sum |x_i + y_i|^p)^{1-1/p} (\sum |y_i|^p)^{1/p}$$

by application of Holder's inequality. Combining these, we get Minkowski's inequality.

[48]. Suppose (X, d) is a metric space. Show that this can be completed in the following sense. There exists a complete metric space $(\hat{X}, \hat{d})$ and a dense subset W of $\hat{X}$ such that (X, d) and $(W, \hat{d})$ are isometric, ie, there exists a bijection $T : X \to W$ such that $d(x, y) = \hat{d}(T(x), T(y))$ for all $x, y \in X$.

hint: Say that two Cauchy sequences $(x_n), (y_n)$ in (X, d) are equivalent if $lim_n d(x_n, y_n) = 0$. Let $\hat{X}$ denote the set of all such equivalence classes of Cauchy sequences in X. If $\hat{x}, \hat{y} \in \hat{X}$, then choose $(x_n) \in \hat{x}, (y_n) \in \hat{y}$ and then define $\hat{d}(\hat{x}, \hat{y}) = lim_n d(x_n, y_n)$. Show that $\hat{d}$ does not depend on the choice of the representatives $(x_n), (y_n)$ in $\hat{x}$ and $\hat{y}$ respectively. Define W to be the set of all elements in $\hat{X}$ of the form $\hat{x}$ where $\hat{x}$ is the equivalence class containing $(x, x, x, ...)$ with $x \in X$. Define $T : X \to W$ so that $T(x) = \hat{x}$. Then show that T is an isometry and that W is dense in $\hat{X}$. We call $\hat{X}$ a completion of X. Show that the completion is unique in the following sense. If (Y, ρ) is any complete metric space that contains a dense subset Z isometric to X, then Y is isometric to $\hat{X}$.

[49]. State the difference between the Wiener and Kalman filters.

hint:The Wiener filter does not rely on any specific model for the random process to be filtered whereas the Kalman filter assumes that the signal model is defined by a set of state variable and output equations, ie, a differential equation model for the process is assumed. Specifically, if $\mathbf{B}(t)$ is a vector valued Brownian motion process, then in the Kalman filter formalism, the state $\mathbf{x}(t)$ of the system satisfies an Ito stochastic differential equation

$$dx(t) = \mathbf{A}(t)|bfx(t)dt + \mathbf{G}(t)d\mathbf{B}(t)$$

and the output, ie the observations are defined by

$$\mathbf{y}(t) = \mathbf{C}(t)\mathbf{x}(t)dt + \mathbf{D}(t)d\mathbf{V}(t)$$

where $\mathbf{V}(.)$ is a vector valued Brownian motion process independent of the process $\mathbf{B}(.)$. In the Wiener filter formalism, we are given two random processes $\mathbf{x}(t) \in \mathbb{R}^n$ and $\mathbf{y}(t) \in \mathbb{R}^m$ and a matrix filter having impulse response $\mathbf{H}(t,s) \in \mathbb{R}^{m \times n}$ is to be designed so that

$$\mathbb{E}(\| \mathbf{x}(t) - \int_E \mathbf{H}(t,s)\mathbf{y}(s)ds \|^2)$$

is to be minimized. Here, E is a subset of the time axis $\mathbb{R}$. If we assume that $E = [0,\infty)$ and $\mathbf{H}(t,s) = 0$ for $s > t$, then what results is the causal Wiener filter. If we do not put any such restrictions on $\mathbf{H}(t,s)$ and assume that $E = \mathbb{R}$, then what results is the non-causal Wiener filter.

[50]. Let V be a finite dimensional real vector space and $B : V \times V \to \mathbb{R}$ be a non-degenerate symmetric bilinear map. Let $f : V \to \mathbb{C}$ be an integrable mapping, ie, we choose a basis $\{e_1, ..., e_n\}$ for V and assume that

$$\int \mathbb{R}^n |f(t_1 e_1 + ... + t_n e_n)| dt_1 ... dt_n < \infty$$

Show that this condition is independent of the basis chosen. Specifically, if $\{f_1, ... f_n\}$ is another basis for V, then we can write $f_i = \sum_{j=1}^n a_{ij} e_j, i = 1, 2, ..., n$ where $A = (a_{ij})$ is a non-singular matrix, Then,

$$\int f(t_1 f_1 + ... + t_n f_n) dt_1 ... dt_n = \int f(\sum_{i,j} t_i a_{ij} e_j) dt_1 ... dt_n = \int f(\sum_j (A^T t)_j e_j) dt_1 ... dt_n$$

$$= \int f(\sum_j t_j e_j) det(A)^{-1} dt_1 ... dt_n$$

using the familiar Jacobian relation for transformations of multiple integrals. Now define the Fourier transform of f so that

$$\hat{f}(x) = \int_V f(y) exp(iB(x,y)) dy$$

This is given meaning by choosing a basis for V and fixing it. Thus,

$$\hat{f}(x) = \int f(\sum_k t_k e_k) exp(i \sum_k t_k B(x, e_k)) dt_1 ... dt_n$$

Thus,

$$\hat{f}(\sum_m s_m e_m) = \int f(\sum_k t_k e_k) exp(i \sum_{k,m} B(e_k, s_m) t_k s_m) dt_1 ... dt_n$$

Determine the Fourier transform of

$$f(x) = exp(- \| x \|^2)$$

assuming that the bilinear form $B(.,.)$ is positive definite. Note that the norm is defined in terms of the basis:

$$\| x \|^2 = \sum_i t_i^2, x = \sum_i t_i e_i$$

hint: diagonalize the form $B(.,.)$ by choosing an appropriate orthonormal basis and transform the multiple integral using the usual Jacobian rule.

[51]. Suppose in the above problem, B is invariant under a non-singular linear transformation T of V, ie, $B(Tx, Ty) = B(x, y)$ for all $x, y \in V$. Then

$$\hat{f}(Tx) = \int_V f(y).exp(iB(Tx, y))dy = \int_V f(y).exp(iB(x, T^{-1}y))dy = \int f(Ty)|T|exp(iB(x, y))dy$$

In other words, the Fourier transform of f evaluated at Tx equals $|T|$ times the Fourier transform of foT where $|T| = |det(T)|$.

[52]. We wish to solve the system of linear equations $Ax = y$ where A is any rectangular or square matrix. If $y \notin R(A)$, then this system has no solution. If $y \in R(A)$, we can write $y = Au$ for some vector u of size equal to the number of columns of A. Then suppose for every u, $x = Gy = GAu$ is a solution. Then $AGAu = AGy = Ax = y = Au$. Since this must hold for every u, it follows that G must satisfy the equation $AGA = A$. In other words, G is a generalized inverse of A.

[53]. Suppose we wish to determine all the vectors x for which for a given y, $\| Ax - y \|$ is a minimum. Explain how you can do this using the notion of generalized inverse.

[54]. Suppose for a fixed y, we wish to determine all the solutions x of the system $Ax = y$ for which $\| x \|$ is a minimum. Explain how you can do this using the notion of generalized inverse.

[55] Problems in detection and estimation theory.

[1] State the most general form of statistical hypothesis testing theory based on minimization of the Bayes risk. Explain how the problem of choosing the decision rule so as to minimize the probability of error is special case of the Bayes risk minimization method.

hint: Let $H_1, ..., H_m$ be the m hypotheses to be tested and let X be the observed random vector. Assume that X takes values in the observation space Ω. Partition Ω into m disjoint regions $Z_i, i = 1, 2, ..., m$ so that if $X \in Z_i$, we decide in favour of H_i. The probability density of X given that H_i is true is $p(X|H_i)$. Let C_{ij} be the cost incurred in deciding in favour of H_i given that H_j is the true hypothesis. Then the average Bayes cost is given by

$$R_B = \sum_{i,j} C_{ij} P_j \int_{Z_i} p(X|H_j)dX = \sum_i W_i$$

where

$$W_i(X) = \sum_j C_{ij} P_j p(X|H_j)$$

Here, P_j is the apriori probability of H_j being true. R_B is minimized provided that Z_i is chosen to be the set of all points X for which $W_i(X) \geq W_j(X)$ for all $j \neq i$. In the special case when $C_{ij} = 1 - \delta_{ij}$, R_B equals the probability of error.

[2] Consider the binary Gaussian problem in which there are two hypotheses H_1, H_0 with apriori probabilities P_1, P_0 respectively. Under H_1, the observation X is a normally distributed random vector with mean vector $\mathbf{m}_1$ and covariance Σ_1 while under H_0, the observation X is a normally distributed random vector with mean vector $\mathbf{m}_0$ and covariance Σ_0. The probability of error is minimized when the decision is to choose H_1 if $\Lambda(X) \geq P_0/P_1$ and to choose H_0 otherwise. Here,

$$\Lambda(X) = exp(\frac{1}{2}((\mathbf{X} - \mathbf{m}_0)^T \Sigma_0^{-1} (\mathbf{X} - \mathbf{m}_0) - (\mathbf{X} - \mathbf{m}_1)^T \Sigma_1^{-1} (\mathbf{X} - \mathbf{m}_1)))$$

Note that

$$L(X) = log\Lambda(X) = \frac{1}{2}\mathbf{X}^T (\Sigma_0^{-1} - \Sigma_1^{-1})\mathbf{X} + (\mathbf{m}_1^T \Sigma_1^{-1} - \mathbf{m}_0^T \Sigma_0^{-1})\mathbf{X}$$

$$+\frac{1}{2}(\mathbf{m}_0^T \Sigma_0^{-1} \mathbf{m}_0 - \mathbf{m}_1^T \Sigma_1^{-1} \mathbf{m}_1)$$

[3] Consider the following binary hypothesis testing problem: Under hypothesis H_1, the observation X has a probability density $p(X|H_1, \theta)$ where θ is a random variable with probability density $f(\theta)$ and under H_0, the observation X has a probability density $p(X|H_0, \theta')$, where θ' is a random variable with probability density $g(\theta')$. Show that the likelihood ratio test can be expressed as:Decide H_1 if $\Lambda(X) > \eta$ and decide H_0 otherwise where

$$\Lambda(X) = \frac{p(X|H_1, \theta)f(\theta)d\theta}{\int p(X|H_0, \theta')g(\theta')d\theta'}$$

[4] Consider the following binary hypothesis testing problem:Under H_1, the observed signal $x(t) = s_1(t) + w(t)$ where $t \in [0, T]$ and $s_1(.)$ is a deterministic signal while $w(.)$ is white noise with spectral density $N_0/2$. Under H_0, the observed signal $x(t) = s_0(t) + w(t)$ where $s_0(t)$ is a deterministic signal while $w(.)$ is white noise with spectral density $N_0/2$. Show that the likelihood ratio test can be expressed as: Decide H_1 if $L(x) > \eta'$ and decide H_0 otherwise where

$$L(x) = \int_0^T x(t)(s_1(t) - s_0(t))dt$$

is the temporal correlation between the observed signal $\{x(t)\}$ and the deterministic signal $\{s_1(t) - s_0(t)\}$.

[5] State the spectral theorem for positive definite kernel $K(t, s), t, s \in [0, T]$ and explain how it can be used in detection theory.

hint:Suppose under H_1, the observed signal $x(t)$ is Gaussian with mean $s_1(t)$ and autocovariance

$$K_1(t_1, t_2) = \mathbb{E}(x(t_1) - s_1(t_1))(x(t_2) - s_1(t_2)|H_1)$$

while under H_0, the observed signal $x(t)$ is Gaussian with mean $s_0(t)$ and autocovariance

$$K_0(t_1, t_2) = \mathbb{E}(x(t_1) - s_0(t_1))(x(t_2) - s_0(t_2)|H_0)$$

Then apply the spectral theorem to the kernels K_1, K_0 and use this to derive the likelihood ratio test.

[6]. Under hypothesis H_1, the signal $x(t) = A.cos(\omega t + \phi) + w(t), 0 \leq t \leq T$ where ϕ is uniformly distributed over $[0, 2\pi)$ while under H_0, $x(t) = w(t)$ where $w(t)$ is white Gaussian noise with spectral density $N_0/2$. Determine the likelihood ratio for this problem.

[7] Under Show that the probability of error in the binary hypothesis testing problem corresponding to the likelihood ratio test $\Lambda(X) \geq \lambda$ implies decide H_1 and $\Lambda(X) \leq \lambda$ implies decide H_0, can be expressed as

$$Pr\{error\} = P_1 \int_{\Lambda(X)\leq\lambda} p(X|H_1)dX + P_0 \int_{\Lambda(X)>\lambda} p(X|H_0)dX$$

$$= P_1 \int_{\Lambda(X)\leq\lambda} \Lambda(X)p(X|H_0)dX + P_0 \int_{\Lambda(X)>\lambda} p(X|H_0)dX$$

[8] Compute the probability of error for the binary Gaussian problem using the likelihood ratio test where under H_1, the observation X is $N(\mathbf{m}_1, \Sigma_1)$ while under H_0 ,the observation X is $N(\mathbf{m}_0, \Sigma_0)$. The apriori probabilities are P_1, P_0 respectively with $P_1 + P_0 = 1$.

[9] In a binary decision problem, let $p(X|H_1)$ and $p(X|H_0)$ denote the respective probability densities of the observation X under the two hypotheses. The likelihood ratio is $\Lambda(X) = \frac{p(X|H_1)}{p(X|H_0)}$. Show that the conditional moments of the likelihood ratio under the two hypotheses satisfy

$$\mathbb{E}(\Lambda(X)^m|H_1) = \mathbb{E}(\Lambda(X)^{m+1}|H_0)$$

for any real number m for which either one side is defined.

[10] Let the observation be a vector $\mathbf{x}$ described by the model

$$\mathbf{x} = \mathbf{As} + \mathbf{w}$$

where $\mathbf{A}$ is a matrix and $\mathbf{s}$ is a nonrandom vector. $\mathbf{w}$ is a complex Gaussian random vector with correlation $\mathbf{R} = \mathbb{E}(\mathbf{x}.\mathbf{x}^*) = \sigma^2\mathbf{I}$ with star denoting conjugate transpose. Show that the maximum likelihood estimate of $\mathbf{s}$ is obtained by minimizing $\| \mathbf{x} - \mathbf{As} \|^2$ and show that it is given by

$$\hat{\mathbf{s}_{\mathbf{ml}}}(\mathbf{x}) = (\mathbf{A}^*\mathbf{A})^{-1}\mathbf{A}^*\mathbf{x}$$

Show that $\mathbf{A}\hat{\mathbf{s}}_{ml}$ equals the orthogonal projection of $\mathbf{x}$ onto $\mathcal{R}(\mathbf{A})$.

[11] Suppose the observations are given by the model

$$x_n = a_n \xi + w_n, n = 1, 2, ..., N$$

where $a_n's$ are constants and ξ is an unknown nonrandom number. $\{w_n\}$ is a sequence of independent zero mean Gaussian random variables with variance $\mathbb{E}x_n^2 = \sigma_n^2$. Show that the maximum likelihood estimate of ξ is obtained by minimizing

$$F(\xi) = \sum_{n=1}^{N}(x_n - a_n\xi)^2/\sigma_n^2$$

Give a recursive algorithm for obtaining the maximum likelihood estimate of ξ as the number of observations increases.

[12] Consider the sequence of observations

$$\mathbf{x}_n = \mathbf{A}_n\mathbf{s} + \mathbf{w}_n, n = 1, 2, ..., N$$

where $\mathbf{w}_n$ are iid zero mean Gaussian random vectors with autocorrelation $\mathbf{R}$. Show that the maximum likelihood estimate of $\mathbf{s}$ is obtained by minimizing

$$\sum_{n=1}^{N}(\mathbf{x}_n - \mathbf{A}_n\mathbf{s})^*\mathbf{R}^{-1}(\mathbf{x}_n - \mathbf{A}_n\mathbf{s})$$

with respect to $\mathbf{s}$.

[13] Let $B(t), t \geq 0$ be a standard Brownian motion process. Consider the observations

$$X(t) = a + b.B(t_k), k = 1, 2, ..., N$$

where a, b are non-random scalars. Obtain the maximum likelihood estimate of a, b. Here assume that $0 \leq t_0 < t_1 < ... < t_N$.

[56] Problems in complex analysis.

[1] State and prove the Cauchy integral theorem for functions of a complex variable. First define an analytic function in a domain, then obtain the Cauchy-Riemann equations for analytic functions and finally using Green's theorem and the Cauchy-Riemann equations, prove the Cauchy integral theorem.

[2] Suppose A is an $n \times n$ complex matrix. Let λ be an eigenvalue of A. and let Γ be a closed contour that encloses λ but not any of the other eigenvalues of A. Then evaluate $\int_{\Gamma}(z\mathbf{I} - \mathbf{A})^{-1}dz$ using the Jordan decomposition theorem of A.

[3] Suppose A is a diagonable matrix and p a polynomial. Show that if Γ is any closed contour that encloses all the eigenvalues of A, then

$$p(A) = \frac{1}{2\pi i}\int_{\Gamma} p(z)(z\mathbf{I} - \mathbf{A})^{-1}dz$$

[4] Suppose $f(z) = u(x,y) + iv(x,y)$ is an analytic function of a complex variable. Then show that the curves $u(x,y) = constt.$ and $v(x,y) = constt.$ are orthogonal to each other at each intersection point.

hint:Make use of the Cauchy-Riemann equations and the fact that $\frac{\partial u}{\partial x}\hat{x} + \frac{\partial u}{\partial y}\hat{y}$ is normal to the curve $u(x,y) = constt.$ at each point and likewise for $v(x,y)$.

[5] Let T be an arbitrary linear operator in a finite dimensional complex vector space. Prove that $T = D + N$ where D diagonable and N nilpotent and D and N commute. Show that D, N are uniquely determined by these requirements. Show that if $c_1, ..., c_r$ are the distinct eigenvalues of T then there exists a resolution $\{E_i : i = 1, 2, ..., r\}$ of the identity such that $D = \sum c_i E_i$ and $N = \sum (T - c_i)E_i$. The $E_i's$ commute with T. Show that $T = \sum_i (c_i E_i + N_i)$ where $N_i^{m_i} = 0$ with m_i being the multiplicity of c_i as a root of the minimal polynomial of T. Here, $N_i = (T - c_i)E_i$. Deduce that $E_i N_j = 0$ if $i \neq j$ and $E_i N_i = N_i$. Suppose f is an analytic function. Then show that

$$f(T) = f(D) + f'(D)N/1! + ... + f^{(p-1)}(D)N^{p-1}/(p-1)!$$

for some positive integer p. Now show that

$$f(D) = \sum c_i E_i, f'(D)N = \sum f'(c_i)N_i, ..., f^{(p-1)}(D)N^{p-1} = \sum f^{(p-1)}(c_i)N_i^{p-1}$$

so that

$$f(T) = \sum_i f(c_i)E_i + f'(c_i)N_i + ... + f^{(m_i-1)}(c_i)N_i^{m_i-1}/(m_i - 1)!$$

Deduce that if Γ is a closed contour enclosing c_i but not c_j for $j \neq i$, then

$$E_i = \int_\Gamma (zI - T)^{-1}dz$$

Give a procedure for computing N_i as a contour integral.

Reference:T.Kato, Perturbation theory for linear operators.

[6] Determine the complex potential for a source emitting fluid.

hint: Taking $\phi(z) = C.log(z)$, where C is a real constant, we find that $\phi(z) = u(r,\theta) + iv(r,\theta)$, where $u = C.log(r)$ and $v = C\theta$. $u = constant$ curves are therefore circles and the stream lines $v = constant$ are radial lines emanating from the origin. Verify by actual calculation that the real and imaginary parts of the complex potential individually satisfy Laplace's equation.

[7] Determine the complex potential for flow corresponding to circular flow (vortices) around a point.

hint: $\phi(z) = i.C.log(z)$ where C is a real constant gives $\phi(z) = u + iv$ where $u = -C\theta$ and $v = C.log(r)$. The stream lines $v = constant$ are circles around the origin.

[57] (Discussion with Dr.Shambunath Sharma on generalized inverses)

[1] Consider the linear system of equations $Ax = b$ where A is an $m \times n$ matrix of rank n. Thus $m \geq n$. Then $R(A)$ is an n dimensional subspace of $\mathbb{C}^m$. It follow that if $m > n$, then there exist

vectors $b \in \mathbb{C}^m$ that are not contained in $R(A)$ and hence there may be no solution to the system. On the other hand suppose $b \in R(A)$. Then writing $b = Ax_0$ for some $x_0 \in \mathbb{C}^n$, it follows that the system $Ax = b = Ax_0$ has the general solution $x \in x_0 + \mathcal{N}(A)$. The dimension of $\mathcal{N}(A)$, namely the nullity of A is however zero by the rank nullity theorem. Hence, in this case, the system of equations has a unique solution x_0.

[2] Generalized inverse: Let A be any rectangular or square matrix and let $b \in R(A)$. Then we have $b = A\xi$ for some vector ξ having size equal to the number of columns of A. Suppose a solution to the system $Ax = b$ is given by $x = Gb$ for some matrix G and for all $b \in R(A)$. Then, $AGb = b$ for all $b \in R(A)$ and hence $AGAx = Ax$ for all vectors x and hence $AGA = A$.

[3] Suppose $AGA = A, (AG)^* = AG$. Then define $P = AG$. It follows that $P^* = P$ and $P^2 = AGAG = AG = P$. Also $R(P) \subset R(A)$ and since $A = AGA = PA$, it follows that $R(A) \subset R(P)$. In other words, $R(P) = R(A)$. Thus, P is the orthogonal projection onto $R(A)$. Let $x = Gb$ be a vector for which $\| b - Ay \|$ is a minimum when $y = x = Gb$. Then $b - AGb \perp R(A)$ for all b and hence $A^*(I - AG) = 0$ or $A^*AG = A^*$. This implies $G^*A^*AG = G^*A^*$ so that if $P = AG$, then $P^2 = P^* = P$. This also implies $A^*AGA = A^*A$ and hence $AGA = A$. On the other hand $AGA = A$ and $G^*A^* = AG$ imply $A^*AG = A^*G^*A^* = A^*$. This argument shows that a necessary and sufficient condition for a matrix G to be such that Gb minimizes $\| b - Ax \|$ for all b is that $AGA = A$ and $G^*A^* = AG$.

[58] Polyphase based estimation of parameters of LTI systems. The input sequence is $\{x(n)\}$ and the output is

$$y(n) = \sum_k h(k, \theta)x(n - k)$$

The parameter θ is to be estimated from measurements on $\{x(n)\}$ and $\{y(n)\}$. We have

$$y(2n) = \sum_k h(k, \theta)x(2n - k) = \sum_k h(2k, \theta)x(2n - 2k) + \sum_k h(2k + 1, \theta)x(2n + 1 - 2k)$$

$$y(2n + 1) = \sum_k h(k, \theta)x(2n + 1 - k) = \sum_k h(2k, \theta)x(2n + 1 - 2k) + \sum_k h(2k + 1, \theta)x(2n - 2k)$$

Define

$$x_1(n) = x(2n), x_2(n) = x(2n + 1), y_1(n) = y(2n), y_2(n) = y(2n + 1),$$

$$h_1(k, \theta) = h(2k, \theta), h_2(k, \theta) = h(2k + 1, \theta)$$

so that

$$y_1(n) = h_1(n, \theta) * x_1(n) + h_2(n, \theta) * x_2(n),$$

$$y_2(n) = h_1(n, \theta) * x_2(n) + h_2(n, \theta) * x_1(n)$$

This system of equations is the polyphase representation of the LTI system. We can estimate θ by minimizing a weighted average of the two errors squared, ie, by minimizing

$$E(\theta) = A\sum_n (y_1(n) - h_1(n, \theta) * x_1(n) - h_2(n, \theta) * x_2(n))^2 + B\sum_n (y_2(n) - h_1(n, \theta) * x_2(n) - h_2(n, \theta) * x_1(n))^2$$

Suppose we filter the output $y(n)$ by a lowpass filter having impulse response $h_1(n)$ and by a highpass filter having impulse response $h_2(n)$. The corresponding outputs are then

$$y_1(n) = h_1(n) * y(n) = \sum_k h_1(k)y(n-k), \, y_2(n) = h_2(n) * y(n) = \sum_k h_2(k)y(n-k)$$

We then decimate the two outputs by a factor of 2 giving the outputs

$$z_1(n) = y_1(2n) = \sum_k h_1(2n-k)y(k) = \sum_k p_1(n-k)y(2k) + \sum_k p_2(n-k)y(2k+1)$$

$$z_2(n) = y_2(2n) = \sum_k h_2(2n-k)y(k) = \sum_k q_1(n-k)y(2k) + \sum_k q_2(n-k)y(2k+1)$$

where

$$p_1(n) = h_1(2n), p_2(n) = h_1(2n+1), q_1(n) = h_2(2n), q_2(n) = h_2(2n+1)$$

We then process the signals z_1 and z_2 by two noise removing filters resulting in the outputs $\tilde{z}_1(n)$ and $\tilde{z}_2(n)$ respectively. We then devise an algorithm to estimate the parameter θ from these two outputs. We define

$$g_1(n, \theta) = h_1(n) * h(n, \theta), g_2(n, \theta) = h_2(n) * h(n, \theta)$$

Then

$$y_1(n) = g_1(n, \theta) * x(n), y_2(n) = g_2(n, \theta) * x(n), z_1(n) = y_1(2n), z_2(n) = y_2(2n)$$

After noise removal, z_1 becomes $\tilde{z}_1$ and z_2 becomes $\tilde{z}_2$. We use the notation z_1, z_2 for $\tilde{z}_1, \tilde{z}_2$. Then, defining

$$p_1(n, \theta) = g_1(2n, \theta), p_2(n, \theta) = g_1(2n+1, \theta), q_1(n, \theta) = g_2(2n, \theta), q_2(n, \theta) = g_2(2n+1, \theta)$$

we have

$$z_1(n) = p_1(n, \theta) * x_1(n) + p_2(n, \theta) * x_2(n), z_2(n) = q_1(n, \theta) * x_1(n) + q_2(n, \theta) * x_2(n)$$

These equations can be used to estimate the parameter θ using a least squares method, ie, by minimizing the energy function

$$E(\theta) = A. \sum_n (z_1(n) - p_1(n, \theta) * x_1(n) - p_2(n, \theta) * x_2(n))^2 + B. \sum_n (z_2(n) - q_1(n, \theta) * x_1(n) - q_2(n, \theta) * x_2(n))^2$$

where $A, B > 0$. The advantage with this method of estimating θ is that if noise of different statistics corrupts the low frequency and high frequency bands of the output signal, then these two bands can be processed by different kinds of Wiener filters and then the parameter θ estimated from these outputs leading to more accurate parameter estimates. Thius method can be generalized to an arbitrary DFT filter bank.

[59] Define a Markov process with the non-negative reals as the time axis. Suppose $P(s, x; t, E)$ is the transition probability distribution of a Markov process for $s < t$. Using the Markov property, establish the Chapman-Kolmogorov equation

$$\int P(s, x; t, dy).P(t, y; u, E) = P(s, x; u, E), s < t < u$$

[60] Suppose $f : [0,1] \to \mathbb{C}$ is a continuous function and $B(.)$ is a standard Brownian motion on a probability space. Define the random variable $B(f) = \int_0^1 f(t)dB(t)$. Show that if g is another continuous function on $[0,1]$. Then show that the map $f \to B(f)$ is an isometry in the sense that

$$\mathbb{E}(B(f)\bar{B}(g)) = \int_0^1 f(t)\bar{g}(t)dt$$

Deduce how the map $f \to B(f)$ can be extended to an isometry from $L^2[0,1]$ into $L^2(\Omega, \mathcal{F}, P)$.

[61] Show that if the integral $\int f dB$ for adapted processes f with respect to Brownian motion is defined as a limit of sums of the form $\sum_k f(t_k)(B(t_{k+1}) - B(t_k))$, then $\mathbb{E}\int f dB = 0$, while on the other hand if it is defined as a limit of sums of the form $\sum f(\theta t_k + (1-\theta)t_{k+1})(B(t_{k+1}) - B(t_k))$ with $0 \le \theta < 1$, then $\mathbb{E}\int f dB$ is not necessarily zero.

[62] Let $X(t)$ be a diffusion process with $b(t,x)$ as the drift coefficient and $\sigma(t,x)$ as the diffusion coefficient. Thus, $X(.)$ satisfies the sde

$$dX(t) = b(t, X(t))dt + \sigma(t, X(t))dB(t)$$

For a smooth function $f(t,x)$, define

$$M_f(t) = f(t, X(t)) - \int_0^t \left(\frac{\partial f(s, X(s))}{\partial s} + L_s f(s, X(s))\right)ds$$

where

$$L_t = b(t,x)\frac{\partial}{\partial x} + \frac{1}{2}\sigma^2(t,x)\frac{\partial^2}{\partial x^2}$$

Then show that M_f is a martingale and in fact,

$$dM_f(t) = \sigma(t, X(t))\frac{\partial f(t, X(t))}{\partial x}dB(t)$$

or equivalently,

$$M_f(t) - M_f(s) = \int_s^t \sigma(u, X(u))\frac{\partial f(u, X(u))}{\partial x}dB(u)$$

[63] Obtain an algorithm for numerically solving the Fokker-Planck equation for a specified drift and diffusion coefficient in one dimension.

hint: Assume that the diffusion process $\xi(t)$ satisfies the sde

$$d\xi(t) = \mu(\xi(t))dt + \sigma(\xi(t))dB(t)$$

Then the probability density $p(t, x)$ of $\xi(t)$ satisfies the Fokker-Planck equation

$$p_{,t} = (-\mu.p)_{,x} + \frac{1}{2}(\sigma^2 p)_{,xx}$$

Now discretize this equation in both the variables t, x and obtain a partial difference equation.

[64] High resolution Fourier analysis: Suppose $x(n)$ is a random signal having the form

$$x(n) = \sum_{k=1}^{p} A_k.exp(j\omega_k n) + w(n)$$

where $\{A_k\}$ are random signal amplitudes and $\{w(n)\}$ is noise. The correlation structure of this signal is given by

$$\mathbb{E}(A_k \bar{A}_m) = P_k \delta_{km}, \mathbb{E}(A_k \bar{w}(n)) = 0, \mathbb{E}(w(n)\bar{w}(m)) = \sigma^2 \delta_{n,m}$$

Then the autocorrelation of the noisy signal $\{x(n)\}$ is given by

$$R_{xx}(m) = \mathbb{E}(x(n)\bar{x}(n - m)) = \sum_{k=1}^{p} P_k.exp(j\omega_k m) + \sigma^2 \delta_{m,0}$$

Let $N > p$ and set

$$\mathbf{R}_{xx} = ((R_{xx}(u - v)))_{0 \leq u,v \leq N}$$

Then

$$\mathbf{R}_{xx} = \mathbf{EDE}^* + \sigma^2 \mathbf{I}$$

where $\mathbf{E}$ is a Vandermonde matrix of size $N \times p$ whose k^{th} column is given by $((exp(j\omega_k m))_{0 \leq m \leq N-1}$ and $k = 1, 2, ..., p$ and $\mathbf{I}$ is an $N \times N$ identity matrix. Further,

$$\mathbf{D} = diag[P_1, P_2, ..., P_p]$$

The positive semidefinite matrix $\mathbf{EDE}^*$ has p positive eigenvalues $\lambda_1, ..., \lambda_p$ and a zero eigenvalue of multiplicity $N - p$. Let $\mathbf{v}_1, ..., \mathbf{v}_p$ denote the eigenvectors of $\mathbf{EDE}^*$ corresponding to the positive eigenvalues $\lambda_1, ..., \lambda_p$ respectively and let $\mathbf{v}_{p+1}, ..., \mathbf{v}_N$ denote the eigenvectors of $\mathbf{EDE}^*$ corresponding to the zero eigenvalue. Then we may assume that $\{\mathbf{v}_1, ..., \mathbf{v}_N\}$ is an orthonormal set and this is a set of eigenvectors of $\mathbf{R}_{xx}$ corresponding to the eigenvalues $\lambda_1 + \sigma^2, ..., \lambda_p + \sigma^2$ and σ^2 repeated $N - p$ times. Then it is clear that if ω is a real number and $\mathbf{e}(\omega) = (exp(j\omega m))_{0 \leq m \leq N-1}$, then $\omega \in \{\omega_1, ..., \omega_p\}$ iff $\mathbf{e}(\omega) \perp \mathbf{v}_{p+1}, ..., \mathbf{v}_N$. it follows that the function

$$P(\omega) = (\sum_{k=p+1}^{N} |\mathbf{v}_k^* \mathbf{e}(\omega)|^2)^{-1}$$

shows peaks in $[0, 2\pi)$ precisely at the points $\omega_1, ..., \omega_p$. This method of estimating the frequencies of the signal is called the MUSIC (Mulitiple Signal Classification) algorithm. In practice, we would replace $\mathbf{R}_{xx}$ by its finite data estimate and then this matrix would have p large eigenvalues and $N-p$ small eigenvalues with the small eigenvalues close to σ^2. The eigenvectors corresponding to the small eigenvalues are used to construct the function $P(\omega)$ whose peaks yield the approximate frequencies in the signal.

[65] Find the Z-transform and the region of convergence of the Z-transform of the following signals:(a) $x[n] = a^n u(n)$, (b) $x[n] = a^{|n|}$ where $|a| < 1$, (c) $x[n] = a^n cos(\omega n)u(n)$, (d) $x[n] = na^n u(n)$, (e) $x[n] = p(n)a^n u(n)$ where $p(x)$ is a polynomial, ie $p(x) = c_0 + c_1 x + ... + c_p x^p$.

[66] Define the following terms in the theory of stochastic processes:(a) Filtration, (b) Adapted process, (c) Measurable process, (d) Progressively measurable process, (e) right continuous filtration, (f) Completion of a filtration, (g) stoptime relative to a filtration.

[67] Let $x(t)$ be a diffusion process with drift $b(x)$ and diffusion coefficient $\sigma(x)$. Thus, $x(t)$ satisfies the sde

$$dx(t) = b(x(t))dt + \sigma(x(t))dB(t)$$

Let f be a smooth function and define

$$M(t) = f(x(t)) - \int_0^t Lf(x(s))ds$$

where

$$Lf(x) = b(x)f'(x) + sigma^2(x)f''(x)/2$$

Show that

$$dM(t) = \sigma(x(t))f'(x(t))dB(t)$$

and hence M is a Brownian martingale. Show that

$$Z(t) = exp(\theta(x(t) - \int_0^t b(x(s))ds) - \frac{\theta^2}{2}\int_0^t \sigma^2(x(s))ds)$$

is a Brownian martingale.

hint: Apply the Ito differential rule to conclude that

$$dZ(t) = \theta.Z(t)\sigma(x(t))dB(t)$$

More generally, prove that

$$Z_f(t) = exp(f(x(t)) - \int_0^t b(x(s))f'(x(s))ds - \frac{1}{2}\int_0^t \sigma^2(x(s))f''(x(s))ds)$$

hint:Show using the Ito rule that

$$dZ_f(t) = Z_f(t)f'(x(t))dB(t)$$

Reference:Multidimensional Diffusion Processes by Stroock and Varadhan.

[68] Cameron-Martin-Girsanov formula.

Let $(\Omega, \mathcal{F}, P)$ be a probability space and $x(t)$ a diffusion process on this space with drift coefficient $b(x)$ and diffusion coefficient $\sigma(x)$. Let $\theta(t)$ be a progressively measurable function on this space. Then

$$M(t) = exp(\int_0^t \theta(u)(dx(u) - b(x(u))du) - \frac{1}{2}\int_0^t \sigma^2(x(u))\theta^2(u)du)$$

is a martingale as follows from the Ito differential rule:

$$dM(t) = M(t)\theta(t)\sigma(x(t))dB(t)$$

Let $c(t)$ be a progessively measurable process with respect to the Brownian filtration $\{\mathcal{F}_t\}$ and define a probability measure Q_t on $\mathcal{F}_t$ so that

$$dQ_t = exp(\int_0^t c(u)dx(u) - \int_0^t (c(u)b(x(u)) + \sigma^2(x(u))c^2(u)/2)du).dP$$

Let $t > s$ and let $A \in \mathcal{F}_s$. Then,

$$Q_t(A) = \mathbb{E}_{Q_t}(\chi_A) = \mathbb{E}_P(\chi_A.exp(\int_0^t c(u)dx(u) - \int_0^t (c(u)b(x(u)) + \sigma^2(x(u))c^2(u)/2)du)$$

$$= \mathbb{E}_P(\chi_A.exp(\int_0^s c(u)dx(u) - \int_0^s (c(u)b(x(u)) + \sigma^2(x(u))c^2(u))du/2)) = \mathbb{E}_{Q_s}(\chi_A) = Q_s(A)$$

from the above martingale property. It follows that Q_t restricted to $\mathcal{F}_s$ is Q_s. Let $Q = Q_T$ where $T > 0$ is fixed. We have for $0 \le s < t \le T$, and $\theta(t)$ progressively measurable, and $A \in \mathcal{F}_s$,

$$\mathbb{E}_P[exp(\int_0^t \theta(u)dx(u) - \int_0^t (\theta(u)b(x(u)) + \sigma^2(x(u))\theta^2(u)/2)du)\chi_A]$$

$$= \mathbb{E}_Q[exp(\int_0^t (\theta(u)dx(u) - \int_0^t (\theta(u)b(x(u)) + \sigma^2(x(u))\theta^2(u)/2)du.$$

$$\times exp(-\int_0^t c(u)dx(u) + \int_0^t (c(u)b(x(u)) + \sigma^2(x(u))c^2(u)/2)du)\chi_A]$$

$$= \mathbb{E}_Q[exp(\int_0^t (\theta(u) - c(u))(dx(u) - (b(x(u)) + \sigma^2(u)c(u))du) - \frac{1}{2}\int_0^t \sigma^2(x(u))(\theta(u) - c(u))^2 du$$

It follows from this equation and the fact that

$$exp(\int_0^t \theta(u)dx(u) - \int_0^t (\theta(u)b(x(u)) + \sigma^2(x(u))\theta^2(u)/2)du)$$

is a martingale relative to P that

$$exp(\int_0^t (\theta(u) - c(u))(dx(u) - (b(x(u)) + \sigma^2(u)c(u))du) - \frac{1}{2}\int_0^t \sigma^2(x(u))(\theta(u) - c(u))^2 du)$$

is a martingale relative to Q.

[69] Numerical techniques for simultating a stochastic differential equation and the equation of evolution of the corresponding probability density:

$$dx(t) = \mu(x(t))dt + \sigma(x(t))dB(t)$$

On discretizing this sde with a time step Δ, we get

$$x[n+1] - x[n] = \mu(x[n])\Delta + \sigma(x[n])W[n+1], n \geq 0$$

where $\{W[n]\}$ is an iid Gaussian sequence with zero mean and variance Δ. The Fokker-Planck equation satisfied by the probability density $p(t,x)$ of $x(t)$ is

$$p_{,t} = -(\mu p)_{,x} + \frac{1}{2}(\sigma^2 p)_{,xx}$$

This evolution equation can be simulated by discretizing it with respect to both the time and spatial variables. If we adopt a time discretization step of τ and a spatial discretization step of Δ, then the Fokker-Planck equation discretizes to

$$(p[n+1,k] - p[n,k])/\tau = -(\mu[k+1]p[n,k+1] - \mu[k]p[n,k])/\Delta$$

$$+\frac{1}{2\Delta^2}(a[k+2]p[n,k+2] - 2a[k+1]p[n,k+1] + a[k]p[n,k])$$

where $a(x) = \sigma^2(x)$.

[70] Let $X[n], n = 0,1,2,...$ be a Markov process with one step transition probability density $\pi(x,y)$, ie,

$$P(X[n+1] \in E|X[n] = x) = \int_E \pi(x,y)dy$$

Then show that the p-step transition probability density is given by

$$\pi^{(n)}(x,y) = \int \pi(x,y_1)dy_1\pi(y_1,y_2)dy_2...dy_{n-1}\pi(y_{n-1},y)$$

Define the operator A acting on functions f defined on the state space of the Markov process by

$$Af(x) = \int \pi(x,y)f(y)dy - f(x) = \mathbb{E}(f(X[n+1])|X[n] = x) - f(x)$$

Consider the process

$$Y[n] = f(X[n]) - \sum_{k=0}^{n-1} Af(X[k])$$

Show that

$$Y[n] - Y[n-1] = f(X[n]) - f(X[n-1]) - Af(X[n-1])$$

and hence

$$\mathbb{E}(Y[n] - Y[n-1]|X[0], ..., X[n-1]) = 0$$

Thus $\{Y[n]\}$ is a martingale with respect to the filtration generated by $\{X[n]\}$.

Reference:Multidimensional diffusion processes by Stroock and Varadhan.

[71] Consider the Ito stochastic differential equation for a damped harmonic oscillator in noise.

$$dx(t) = v(t)dt, dv(t) = -\gamma v(t)dt - \omega^2 x(t)dt + \sigma.dB(t)$$

where $B(.)$ is a Brownian motion process. Solve this equation and hence obtain a formula for the transition probability density of the Markov process $\{(x(t), v(t))\}$. Show that this transition probability density satsifies the Fokker-Planck equation.

[72] Prove the multidimensional version of the Parseval relation:Suppose $f(\mathbf{x})$ and $g(\mathbf{x})$ are both integrable as well as square integrable functions on $\mathbb{R}^n$. Define their Fourier transforms

$$\hat{f}(\mathbf{t}) = \int f(\mathbf{x})exp(i<\mathbf{t},\mathbf{x}>)d^n x, \hat{g}(\mathbf{t}) = \int g(\mathbf{x})exp(i<\mathbf{t},\mathbf{x}>)d^n x, \mathbf{t} \in \mathbb{R}^n$$

Then,

$$<f,g>=\int_{\mathbb{R}^n} f(\mathbf{x})g^*(\mathbf{x})d^n x = (2\pi)^{-n}<\hat{f},\hat{g}>= (2\pi)^{-n}\int_{\mathbb{R}^n} \hat{f}(\mathbf{t})\hat{g}^*(\mathbf{t})d^n t$$

where star means complex conjugate.

[73] Define the autocorrelation function of a multidimensional random field and hence its Power spectral density.

hint: Let $F(\mathbf{x})$ be a stationary random field defined on $\mathbb{R}^n$. If this field is Wide sense stationary, then its autocorrelation function is given by

$$R_{FF}(\mathbf{z}) = \mathbb{E}(F(\mathbf{x}+\mathbf{z})F(\mathbf{x})), \mathbf{x}, \mathbf{z} \in \mathbb{R}^n$$

Its power spectral density is the Fourier transform of the autocorrelation function:

$$S_{FF}(\mathbf{t}) = \int R_{FF}(\mathbf{x})exp(i<\mathbf{t},\mathbf{x}>)d^n x$$

The significance of the power spectral density is as follows. Suppose that the random field F is passed through a linear shift invariant multidimensional system having impulse response $h(\mathbf{x})$. The output is then given by

$$G(\mathbf{x}) = h(\mathbf{x}) * F(\mathbf{x}) = \int h(\mathbf{x}-\mathbf{z})F(\mathbf{z})d^n z$$

and the power spectral density of the output G is then given by

$$S_{GG}(\mathbf{t}) = |H(\mathbf{t})|^2 S_{FF}(\mathbf{t})$$

[74] Let X, Y be random variables and let $p, q > 1$ be such that $1/p + 1/q = 1$. Then deduce Holder's inequality

$$\mathbb{E}|XY| \leq (\mathbb{E}|X|^p)^{1/p}/(\mathbb{E}|Y|^q)^{1/q}$$

hint:Start with the fact that if α, β are nonnegative real numbers, then

$$\alpha^{1/p}\beta^{1/q} \leq \alpha/p + \beta/q$$

and hence

$$\alpha\beta \leq \alpha^p/p + \beta^q/q$$

(Arithmetic mean$\geq$ Geometric mean). Now set

$$\alpha = |X|/(\mathbb{E}|X|^p)^{1/p}, \beta = |Y|/(\mathbb{E}|Y|^q)^{1/q}$$

in this inequality to get

$$|XY|/[(\mathbb{E}|X|^p)^{1/p}(\mathbb{E}|Y|^q)^{1/q}] \leq \mathbb{E}(|X|^p)/p(E|X|^p) + \mathbb{E}(|Y|^q)/q(\mathbb{E}|Y|^q)^{1/q}$$

Take the expected value on both sides of this inequality to end up with Holder's inequality.

[75] Suppose $x(n)$ is a sinusoid in discrete time with frequency ω and is corrupted by additive white Gaussian noise. Derive a formula for the maximum likelihood estimator of the frequency ω based on N samples of the noisy signal. How is the algorithm modified if the noise is coloured Gaussian with a specified autocorrelation function.

[76] Let $f(x_1, ..., x_n; t_1, ..., t_n)$ be the joint probability density of random variables $X(t_1), ..., X(t_n)$ where $\{X(t)\}$ is a random process. The moments $\mathbb{E}(X(t_1)...X(t_n))$ can be expressed as

$$\mathbb{E}(X(t_1)...X(t_n)) = \int_{\mathbb{R}^n} x_1...x_n f(x_1, ..., x_n; t_1, ..., t_n)dx_1...dx_n$$

More generally, if $g : \mathbb{R}^n \to \mathbb{R}$ is a bounded Borel measurable function, then

$$\mathbb{E}(g(X(t_1), ..., X(t_n)) = \int g(x_1, ..., x_n)f(x_1, ..., x_n; t_1, ..., t_n)dx_1...dx_n$$

[77] Suppose $X(t)$ is a zero mean Gaussian random process with autocorrelation function $\mathbb{E}(X(t)X(s)) = R(t, s)$. Show that if ϕ is a nonramdom function, then

$$\mathbb{E}exp(\int_0^T \phi(t)X(t)dt) = exp(\frac{1}{2}\int_0^T \int_0^T \phi(t)\phi(s)R(t, s)dtds)$$

Use this identity to derive a formula for the moments $\mathbb{E}(X(t_1)...X(t_n))$.

[78] Compute the volume of the n-dimensional unit sphere in $\mathbb{R}^n$. Note that this sphere is defined by

$$\{(x_1, ..., x_n) : \sum_{i=1}^{n} x_i^2 \leq 1\}$$

[79] Let $(\Omega_i, \mathcal{F}_i, P_i), i = 1, 2, ..., n$ be probability spaces For $E_i \in \mathcal{F}_i, i = 1, 2, ..., n$, define $P(E_1 \times ... \times E_n) = \Pi_{i=1}^{n} P_i(E_i)$. Show how P can be extended to a measure on the product σ-field $\times_{i=1}^{n} \mathcal{F}_i$.

[80] Using Apollonius' theorem in a Hilbert space, explain how one can define the orthogonal projection operator in an infinite dimensional Hilbert space.

[81] If $X_n, n = 1, 2, ...$ is a sequence of random variables in a probability space $(\Omega, \mathcal{F}, P)$, then show that $sup X_n, inf X_n, limsup X_n, liminf X_n$ are all random variables in the same space and hence that $\{lim X_n exists\}$ is an event, ie, an element of $\mathcal{F}$.

[82] Suppose X is a normed linear space and A is a closed subspace of X. If $x \in X - A$, then show that $d(x, A) = inf\{\| x - y \| : y \in A\}$ is positive. If A is not closed, then show that this may not be true.

[83] Suppose P is an operator in a finite dimensional vector space V such that $P^2 = P$. Show that $V = \mathcal{R}(P) \otimes \mathcal{N}(P)$ where $\mathcal{R}(P)$ is the range of P and $\mathcal{N}(P)$ is the nullspace of P.

[84] Let X be a random variable defined on a probability space $(\Omega, \mathcal{F}, P)$ and let $g : \mathbb{R} \to \mathbb{R}$ be a non-negative and non-decreasing measurable function such that $\mathbb{E}g(X) < \infty$. Show that for $a \in \mathbb{R}$

$$P(X > a) \leq \mathbb{E}[g(X)]/g(a)$$

[85] Let X_t be a diffusion process on $\mathbb{R}$ with drift coefficient $b(x)$ and diffusion coefficient $\sigma(x)$. Let $a(x) = \sigma^2(x)$ and define the partial differential operator

$$L = b(x)\frac{\partial}{\partial x} + \frac{1}{2}a(x)\frac{\partial^2}{\partial x^2}$$

Let

$$Z_t = exp(-\lambda t)\phi(X_t)$$

Show that

$$dZ_t = exp(-\lambda t)(-\lambda.\phi(X_t)dt + \phi'(X_t)(b(X_t)dt + \sigma(X_t)dB_t) + \frac{1}{2}\phi''(X_t)\sigma^2(X_t)dt)$$

$$= -\lambda exp(-\lambda t)\phi(X_t)dt + exp(-\lambda t)L\phi(X_t)dt + \sigma(X_t)exp(-\lambda t)dB_t$$

and hence deduce t hat

$$Z_t + \int_0^t exp(-\lambda u)(\lambda\phi(X_u) - L\phi(X_u))du$$

Is a martingale. Deduce that if $L\phi(x) \leq \lambda\phi(x)$ for all x, then

$$\mathbb{E}(Z_t|X_s = x) \leq exp(-\lambda s)\phi(x)$$

[86] Explain how you would simulate the Fokker-Planck equation for the evolution of the probability density of a diffusion process using the method of moments.

hint: The Fokker-Planck equation is given by

$$p_{,t}(t,x) = -(b(x)p(t,x))_{,x} + \frac{1}{2}(a(x)p(t,x))_{,xx}$$

To simulate this equation, we expand $p(t,x)$ in terms of test functions $\phi_i(x), i = 1, 2, ..., N$ so that

$$p(t,x) \approx \sum_{i=1}^{N} \psi_i(t)\phi_i(x)$$

We substitute this into the Fokker-Planck equation, form inner products on both sides with the test functions $\phi_j(x)$ and thereby derive ordinary differential equations for the functions $\psi_i(t)$.

[87] linear prediction of stationary vector valued time series. $\mathbf{X}(n) = (X_1(n), X_2(n), ..., X_p(n))^T, n = 0, 1, 2, ...$ is the given vector valued time series. Its autocorrelation function $\mathbf{R}(k)$ is matrix valued with entries

$$R_{ij}(k) = \mathbb{E}(X_i(n)X_j(n-k))$$

or equivalently,

$$\mathbf{R}(k) = \mathbb{E}(\mathbf{X}(n)\mathbf{X}(n-k)^T)$$

where

$$\mathbf{R}(k) = ((R_{ij}(k)))_{1 \leq i,j \leq p}$$

The linear predictor of order m has the form

$$\hat{\mathbf{X}}(n) = -\sum_{k=1}^{m} \mathbf{A}(k)\mathbf{X}(n-k)$$

with the prediction error given by

$$\mathbf{e}(n) = \mathbf{X}(n) - \hat{\mathbf{X}}(n) = \mathbf{X}(n) + \sum_{k=1}^{m} \mathbf{A}(k)\mathbf{X}(n-k) = \sum_{k=0}^{m} \mathbf{A}(k)\mathbf{X}(n-k)$$

where

$$\mathbf{A}(0) = \mathbf{I}$$

The matrices $\mathbf{A}(k), k = 1, 2, ..., m$ are chosen so that $\mathbb{E}(\| \mathbf{e}(n) \|^2)$ is a minimum. This leads us at once to the normal equations

$$\mathbb{E}(\mathbf{e}(n)\mathbf{X}(n - k)^T) = \mathbf{0}, k = 1, 2, ..., p$$

or

$$\mathbb{E}[(\mathbf{X}(n) + \sum_{r=1}^{m} \mathbf{A}(r)\mathbf{X}(n - r))\mathbf{X}(n - k)^T] = 0, k = 1, 2, ..., p$$

or

$$\sum_{r=1}^{m} \mathbf{A}(r)\mathbf{R}(k - r) = -\mathbf{R}(k), k = 1, 2, ..., m$$

The solution to this system of linear equations gives the desired predictor.

Least squares method:Here, we do not assume any statistics for the time series but choose the prediction coefficient matrices $\mathbf{A}(k), k = 1, 2, ..., m$ so that $\sum_{n=m}^{N-1} \| \mathbf{e}(n) \|^2$ is a minimum. In other words, instead of minimizing the ensemble averaged mean square prediction error, we minimize the time averaaged mean square error. The optimal equations are

$$\sum_{n=m}^{N-1} \mathbf{e}(n)\mathbf{X}(n - k)^T = \mathbf{0}$$

[88] Solve the Bessel differential equation

$$x^2 f''(x) + x f'(x) + (x^2 - \nu^2)f(x) = 0$$

using the power series method.

[89] Prove that if $\sum_n |x(n)|^2 < \infty$ and $\sum_n |x(n)| < \infty$, then with $X(\omega) = \sum_n x(n)exp(-i\omega n)$, we have that

$$\sum_n |x(n)|^2 = \frac{1}{2\pi} \int_{-\pi}^{\pi} |X(\omega)|^2 d\omega$$

[90] Define the following terms in martingale calculus and give examples.

1. Stoptime, Martingale, continuous martingale, local martingale, uniformly integrability of a family of random variables.

2. Submartingale, supermartingale, potential, natural process, predictable process.

3.Variation of a process, quadratic variation of a process.

4. Levy process, decomposition of a Levy process as a superposition of Brownian motion and Poisson processes with varying intensities.

[91] Let $N(t), t \geq 0$ be a Poisson process with intensity λ. Let $t_1 < t_2 < \ldots$ be the jump times of this process. Let $X_1, X_2, \ldots$ be identically distributed independent random variables independent of the process $N(.)$. Define the processes (1)

$$X(t) = \sum_{i=1}^{N(t)} X_i$$

(2)

$$N(t, B) = \sum_{i=1}^{N(t)} \chi_B(X_i)$$

where B is a Borel subset of $\mathbb{R}$ and χ_B is the indicator of B. Compute the characteristic functions of $X(t)$ and $N(t, B)$. Verify these by numerical simulations.

[92] Suppose $B(t)$ is Brownian motion and f is a twice differentiable function. Show that the process

$$M_f(t) = f(B(t)) - \frac{1}{2} \int_0^t f''(B(s))ds$$

is a martingale. Take $f(x) = x^2$ and verify the martingale property by simulations. Let $M(t)$ and $N(t)$ be two martingales on the same probability space and let $X(t)$ be an adapted process. Consider the stochastic integrals $I_M(X) = \int_0^T X(t)dM(t)$ and $I_N(X) = \int_0^T X(t)dN(t)$. Simulate the process $[M, N]_t$ by approximating using partial sums, ie, if $0 = t_0 < t_1 < \ldots < t_n = t$, then

$$[M, N]_t \approx \sum_{k=0}^{n-1} (M(t_{k+1}) - M(t_k))(N(t_{k+1}) - N(t_k))$$

Verify via simulations that

$$\left[\int X dM, \int X dN \right] = \int X d[M, N]$$

where X is an adapted process. Verify more generally that if X, Y are two adapted processes, then

$$\left[\int X dM, \int Y dN \right] = \int XY d[M, N]$$

[93] Verify via simulations that if $B(t)$ is Brownian motion, then the process

$$M_\lambda(t) = exp(\lambda.B(t) - \lambda^2 t/2)$$

satisfies the stochastic differential equation

$$dM_\lambda(t) = \lambda.M_\lambda(t)dB(t)$$

[94] Assume that the position of a particle at any time $n = 0, 1, 2, ...$ is $k, k = 0 \pm 1, \pm 2, ...$ We denote the position of the particle at time n by $X(n)$. The transition probability law is $Pr(X(n+1) = k + 1 | X(n) = k) = p$ and $Pr(X(n+1) = k - 1 | X(n) = k) = q$ where $p + q = 1$. Then show that under the assumption that $\{X(n)\}$ is a Markov process that

$$Pr(X(n) = k) = \binom{n}{(n+k)/2} p^{(n+k)/2} q^{(n-k)/2}$$

provided that $n + k$ is even. If $n + k$ is odd, then this probability is zero. Assume that $X(0) = 0$. If now p is a random variable assuming values in $[0, 1]$, with probability density $f(p)$ and moment generating function $\mathbb{E}(p^x) = \int_0^1 p^x f(p) dp = \phi(x)$, then compute $Pr(X(n) = k)$ in terms of the function $\phi(.)$.

[95] Assume that a particle moves on $\mathbb{Z}$ such that if $X(n)$ is its position at time n, then $Pr(X(n+1) = k + 1 | X(n) = k) = p(n)$ and $Pr(X(n+1) = k - 1 | X(n) = k) = q(n)$ where $p(n) + q(n) = 1$. Then assuming that $\{p(n)\}$ is a random iid sequence with density $f(p)dp$, compute $Pr(X(n) = k)$ by averaging over the sequence $\{p(n)\}$.

hint: First assume that $\{p(n)\}$ is a non-random sequence. Define

$$\mathbb{E}(z^{X(n)}) = \phi(z, n)$$

Then,

$$\phi(z, n+1) = (p(n)z + q(n)z^{-1})\phi(z, n)$$

It follows that

$$\phi(z, n) = \Pi_{m=0}^{n-1}(p(m)z + q(m)z^{-1})$$

and when $\{p(n)\}$ becomes a random sequence, we average this expression over $\{p(n)\}$ to get the moment generating function of $X(n)$.

[96] Problems in signal theory:

[1] Let V denote the vector space of all polynomials in an indeterminate t over the real field $\mathbb{R}$ having degree less than or equal to n where n is a positive integer. Let D denote the differentiation operator acting on V. find $[D]_{\mathcal{B}}$ relative to the basis $\mathcal{B} = \{1, t, t^2, ..., t^n\}$ of V.

[2] Determine the eigenvalues and eigenfunctions of the operator $T = \frac{d^2}{dx^2}$ acting on continuously differentiable functions f on the interval $[0, L]$ with boundary conditions $f(0) = f(L) = 0$.

[3] State the Cayley-Hamilton theorem and verify it for the 2×2 matrix

$$A = \begin{pmatrix} a & b \\ c & d \end{pmatrix}$$

[4] Let T be a linear transformation on a finite dimensional vector space V and let $\mathcal{B}$ and $\mathcal{B}'$ be two bases for V. What is the relationship between $[T]_{\mathcal{B}}$ and $[T]_{\mathcal{B}'}$.

[97] Let $N(t)$ be a Poisson process with mean λt. Prove that $(dN(t))^2 = dN(t)$ and hence deduce the Ito formula for the Poisson process, namely, $df(N(t)) = (f(N(t)+1) - f(N(t))dN(t)$, where f is any function on the natural numbers $\mathbb{Z}_+$. Let $0 < t_1 < t_2 < ...$ be the jump times of the Poisson process $\{N(t)\}$ and let $X_1, X_2, ...$ be iid random variables with distribution F independent of the Poisson process $N(.)$. For any Borel set E, define the process

$$N(t, E) = \sum_{i=1}^{N(t)} \chi_E(X_i)$$

Then show that $N(., E)$ is a Poisson process with mean $\lambda.F(E)t$. Show that if g is a bounded Borel measurable function, then

$$\xi(t) = \int g(x)N(t, dx) = \sum_{i=1}^{N(t)} g(X_i)$$

is an independent increment process satifying the differential rule

$$d\xi(t) = (g(X_{N(t)+1}) - g(X_{N(t)}))dN(t)$$

Calculate the moment generating function of $\xi(t)$.

hint:

$$\mathbb{E}(exp(s\xi(t))|N(t) = n) = (\int exp(sg(x))dF(x))^n$$

and hence,

$$\mathbb{E}exp(s\xi(t)) = exp(\lambda t(\int exp(sg(x))dF(x) - 1))$$

[98] Solve the following stochastic differential equation.

$$dX(t) = (aX(t) + b)dt + (cX(t) + f)dB(t)$$

where $B(t)$ is standard Brownian motion and a, b, c, f are constants.

[99] Let $X_1, ..., X_n$ be iid $N(0, 1)$ random variables on a probability space $(\Omega, \mathcal{F}, P)$. Define the probability measure Q on the same space by

$$dQ = exp(\sum_i a_i X_i - \sum_i a_i^2/2)dP$$

Check that

$$Q(\Omega) = \int_\Omega dQ = \mathbb{E}_P(exp(\sum_i (a_i X_i - a_i^2/2)) = 1$$

so that Q as defined above is indeed a probability measure. Observe that if $Y_i = X_i - a_i$ then

$$\mathbb{E}_Q exp(\sum_i b_i Y_i) = \mathbb{E}_P(exp(\sum_i (b_i(X_i - a_i) + a_i X_i - a_i^2/2)) = \mathbb{E}_P exp(\sum_i (b_i + a_i)X_i - a_i b_i - a_i^2/2)$$

$$= exp(\sum_i (b_i + a_i)^2/2 - a_i b_i - a_i^2/2) = exp(\sum_i b_i^2/2)$$

This implies that under the probability measure Q, the random variables $X_1, ..., X_n$ are independent with means $a_1, ..., a_n$ and unit variance. This result can be generalized to the Girsanov theorem.

[100] State and prove the Martingale up-crossing inequality for discrete time martingales and show how it can be used to prove the Martingale convergence theorem.

[101] Suppose $B(t), t \geq 0$ is standard Brownian motion starting at zero. For $b > 0$, let $T_b = inf\{t \geq 0 : B(t) = b\}$. Define the reflected Brownian motion by

$$B_b(t) = B(t)\chi_{t \leq T_b} + (2b - B(t))\chi_{t > T_b}$$

Prove that $B_b(t), t \geq 0$ is also a Brownian motion process and hence derive the probability distribution of T_b. Also derive the joint probability distribution of $(M(t), B(t))$ where $M(t) = max(B(s) : s \leq t)$.

[102] Let E, F be two subsets of a set Ω. With χ_E denoting the indicator of the set E, show that $|\chi_E - \chi_F| = \chi_{E \Delta F}$, where the symmetric difference $E \Delta F$ between E and F is defined by

$$E \Delta F = E F^c \cup F E^c$$

[103] What do you understand by uniform integrability of a collection of random variables. Deduce the generalization of the Lebesgue dominated convergence theorem for a sequence of uniformly integrable random variables.

[104] Derive the central limit theorem for the random variables $(N(t) - \lambda t)/\sqrt{\lambda t}$ where $N(t), t \geq 0$ is a Poisson process with mean rate λ. Prove this using moment generating functions.

[105] Suppose $X_1, X_2, ...$ is a sequence of independent random variables defined on a probability space $(\Omega, \mathcal{F}, P)$. Let X be a random variable on the same space that is measurable with respect to the σ-field $\bigcap_{n=1}^{\infty} \sigma(X_{n+1}, X_{n+2}, ...)$. Then show that X is almost surely equal to a constant. This is known as the Kolmogorov zero-one law.

hint:Let E be a set that is an element of $\bigcap_{n=1}^{\infty} \sigma(X_{n+1}, X_{n+2}, ...)$. Then for every $n \geq 1$, E is independent of $\sigma(X_1, ..., X_n)$. It follows that E is independent of $\sigma(X_1, X_2, ...)$. However since $E \in \sigma(X_1, X_2, ...)$, it follows that $P(E) = P(E \cap E) = P(E)^2$ and hence $P(E) = 0$ or $P(E) = 1$.

[106] Consider the motion of a pollen particle in a one dimensional fluid. Let $p(t, x)$ be the probability density of the particle at time t. A molecular kick displaces the particle by a random amount Δ which has probability density $\phi(\Delta)$. The duration of a molecular kick is τ. Justify the conditions under which the Chapman-Kolmogorov equation holds

$$p(t + \tau, x) = \int_{-\infty}^{\infty} p(t, x - \Delta)\phi(\Delta)d\Delta$$

Show by making appropriate approximations and imposing appropriate conditions that this equation becomes the diffusion equation

$$\frac{\partial p(t,x)}{\partial t} = D\frac{\partial^2 p(t,x)}{\partial x^2}, D = \int_{-\infty}^{\infty}\Delta^2.\phi(\Delta)d\Delta$$

[107] Consider the motion of two pollen particles moving under an interaction potential $U(|x_1-x_2|)$. Formulate the system of stochastic differential equations that describe the motion of this pair when noisy forces and damping forces are present. Simulate the resulting motion using MATLAB.

[108] Derive the Cauchy-Riemann equations for the real and imaginary parts of an analytic function of a complex variable. Start with the definition of an analytic function.

[109] Let z be a complex number having real part greater than one. Show that the series $\zeta(z) = \sum_{n=1}^{\infty}\frac{1}{n^z}$ is absolutely convergent and express it as an integral.

hint:

$$\int_0^{\infty} exp(-nt)t^{z-1}dt/\Gamma(z) = n^{-z}$$

and so

$$\zeta(z) = \frac{1}{\Gamma(z)}\int_0^{\infty}\frac{t^{z-1}}{e^t-1}dt$$

[110] Consider an infinite series of the form $\sum_{n=1}^{\infty} z_n$ where $\{z_n\}$ is a sequence of complex numbers. Suppose $limsup|z_n|^{1/n} = a$ is such that $a < 1$. Then show that the series is absolutely convergent.

hint:

There exists a positive integer N such that for all $n > N$, $|z_n|^{1/n} \le (1+a)/2$.

[111] Formulate and prove the general theorem on existence and uniqueness of solutions to the Ito stochastic differential equation under Lipschitz conditions satisfied by the drift and diffusion coefficients.

[112] Define the shot noise process in terms of the arrival times of a Poisson process and calculate its power spectral density.

[113] Two Brownian particles moving under a mutual interaction. We assume that the motion is one dimensional. At time t, $x_1(t), x_2(t)$ are the positions of the two particles and $v_1(t) = dx_1(t)/dt, v_2(t) = dx_2(t)/dt$ are their respective velocities. The interaction potential between the two particles is $U(|x_1-x_2|)$. It follows that the force experienced by the first particle is given by

$$F_1 = -\frac{\partial}{\partial x_1}U(|x_1-x_2|) = -sgn(x_1-x_2)U'(|x_1-x_2|),$$

and that experienced by the second particle is

$$F_2 = -\frac{\partial}{\partial x_2}U(|x_1 - x_2|) = sgn(x_1 - x_2)U'(|x_1 - x_2|)$$

The equations of motion of the system of two particles moving inside a liquid involves in addition to the above interaction forces, the fluctuating forces due to molecular collision on the pollen particles and the dissipative forces due to the viscosity of the liquid. The equations of motion are therefore given by

$$dx_1(t) = v_1(t)dt, dx_2(t) = v_2(t)dt,$$

$$dv_1(t) = -sgn(x_1(t) - x_2(t))U'(|x_1(t) - x_2(t)|)dt - \gamma_1 v_1(t)dt + \sigma_1 dB_1(t),$$

$$dv_2(t) = sgn(x_1(t) - x_2(t))U'(|x_1(t) - x_2(t)|)dt - \gamma_2 v_2(t)dt + \sigma_2 dB_2(t)$$

The first problem is to derive the mean and covariance propagation equations for this system of stochastic differential equations. The second problem is to estimate the trajectories based on measurements of the energy and if possible, other observables at different times. Note that the energy observable is given by

$$E(x_1, x_2, v_1, v_2) = v_1^2/2 + v_2^2/2 + U(|x_1 - x_2|)$$

[114] Show that the function $u(t,x) = \mho(x + B(t))$, where $\{B(t)\}$ is standard Brownian motion, satisfies the heat equation $u_{,t} = \frac{1}{2}u_{,xx}$ with initial condition $u(0,x) = f(x)$.

[115] Let $\mathbf{B}(t)$ be standard three dimensional Brownian motion starting at the origin and let D be a bounded open subset of $\mathbb{R}^3$ with boundary ∂D. For $x \in D$, define $\tau_x = inf\{t \geq 0 : x + \mathbf{B}(t) \in \partial D\}$. Then show that if f is a smooth function on ∂D, and $u(x) = \mathbb{E}f(x + \mathbf{B}(\tau_x))$, then $\Delta u(x) = 0$ for $x \in D$ and $u(x) = f(x)$ for $x \in \partial D$. Here,

$$\Delta = \sum_{i=1}^{3}\frac{\partial^2}{\partial x_i^2}$$

is the Laplacian operator.

[116] Generalize the result of the preceding problem to an arbitrary three dimensional diffusion process with specified drift and diffusion coefficients.

[117] What does the discretized form of the Green's identity

$$\nabla.(U\nabla V - V\nabla U) = U\nabla^2 V - V\nabla^2 U$$

look like where U, V are two scalar fields in three dimensional space.

[118] Let $A(t), t \geq 0$ be a $d \times d$ matrix valued function of time and let $\Phi(t), t \geq 0$ be the unique $d \times d$ matrix valued function of time satisfying

$$\Phi'(t) = A(t)\Phi(t), t \geq 0, \Phi(0) = I$$

Then if $\sigma(t)$ is a $d \times p$ matrix valued function of time and $B(t), t \geq 0$ is a standard p dimensional Brownian motion, show that the solution to the stochastic differential equation

$$dX(t) = A(t)X(t)dt + \sigma(t)dB(t), t \geq 0, X(0) = x$$

where $X(t), x \in \mathbb{R}^d$ can be expressed as

$$X(t) = \Phi(t)x + \int_0^t \Phi(t)\Phi(s)^{-1}\sigma(s)dB(s)$$

Defining $\Phi(t, s) = \Phi(t)\Phi(s)^{-1}$, show that if $x = 0$, then

$$\mathbb{E}(X(t_1)X(t_2)^T) = \int_0^{min(t_1,t_2)} \Phi(t_1, s)\sigma(s)\sigma(s)^T\Phi(t_2, s)^T ds$$

Derive using this equation, the differential equation satisfied by $R(t) = \mathbb{E}(X(t)X(t)^T)$. In fact show that

$$R'(t) = A(t)R(t) + R(t)A(t)^T + \sigma(t)\sigma(t)^T$$

[119] We consider the linear coupled differential equations

$$dX(t)/dt = A(t)X(t) + \sigma(t)v(t), t \geq 0$$

Here $A(t)$ is $d \times d$, $X(t)$ is $d \times 1$, $\sigma(t)$ is $d \times p$ and $v(t)$ is $p \times 1$. The solution starting at $t = 0, X(0) = x$ is given by

$$X(T) = \Phi(T)(x + \int_0^T \Phi(s)^{-1}\sigma(s)v(s)ds)$$

where $T > 0$. We want to determine conditions under which for each $x, y \in \mathbb{R}^d$, there exists an input process $v(s), 0 \leq s \leq T$ such that $X(T) = y$. When this happens, we say that the given dynamical system is controllable over $[0, T]$. It is clear that a necessary and sufficient condition for this to happen is that as $v(s), 0 \leq s \leq T$ ranges over all input processes, $\int_0^T \Phi(s)^{-1}\sigma(s)v(s)$ should range over the whole of $\mathbb{R}^d$. We want to show this happens iff the positive semidefinite matrix

$$M(T) = \int_0^T \Phi(s)^{-1}\sigma(s)\sigma(s)^T\Phi(s)^{-T}ds$$

is nonsingular. Suppose $M(T)$ is singular. Then there exists a nonzero vector ξ in $\mathbb{R}^d$ such that $\xi^T M(T)\xi = 0$ and this implies that $\xi^T \Phi(s)^{-1}\sigma(s) = 0$ for μ a.e. $s \in [0, T]$ where μ is the Lebesgue measure on $\mathbb{R}$. This in turn implies that for all (measurable) inputs $v(s), 0 \leq s \leq T$, we have

$$\xi^T \int_0^T \Phi(s)^{-1}\sigma(s)v(s)ds = 0$$

which shows that the system is not controllable over $[0, T]$. On the other hand, suppose $M(T)$ is non-singular. For $\xi \in \mathbb{R}^d$, define $v(s) = \sigma(s)^T \Phi(s)^{-1} M(T)^{-1} \xi$. Then

$$\int_0^T \Phi(s)^{-1} \sigma(s) v(s) ds = \xi$$

proving that the system is controllable over $[0, T]$.

Reference:Brownian motion and Stochastic calculus, by Karatzas and Shreve.

[120] Let $x[n] = \sum_{i=1}^p A_i a_i^n u(n)$ where $u(n) = 1$ for $n \geq 0$ and $u(n) = 0$ for $n < 0$. Here, $\{a_i\}$ are complex numbers. Then show that if $|z| > |a_i|$ for all $i = 1, 2, ..., p$, then

$$\sum_n x[n] z^{-n} = \sum_{i=1}^p A_i (1 - a_i z^{-1})^{-1}$$

Using this formula explain how you would calculate the inverse Z transform of the function $\frac{1}{P(z)}$ where $P(z)$ is a polynomial having no repeated roots.

hint: Perform a partial fraction expansion of $\frac{1}{P(z)}$ expressing it as $\sum_{i=1}^p \frac{A_i}{z - b_i}$ where $p = degP(z)$.

[121] What do you understand by a semisimple and a nilpotent linear transformation on a finite dimensional vector space. Prove that every linear operator in a finite dimensional complex vector space can be expressed uniquely as the sum of a semisimple and a nilpotent element, both of which commute.

[122] Define the Z transform of a multidimensional signal $x[n_1, ..., n_r]$ on $\mathbb{Z}^r$ by

$$X(z_1, ..., z_r) = \sum_{n_1, ..., n_r \in \mathbb{Z}} x[n_1, ..., n_r] z_1^{-n_1} ... z_r^{-n_r}$$

Prove the convolution property for such signals, ie, if $y[n_1, ..., n_r]$ is another such signal and

$$f[n_1, ..., n_r] = \sum_{k_1, ..., k_r} x[n_1 - k_1, ..., n_r - k_r] y[k_1, ..., k_r]$$

then,

$$F(z_1, ..., z_r) = X(z_1, ..., z_r) Y(z_1, ..., z_r)$$

provided that the complex numbers $z_1, ..., z_r$ fall in an appropriate domain.

[123] Consider the input output equation

$$y(n) = h(n, \theta) * x(n) = \sum_k h(k, \theta) x(n - k)$$

We want to estimate the parameter θ from input-output measurements. We assume that white noise of different spectral densities corrupts the lowpass and highpass bands. The lowpass band is $|\omega| \leq \pi/2$ and the highpass band is $\pi/2 \leq |\omega| < \pi$. The ideal lowpass filter is defined by $H_1(\omega) = 1$ if $|\omega| < \pi/2$ and $H_2(\omega) = 0$ if $\pi/2 < |\omega| < \pi$. Let $h_1(n)$ denote the corresponding impulse response. The ideal highpass filter is $H_2(\omega) = 1$ if $\pi/2 < |\omega| < \pi$ and $H_2(\omega) = 0$ if $|\omega| < \pi/2$. Let $h_2(n)$ denote the corresponding impulse response. Filtering the output using these two filters yields

$$z_1(n) = h_1(n) * y(n) = g_1(n,\theta) * x(n), \ z_2(n) = h_2(n) * y(n) = g_2(n,\theta) * x(n)$$

where

$$g_1(n,\theta) = h_1(n) * h(n,\theta), g_2(n,\theta) = h_2(n) * h(n,\theta)$$

$z_1(n)$ is a lowpass sequence, ie, its Discrete time Fourier transform $Z_1(\omega)$ is zero for $\pi/2 < |\omega| < \pi$ and likewise, $z_2(n)$ is a highpass sequence, ie, its Discrete time Fourier transform $Z_2(\omega)$ is zero for $|\omega| < \pi/2$. We then decimate the sequence $z_1(n)$ giving $y_1(n) = z_1(2n)$ and likewise $y_2(n) = z_2(2n)$. Then apply Wiener filters $W_1(\omega)$ and $W_2(\omega)$ to the two sequences $y_1(n)$ and $y_2(n)$ for cancelling out the respective noises. Let the corresponding outputs be denoted by $v_1(n)$ and $v_2(n)$. In the ideal noiseless case, we have

$$v_1(n) = \sum_k g_1(k,\theta)x(2n-k), \ v_2(n) = \sum_k g_2(k,\theta)x(2n-k)$$

We write these equations as

$$v_1(n) = \sum_k g_1(2k,\theta)x(2n-2k) + \sum_k g_1(2k+1,\theta)x(2n-2k-1)$$

$$v_2(n) = \sum_k g_2(2k,\theta)x(2n-2k) + \sum_k g_2(2k+1,\theta)x(2n-2k-1)$$

From these equations, the parameter θ can be estimated by a least squares algorithm. This method can be generalized to any number of bands. Suppose that the entire frequency band $[-\pi, \pi)$ is partitioned into N equal parts, each of width $2\pi/N$. Let $h_i(n)$ be the ideal bandpass filter that passes the ith band and rejects the other bands. The output is given by

$$y(n) = \sum_k h(k,\theta)x(n-k)$$

with $x(n)$ as the input. The aim is to estimate the parameter θ. We set

$$z_i(n) = h_i(n) * y(n) = g_i(n,\theta) * x(n), g_i(n,\theta) = h_i(n) * h(n,\theta), i = 1, 2, ..., N$$

Decimate each output by a factor of N giving the sequences

$$z_i(Nn) = \sum_k g_i(k,\theta)x(Nn-k) = \sum_{m \in \mathbb{Z}, 0 \leq p \leq N-1} g_i(Nm+p,\theta)x(Nn-Nm-p)$$

From these equations, we can estimate θ after applying the appropriate Wiener filter to cancel out the respective noise. Another approach is to remove the noise from the respective bands by Wiener filtering, interpolating the different noise removed sequences by the factor of N and then applying a band pass filter to each output and patching up the outputs together to give a noiseless version of $y(n)$ from which θ can be estimated using the original input-output equation $y(n) = h(n,\theta) * x(n)$.

[124] Synopsis of the thesis: The aim of this work is to devise a scheme by which one can, based on past data, predict the quantity and quality of journal publications in technical institutions using the method of time series analysis for vector valued processes. Such a prediction will be useful in devising methods for improving the quantity and quality of publications. The idea in brief is as follows. Suppose we have a record of the successive years numbered $0, 1, 2, ..., N$. In the ith year the record gives us the number of publications in the journal numbered j as $X(i, j)$. The journal index j assumes values $1, 2, ..., M$ in increasing order of impact factor. We define a vector

$$\mathbf{X}(i) = (X(i, 1), X(i, 2), ..., X(i, M))^T, i = 0, 1, 2, ..., N$$

Then the sequence $\{\mathbf{X}(i)\}$ defines an M component vector valued time series of length $N + 1$ and we want to use this data to predict the journal publications in the year $N + r$ for some $r \geq 1$. Note that for a fixed year index i, $\mathbf{X}(i)$ gives the number of publications in the year i in all the journals. $\mathbf{X}(i)$ will in general be a correlated process with the autocorrelation matrix defined by

$$\mathbf{R}(p, q) = \sum_i \mathbf{X}(p + i)\mathbf{X}(q + i)^T$$

The information contained in these correlations can be used to construct a linear predictor for the process. For example, suppose that we predict r years into the future using the linear model

$$\hat{\mathbf{X}}(n + r) = \sum_{k=0}^{p} \mathbf{A}(k)\mathbf{X}(n - k)$$

where $\mathbf{A}(k)$ are matrix coefficients to be determined by a least squares criterion in which we minimize the sum of squared errors

$$\sum_n \| \mathbf{X}(n + r) - \sum_{k=0}^{p} \mathbf{A}(k)\mathbf{X}(n - k) \|^2$$

with respect to the matrices $\mathbf{A}(k), k = 0, 1, ..., p$. The minimization leads to the optimal normal equations

$$\sum_n \mathbf{X}(n + r)\mathbf{X}(n - k)^T = \sum_{m=0}^{p} \mathbf{A}(m) \sum_n \mathbf{X}(n - m)\mathbf{X}(n - k)^T$$

or in terms of the correlation matrices introduced above,

$$\mathbf{R}(r, -k) = \sum_{m=0}^{p} \mathbf{A}(m)\mathbf{R}(-m, -k), k = 0, 1, ..., p$$

This is a system of linear equations and can be solved to yield the coefficients $\mathbf{A}(k)$. This is the simplest possible prediction algorithm. This procedure can be made recursive so that prediction can be carried out on a real time basis. The idea in brief is to express the solution to the above system of linear equations as $\mathbf{R}_N^{-1}\mathbf{r}_N$ where $\mathbf{R}_N$ is a data correlation matrix constructed using the data collected uptill the year N and likewise $\mathbf{r}_N$. Then when the data for the year $N + 1$ arrives, we update $\mathbf{R}_N$ by adding to it the new data and so also for $\mathbf{r}_N$. Then using the matrix inversion lemma, we can make the solution of this system recursive. Such a recursive approach can be used to program a computer to keep providing us on a real time basis with new forecasts. Another useful tool for statistical inference is the power spectral density. This enables us to determine the approximate frequency of publications in each journal over a given time frame. For example, we consider the publications in journal number k $X(i,k), i = 0, 1, ..., N$. The periodogram constructed using this data is

$$P_k(\omega) = \| \sum_{m=0}^{N} X(m,k)exp(-j\omega m) \|^2$$

A plot of $P_k(\omega)$ gives us peaks at certain frequencies which we can call the dominant frequency for journal number k. These dominant frequencies can be used to infer about the maximum variation in the number of publications in a given journal. There are more accurate methods like the high resolution eigensubspace algorithms for determining the dominant frequencies more accurately. Suppose that we have isolated a single dominant frequency $\omega(k)$ for journal number k. Then we can use this information to assess in which future years will a certain journal receive a low number of publications and we can take appropriate steps to improve this. For example suppose $\omega(k)$ is the dominant frequency for the journal numbered k. if in a certain year i we have a very low number of publications in this journal, then we can say that approximately afer $2\pi/\omega(k)$ years, this low number of publications will recur and we can thus take appropriate remedial steps. Another useful index is the cross spectral density between publications in two different journals numbered k and r. This is given by

$$P_{k,r}(\omega) = (\sum_m X(m,k)exp(-j\omega m))(\sum_n X(m,r)exp(j\omega r))^T$$

This function can be used to determine in which future years will publications in the two journals be maximally correlated. These auto and cross power spectral densities can be combined into a single matrix valued power spectral density:

$$\mathbf{P}(\omega) = (\sum_m \mathbf{X}(m)exp(-j\omega m))(\sum_n \mathbf{X}(n)^T exp(j\omega n))$$

This power spectral density matrix gives us the dominant frequencies of each journal over the entire time duration. However, if we wish to determine the dominant frequencies over a short time interval, we can use the short time Fourier transform as follows. Let $h(n)$ be a window function having maximum at $n = 0$ and tapering towards zero on both the sides. Then the short time Fourier transform of the time series $\mathbf{X}(n)$ is the function

$$S(\omega, m) = \sum_n \mathbf{X}(n)w(n-m)exp(-j\omega n)$$

whose journal components are given by

$$S(\omega, m, k) = \sum_n X(n, k)w(n-m)exp(-j\omega n)$$

Suppose this function shows a peak for a given m and k at the frequency $\omega(m,k)$. Then we can infer that during the years (m-r,m+r) where $2r$ is the window duration, the frequency of publication in the journal numbere k is approximately $\omega(m,k)$. This local Fourier transform gives us a method of computing in the vicinity of a given year, when will a low or high publication record be repeated. In summary, we propose in this thesis to use signal processing tools for predicting journal publications, and for analyzing the frequency of journal publications and developing remedial measures to improve both their quantity and quality. Suppose that the impact factor of journal number k is $I(k)$. Then, a good measure of the overall performance index in a given year i is the weighted average $\sum_k I(k)X(i,k)$. This index can be used to determine in which years the publications have been the best.

[125] Prove that convergence in probability implies convergence in distribution, but the converse is false.

[126] Let X and Y be two random variables defined on the same probability space. Suppose that for all bounded Borel measurable functions $f : \mathbb{R}^2 \to \mathbb{R}$, we have $\mathbb{E}f(X,Y) = \mathbb{E}f(X,X)$, then $P(X = Y) = 1$ and the converse is also true. Simply take $f = \chi_E$, where $E = \{(x,y) : x \neq y\}$.

[127] A Gaussian random vector of size n is a set of n random variables all defined on the same probability space such that every linear combination of these variables is a Gaussian random variable. Deduce using this the general form of the joint characteristic function and hence the joint probability density of a Gaussian random vector.

[128] Let $(V, < .,. >)$ be a finite dimensional inner product space. Show by making use of the Gram-Schmidt orthonormalization process that there exists an orthonormal basis for this space, ie, a basis $\{e_1, ..., e_n\}$ for V satisfying $< e_i, e_j >= \delta_{ij}, 1 \leq i, j \leq n$.

[129] Let $(\Omega, \mathcal{F}, \{\mathcal{F}_t\}, P)$ be a filtered probability space. Show that $\mathcal{F}_{t+} = \bigcap_{s>t} \mathcal{F}_s$ defines a right continuous filtration. Give an example in which $\mathcal{F}_{t+} \neq \mathcal{F}_t$ for some $t \geq 0$.

[130] Let $\{X_n : n \geq 0\}$ be a discrete time Markov process. Show that conditioned on the present, the past and future are independent. Specifically, if $0 < n < N$, then

$$P(X_k \in dx_k, 0 \leq k \leq N, k \neq n | X_n) = P(X_k \in dx_k, 0 \leq k \leq n-1 | X_n).P(X_k \in dx_k, n+1 \leq k \leq N | X_n)$$

[131] Prove that if f is a bounded twice continuously differentiable function of n variables $f(x_1, ..., x_n)$ and $B_1(t), ..., B_n(t)$ are n independent standard Brownian motion processes, then, for $s < t$

$$f(B_1(t), ..., B_n(t)) = f(B_1(s), ..., B_n(s)) + \sum_{k=1}^{n} \int_s^t \frac{\partial f(B_1(u), ..., B_n(u))}{\partial x_i} dB_i(u)$$

$$+ \sum_{k=1}^{n} \int_s^t \frac{\partial^2 f(B_1(u), ..., B_n(u))}{\partial x_i^2} du$$

where the integrals in the second term on the right are Ito integrals.

[132] If $x_1, ..., x_n$ are real numbers and $p \geq 1$, then

$$|\sum_{i=1}^{n} x_i|^p \leq a(n,p) \sum_{i=1}^{n} |x_i|^p$$

where the constants $a(n,p)$ do not depend on $x_1, ..., x_n$, ie, they are functions of only n, p.

hint. It follows from Holder's inequalty for $p > 1$ and for $p = 1$, it is simply the triangle inequality. For $p > 1$, we have in fact

$$|\sum_{i=1}^{n} x_i y_i| \leq (\sum |x_i|^p)^{1/p} (\sum |y_i|^q)^{1/q}$$

where $1/p + 1/q = 1$. Take $y_i = 1$ for all i and conclude that

$$|\sum x_i| \leq n^{1/q} (\sum |x_i|^p)^{1/p}$$

and hence

$$|\sum x_i|^p \leq n^{p/q} \sum |x_i|^p$$

so that we can take

$$a(n,p) = n^{p/q} = n^{p-1}$$

Now let M_t be a bounded continuous martingale and let $p \geq 2$. We set $\phi(x) = |x|^p$. Then $\phi'(x) = sgn(x).p.|x|^{p-1}$ and

$$\phi''(x) = 2p|x|^{p-1}\delta(x) + p(p-1)|x|^{p-2} = p(p-1)|x|^{p-2}$$

Thus, Ito's formula gives

$$|M_t|^p = p \int_0^t sgn(M_s)|M_s|^{p-1} dM_s + p(p-1) \int_0^t |M_s|^{p-2} d < M, M >_s$$

Thus,

$$\mathbb{E}(|M_\infty|^p) = p(p-1)\mathbb{E} \int_0^\infty |M_s|^{p-2} d < M, M >_s \leq p(p-1)\mathbb{E}(M_\infty^{*(p-2)} < M, M >_\infty)$$

where

$$M_\infty^* = sup_{0 \leq t < \infty} |M_t|$$

Application of Holder's inequality gives

$$\mathbb{E}(M_\infty^{*(p-2)} < M, M >_\infty) \leq (\mathbb{E} < M, M >^{p/2})^{2/p} (\mathbb{E} M_\infty^{*p})^{(1-2/p)}$$

Also Doob's inequality gives

$$\mathbb{E}(M_\infty^{*p}) \leq K_p \mathbb{E}(|M_\infty|^p)$$

for some constant K_p. Combining this with the previous inequalities gives us

$$\mathbb{E}(|M_\infty|^p) \leq (\mathbb{E}(< M, M >^{p/2}))^{2/p} (K_p \mathbb{E}|M_\infty|^p)^{1-2/p}$$

so that

$$\mathbb{E}(|M_\infty|^p) \leq C_p \mathbb{E} < M, M >^{p/2}$$

for some constant C_p. Again combining this with Doob's inequality gives

$$\mathbb{E}(M_\infty^{*p}) \leq A_p \mathbb{E}(< M, M >^{p/2})$$

for some constant A_p. This is one of the Burkholder-Davies-Guindy (BDG) inequalities.

Reference: Daniel Revuz and Marc Yor, Continuous Martingales and Brownian Motion.

[133] Use the BDG inequalities to derive the general theory of existence and uniqueness of solutions to stochastic differential equations.

[134] Let $\mathbf{a}, \mathbf{b}, \mathbf{c}$ be three vectors in three dimensional Euclidean space. Show that the volume of the parallelopiped formed using these three vectors is given by $\mathbf{a} \times \mathbf{b}.\mathbf{c}$.

[135] Given two polynomials f, d in one variable, show that there exists a polynomial q such that either $f = d.q$ or else $deg(f - d.q) < deg(d)$.

[136] Given two sequences $\{x_n\}$ and $\{y_n\}$ in a metric space (X, d), show that if x_n converges to a limit x and $d(x_n, y_n) \to 0$, then y_n also converges to x.

hint:Use the triangle inequality.

[137] Given a topological space (X, τ) (τ is the topology on X and the elements of τ are subsets of X called the open sets) we say that a sequence $\{x_n\} \subset X$ converges to $x \in X$ if given any $U \in \tau$ such that $x \in U$, there exists an integer N such that $x_n \in U$ for all $n > N$.

[138] Let $T : \mathbb{R}^n \to \mathbb{R}^n$ be any transformation. Let $< .,. >$ denote the standard inner product on $\mathbb{R}^n$, ie, $< x, y >= y^T x = x^T y$. Show that if $\| T(x) - T(y) \|=\| x - y \|$ for all $x, y \in \mathbb{R}^n$, then T is an affine linear transformation, ie, there exists an $n \times n$ matrix S and a vector $z \in \mathbb{R}^n$ such that $T(x) = S(x) + z$ for all $x \in \mathbb{R}^n$.

[139] Let $B(t), t \geq 0$ be a standard Brownian motion process and μ a probability distribution on $\mathbb{R}$. Under what conditions does there exist a stoptime T for the Brownian filtration such that $B(T)$ has the distribution μ.

[140] Let $x(t)$ be a diffusion processs on $\mathbb{R}$ satisfying the sde

$$dx(t) = \mu(x(t))dt + \sigma(x(t))dB(t)$$

Let $\xi(0) = x$ and let $a < x < b$. Let $T_a = min(t \geq 0 : x(t) = a\}$ and $T_b = min(t \geq 0 : x(t) = b\}$. Give a method for determining $P(T_a \leq T_b)$. Specialize to the case when $x(t)$ is Brownian motion.

hint: Set

$$u(x) = P(T_a \leq T_b | x(0) = x), a \leq x \leq b$$

Then $u(a) = 1, u(b) = 0$ and

$$u(x) = \mathbb{E}(u(x(h))|x(0) = x) + o(h), h \to 0$$

Application of Ito's formula then gives

$$\mu(x)u'(x) + \frac{\sigma^2(x)}{2}u''(x) = 0$$

[141] Let $x(t)$ be a diffusion process satifying the sde

$$dx(t) = \mu(x(t))dt + \sigma(x(t))dB(t)$$

and let $T_a = min(t \geq 0 : x(t) = a\}$. Give a method for determining $u(x) = \mathbb{E}(\phi(x(T_a))|x(0) = x)$ where ϕ is continuous and bounded and has continuous and bounded first and second order derivatives.

hint: $u(a) = \phi(a)$ and further

$$u(x) = \mathbb{E}(u(x(h))|x(0) = x) + o(h), h \to 0$$

[142] State Ito's formula for Cadlag semimartingales, ie, semimartingales which are right continuous with left limits.

hint: Let $X(t)$ be a Cadlag semimartingale. Then if f has bounded derivatives upto order two, We set

$$df(X(t)) = f(X(t+dt)) - f(X(t-))$$

Write $X(t) = X_c(t) + X_J(t)$ where X_c is the continuous part of X and X_J is the pure jump part of X. Then, $dX(t) = dX_c(t) + \Delta X(t)$ where $\Delta X(t) = X(t) - X(t-)$. Then,

$$f(X(t+dt)) = f(X(t-) + dX(t)) = f(X(t-)) + \Delta X(t) + dX_c(t)) = f(X(t) + dX_c(t))$$

$$= f(X(t)) + f'(X(t))dX_c(t) + \frac{1}{2}f''(X(t))d < X, X >_c (t)$$

$$= f(X(t-)) + \Delta f(X(t)) + f'(X(t-))dX_c(t) + \frac{1}{2}f''(X(t-))d < X, X >_c (t)$$

where $o(dt)$ terms have been neglected. Thus,

$$df(X(t)) = f(X(t+dt)) - f(X(t-))$$

$$= \Delta f(X(t)) - f'(X(t-))\Delta X(t) + f'(X(t-))dX(t) + \frac{1}{2}f''(X(t-))d < X, X >_c (t)$$

Note that

$$\Delta f(X(t)) = f(X(t)) - f(X(t-))$$

[143] The equation of a circle in cartesian coordinates is given by

$$x^2 + y^2 + 2gx + 2fy + c = 0$$

Determine the centre and radius of the circle.

[144] The equation of a sphere in cartesian coordinates is given by

$$x^2 + y^2 + z^2 + 2gx + 2fy + 2hz + c = 0$$

Determine the centre and radius of the sphere.

[145] Solve the heat equation with source.

$$D.\nabla^2 u(t, \mathbf{x}) - u_{,t}(t, \mathbf{x}) = s(t, \mathbf{x}), \mathbf{x} \in \mathbb{R}^3, t \geq 0$$

hint: Use Fourier analysis. Specifically, derive an ordinary differential equation in t satisfied by

$$\hat{u}(t, \mathbf{k}) = \int u(t, \mathbf{x}) exp(-i\mathbf{k}.\mathbf{x}) d^3 x$$

[146] Let $\mathcal{L}$ denote the Laplace transform, ie, for a function $f : [0, \infty) \to \mathbb{C}$, we have

$$F(s) = \mathcal{L}(f)(s) = \int_0^\infty exp(-st) f(t) dt$$

Show that under sufficiently general conditions on f,

$$f(0) = lim_{s \to \infty} sF(s), f(\infty) = lim_{s \to 0} sF(s)$$

hint:

$$\mathcal{L}(f')(s) = sF(s) - f(0)$$

and letting $s \to \infty$ on both sides gives $f(0) = lim_{s \to \infty} sF(s)$. Again, letting $s \to 0$ gives

$$f(\infty) - f(0) = \int_0^\infty f'(t) dt = lim_{s \to 0} \mathcal{L}(f')(s) = lim_{s \to 0}(sF(s) - f(0))$$

[147] Let X, Y be two independent random variables on a probability space having densities f_X, f_Y respectively. Show that the density of $Z = X + Y$ is given by

$$f_Z(z) = \int_{-\infty}^\infty f_X(z - y) f_Y(y) dy$$

More generally, let $X_1, ..., X_n$ be n independent random variables on a probability space having densities $f_1, ..., f_n$ respectively. Show that the density of $X_1 + ... + X_n$ is given by

$$f(z) = \int_{\mathbb{R}^{n-1}} f_1(z - x_2 - ... - x_n) f_2(x_2)...f_n(x_n) dx_2...dx_n$$

[148] Let $X_1, ..., X_n$ be n random variables on a probability space having joint density $f(x_1, ..., x_n)$. Show that the density of $X_1 + ... + X_n$ is given by

$$f(z) = \int_{\mathbb{R}^{n-1}} f(z - x_2 - ... - x_n, x_2, ..., x_n) dx_2...dx_n$$

[149] Let X and Y be independent random variables on a probability space assuming values in a topological group G. Let the probability distribution of X be μ_X and that of Y be μ_Y. The σ algebra on G is the Borel σ algebra, ie, that generated by the open sets. Show that if E is any Borel subset of G, then

$$\Pr(XY \in E) = \int_G P(X \in Ey^{-1}) d\mu_Y(y) = \int_G \mu_X(Ey^{-1}) d\mu_Y(y)$$

$$= \int Pr(Y \in x^{-1}E) d\mu_X(x) = \int_G \mu_Y(x^{-1}E) d\mu_X(x)$$

[150] Suppose $X_1, ..., X_n$ are independent random variables with means $\mathbb{E}X_i = \mu_i$ and variances $V(X_i) = \sigma_i^2$. Then show that $S = X_1 + ... + X_n$ is a random variable with mean $\mathbb{E}S = \sum_{i=1}^n \mu_i$ and variance $V(S) = \sum_i \sigma_i^2$. Prove this by assuming that the $X_i's$ have probability densities.

[151] Suppose $X(t), t \in [0, T]$ is a random process with characteristic functional

$$\mathbb{E}[exp(i \int_0^T \phi(t) X(t) dt)] = \psi(\phi)$$

Show that if $t_1, ..., t_n$ are distinct points in $[0, T]$, then

$$\frac{\partial^n}{\partial \phi(t_1)...\partial \phi(t_n)} \psi(\phi)|_{\phi=0} = i^n \mathbb{E}(X(t_1)...X(t_n))$$

[152] Explain how you would obtain the general solution to the difference equation

$$x(n) + a_1 x(n-1) + ... + a_p x(n-p) = 0$$

hint:Substitute $x(n) = z^n$.

[153] Prove that the Fourier transform on $\mathbb{R}^n$ is a unitary operator when restricted to functions in $L^2(\mathbb{R}^n)$.

[154] Let $B(t), t \geq 0$ be standard one dimensional Brownian motion and let τ_a be the first time at which this Brownian motion starting at the origin hits the level $a > 0$, ie,

$$\tau_a = inf\{t > 0, B(t) = a\}$$

Then compute the probability distribution of τ_a using the reflection principle and compute the moment generating function of τ_a using the optional stopping theorem for the exponential martingale and from these two deduce the formula for the Laplace transform of the heat kernel with respect to the time argument.

[155] Suppose F is a totally finite measure on metric group G. Let F^{*n} denote the n-fold convolution of the distribution of F with itself and define

$$e(F) = exp(-F(X)) \sum_{n=0}^{\infty} \frac{F^{*n}}{n!}$$

Then show that $e(F)$ is a probability distribution on X and that its characteristic function is given by

$$\hat{e}(F)(y) = \int_X (x, y) de(F)(x) = exp(\int_X ((x, y) - 1) dF(x))$$

where $y \in \hat{X}$, the character group of X. Show further that $e(F)$ is infinitely divisible and in fact, $e(F) = e(F/n)^{*n}$ for every positive integer n. More generally, show that if F, G are totally finite measures on X, then $e(F + G) = e(F) * e(G)$. This is most easily proved using Fourier transforms.

[156] Let $B(t), t \geq 0$ be standard Brownian motion with $B(0) = x$. For $a < x < b$, let $\tau_{a,b}$ be the first time at which the process hits either the level a or the level b, ie,

$$\tau_{a,b} = min(t \geq 0 : B(t) = a \, or \, b\}$$

Then compute using the optional stopping theorem for the exponential martingale, the moment generating function of $\tau_{a,b}$.

hint: $B(\tau_{a,b}) \in \{a, b\}$ so $|B(\tau_{a,b}) - (a+b)/2| = (b-a)/2$. Now, the optional stopping theorem gives

$$\mathbb{E}[exp(\lambda B(\tau_{a,b}) - \lambda^2 \tau_{a,b}/2)] = 1, \lambda \in \mathbb{R}$$

and hence

$$\mathbb{E}[exp(\lambda(B(\tau_{a,b}) - (a+b)/2) - \lambda^2 \tau_{a,b}/2)] = exp(-\lambda(a+b)/2)$$

Change λ to $-\lambda$ in this equation and add the two equations to get

$$\mathbb{E}[exp(-\lambda^2 \tau_{a,b}/2)] = \frac{cosh(\lambda(a+b)/2)}{cosh(\lambda(b-a)/2)}$$

[157] Draw the circuit diagram of an integrator and a differentiator using the ideal operational amplifier. Determine the exact transfer functions of these assuming that the operational amplifier has a finite input impedance, a finite output impedance and a frequency dependent voltage gain of the form $\frac{A_{v0}}{1+\tau.s}$.

[158] Assume that the input signal to a discrete time system is $x(n)$ and that the output is given by a second order Volterra system as

$$y(n) = \sum_k h(k)x(n-k) + \sum_{k,m} g(k,m)x(n-k)x(n-m)$$

Design a linear time invariant filter having impulse response $q(n)$ that takes the signal $y(n)$ as input and outputs an estimate $\hat{x}(n)$ of the input so that

$$\sum_n (x(n) - \hat{x}(n))^2$$

is a minimum.

[159] Repeat the previous problem assuming that the signals under consideration are two dimensional signals, ie, the input is $x(n,m)$, the output is $y(n,m)$ given by

$$y(n,m) = \sum_{k,l} h(k,l)x(n-k,m-l) + \sum_{k,l,r,s} g(k,l,r,s)x(n-k,m-l)x(n-r,m-s)$$

and that the estimate of the input is given by

$$\hat{x}(n,m) = \sum_{k,l} q(k,l)y(n-k,m-l)$$

with the filter coefficients $q(k,l)$ chosen so that

$$\sum_{n,m} (x(n,m) - \hat{x}(n,m))^2$$

is a minimum.

[160] Determine the matrix of a reflection about a plane passing through the origin in $\mathbb{R}^3$. What is the determinant of this matrix.

[161] Suppose that the diffusion coefficient is a function of the particle concentration, ie, with $N(t,x)$ denoting the concentration and D the diffusion coefficient, we have $D = D(N)$. The diffusion equation reads

$$(D(N)N_{,x})_{,x} - N_{,t} = 0$$

We are assuming that diffusion takes place in one dimension. If it were taking place in three dimensions, the diffusion equation would read

$$(\nabla, D(N)\nabla N) = N_{,t}$$

Expanding the terms on the left, we have

$$D(N)N_{,xx} + D'(N)N_{,x}^2 = N_{,t}$$

We write

$$D(N) = D_0 + \epsilon . D_1(N)$$

where D_0 is a constant and ϵ is a small parameter. We seek the solution to the diffusion equation upto linear terms in ϵ. Retaining terms upto $O(\epsilon)$ in the above diffusion equation gives us

$$(D_0 + \epsilon . D_1(N))N_{,xx} + \epsilon . D_1'(N)N_{,x}^2 = N_{,t}$$

or

$$D_0 N_{,xx} - N_{,t} + \epsilon . (D_1(N)N_{,xx} + D_1'(N)N_{,x}^2) = 0$$

Write

$$N = N^{(0)} + \epsilon . N^{(1)}$$

Then upto $O(\epsilon)$, we get two equations,

$$D_0 N_{,xx}^{(0)} - N_{,t}^{(0)} = 0,$$

$$D_0 N_{,xx}^{(1)} - N_{,t}^{(1)} + D_1(N^{(0)})N_{,xx}^{(0)} + D_1'(N^{(0)})N_{,x}^{(0)2} = 0$$

Since $N^{(0)}$ is known by solving the previous equation, this is a linear partial differential equation for $N^{(1)}$ with a source term and the solution for $N^{(1)}$ can be determined using standard Fourier methods.

[162] Solve

$$DN_{,xx}(t,x) - N_{,t}(t,x) = s(t,x)$$

where D is a constant and s is a known function.

hint: Take the Fourier transform with respect to the spatial variable x, so that

$$\hat{N}(t,k) = \int_\infty^\infty N(t,x)exp(-ikx)dx$$

and

$$\hat{s}(t,k) = \int_\infty^\infty s(t,x)exp(-ikx)dx$$

Then,

$$-Dk^2\hat{N}(t,k) - \hat{N}_{,t}(t,k) = \hat{s}(t,k)$$

or

$$\hat{N}_{,t}(t,k) + Dk^2\hat{N}(t,k) = -\hat{s}(t,k)$$

The solution is

$$\hat{N}(t,k) = exp(-tDk^2)\hat{N}(0,k) - \int_0^t exp(-(t-u)Dk^2)\hat{s}(u,k)du$$

Now take the inverse Fourier transform on both sides to obtain the solution for any initial condition $N(0,x)$.

[163] Show from first principles that the exact equation of a vibrating string having constant mass per unit length and length L with endpoints fixed is given by

$$(1+u_x^2)^{1/2}u_{tt} - c^2 u_{xx} = 0, 0 \le x \le L, t \ge 0$$

and under the small amplitude assumption, this approximates to

$$u_{tt} - c^2 u_{xx} = 0$$

Here, $u(t,x)$ is the amplitude of the string and

$$u_x = \frac{\partial u}{\partial x}, u_t = \frac{\partial u}{\partial t}, u_{xx} = \frac{\partial^2 u}{\partial x^2}, u_{tt} = \frac{\partial^2 u}{\partial t^2}$$

Assuming that there is an external force $f(t,x)$ acting on the string, the equation of motion is

$$(1+u_x^2)^{1/2}u_{tt} - c^2 u_{xx} = f(t,x), 0 \le x \le L, t \ge 0$$

Approximating the nonlinear term $(1+u_x^2)^{1/2}$ by $1+u_x^2/2$ when $(|u_x| << 1)$ and treating the nonlinear term $u_x^2 u_{tt}/2$ in the resulting equation as a small perturbation, obtain an approximate solution to this equation based on first order perturbation theory.

[164] Let $X_1,...,X_n$ be jointly Gaussian random variables with second moments $R(i,j) = \mathbb{E}(X_i X_j), 1 \le i,j \le n$. Derive a formula for the moment $\mathbb{E}(\Pi_{i=1}^n X_i^{r_i})$ where $r_1,...,r_n$ are non-negative integers.

[165] Suppose $x(n), n \in \mathbb{Z}$ is a Gaussian process with mean $\mu(n) = \mathbb{E}x(n)$ and autocorreltation $R(n,m) = \mathbb{E}(x(n)x(m))$. The compute the mean and variance of the random variable

$$y(n) = \sum_k h(n,k)x(k) + \sum_{k,m} g(n,k,m)x(k)x(m)$$

where $h(n,k)$ and $g(n,k,m)$ are nonrandom coefficients. Also compute the autocorrelation function of $y(n)$, ie, $\mathbb{E}(y(n)y(m))$.

[166] Obtain the general solution to the n dimensional wave equation

$$c^2 \sum_{i=1}^n \frac{\partial^2 \psi(t,x_1,...,x_n)}{\partial x_i^2} - \frac{\partial^2 \psi(t,x_1,...,x_n)}{\partial t^2} = 0$$

hint. Make use of the n-dimensional Fourier transform with respect to the variables $x_1, ..., x_n$, ie, show that if

$$\hat{\psi}(t, k_1, ..., k_n) = \int_{\mathbb{R}^n} \psi(t, x_1, ..., x_n) exp(-i(k_1 x_1 + ... + k_n x_n)) dx_1 ... dx_n$$

then

$$\hat{\psi}_{,tt} + k^2 c^2 \hat{\psi} = 0$$

where

$$k^2 = \sum_{i=1}^{n} k_i^2$$

[167] Let $B_1(t), ..., B_\delta(t)$ be independent standard Brownian motions. Define

$$X(t) = (\sum_{i=1}^{\delta} B_i(t)^2)^{1/2}$$

We wish to determine the stochastic differential equation satisfied by $X(t)$. We have with $Y(t) = X^2(t)$,

$$dY(t) = d(\sum_{i=1}^{d} B_i^2(t)) = 2 \sum_i B_i(t) dB_i(t) + \delta.dt$$

It follows that

$$dX(t) = d\sqrt{Y(t)} = dY/2\sqrt{Y} - \frac{1}{8Y^{3/2}}(dY)^2$$

$$(dY)^2 = 4 \sum_i B_i^2 dt = 4X^2 dt$$

$$dY/2\sqrt{Y} = \sum B_i.dB_i/X + \delta.dt/2X$$

Define the process β so that

$$d\beta = \sum B_i.dB_i/X$$

Then β is a continuous martingale and $(d\beta)^2 = dt$. Thus, β is a Brownian motion. We have

$$dX = d\beta + \delta.dt/2X - \frac{1}{2X}dt = d\beta + ((\delta - 1)/2X)dt$$

or in other words, if $X(0) = x$, then

$$X(t) = x + \beta(t) + ((\delta - 1)/2) \int_0^t \frac{ds}{X(s)}$$

$X(t)$ is called a Bessel process of order δ and denoted by BES^δ.

[168] Let $B(t)$ be standard Brownian motion. Show that if $x \in \mathbb{R}$, then $M_x(t) = exp(x.B(t) - x^2 t/2)$ is a martingale. Apply the optional stopping theorem to this martingale and derive the moment generating function of T_a, the first time at which the Brownian motion hits the level a.

[169] Suppose $t_1, ..., t_n$ are n distinct complex numbers. Define the polynomials

$$L_i(t) = \frac{\Pi_{1 \leq j \leq n, j \neq i}(t - t_j)}{\Pi_{1 \leq j \leq n, j \neq i}(t_i - t_j)}$$

Show that

$$L_i(t_j) = \delta_{ij}, 1 \leq i, j \leq n$$

and hence deduce that if $f(t)$ is any polynomial of degree at most $n - 1$, then

$$f(t) = \sum_{i=1}^{n} f(t_i) L_i(t)$$

[170] Suppose $\{X_n, n = 0, 1, 2, ...\}$ is a martingale with respect to a filtration $\mathcal{F}_n, n = 0, 1, 2, ...$ Let $\tau_x = min(k \geq 0 : |X_k| = x\}$. Then,

$$P\{\tau_x \leq n\} = P(max_{0 \leq k \leq n}|X_k| \geq x\}$$

Then,

$$\{\tau_x = k\} = \{|X_m| < x, m = 0, 1, ..., k - 1, |X_k| \geq x\} \in \mathcal{F}_\kappa$$

Thus,

$$\mathbb{E}[|X_n|\chi_{\tau_x \leq n}] = \sum_{k=0}^{n} \mathbb{E}[|X_n|\chi_{\tau_x = k}]$$

$$\geq \sum_{k=0}^{n} \mathbb{E}[|X_k|\chi_{\tau_x = k}] \geq x. \sum_{k=0}^{n} P\{\tau_x = k\} = x.P\{\tau_x \leq n\}$$

Thus,

$$P\{max_{0 \leq k \leq n}|X_k| \geq x\} \leq x^{-1}\mathbb{E}[|X_n|]$$

This is Kolmogorov's maximal inequality for martingales. Under sufficiently general conditions, it can be extended to include continuous martingales. Thus, let $B(t)$ be standard Brownian motion. Then,

$$\{max_{s \leq t}(B(s) - \alpha.s) \geq \beta\}$$

$$= \{max_{s \leq t}(x.B(s) - x^2 s/2 + x^2 s/2 - \alpha x s) \geq x.\beta\}$$

$$\subset \{max_{s \leq t}(x.B(s) - x^2 s/2) + x^2 t/2 - \alpha x t \geq x.\beta\}$$

provided

$$x \geq 2\alpha$$

It follows that in this case, if we define the exponential martingale

$$M_x(s) = exp(x.B(s) - x^2 s/2), s \geq 0$$

then,

$$P\{max_{s \leq t}(B(s) - \alpha s) \geq \beta\}$$

$$\leq P\{max_{s\leq t}M_x(s) \geq exp(\beta x + \alpha x t - x^2 t/2)\}$$

$$\leq exp(x^2 t/2 - \beta x - \alpha x t)$$

in view of Kolmogorov's maximal inequality for martingales. We are assuming here that $x \geq 2\alpha$. Putting $x = 2\alpha$ in this inequality gives

$$P\{max_{s\leq t}(B(s) - \alpha s) \geq \beta\} \leq exp(-2\alpha\beta)$$

This inequality is useful only if either both α and β are positive or both are negative.

[171] Let $B(t)$ be standard Brownian motion. Then we have by formal application of the Ito rule to the function $|x|$,

$$d|B(t)| = sgn(B(t))dB(t) + \delta(B(t))dt$$

or under the assumption that $|B(0)| = x$,

$$|B(t)| = x + W(t) + L(t)$$

where

$$W(t) = \int_0^t sgn(B(s))dB(s)$$

is another Brownian motion process and

$$L(t) = \int_0^t \delta(B(s))ds$$

is the local time of the Brownian motion $B(.)$ at zero. More generally, the local time of the Brownian motion $B(.)$ at some $x \in \mathbb{R}$ is defined by

$$L(t, x) = \int_0^t \delta(B(s) - x)ds$$

We have under the assumption $B(0) = x$,

$$|B(t) - a| = |x - a| + W(t, a) + L(t, a)$$

where

$$W(t, a) = \int_0^t sgn(B(s) - a)dB(s)$$

is another Brownian motion process. Clearly,

$$\int_{-\infty}^{\infty} \phi(x)L(t, x)dx = \int_0^t \phi(B(s))ds$$

In particular, if E is any Borel subset of $\mathbb{R}$,

$$\int_E L(t, x)dx = \int_0^t \chi_E(B(s))ds$$

is the total time spent by the process $B(.)$ in the set E during the time interval $[0, t]$. We have for a twice differentiable function ϕ,

$$\phi(B(t)) - \phi(B(0)) = \int_0^t \phi'(B(s))dB(s) + \frac{1}{2}\int_0^t \phi''(B(s))ds$$

and

$$\int_0^t \phi''(B(s))ds = \int_{\mathbb{R}} \phi''(x)L(t,x)dx$$

so that the Ito rule can be expressed as

$$\phi(B(t)) - \phi(B(0)) = \int_0^t \phi'(B(s))dB(s) + \frac{1}{2}\int_{\mathbb{R}} \phi''(x)L(t,x)dx$$

For a rigorous discussion of the Brownian local time, see the book "Continuous Martingales and Brownian Motion" by Daniel Revuz and Marc Yor.

[172] Obtain particular solutions to the two dimensional wave equation with source using, Cartesian coordinates and polar coordinates. Note that in rectangular coordinates, this equation is

$$\psi_{,xx} + \psi_{,yy} - \frac{1}{c^2}\psi_{,tt} = -f$$

where $\psi(x,y,t)$ is the wave amplitude and $f(x,y,t)$ is the source. In polar coordinates, this equation is

$$\frac{1}{r}(r\psi_{,r})_{,r} + \frac{1}{r^2}\psi_{,\phi\phi} - \frac{1}{c^2}\psi_{,tt} = -f$$

or equivalently,

$$\psi_{,rr} + \frac{1}{r}\psi_{,r} + \frac{1}{r^2}\psi_{,\phi\phi} - \frac{1}{c^2}\psi_{,tt} = -f$$

Where (r, ϕ) are the polar coordinates related to the Cartesian coordinates by

$$x = r.cos(\phi), y = r.sin(\phi)$$

$\psi(r, \phi, t)$ is the wave amplitude and $f(r, \phi, t)$ is the source.

[173] Derive a least mean square algorithm for estimating the velocity of a one dimensional wave based on measurements of the wave amplitude over space at a sequence of time points. The algorithm should be a real time implementable one for situations in which the wave velocity is constant over space but varies slowly with time.

hint: Assume that the wave motion is governed by the partial differential equation

$$\psi_{,tt} - c^2\psi_{,xx} = 0$$

Let $c^2 = \theta$ so that the wave equation can be expressed as

$$\psi_{,tt} - \theta\psi_{,xx} = 0$$

The LMS algorithm for estimating θ iteratively in time reads

$$\hat{\theta}(t+\delta t) = \hat{\theta}(t) - \mu.\frac{\partial}{\partial\theta}\int_D (\psi_{,tt}(x,t) - \theta\psi_{,xx}(x,t))^2 dx\big|_{\theta=\hat{\theta}(t)}$$

where D is the region of space over which the wave amplitude is measured.

[174] What do you understand by an infinitely divisible distribution. What is the Levy-Kintchine representation for the characteristic function of such a distribution.

hint:Given a probability distribution F, suppose that for each positive integer n, there is a probability distribution F_n such that $F = F_n^{(*n)}$, ie, F is the n-fold convolution of F_n with itself, then F is said to be an infinitely divisible distribution. Equivalently, if ψ is the characteristic function of F, then F is infinitely divisible iff for each positive integer n, there is a characteristic function ψ_n, such that $\psi = \psi_n^n$.

[175] Let $\phi : \mathbb{R}^3 \to \mathbb{R}$ be a smooth function and let V be a volume in $\mathbb{R}^3$ bounded by the closed surface S. Let $\mathbf{n}$ be the unit normal to the surface at the point $\mathbf{r}$ on it. Then derive Gauss law for $\int_S \phi\mathbf{n}dS$.

hint: Let $\mathbf{a}$ be any constant vector. Then, by Gauss' theorem,

$$\mathbf{a}.\int_S \phi\mathbf{n}dS = \int_S \phi\mathbf{a}.\mathbf{n}dS = \int_V div(\mathbf{a}\phi)dV$$

$$= \int_V \mathbf{a}.\nabla\phi dV = \mathbf{a}.\int_V \nabla\phi dV$$

Since this holds for all vectors $\mathbf{a}$, we conclude that

$$\int_S \phi\mathbf{n}dS = \int_V \nabla\phi dV$$

[176] Let $B_t(\omega), t \geq 0$ be a standard Brownian motion process. Then we claim that the set $\gamma(\omega) = \{B_t(\omega) : 0 \leq t \leq 1\}$ is a measurable subset of $\mathbb{R}$ for almost all ω. This follows from the fact that the Brownian paths are almost surely continuous and that the continuous image of a compact set is compact and is therefore a Borel set. An alternate proof can be given as follows. Let N be a null set such that for all $\omega \in N^c$, $t \to B_t(\omega)$ is continuous. Then for $\omega \in N^c$, we have

$$\gamma(\omega) = \{z \in \mathbb{R} : inf_{t\in[0,t]}|z - B_t(\omega)| = 0\}$$

$$= \{z \in \mathbb{R} : inf_{t\in Q\cap[0,1]}|z - B_t(\omega)| = 0\}$$

where Q is the set of rationals. Since $Q \cap [0, 1]$ is countable, for each $n = 1, 2, ...$ we can find a finite set A_n such that $A_n \subset A_{n+1}$ for all $n = 1, 2, ...$ and $Q \cap [0, 1] = \bigcup_n A_n$. Define the function

$$\phi_n(z,\omega) = min\{|z - B_t(\omega)| : t \in A_n\}$$

Then, for each n, the map $(z, \omega) \to \phi_n(z, \omega)$ from $\mathbb{R} \times N^c$ into $\mathbb{R}$ is measurable and hence

$$\phi(z, \omega) = inf\{|z - B_t(\omega)| : t \in [0.1]\} = lim_n \phi_n(t, \omega)$$

is also measurable (with respect to the product σ algebra). It follows that $E = \{(z, \omega) \in \mathbb{R} \times N^c : \phi(z, \omega) = 0\}$ is measurable and hence

$$E_\omega = \{z : (z, \omega) \in E\} = \{z : \phi(z, \omega) = 0\} = \gamma(\omega)$$

is measurable for all $\omega \in N^c$. This proves the claim. The same argument works for two dimensional Brownian motion. Suppose $B_t(\omega) = (B_{1t}(\omega), B_{2t}(\omega))$ is standard two dimensional Brownian motion. Then the image of $[0, 1]$ under B is almost surely a compact subset of $\mathbb{R}^2$ and hence almost surely measurable. It follows that if λ denotes the Lebesgue measure on $\mathbb{R}^2$, then $\omega \to \lambda(\gamma_1(\omega))$ is a random variable where

$$\gamma_1(\omega) = \{B_t(\omega) : t \in [0.1]\}$$

We claim that $\lambda(\gamma_1(\omega)) = 0$ almost surely. To prove this, we use the fact that $\frac{1}{\sqrt{2}}B(2t), 0 \leq t \leq 1$ is a standard Brownian motion and hence, if

$$\gamma_2(\omega) = \{B_t(\omega); 0 \leq t \leq 2\}$$

then

$$\mathbb{E}\lambda(\gamma_2(\omega)) = 2\mathbb{E}\lambda(\gamma_1(\omega))$$

[177] Let $(\Omega_i, \mathcal{F}_i), i = 1, 2$ be measurable spaces. We define $\Omega = \Omega_1 \times \Omega_2$ and $\mathcal{F} = \mathcal{F}_1 \otimes \mathcal{F}_2$, the latter meaning that $\mathcal{F}$ is the σ algebra on Ω generated by all measurable rectangles $E_1 \times E_2$ with $E_i \in \mathcal{F}_i, i = 1, 2$. Let $E \in \mathcal{F}$. We want to show that for each $\omega' \in \Omega_2$, the set $E_{\omega'} = \{\omega \in \Omega_1 : (\omega, \omega') \in E\}$ is in $\mathcal{F}_1$. Indeed, let

$$\mathcal{C} = \{E \in \mathcal{F} : E_\omega \in \mathcal{F}_1\}$$

where $\omega \in \Omega_2$ is given. For $E = E_1 \times E_2$ with $E_i \in \mathcal{F}_i, i = 1, 2$, it is clear that $E_\omega = E_1$ if $\omega \in E_2$ and $E_\omega = \phi$ if $\omega \notin E_2$. Hence in both cases $E_\omega \in \mathcal{F}_1$ and so $E \in \mathcal{C}$.

[178] Let $B(t), t \geq 0$ be standard Brownian motion. Prove that if $t > 0$ and $P_n : 0 = t_0 < ... < t_n = t, n = 1, 2, ...$ are partitions of $[0, t]$ such that $|P_n| = max_{0 \leq k \leq n-1}(t_{k+1} - t_k) \to 0$ as $n \to \infty$, then $\sum_{k=0}^{n-1}(B(t_{k+1}) - B(t_k))^2$ converges in L^2 to t as $n \to \infty$.

[179] Suppose $X(t)$ is a continuous martingale and $X^2(t) - t$ is also a martingale. Then show that $X(t)$ is a Brownian motion process.

[180] Suppose $X(t)$ is a stochastic process having stationary independent increments. Assume that $X(0) = 0$. Show that its characteristic function must be of the form

$$\mathbb{E}[exp(iuX(t))] = exp(t\psi(u)), t \geq 0, u \in \mathbb{R}$$

hint:

$$\mathbb{E}exp(iuX(t+s)) = \mathbb{E}exp(iu(X(t+s) - X(t)))\mathbb{E}exp(iuX(t))$$

Let

$$\phi(u,t) = \mathbb{E}exp(iuX(t))$$

Conclude that

$$\phi(u,t+s) = \phi(u,s)\phi(u,t), t,s \geq 0, u \in \mathbb{R}$$

[181] Suppose $f(t,x)$ is a random field on $\mathbb{R}^2$. Assume that f satisfies the partial differential equation

$$f_{,t}(t,x) = D.f_{,xx}(t,x) + s(t,x)$$

where s is another random field. Then show that if $< f(t,x)s(u,y) >= R_{fs}(t,x,u,y)$ and $< s(t,x)s(u,y) >= R_{ss}(t,x,u,y)$, then

$$R_{fs,t}(t,x,u,y) - D.R_{fs,xx}(t,x,u,y) = R_{ss}(t,x,u,y)$$

Show further that if $< f(t,x)f(u,y) >= R_{ff}(t,x,u,y)$, then

$$R_{ff,t}(t,x,u,y) - D.R_{ff,xx}(t,x,u,y) = R_{fs}(u,y,s,x)$$

[182] Prove that the Fourier transform of $1/(\pi t)$ taken in the principal value sense is $-i.sgn(\omega)$. An LTI system with $1/(\pi t)$ as the impulse response is called a Hilbert transformer. What is the effect of the Hilbert transformer on sinusoidal signals.

[183] Let $M_t, t \geq 0$ be a non-negative continuous martingale such that $lim_{t\to\infty} M_t = 0$. Let $\mathcal{F}_t$ be the given filtration with respect to which M is a martingale. Fix $0 < A < \infty$ and define for $x \geq 0$

$$T_x = inf\{t \geq 0 : M_t > x\}$$

if $M_t > x$ for some $t \in [0, A]$ and define $T_x = A$ if $M_t \leq x$ for all $t \in [0, A]$. Then we have by the optional stopping theorem,

$$\mathbb{E}(M_{T_x}|\mathcal{F}_0) = M_0$$

If $M_0 > x$, then $T_x = 0$ and $M_{T_x} = M_0$. In this case, the above equation is trivial. Suppose $M_0 \leq x$. Then, from the above equation,

$$x.P(T_x < A|\mathcal{F}_0) + \mathbb{E}(M_A.\chi_{T_x=A}|\mathcal{F}_0) = M_0$$

Define

$$M_t^* = sup\{M_s : s \leq t\}$$

Then, from the above equation,

$$M_0 - x.P(M_A^* > x|\mathcal{F}_0) = \mathbb{E}(M_A.\chi_{T_x=A}|\mathcal{F}_0)$$

Now,

$$\mathbb{E}(M_A \cdot \chi_{T_x = A} | \mathcal{F}_0) = \mathbb{E}(M_A \cdot chi_{M_A^* \leq x} | \mathcal{F}_0)$$

Define the random variable

$$N_A = M_A \cdot \chi_{M_A^* \leq x}$$

Obviously, $N_A \leq x$ for all $A > 0$ and hence by Fatou's lemma,

$$limsup_{A \to \infty} \mathbb{E}(N_A | \mathcal{F}_0) \leq \mathbb{E}(limsup_{A \to \infty} N_A | \mathcal{F}_0) = 0$$

since $lim_{A \to \infty} M_A = 0$. Thus, we get on letting $A \to \infty$,

$$P(M_\infty^* > x | \mathcal{F}_0) = M_0 / x$$

We thus have for all $x > 0$,

$$P(M_\infty^* > x | \mathcal{F}_0) = min(1, M_0 / x)$$

Reference: D.Revuz and M.Yor, Continuous Martingales and Brownian Motion.

[184] Let $B_1(t)$ and $B_2(t)$ be two independent standard Brownian motion processes starting at zero. Let $a > 0$ and

$$\tau_{1a} = min(t \geq 0 : B_1(t) = a\}$$

Compute the probability distribution of $B_2(\tau_{1a})$.

hint:Using the reflection principle, first compute the probability density of τ_{1a} and then

$$P(B_2(\tau_{1a}) \leq x) = \int_0^\infty P(\tau_{1a} \in [t, t+dt]) P(B_2(t) \leq x)$$

[185] If X is a random variable assuming values $\{1, 2, ..., n\}$ with probabilities $\{p_1, ..., p_n\}$ so that $p_i \geq 0$ and $\sum_i p_i = 1$, then the entropy of X is defined by

$$H(X) = -\sum_i p_i log(p_i)$$

If $x = 0$, then $x.log(x)$ is taken to be zero. Then if n is fixed, $H(X)$ attains its maximum value of $log(n)$ when $p_i = 1/n$ for each $i = 1, 2, ..., n$. Suppose X, Y are two random variables defined on the same probability space such that $P(X = a_i, Y = b_j) = p(i, j), 1 \leq i \leq n, 1 \leq j \leq m$ so that $p(i, j) \geq 0$ and $\sum_{i,j} p(i, j) = 1$. Then, the joint entropy of the pair (X, Y) is given by

$$H(X, Y) = -\sum_{i,j} p(i, j).log(p(i, j))$$

We write

$$p(i, j) = p(i)q(j|i), p(i) = \sum_j p(i, j)$$

Then,

$$q(j|i) = P(Y = b_j | X = a_i)$$

Then,

$$H(X,Y) = -\sum_{i,j} p(i)q(j|i).log(p(i)q(j|i)) = -\sum_{i,j} p(i)q(j|i)(log(p(i)) + log(q(j|i)))$$

$$= -\sum p(i)log(p(i)) - \sum_{i,j} p(i)q(j|i)log(q(j|i))$$

since

$$\sum_j q(j|i) = 1$$

Define

$$H(Y|X = i) = \sum_j q(j|i)log(q(j|i))$$

and define

$$H(Y|X) = \sum_i p(i)H(Y|X = i)$$

Then the above formula can be expressed as

$$H(X,Y) = H(X) + H(Y|X)$$

Suppose $X_1, X_2, ..., X_n$ are random variables assuming a finite number of values and all are defined on the same probability space. Then their joint entropy is given by

$$H(X_1, ..., X_n) = H(X_n|X_1, ..., X_{n-1}) + H(X_1, ..., X_{n-1}) = H(X_1) + \sum_{i=2}^{n} H(X_i|X_{i-1}, X_{i-2}, ..., X_1)$$

Suppose $X_n, n = 0, \pm 1, \pm 2, ...$ is a stationary sequence. Then, it follows that

$$lim_{n \to \infty} H(X_n|X_{n-1}, ..., X_1) = lim_{n \to \infty} H(X_0|X_{-1}, X_{-2}, ..., X_{-n+1})$$

exists ($H(X_0|X_{-1}, ..., X_{-n+1}), n = 2, 3, ...$ is a non-increasing sequence) and we denote this limit by

$$H(X_0|X_{-1}, X_{-2}, ...)$$

Then by the basic theorem on Cesaro sums, it follows that

$$\bar{H} = lim_{n \to \infty} \frac{1}{n} H(X_1, ..., X_n) = H(X_0|X_{-1}, X_{-2}...,)$$

$\bar{H}$ is called the entropy rate of the stationary process (X_n). If X_n is a stationary Markov process, then it is easily seen that its entropy rate is given by

$$\bar{H} = H(X_1|X_0)$$

[186] For $0 \leq 1$, define

$$H(p) = -p.log(p) - (1-p).log(1-p)$$

$H(p)$ is thus the entropy of a random variable X assuming two values with probabilities p and $1-p$ respectively. Consider the function $\phi(x) = -x.log(x)$. We have $\phi'(x) = -log(x) - 1$ and $\phi''(x) = -1/x$. Thus, for $x > 0$, ϕ is a concave function. Hence, if p, q are in $[0,1]$ and $\alpha \in [0,1]$, it follows that

$$H(\alpha.p + (1-\alpha).q) \geq \alpha.H(p) + (1-\alpha).H(q)$$

This result can be generalized to a random variables assuming n distinct values. In other words, if $\{p_1, ..., p_n\}$ and $\{q_1, ..., q_n\}$ are two probability distributions and $0 \leq \alpha \leq 1$, then

$$H(\alpha.p_1 + (1-\alpha).q_1, ..., \alpha.p_n + (1-\alpha).q_n) \geq \alpha.H(p_1, ..., p_n) + (1-\alpha).H(q_1, ..., q_n)$$

[187] Let $X_1, X_2, ...$ be iid random variables assuming values in the finite set $A = \{a_1, ..., a_m\}$ with probabilities $p(a) = P(X_1 = a), a \in A$. Define

$$H(X) = -\sum_{i=1}^{m} p_i log(p_i)$$

By the weak law of large numbers, the random variable $-\sum_{i=1}^{n} log(p(X_i))$ converges in probability to $H(p) = -\mathbb{E}[log(p(X_1))]$. Define for $\epsilon > 0$

$$E(n, \epsilon) = \{(x_1, ..., x_n) \in A^n : |n^{-1} \sum_{i=1}^{n} log(p(x_i)) + H(p)| < \epsilon\}$$

Then by the Chebyshev inequality,

$$P(E(n, \epsilon)) = \sum_{(x_1, ..., x_n) \in E(n, \epsilon)} p(x_1)...p(x_n) \geq 1 - Var(log(X_1))/n\epsilon^2$$

Let $\mu(E)$ denote the number of elements in a finite set E. By the definition of $E(n, \epsilon)$,

$$exp(-n(H(p) + \epsilon)) \leq p(x_1)...p(x_n) \leq exp(-n(H(p) - \epsilon)), (x_1, ..., x_n) \in E(n, \epsilon)$$

Summing over $(x_1, ..., x_n) \in E(n, \epsilon)$ gives us

$$exp(-n(H(p) + \epsilon))\mu(E(n, \epsilon)) \leq P(E(n, \epsilon)) \leq exp(-n(H(p) - \epsilon))\mu(E(n, \epsilon))$$

so that

$$(1 - Var(log(X_1))/n\epsilon^2)exp(n(H(p) - \epsilon)) \geq \mu(E(n, \epsilon)) \leq exp(n(H(p) + \epsilon))$$

Thus,

$$H(p) - \epsilon \leq liminf_{n \to \infty} \frac{log(\mu(E(n, \epsilon)))}{n} \leq limsup_{n \to \infty} \frac{log(\mu(E(n, \epsilon)))}{n} \leq H(p) + \epsilon$$

[188] Define a martingale, submartingale and supermartingale. State and prove the down crossings inequality for submartingales. Use this inequality to prove the martingale convergence theorem.

[189] Suppose $f : [a, b] \rightarrow \mathbb{R}$ is a measurable function. We approximate its integral $\int_a^b f(x)dx$ by the partial sum $\sum_{k=0}^{N-1} f(a+k(b-a)/N)(b-a)/N$. Assuming that f is continuously differentiable, derive a bound for the approximation error.

hint: Let $\Delta = (b-a)/N$ and for $x \in [a+k\Delta, a+(k+1)\Delta]$, apply the Lagrange mean value theorem.

$$f(x) = f(a + k\Delta) + (x - a - k\Delta)f'(a + k\Delta + \theta(x - a - k\Delta)), \theta \in [0, 1]$$

Let $M = sup_x|f'(x)|$. Then,

$$-M(x - a - k\Delta) \leq f(x) - f(a + k\Delta) \leq M(x - a - k\Delta), x \in [a + k\Delta, a + (k + 1)\Delta]$$

Integrating over $[a + k\Delta, a + (k + 1)\Delta]$ gives

$$-M\Delta^2/2 \leq \int_{a+k\Delta}^{a+(k+1)\Delta} f(x)dx - f(a + k\Delta)\Delta \leq M\Delta^2/2$$

and summing over k, we get

$$-M(b - a)^2/2N = -MN\Delta^2/2 \leq \int_a^b f(x)dx - \sum_{k=0}^{N-1} f(a + k\Delta)\Delta \leq M(b - a)^2/2N$$

[190] Define uniform integrability of a family of random variables. Show that if X is an integrable random variable on a probability space $(\Omega, \mathcal{F}, P)$ and if $\{\mathcal{B}_\alpha : \alpha \in I\}$ is a family of sub-σ-algebras of $\mathcal{F}$, then $\{\mathbb{E}(X|\mathcal{B}_\alpha), \alpha \in I\}$ is a uniformly integrable family of random variables.

hint: A family $\{X_\alpha, \alpha \in \Gamma\}$ of random variables, all defined on the same probability space $(\Omega, \mathcal{F}, P)$ is said to be uniformly integrable if (a) $sup_{\alpha \in \Gamma}\mathbb{E}(|X_\alpha|) < \infty$ and (b)

$$lim_{c \rightarrow \infty} sup_{\alpha \in \Gamma}\mathbb{E}(|X_\alpha|\chi_{|X_\alpha|>c}) = 0$$

Suppose $\{X_\alpha : \alpha \in \Gamma\}$ is a uniformly integrable family of random variables. Let $\epsilon > 0$ be given and choose $0 < c = c(\epsilon) < \infty$ such that $\mathbb{E}(|X_\alpha|\chi_{|X_\alpha|>c}) < \epsilon$ for all $\alpha \in \Gamma$. Suppose $\delta > 0$ and $A \in \mathcal{F}$ is such that $P(A) < \delta$. Then,

$$\mathbb{E}(|X_\alpha|\chi_A) = \mathbb{E}(|X_\alpha|\chi_{|X_\alpha|>c}\chi_A) + \mathbb{E}(|X_\alpha|\chi_{|X_\alpha|\leq c}\chi_A)$$

$$\leq \mathbb{E}(|X_\alpha|\chi_{|X_\alpha|>c}) + c.P(A) \leq \delta.c + \epsilon, \forall \alpha \in \Gamma$$

Taking $\delta = \epsilon/c$ shows that uniform integrability of $\{X_\alpha, \alpha \in \Gamma\}$ implies that given any $\epsilon > 0$ there exists a $\delta > 0$ such that $P(A) < \delta$ implies $sup_{\alpha \in \Gamma}\mathbb{E}(|X_\alpha|\chi_A) < \epsilon$. Conversely, suppose $X_\alpha, \alpha \in \Gamma$ is a family of random variables defined on $(\Omega, \mathcal{F}, P)$ such that (a) $sup_{\alpha \in \Gamma}\mathbb{E}(|X_\alpha|) < \infty$ and (b) For every $\epsilon > 0$, there exists a $\delta > 0$ such that $A \in \mathcal{F}$ and $P(A) < \delta$ imply $sup_{\alpha \in \Gamma}\mathbb{E}(|X_\alpha|\chi_A) < \epsilon$. Then, we claim that

$X_\alpha, \alpha \in \Gamma$ is a uniformly integrable family. Indeed, let $\epsilon > 0$ and choose $\delta > 0$ such that $\mathbb{E}(|X_\alpha|\chi_A) < \epsilon$ whenever $P(A) < \delta$. Then, let $\epsilon > 0$ and let $\delta > 0$ be such that $P(A) < \delta$ implies $\mathbb{E}(|X_\alpha|\chi_A) < \epsilon$ for all $\alpha \in \Gamma$. Let $\alpha \in \Gamma$ be arbitrary. Since $sup_{\alpha \in \Gamma}\mathbb{E}|X_\alpha| = K < \infty$, we have by Chebyshev's inequality,

$$P|X_\alpha| > c) \le K/c, c > 0, \alpha \in \Gamma$$

and hence taking $c > K/\delta$, it follows that $P(|X_\alpha| > c) < \delta$ for all $\alpha \in \Gamma$. Then, by hypothesis,

$$\mathbb{E}(|X_\alpha|\chi_{|X_\alpha|>c}) < \epsilon$$

for all $\alpha \in \Gamma$ and this proves that $X_\alpha, \alpha \in \Gamma$ is a uniformly integrable family of random variables. We have thus proved that $X_\alpha, \alpha \in \Gamma$ is a uniformly integrable family of random variables iff $sup_{\alpha \in \Gamma}\mathbb{E}(|X_\alpha|) < \infty$ and for every $\epsilon > 0$, there is a $\delta > 0$ such that $P(A) < \delta$ implies $sup_{\alpha \in \Gamma}\mathbb{E}(|X_\alpha|\chi_A) < \epsilon$.

[191] Let $B(t), t \ge 0$ be standard Brownian motion with $B(0) = 0$. Consider the operator P_t so that

$$P_t f(x) = \mathbb{E}f(x + B(t)) = \mathbb{E}exp(B(t)D)f(x) = exp(tD^2/2)f(x)$$

where $D = d/dx$. We have

$$\int_0^1 (DP_{1-t})f(x + B(t))dt = \int_0^1 D.exp((1-t)D^2/2).exp(B(t)D)f(x)dB(t)$$

$$= exp(D^2/2)(\int_0^1 exp(B(t)D - tD^2/2).D.dB(t))f(x) = exp(D^2/2)\int_0^1 d(exp(B(t)D - tD^2/2))f(x)$$

$$= exp(D^2/2)(exp(B(1)D - D^2/2) - 1)f(x) = (exp(B(1)D) - exp(D^2/2))f(x) = f(x + B(1)) - \mathbb{E}(f(x + B(1)))$$

[192] Prove that if V is a finite dimensional complex vector space and $\{T_\alpha, \alpha \in I\}$ is a family of diagonable linear operators in V such that for all $\alpha, \beta \in I$, $[T_\alpha, T_\beta] = 0$, then the family $\{T_\alpha\}$ is simultaneously diagonable, ie, there exists a basis $\mathcal{B}$ for V such that $[T_\alpha]_\mathcal{B}$ is diagonable for all $\alpha \in I$.

[193] Let V be a finite dimensional complex vector space. Prove that any linear operator T in V is triangulable, ie, there exists a basis $\mathcal{B}$ for V, such that $[T]_\mathcal{B}$ is upper triangular.

[194] Prove Holder's inequality in the following form.

$$|\int_{-\infty}^{\infty} f(t)\bar{g}(t)dt| \le (\int_{-\infty}^{\infty} |f(t)|^p dt)^{1/p}.(\int_{-\infty}^{\infty} |g(t)|^q)^{1/q}$$

where f, g are complex valued measurable functions on $\mathbb{R}$, $p, q > 1, 1/p + 1/q = 1$ and

$$\int_{-\infty}^{\infty} |f(t)|^p dt < \infty, \int_{-\infty}^{\infty} |g(t)|^q dt < \infty$$

[195] Suppose $B(t), t \geq 0$ is a Brownian motion process. Show that $|B(t)|, t \geq 0$ is a Markov process and compute its transition probability density.

hint: for $s < t$

$$P(|B(t)| \leq y|B(u), u \leq s) = P(|B(t)| \leq y|B(s)) = P(-y \leq B(t) \leq y|B(s))$$

$$= P(-y - B(s) \leq B(t) - B(s) \leq y - B(s)|B(s)) = \Phi((y - B(s))/\sqrt{t-s}) + \Phi((y + B(s))/\sqrt{t-s}) - 1$$

$$= \Phi((y - |B(s)|)/\sqrt{t-s}) + \Phi((y + |B(s)|)/\sqrt{t-s}) - 1$$

and hence

$$P(|B(t)| \leq y||B(u)|, u \leq s) = \psi(t - s, |B(s)|, y)$$

where

$$\psi(t, x, y) = \Phi((y - x)/\sqrt{t}) + \Phi((y + x)/\sqrt{t}) - 1$$

[196] Nonlinear heat equation.

$$(D(u)u_{,x})_{,x} - u_{,t} = 0$$

This expands to give

$$D(u)u_{,xx} + D'(u)u_{,x}^2 - u_{,t} = 0$$

Set

$$D(u) = D_0 + \epsilon.D_1(u)$$

Then upto $O(\epsilon)$ the solution $u = u_0 + \epsilon u_1$ can be obtained by substitution. Equating coefficients of ϵ^0 gives

$$D_0 u_{0,xx} - u_{0,t} = 0$$

Equating coefficients of ϵ^1 gives

$$D_0 u_{1,xx} + D_1(u_0)u_{0,xx} + D_1'(u)u_{0,x}^2 - u_{1,t} = 0$$

u_1 thus satisfies a linear differential equation with a source depending on u_0.

[197] Let $\mathbf{X}, \mathbf{Y}$ be jointly Gaussian random vectors with

$$\mathbb{E}(\mathbf{X}) = \mathbf{m_X}, \mathbb{E}(\mathbf{Y}) = \mathbf{m_Y},$$

$$Cov(\mathbf{X}, \mathbf{Y}) = \mathbb{E}[(\mathbf{X} - \mathbf{m_X})(\mathbf{Y} - \mathbf{m_Y})^T] = \mathbf{R_{XY}},$$

$$Cov(\mathbf{X}) = \mathbb{E}[(\mathbf{X} - \mathbf{m_X})(\mathbf{X} - \mathbf{m_X})^T] = \mathbf{R_{XX}},$$

$$Cov(\mathbf{Y}) = \mathbb{E}[(\mathbf{Y} - \mathbf{m_Y})(\mathbf{Y} - \mathbf{m_Y})^T] = \mathbf{R_{YY}}$$

Then compute

$$\mathbb{E}(\mathbf{X}|\mathbf{Y}), Cov(\mathbf{X}|\mathbf{Y}) = \mathbb{E}((\mathbf{X} - \mathbb{E}(\mathbf{X}|\mathbf{Y}))(\mathbf{X} - \mathbb{E}(\mathbf{X}|\mathbf{Y}))^T|\mathbf{Y})$$

[198] Give an example of a discontinuous martingale whose quadratic variation is t. Show that if the martingale is continuous and has quadratic variation t, then the martingale is a Brownian motion process.

[199] Let T, S be two linear operators in a finite dimensional vector space V. Suppose W is the eigensubspace of T corresponding to the eigenvalue λ and $TS - ST = 0$. Then show that W is invariant under S. Using this result show that if T, S are diagonable and $TS - ST = 0$, then T and S are simultaneously diagonable.

[200] Let $B(t), t \geq 0$ be standard Brownian motion and let $c > 0$. Show that $X(t) = c.B(t/c^2), t \geq 0$ is also a standard Brownian motion process.

hint: You can either use Ito's formula or show that X has a.s. continuous sample paths and the same mean and covariance as that of B.

[201] Let $B(t)$ be standard Brownian motion and $c > 0$. Consider the Brownian motion process $X(t) = c.B(t/c^2)$. Then,

$$T_a(X) = inf\{t \geq 0 : X(t) = a\} = inf\{t \geq 0 : c.B(t/c^2) = a\}$$

$$= c^2.\inf\{t/c^2 : t \geq 0, c.B(t/c^2) = a\} = c^2.inf\{t/c^2 : t \geq 0, B(t/c^2) = a/c\} = c^2.T_{a/c}(B)$$

Deduce that if T_a is the first hitting time of a Brownian motion process at level a when it starts at zero, then T_a and $c^2.T_{a/c}$ have the same distribution and in fact, the processes $\{T_a : a \geq 0\}$ and $\{c^2.T_{a/c} : a \geq 0\}$ have the same laws.

[202] Sketch the following discrete time signals. (1) $x[n] = sin(2\pi n/N), n \in \mathbb{Z}$ where N is a positive integer, (2) $x[n] = a^{|n|}, n \in \mathbb{Z}$ where $0 < a < 1$, (3) $x[n] = u[n - a] - u[n - b]$ where a, b are integers with $a < b$ and $u[n] = 1$, for $n \geq 0$ and $u[n] = 0$ for $n < 0$.

[203] Explain how you would graphically convolve two finite duration discrete time signals. Specifically, assume $h[n]$ is non-zero only for $0 \leq n \leq N - 1$ and $x[n]$ is non-zero only for $0 \leq n \leq N_2 - 1$ where N_1, N_2 are positive integers. Then plot $\sum_{k \in \mathbb{Z}} h[k]x[n - k]$.

[204] Compute $E_x = \sum_{n=0}^{N-1} x^2[n]$ where $x[n] = sin(2\pi kn/N)$ with k a positive integer. Also compute $E_y = \int_0^T y^2(t)dt$ where $y(t) = sin(2\pi t/T)$.

[205] If $x[n, m], n, m \in \mathbb{Z}$ is a two dimensional signal, then define

$$X(\omega_1, \omega_2) = \sum_{n,m \in \mathbb{Z}} x[n, m]exp(-j(n\omega_1 + m\omega_2))$$

Prove that

$$x[n, m] = \frac{1}{4\pi^2} \int_0^{2\pi} \int_0^{2\pi} X(\omega_1, \omega_2)exp(j(n\omega_1 + m\omega_2))d\omega_1 d\omega_2$$

Deduce that

$$\sum_{n,m\in\mathbb{Z}} |x[n,m]|^2 = \frac{1}{4\pi^2} \int_0^{2\pi} \int_0^{2\pi} |X(\omega_1,\omega_2)|^2 d\omega_1 d\omega_2$$

[206] Obtain an approximate solution to the first order linear differential equation

$$y'(t) + ay(t) = x(t), t \geq 0$$

by discretization. Here, x is a known function and y is the solution. Compare the approximate solution with the exact solution.

hint: Discretization with time step Δ gives on setting $y[n] = y(n\Delta), x[n] = x(n\Delta)$, the approximate equation

$$(y[n+1] - y[n])/\Delta + ay[n] = x[n], n = 0, 1, 2, ...$$

or

$$y[n+1] = (1 - a\Delta)y[n] + \Delta.x[n], n = 0, 1, 2, ...$$

[207] Give an example of a linear two dimensional difference equation that does not admit recursive computation of the output.

hint.

$$y[n,m] = y[n-1,m] + y[n+1,m] + y[n,m-1] + y[n,m+1] + x[n,m]$$

[208] Let z be a complex number with $|z| < 1$. If r is a positive integer, then compute $\sum_{n=0}^{\infty} n^r z^n$

hint. Apply the operator $(z\frac{d}{dz})^r$ to the function $\frac{1}{1-z}$.

[209] Let $B_t, t \geq 0$ be standard Brownian motion and let $S_t = sup_{s\leq t}B_s$. Show that if S increases at t then $S_t = B_t$. Deduce that if $dS_t > 0$, then $S_t - B_t = 0$ and hence $\int_0^{\infty}(S_t - B_t)dS_t = 0$. This result is also valid if B is replaced by any continuous martingale M.

[210] Let M_t be a continuous martingale with quadratic variation $< M, M >$ and assume that $< M, M >_{\infty} = \infty$. Define the stop time

$$T_s = inf\{t \geq 0 :< M, M >_t > s\}$$

Then show that $B_t = M_{T_t}, t \geq 0$ is a standard Brownian motion process.

hint. Note that

$$\{T_t > s\} = \{< M, M >_s \leq t\}$$

If $\mathcal{F}_t$ is the underlying filtration, then it is clear from the above equation that $\{T_t > s\} \in \mathcal{F}_s$ and hence $\{T_t \leq s\} \in \mathcal{F}_s$ which means that for each t, T_t is a stop-time. Define $\tilde{\mathcal{F}}_t = \mathcal{F}_{T_t}$. Then B_t is adapted to the filtration $\tilde{\mathcal{F}}_t$. Show that B is a continuous martingale with respect to the filtration $\tilde{\mathcal{F}}_t$ and that its quadratic variation is given by

$$< B, B >_t = < M, M >_{T_t} = t$$

Deduce using Levy's characterization of Brownian motion that B is a Brownian motion process with respect to the filtration $\tilde{F}_t$.

[211] Let $B_1(t), ..., B_n(t)$ be standard independent Brownian motion processes and let $f : \mathbb{R}^n \to \mathbb{R}$ be a twice continuously differentiable function. Consider the process

$$X(t) = f(B_1(t), ..., B_n(t))$$

Show that

$$X(t) - X(0) = \int_0^t \sum f_{,i}(B_1(s), ..., B_n(s)) dB_i(s) + \frac{1}{2} \int_0^t \sum_{i=1}^n f_{,ii}(B_1(s), ..., B_n(s)) ds$$

Deduce that if $\sum_i f_{,ii} = 0$, ie f satisfies Laplace's equation in n dimensions, then $X(t)$ is a Martingale.

[212] Solve the n dimensional heat equation

$$u_{,t}(t, x_1, ..., x_n) = D . \sum_{i=1}^n u_{,ii}(t, x_1, ..., x_n)$$

hint:

Derive the equation satisfied by the spatial Fourier transform of u, ie,

$$\hat{u}(t, k_1, ..., k_n) = \int u(t, x_1, ..., x_n) exp(-i(k_1 x_1 + ... + k_n x_n)) dx_1 ... dx_n$$

Note: Suppose $B_1(t), ..., B_n(t)$ are n independent standard Brownian motion processes and $f : \mathbb{R}^n \to \mathbb{R}$ is twice continuously differentiable. Define

$$u(t, x_1, ..., x_n) = \mathbb{E} f(x_1 + B_1(t), ..., x_n + B_n(t))$$

Then u satisfies the n dimensional heat equation with initial condition $u(0, .) = f$.

[213] Let X be a non-negative random variable with distribution F. We have

$$\int_0^T P(X > x) dx = \int_0^T (1 - F(x)) dx = T - \int_0^T F(x) dx$$

$$= T - \int_0^T (d(x . F(x)) - x . dF(x)) = T - TF(T) + \int_0^T x . dF(x)$$

Suppose

$$lim_{T \to \infty} T . (1 - F(T)) = 0$$

Then we get

$$\int_0^\infty P(X > x)dx = \int_0^\infty x.dF(x) = \mathbb{E}(X)$$

Suppose X is integrable, ie, it has a finite expectation. Then

$$T(1 - F(T)) = T.\int_T^\infty dF(x) \le \int_T^\infty x.dF(x) \to 0, T \to \infty$$

[214] Let $B(t), t \ge 0$ be a Brownian motion process. Define the process $X(t) = t.B(1/t), t > 0$ and $X(0) = 0$. Then $X(t)$ is a continuous Gaussian process and if $0 < s \le t$, then

$$\mathbb{E}(X(s)X(t)) = st.min(1/s, 1/t) = s$$

Hence, $X(t), t \ge 0$ is also a Brownian motion process.

[215] Let $X(t)$ be a semimartingale with quadratic variation $< X, X > (t)$. Then

$$d|X(t)| = sgn(X(t))dX(t) + \delta(X(t))d < X, X > (t)$$

Define the local time of the process X at a as

$$L_t^a(X) = \int_0^t \delta(X(s) - a)d < X, X > (s)$$

Then

$$\int_{\mathbb{R}} f(a)L_t^a(X)da = \int_0^t f(X(s))d < X, X > (s)$$

We have

$$|X(t)| = \int_0^t sgn(X(s))dX(s) + L_t^0(X)$$

[216] Let $\phi(t), t \in \mathbb{R}$ be a smooth function such that $\int_{-\infty}^\infty \phi(t)dt = 1$. Then if $f(t)$ is any function, we have

$$n \int_{-\infty}^\infty f(t)\phi(nt)dt = \int_{-\infty}^\infty f(t/n)\phi(t)dt \to f(0) \int_{-\infty}^\infty \phi(t)dt = f(0),$$

as $n \to \infty$. We have assumed that dominated convergence can be applied. This result formally means that the sequence of functions $\phi_n(t) = n.\phi(nt), n = 1, 2, 3, ...$ converges to the Dirac delta function $\delta(t)$ as $n \to \infty$.

[217] Let $B_1(t)$ and $B_2(t)$ be two standard independent Brownian motion processes and define the complex Gaussian process $Z(t) = B_1(t) + i.B_2(t)$. Let $f(z)$ be a holomorphic function of the complex variable z and consider the process $X(t) = f(Z(t)) = f(B_1(t) + iB_2(t))$. We have

$$dX(t) = f'(Z(t)dZ(t) + \frac{1}{2}f''(Z(t))(dZ(t))^2 = f'(Z(t))dZ(t)$$

since

$$(dZ(t))^2 = (dB_1(t) + idB_2(t))^2 = (dB_1)^2 + (idB_2)^2 + 2idB_1.dB_2 = 0$$

Hence,

$$X(t) = X(0) + \int_0^t f'(Z(s))dZ(s)$$

and this proves that $X(t)$ is a martinagle with respect to the filtration $\mathcal{F}_t = \sigma(B_1(s), B_2(s), s \leq t)$.

[218] Let $B(t)$ be a standard Brownian motion process and let $T_a = inf(t \geq 0 : B(t) = a)$ where $a \in \mathbb{R}$. Let $f(x)$ be a smooth function and define

$$u(x) = \mathbb{E}[\int_0^{T_a \wedge T_b} f(B(t))dt | B(0) = x], a \leq x \leq b$$

Then $u(a) = u(b) = 0$ and by the Markov property,

$$u(x) = f(x)h + \mathbb{E}(u(B(h))|B(0) = x) + o(h) = f(x)h + u(x) + \frac{h}{2}u''(x) + o(h)$$

so that

$$u''(x) + f(x) = 0$$

[219] Calculate the sum of the following series. (a) $\sum_{n=1}^{\infty} \frac{1}{n^2}$, (b) $\sum_{n=1}^{\infty} \frac{1}{n^4}$, (c) $\sum_{n=1}^{\infty} \frac{1}{z^2+n^2}$ where z is any complex number.

[220] Show that if $f(x,y)$ is a function of two variables and $a(x)$ and $b(x)$ are two functions of one variable, then

$$\frac{d}{dx} \int_{a(x)}^{b(x)} f(x,y)dy = b'(x)f(x,b(x)) - a'(x)f(x,a(x)) + \int_{a(x)}^{b(x)} \frac{\partial f(x,y)}{\partial x}dy$$

[221] Let $m_1, ..., m_N$ be random variables with joint probability density $f(m_1, ..., m_N)$ and let $\mathbf{r}_1, ..., \mathbf{r}_N$ be random vectors in $\mathbb{R}^3$ with joint density $g(\mathbf{r}_1, ..., \mathbf{r}_N)$. Assume that the set of random variables $\{m_1, ..., m_N\}$ is independent of the set of random vectors $\{\mathbf{r}_1, ..., \mathbf{r}_N\}$ so that the joint density of $\{m_1, ..., m_N, \mathbf{r}_1, ..., \mathbf{r}_N\}$ is given by $f(m_1, ..., m_N)g(\mathbf{r}_1, ..., \mathbf{r}_N)$. Let $V(r)$ be a spherically symmetric potential and let M be a positive constant. Consider the equation of motion

$$\mathbf{r}''(t) = -MV'(|\mathbf{r}|)\mathbf{r}/|\mathbf{r}| - \epsilon \sum_{k=1}^{N} m_k V'(|\mathbf{r} - \mathbf{r}_k|)(\mathbf{r} - \mathbf{r}_k)/|\mathbf{r} - \mathbf{r}_k|$$

where ϵ is a small parameter. Determine the trajectory $\mathbf{r}(t)$ upto $O(\epsilon)$ assuming that the solution $\mathbf{r}_0(t)$ to the unperturbed problem is known. Explain how you can determine the approximate statistics of the perturbed trajectory (specifically the moments $\mathbb{E}[\xi_{i_1}(t_1)...\xi_{i_m}(t_m)]$ where $\mathbf{r}(t) = (\xi_1(t), \xi_2(t), \xi_3(t))$. Consider the special case $V(r) = -K/r$. This problem can be viewed as computing the fluctuations in the trajectory of a particle moving under the influence of a central forcefield and subject to perturbing forces due to the presence of particles of random masses and whose locations are random points in space. (perturbation by randomly distributed dust).

[222] (The Cramer-Rao bound) Suppose θ is a parameter assuming real values and X is a random vector whose probability density depends on θ. We may thus write the probability density of X as $p(X|\theta)$. Let $\hat{\theta}(X)$ be a function of X and an estimator of θ. We denote by $B(\theta)$ the bias in this estimator, ie,

$$\mathbb{E}(\hat{\theta}(X) - \theta) = \int (\hat{\theta}(X) - \theta).p(X|\theta)dX = B(\theta)$$

Differentiating both sides with respect to θ gives

$$1 + B'(\theta) = \int (\hat{\theta}(X) - \theta)\frac{\partial p(X|\theta)}{\partial \theta}dX = \mathbb{E}[(\hat{\theta}(X) - \theta)\frac{\partial log(p(X|\theta))}{\partial \theta}]$$

and using the Cauchy-Schwarz inequality, we get from this,

$$(1 + B'(\theta))^2 \leq \mathbb{E}[(\hat{\theta}(X) - \theta)^2].\mathbb{E}[(\frac{\partial log(p(X|\theta))}{\partial \theta})^2]$$

or

$$\mathbb{E}[(\hat{\theta}(X) - \theta)^2] \geq \frac{(1 + B'(\theta))^2}{\mathbb{E}[(\frac{\partial log(p(X|\theta))}{\partial \theta})^2]}$$

[223] The Bhattacharya bound. Let X be the observed random vector whose probability density depends on a real valued parameter θ, ie, $p(X|\theta)$. Let $\hat{\theta}(X)$ be an estimator of θ based on the observation X and let $B(\theta)$ be the bias of this estimator. Thus,

$$\int (\hat{\theta}(X) - \theta).p(X|\theta)dX = B(\theta)$$

Differentiating with respect to θ gives

$$1 + B'(\theta) = \int (\hat{\theta}(X) - \theta)p'(X|\theta)dX$$

Differentiating this expression $k - 1$ times with respect to θ gives

$$\delta_{k,1} + B^{(k)}(\theta) = \int (\hat{\theta}(X) - \theta)p^{(k)}(X|\theta)dX, k = 1, 2, ..., n$$

where the notation $p'(X|\theta)$ stands for $\frac{\partial p(X|\theta)}{\partial \theta}$ and $p^{(k)}(X|\theta)$ stands for $\frac{\partial^k p(X|\theta)}{\partial \theta^k}$. Thus if $\mathbf{a} = (a_1, ..., a_n)^T$ is any two vector in $\mathbb{R}^n$, we have

$$\mathbf{b}^T\mathbf{a} = \mathbb{E}[(\hat{\theta}(X) - \theta)\mathbf{a}^T\mathbf{p}]$$

where

$$\mathbf{b} = (1 + B', B'', ..., B^{(n)})^T, \mathbf{p} = (p'/p, p''/p, ..., p^{(n)}/p)^T, p = p(X|\theta)$$

Thus,

$$(\mathbf{a}^T\mathbf{b})^2 \leq \mathbf{a}^T\mathbf{Ja}.\mathbb{E}[(\hat{\theta}(X) - \theta)^2]$$

where

$$J = \mathbb{E}(\mathbf{pp}^T)$$

Equivalently, for all $\mathbf{a} \in \mathbb{R}^n$, we have

$$\mathbb{E}[(\hat{\theta}(X) - \theta)^2] \geq \frac{(\mathbf{a}^T\mathbf{b})^2}{\mathbf{a}^T\mathbf{Ja}}$$

[224] Design an integrator using an ideal operational amplifier, a resistance and a capacitance.

[225] Let A, B be two $n \times n$ non-singular Hermitian matrices with B positive definite. Calculate the points $x \in \mathbb{C}^n$ at which the function $\phi : \mathbb{C}^n - \{0\} \to \mathbb{R}$ defined by

$$\phi(x) = \frac{< Ax, x >}{< Bx, x >}$$

is an extremum.

[226] Suppose $\mathbf{s}, \mathbf{w}$ are two random vectors in $\mathbb{R}^n$ such that $\mathbf{R}_{ww} = \mathbb{E}(\mathbf{w}.\mathbf{w}^T)$ is non-singular. Let $\mathbf{a} \in \mathbb{R}^n$ and let $\mathbf{x} = \mathbf{s} + \mathbf{w}$. We view $\mathbf{x}$ as the observed random vector and $\mathbf{s}$ and $\mathbf{w}$ respectively as the signal and noise vectors. After processing $\mathbf{x}$ by the filter $\mathbf{a}$, the output can be expressed as $y = \mathbf{a}^T\mathbf{x} = y_s + y_w$ where $y_s = \mathbf{a}^T\mathbf{s}$ and $y_w = \mathbf{a}^T\mathbf{w}$. The signal component y_s of the output y has power $P_s = \mathbb{E}(y_s)^2 = \mathbf{a}^T\mathbf{R}_{ss}\mathbf{a}$ where $\mathbf{R}_{ss} = \mathbb{E}(\mathbf{ss}^T)$. The noise component y_w of the output y has power $P_w = \mathbb{E}(y_w)^2 = \mathbf{a}^T\mathbf{R}_{ww}\mathbf{a}$. The signal to noise ratio at the output is therefore

$$SNR(\mathbf{a}) = \frac{P_s}{P_w} = \frac{\mathbf{a}^T\mathbf{R}_{ss}\mathbf{a}}{\mathbf{a}^T\mathbf{R}_{ww}\mathbf{a}}$$

The problem is to calculate $\mathbf{a}$ for which $SNR(\mathbf{a})$ is a maximum.

[227] For $x \in \mathbb{R}$, let $x^+ = max(x, 0)$. Let $\theta : \mathbb{R} \to \mathbb{R}$ be defined so that $\theta(x) = 1$ if $x > 0$ and $\theta(x) = 0$ if $x \leq 0$. Then $x^+ = x.\theta(x)$ say. Then,

$$d(x^+)/dx = x.\delta(x) + \theta(x) = \theta(x), d^2(x^+)/dx^2 = \delta(x)$$

Thus, if X is a continuous semimartingale, we have by Ito's formula,

$$d(X_t^+) = \theta(X_t)dX_t + \frac{1}{2}\delta(X_t)d < X, X >_t$$

We define

$$L_t^x(X) = \int_0^t \delta(X_s - x)d < X, X >_s$$

and set $L_t(X) = L_t^0(X)$. Then, the above formula can be expressed as

$$X_t^+ - X_0^+ = \int_0^t \theta(X_s)dX_s + \frac{1}{2}L_t(X)$$

Note that

$$\int_{\mathbb{R}} f(x)L_t^x(X)dx = \int_{[0,t]\times\mathbb{R}} f(x)\delta(X_s - x)d < X, X >_s .dx = \int_0^t f(X_s)d < X, X >_s$$

We also note that

$$df(X_t) = f'(X_t)dX_t + \frac{1}{2}f''(X_t)d < X, X >_t$$

so that

$$f(X_t) - f(X_0) = \int_0^t f(X_s)dX_s + \frac{1}{2}\int_0^t f''(X_s)d < X, X >_s$$

$$= \int_0^t f'(X_s)dX_s + \frac{1}{2}\int_{\mathbb{R}} f''(x)L_t^x(X)dx$$

[228] Let $X_t, Y_t, t \geq 0$ be two continuous semimartingales. Consider the semimartingale $Z_t = max(X_t, Y_t)$. We can write

$$Z_t = X_t.\theta(X_t - Y_t) + Y_t.(1 - \theta(X_t - Y_t)) = Y_t + (X_t - Y_t)\theta(X_t - Y_t)$$

It follows that

$$dZ_t = dY_t + (dX_t - dY_t)\theta(X_t - Y_t) + \frac{1}{2}dL_t(X - Y)$$

where

$$dL_t(X - Y) = \delta(X_t - Y_t)d < X - Y, X - Y >_t$$

Consider now the process

$$U_t = max(max(X_t, Y_t), 0) = max(X_t, Y_t, 0) = max(max(X_t, 0), max(Y_t, 0))$$

$$= max(max(X_t, 0) - max(Y_t, 0), 0) + max(Y_t, 0)$$

Equivalently,

$$U_t = max(X_t, Y_t)^+ = max(X_t^+ - Y_t^+)^+ + Y_t^+$$

It follows that on the one hand,

$$dU_t = \theta(max(X_t, Y_t)).dmax(X_t, Y_t) + \frac{1}{2}dL_t(max(X, Y))$$

and on the other hand,

$$dU_t = \theta(X_t^+ - Y_t^+).(dX_t^+ - dY_t^+) + \theta(Y_t)dY_t + \frac{1}{2}dL_t(Y)$$

$$= \theta(X_t^+ - Y_t^+)(\theta(X_t)dX_t + \frac{1}{2}dL_t(X) - \theta(Y_t)dY_t - \frac{1}{2}dL_t(Y)) + \frac{1}{2}dL_t(Y)$$

[229] Let $B(t), t \geq 0$ be standard Brownian motion and Define $X(t) = tB(1) - B(t), 0 \leq t \leq 1$. Then, $X(.)$ is a Gaussian process and $X(0) = 0$ and $X(1) = 0$. Further, for $0 \leq s \leq t \leq 1$, we have

$$\mathbb{E}(X(t)X(s)) = \mathbb{E}((tB(1) - B(t))(sB(1) - B(s))) = ts - ts - st + s = s(1-t)$$

[230] Suppose $X_n, n = 1, 2, 3, \ldots$ is a sequence of non-negative integrable random variables on a probability space. We have

$$inf_{n \geq k} X_n \leq X_m, m \geq k$$

Thus,

$$\mathbb{E}(inf_{n \geq k} X_n) \leq \mathbb{E}X_m, m \geq k$$

It follows that

$$\mathbb{E}(inf_{n \geq k} X_n) \leq inf_{m \geq k} \mathbb{E}(X_m)$$

and hence by the monotone convergence theorem,

$$\mathbb{E}(liminf X_n) \leq liminf \mathbb{E}(X_n)$$

Suppose $X_n, n = 1, 2, \ldots$ are integrable random variables and there exists an integrable random variable X such that $X_n \geq X$ for all n. Then, applying the above result to the sequence $Y_n = X_n - X$, we get

$$\mathbb{E}(liminf X_n) - \mathbb{E}X \leq liminf \mathbb{E}(X_n) - \mathbb{E}X$$

or

$$\mathbb{E}(liminf X_n) \leq liminf \mathbb{E}(X_n)$$

This result is one of the versions of Fatou's lemma.

[231] Show that if $\{X_t, t \geq 0\}$ and $\{Y_t, t \geq 0\}$ are two continuous stochastic processes and $P(X_t = Y_t) = 1$ for all $t \geq 0$, then $P(X_t = Y_t, \forall t \geq 0) = 1$.

[232] Consider a circuit consisting of a diode in series with a resistor R and a voltage source of $v(t)$ volts. Let $x(t)$ be the current through the circuit. Then,

$$f(x(t)) + Rx(t) = v(t)$$

where $V = f(X)$ is the voltage current relation for the diode. Note that $X = f^{-1}(V) = I_0(exp(V/V_T)-1)$, so that $V = f(X) = V_T.log(X/I_0 + 1)$. The above equation can therefore be expressed as

$$V_T.log(x(t)/I_0 + 1) + R.x(t) = v(t)$$

or

$$x(t) = I_0.(exp((v(t) - Rx(t))/V_T) - 1)$$

[233] Consider the parallel connection of a resistor and a capacitor and the resulting in series with a diode and a voltage source $v(t)$. Let $v_o(t)$ be the voltage across the capacitor-resistor combination. Then, the current through the diode is given by

$$x(t) = I_0(exp((v(t) - v_o(t))/V_T) - 1)$$

This current also flows through the parallel combination of the resistor and capacitor. Thus,

$$I_0(exp((v(t) - v_o(t))/V_T) - 1) = Cdv_o(t)/dt + v_o(t)/R$$

The problem is to derive a Volterra representation of the input output relation with $v(t)$ as the input and $v_o(t)$ as the output.

[234] Let $(\Omega, \mathcal{F}, P)$ be a probability space and $\{\mathcal{F}_t : t \geq 0\}$ a filtration on this space, ie, a non-decreasing family of sub-σ-fields of $\mathcal{F}$. A stochastic process $\{X_t(\omega), t \geq 0\}$ is said to be adapted to this filtration if for each $t \geq 0$, the random variable X_t is $\mathcal{F}_t$-measurable. The stochastic process $\{X_t\}$ is said to be measurable if the map $(t, \omega) \rightarrow X_t(\omega)$ from $([0, \infty) \times \Omega, \mathcal{B}[0, \infty) \otimes \mathcal{F})$ into $(\mathbb{R}, \mathcal{B}(\mathbb{R}))$ The stochastic process $\{X_t\}$ is said to be progressively measurable if for every $t \geq 0$, the map $(s, \omega) \rightarrow X_s(\omega)$ from $([0, t] \times \Omega, \mathcal{B}[0, t] \otimes \mathcal{F}_t)$ into $(\mathbb{R}, \mathcal{B}(\mathbb{R}))$ is measurable. A non-negative random variable T on this space is said to be a stop time if for each $t \geq 0$, $\{T \leq t\} \in \mathcal{F}_t$. If T is a stop time, then $\mathcal{F}_T$ is defined to be the class of all events $E \in \mathcal{F}$ such that $E \cap \{T \leq t\} \in \mathcal{F}_t$ for all $t \geq 0$. It is easy to see that $\mathcal{F}_T$ is a σ-field. In fact, if $E_n, n = 1, 2, ...$ is a sequence in $\mathcal{F}_T$ then

$$\left(\bigcup_n E_n\right) \cap \{T \leq t\} = \bigcup_n (E_n \cap \{T \leq t\}) \in \mathcal{F}_t, t \geq 0$$

and further if $E \in \mathcal{F}_T$, then the identity

$$\{T \leq t\} - E \cap \{T \leq t\} = E^c \cap \{T \leq t\}, t \geq 0$$

shows that $E^c \in \mathcal{F}_T$. Suppose $\{X_t\}$ is a progressively measurable process and T is a stoptime. Then X_T is an $\mathcal{F}_T$ measurable random variable.

[235] Let $P_t(x, dy), t \geq 0$ be the transition probability kernel of a time homogeneous Markov process with state space $\mathbb{R}$. Thus, if E is any Borel subset of $\mathbb{R}$ and $t, s \geq 0$, then P_t satisfies the Chapman-Kolmogorov equation

$$P_{t+s}(x, E) = \int \mathbb{R} P_t(x, dy) P_s(y, E)$$

As a linear operator,

$$(P_t f)(x) = \int \mathbb{R} P_t(x, dy) f(y)$$

Then

$$P_{t+s} = P_t P_s, t, s \geq 0$$

Let A be the generator of the Markov process. Thus,

$$A f(x) = \frac{d}{dt} P_t f(x)|_{t=0} = lim_{h \to 0} \frac{P_h f(x) - f(x)}{h}$$

Then formally,

$$P_t = exp(tA)$$

and hence

$$\int_0^\infty exp(-pt) P_t = (p.I - A)^{-1} = U_p$$

U_p is the resolvent of the generator A. Then,

$$U_p U_q f(x) = \int_0^\infty exp(-pt) P_t U_q f(x) dt = \int_0^\infty \int_0^\infty exp(-pt) exp(-qs) P_t P_s f(x) dt ds$$

$$= \int exp(-pt - qs) P_{t+s} f(x) dt ds$$

or equivalently,

$$U_p U_q = \int exp(-pt - qs) P_{t+s} dt ds = \int_0^\infty (\int_0^u exp(-pt - q(u - t)) P_u dt) du$$

$$= \int_0^\infty exp(-qu) P_u du \int_0^u exp((q - p)t) dt = \int_0^\infty exp(-qu) P_u (exp((q - p)u - 1) du/(q - p)$$

or

$$(q - p) U_p U_q = \int_0^\infty (exp(-pu) - exp(-qu)) P_u du = U_p - U_q$$

This equation can also be proved formally as

$$U_q - U_p = (qI - A)^{-1} - (pI - A)^{-1} = (qI - A)^{-1}((pI - A) - (qI - A))(pI - A)^{-1} = (p - q) U_q U_p$$

Note that U_q and U_p commute. Note that

$$\frac{d}{dt} P_t f(x) = lim_{h \to 0} \frac{P_{t+h} f(x) - P_t f(x)}{h}$$

$$= lim_{h \to 0} \frac{(P_h - I) P_t f(x)}{h} = A P_t f(x) = P_t A f(x)$$

which is the reason that $P_t = exp(tA)$. To make these arguments more rigorous one must make use of the theory of unbounded operators in a Banach space.

[236] Find the DTFT of the signal $x[n] = r^n u[n]$ where $|r| < 1$. Note that the DTFT of a summable sequence $x[n]$ is defined as $X(\omega) = \sum_{n=-\infty}^{\infty} x[n] exp(-j\omega n)$.

[237] Give an example of a sequence that is not summable and hence its DTFT does not exist but its Z transform exists in a certain ROC.

hint:

$$x[n] = r^n u[n], r > 1$$

[238] Prediction of a time series. Let $x(n), n = 0, 1, 2, ...$ be a time series in $\mathbb{R}$, (ie, a sequence of real numbers). Its pth order linear predictor has the form

$$\hat{x}(n) = -\sum_{k=1}^{p} a(k)x(n - k)$$

where $a(1), ..., a(p)$ are obtained by minimizing

$$E(a(1), ..., a(p)) = \sum_{n=p}^{N} (x(n) - \hat{x}(n))^2$$

with respect to $a(1), ..., a(p)$. Derive the optimal equations satisfied by the $a(k)'s$ and determine the solution and also the minimum value of the prediction energy $E(a(1), ..., a(p))$. Suppose we use a second order Volterra predictor of order p. This will give a more accurate prediction ie, smaller prediction value of the minimum prediction energy.

[239] Mean ergodicity. Let $x[n], n \in \mathbb{Z}$ be a wide sense stationary random process with mean $m_x = \mathbb{E}x[n]$ and covariance $C_x[k] = \mathbb{E}(x[n]x[n - k]) - m_x^2$. Let

$$\hat{m}_x[N] = \frac{1}{2N + 1} \sum_{n=-N}^{N} x[n]$$

We want to determine the condition on C_x such that $\mathbb{E}(\hat{m}_x[N] - m_x)^2 \to 0$ as $N \to \infty$. Note that $\mathbb{E}(\hat{m}_x[N]) = m_x$ for all $N = 1, 2, ...$ Further

$$\mathbb{E}(\hat{m}_x[N] - m_x)^2 = \frac{1}{(2N + 1)^2} \sum_{n,m=-N}^{N} C_x[n - m]$$

$$= (2N + 1)^{-1} \sum_{k=-2N}^{2N} (1 - |k|/(2N + 1))C_x[k]$$

and hence a sufficient condition for $\mathbb{E}(\hat{m}_x[N] - m_x)^2$ to converge to zero as $N \to \infty$ is that $\sum_{k=-\infty}^{\infty} |C_x[k]| < \infty$.

[240] Let $x[n], n \in \mathbb{Z}$ be a real wide sense stationary Gaussian random process with zero mean and autocorrelation $R_x[k] = \mathbb{E}(x[n]x[n-k])$. Define the periodogram

$$S_N(\omega) = \frac{1}{2N+1} | \sum_{n=-N}^{N} x[n] exp(-j\omega n)|^2$$

Determine $\mathbb{E}(S_N(\omega))$ and $Cov(S_N(\omega_1), S_N(\omega_2))$ in terms of the power spectral density of $x[n]$. Note that the power spectral density of $x[n]$ is given by

$$S(\omega) = \sum_{n=-\infty}^{\infty} R_x[k] exp(-j\omega k)$$

hint:For determining the covariance of $S_N(\omega)$, make use of the fact that

$$\mathbb{E}(x[n_1]x[n_2]x[n_3]x[n_4]) = R_x[n_1 - n_2]R_x[n_3 - n_4] + R_x[n_1 - n_3]R_x[n_2 - n_4] + R_x[n_1 - n_4]R_x[n_2 - n_3]$$

[241] Carleman linearization. Consider a nonlinear first order ordinary differential equation

$$x'(t) = f(x(t), t)$$

Assume that f has a Taylor expansion in the first variable, ie,

$$f(x, t) = \sum_{n=0}^{\infty} c_n(t)x^n$$

Then, set

$$x_n(t) = x(t)^n, n = 0, 1, 2, \ldots$$

we have

$$x'_n(t) = nx'(t)x(t)^{n-1} = n\sum_{k=0}^{\infty} c_k(t)x(t)^{k+n-1} = n\sum_{k=0}^{\infty} c_k(t)x_{k+n-1}(t), n = 0, 1, 2, \ldots$$

This shows that the above nonlinear dynamical system in one state variable can equivalently be described by a system of linear ordinary differential equations in an infinite number of state variables.

[242] Write down the following fundamental equations of mathematical physics and explain the meaning of the different terms in them.

(a) Heat equation in one, two and three dimensions, (b) Wave equation in one, two and three dimensions, (c) Laplace's and Poisson' equation in electrostatics, gravitation and fluid dynamics (d) Navier-Stokes and mass conservation equation for a fluid, (e) Boltzmann's kinetic transport equation, (f) Schrodinger's wave equation in quantum mechanics, (g) Maxwell's equations, Helmholtz equation, (e) Einstein's equation of gravitation.

hint:

Heat equation is $u_{,t} = D.\nabla^2 u$ where for $u(t, x_1, ..., x_n)$,

$$u_{,t} = \frac{\partial u}{\partial t}, \nabla^2 u = \sum_{i=1}^{n} \frac{\partial^2 u}{\partial x_i^2}$$

Wave equation is

$$u_{,tt} = c^2 \nabla^2 u$$

Laplace's equation for $u(x_1, ..., x_n)$ is

$$\nabla^2 u = 0$$

Poisson's equation is

$$\nabla^2 u = -s$$

where $s(x_1, ..., x_n)$ is the source.

Navier-Stokes equation and mass conservation equation are respectively

$$\rho(\mathbf{v}.\nabla \mathbf{v} + \mathbf{v}_{,t}) = -\nabla p + \mathbf{f} + \nu \nabla^2 \mathbf{v}$$

$$\nabla.(\rho \mathbf{v}) + \rho_{,t} = 0$$

Boltzmann's kinetic transport equation for the particle number density $f(t, \mathbf{r}, \mathbf{v})$ in phase space is

$$f_{,t} + \mathbf{v}.\nabla \mathbf{f} + \frac{1}{m}\mathbf{F}.\nabla_{\mathbf{v}} f = C(f)$$

where $C(f)$ is the collision integral, ie, the rate of increase of the number density in phase space due to collisions.

Schrodinger's wave equation in quantum mechanics is

$$\frac{ih}{2\pi}\psi_{,t} = -\frac{h^2}{8\pi^2 m}\nabla^2 \psi + V\psi$$

The Maxwell equations are

$$\nabla.\mathbf{D} = \rho, \nabla \times \mathbf{E} = -\mathbf{B}_{,t}, \nabla \times \mathbf{H} = \mathbf{J} + \mathbf{D}_{,t},$$

$$\nabla.\mathbf{B} = 0$$

with

$$\mathbf{D} = \epsilon \mathbf{E}, \mathbf{B} = \mu \mathbf{H}$$

Helmholtz's equation is

$$(\nabla^2 + k^2)\psi = 0$$

and finally Einstein's equation of gravitation is

$$R_{\mu\nu} - \frac{1}{2}R g_{\mu\nu} = K T_{\mu\nu}$$

where K is a constant, $R_{\mu\nu}$ is the Ricci tensor and can be expressed in terms of the metric tensor $g_{\mu\nu}$ and its first and second order partial derivatives and $T_{\mu\nu}$ is the energy-momentum tensor of matter and radiation.

[243] The Riemann Zeta function. Let z be a complex number so that the series $\zeta(z) = \sum_{n=1}^{\infty} \frac{1}{n^z}$ is absolutely convergent. Then since

$$\int exp(-nt)t^{z-1}dt = \Gamma(z)/n^z$$

it follows that

$$\zeta(z) = \frac{1}{\Gamma(z)} \sum_{n=1}^{\infty} \int_0^{\infty} exp(-nt)t^{z-1}dt = \frac{1}{\Gamma(z)} \int_0^{\infty} \frac{t^{z-1}}{exp(t) - 1}dt$$

Note that if $Re(z) > 1$, then the series for $\zeta(z)$ is absolutely convergent since $|n^{-z}| = n^{-Re(z)}$ and $\int_1^{\infty} x^{-Re(z)}dx$ is finite.

[244] Let E be the set of primes. Then show that

$$\Pi_{p \in E}(1 - 1/p^z)^{-1} = \zeta(z)$$

Hint

$$(1 - 1/p^z)^{-1} = \sum_{n=0}^{\infty} \frac{1}{p^{nz}}$$

Combine this with the fact that every positive real number is uniquely expressible as a product of powers of distinct primes.

[245] Let N be a positive integer and let $p_1, ..., p_k$ be all the prime factors of N. We can write

$$N = p_1^{n_1}...p_k^{n_k}$$

where $n_1, ..., n_k$ are positive integers. For any $i = 1, 2, ..., k$, the number of positive integers x such that $x \le N$ and such that p_i divides x is clearly N/p_i (These numbers are $p_i, 2p_i, 3p_i, ...Np_i/p_i$. Let E_i denote this set. It follows that the number of positive integers x such that $x \le N$ and $gcd(x, N) = 1$ is given by

$$\mu((E_1 \cup ... \cup E_k)^c) = N - \mu(E_1 \cup ... \cup E_k) = N - \sum_i \mu(E_i) + \sum_{i<j} \mu(E_i E_j) - \sum_{i<j<r} \mu(E_i E_j E_r) + ...$$

$$= N(1 - \sum_i 1/p_i + \sum_{i<j} 1/p_i p_j - \sum_{i<j<r} 1/p_i p_j p_r + ...) = N\Pi_{m=1}^{k}(1 - 1/p_m)$$

Here, $\mu(E)$ denotes the number of elements in the set E.

[246] The Ornstein-Uhlenbeck process $X(t)$ satisfies the stochastic differential equation

$$dX(t) = -aX(t)dt + dB(t)$$

whre $\{B(t)\}$ is the standard Brownian motion and $a > 0$. The solution is

$$X(t) = exp(-at)X(0) + \int_0^t exp(-a(t-s))dB(s)$$

Assuming that $X(0)$ is a zero mean Gaussian random variable independent of $\{B(t)\}$, it follows that for $s \leq t$,

$$\mathbb{E}X(t) = 0,$$

$$\mathbb{E}(X(t)X(s)) = exp(-a(t+s))\mathbb{E}(X(0)^2) + \int_0^s exp(-a(t+s-2u))du$$

$$= exp(-a(t+s))\mathbb{E}(X(0)^2) + exp(-a(t+s))(exp(2s)-1)/2a$$

$$= exp(-a(t+s))(\mathbb{E}(X(0)^2) - 1/2a) + \frac{1}{2a}exp(-a(t-s))$$

It follows that the process $\{X(t)\}$ is stationary iff $\mathbb{E}(X(0)^2) = 1/2a$.

[247] Suppose $f(x)$ is a function on $\mathbb{R}$ such that for all $x \in \mathbb{R}$, $g(x) = \sum_{n\in\mathbb{Z}} f(x-n)$ is absolutely convergent. g is then periodic with period 1. It therefore has a Fourier series

$$g(x) = \sum_k a_k exp(2\pi i k x)$$

where

$$a_k = \int_0^1 g(x)exp(-2\pi i k x)dx = \sum_n \int_0^1 f(x-n)exp(-2\pi i k x)dx$$

$$= \sum_n \int_{-n}^{-n+1} f(x)exp(-2\pi i k x)dx = \int_{-\infty}^{\infty} f(x)exp(-2\pi i k x)dx = F(k)$$

where F is the Fourier transform of f.

$$F(y) = \int_{-\infty}^{\infty} f(x)exp(-2\pi i y x)dx$$

We are assuming here that f is integrable on $\mathbb{R}$. We have thus proved Poisson's summation formula.

$$\sum_n f(x-n) = \sum_k F(k)exp(2\pi i k x)$$

Consider

$$f(x) = exp(-a|x|)$$

where $a > 0$. Its Fourier transform is

$$F(y) = \int_0^{\infty} exp(-(a+2\pi i y)x)dx + \int_{-\infty}^0 exp((a-2\pi i y)x)dx$$

$$= \frac{1}{a-2\pi i y} + \frac{1}{a} - 2\pi i y = \frac{2a}{a^2 + 4\pi^2 y^2}$$

Applying the Poisson summation formula

$$\sum_n f(n) = \sum_n F(n)$$

we get

$$\sum_n \frac{2a}{a^2 + 4\pi^2 n^2} = \sum_n exp(-a|n|)$$

Now,

$$\sum_n exp(-a|n|) = \sum_{n=0}^{\infty} exp(-an) + \sum_{n=-\infty}^{-1} exp(an)$$

$$= \frac{1}{1 - exp(-a)} + \frac{exp(-a)}{1 - exp(-a)} = \frac{exp(a) + 1}{exp(a) - 1} = coth(a/2)$$

Thus,

$$\sum_n \frac{2a}{a^2 + 4\pi^2 n^2} = coth(a/2)$$

[248] Consider the series

$$F(x) = \sum_{n=1}^{\infty} \frac{a_n}{n^x}$$

We have

$$\Gamma(x)F(x) = \sum_{n=1}^{\infty} a_n \int_0^{\infty} exp(-nt) t^{x-1} dt$$

Define

$$A(z) = \sum_{n=1}^{\infty} a_n z^{-n}$$

Then,

$$\Gamma(x)F(x) = \int_0^{\infty} A(exp(t)) t^{x-1} dt$$

[249] Let $x(t)$ and $y(t)$ be respectively the input and output of a nonlinear system satisfying the differential equation

$$\frac{d^n y(t)}{dt^n} = f\left(\frac{d^{n-1} y(t)}{dt^{n-1}}, \ldots \frac{dy(t)}{dt}, x(t)\right)$$

Cast this system in state variable form, ie, as a system of coupled first order differential equations.

[250] Find the Fourier series expansion for the periodic signal

$$x(t) = \sum_{n=-\infty}^{\infty} \delta(t - nT)$$

and deduce the Poisson summation formula from this.

hint:

$$x(t) = \sum_n T^{-1} exp(-jn2\pi t/T)$$

and hence

$$\sum_n f(nT) = T^{-1} \sum_n F(2n\pi/T)$$

where F is the Fourier transform of f.

[251] Let T_1 and T_2 be two linear systems with impulse responses $h_1(n,k)$ and $h_2(n,k)$ respectively. Then the respective input output relations are

$$y_1(n) = \sum_k h_1(n,k)x(k), \, y_2(n) = \sum_k h_2(n,k)x(k)$$

Suppose we pass the signal $x(n)$ through the system T_1T_2 so that $y(n) = T_1T_2\{x(n)\}$ is the output. Then

$$y(n) = \sum_{k,r} h_1(n,k)h_2(k,r)x(r) = \sum_r h(n,r)x(r)$$

where

$$h(n,r) = \sum_k h_1(n,k)h_2(k,r)$$

is the impulse response of the system T_1T_2. Note that in general, $T_1T_2 \neq T_2T_1$ since the impulse response of T_2T_1 is given by

$$g(n,r) = \sum_k h_2(n,k)h_1(k,r)$$

[252] Find the general solution to the ordinary differential equation

$$p(D)y = x$$

where $p(s)$ is a polynomial having the factorization

$$p(s) = \Pi_{i=1}^{r}(s - s_i)^{m_i}$$

with $s_1, ..., s_r$ distinct and $m_1, ..., m_r$ positive integers. Here $x(t)$ is the input, ie, it is known and $y(t)$ is the output, to be expressed in terms of the initial conditions $y^{(k)}(0), k = 0, 1, ..., m - 1$ and the input $\{x(t)\}$. $D = d/dt$.

[253] State and prove the sub-martingale downcrossing inequality and use this to derive the sub-martingale convergence theorem.

[254] Express the Markov property of a stochastic process using the shift operator θ_t on the space of paths.

[255] Determine the possible frequencies of oscillation of an electromagnetic field inside a cylindrical cavity. Assume that the height of the cylinder is L and its radius is R. Consider both the TE and TM modes. Assume that the walls of the cavity are perfect conductors.

[256] Solve the heat equation in three spatial dimensions in cylindrical coordinates inside a cylinder of radius R and length L.

hint: The heat equation

$$u_{,t} = D\nabla^2 u$$

becomes in cylindrical coordinates,

$$u_{,t} = D(\rho^{-1}(\rho u_{,\rho})_{,\rho} + \rho^{-2}u_{,\phi\phi} + u_{,zz})$$

where $u(t, \rho, \phi, z)$ is the temperature inside the disc $0 \leq \rho \leq R, 0 \leq \phi < 2\pi, 0 \leq z \leq L$. Adopt the separation of variables

$$u(t, \rho, \phi, z) = T(t)R(\rho)\Phi(\phi)Z(z)$$

Substituting gives

$$T'/T = D(R''/R + R'/\rho R + \Phi''/\rho^2\Phi + Z''/Z)$$

The left side is a function of only t while the right side is a function of only (ρ, ϕ, z). So both sides must equal a constant, say $-\lambda$. Then,

$$T(t) = exp(-\lambda t)$$

$$R''/R + R'/\rho R + \Phi''/\rho^2\Phi + \lambda/D = -Z''/Z$$

Again both sides must equal a constant say α^2. Then,

$$Z(z) = A_1.cos(\alpha z) + A_2.sin(\alpha z)$$

Further,

$$\rho^2(R''/R + R'/\rho R) + (\lambda/D) - \alpha^2)\rho^2 = -\Phi''/\Phi$$

Thus both sides again equal a constant say m^2. Then m must be an integer for Φ to be single valued and

$$\Phi(\phi) = B_1cos(m\phi) + B_2sin(m\phi)$$

and

$$\rho^2 R'' + \rho R' + ((\lambda/D - \alpha^2)\rho^2 - m^2) = 0$$

so that

$$R(\rho) = J_m(\beta\rho)$$

where

$$\beta = \sqrt{\lambda/D - \alpha^2}$$

and J_m is the Bessel function of order m. The complete solution is a linear superposition of the functions

$$u(t, \rho, \phi, z|m, \alpha, \beta) = J_m(\beta\rho)(A_1cos(\alpha z) + A_2sin(\alpha z))(B_1cos(m\phi) + B_2sin(m\phi))exp(-D\sqrt{\alpha^2 + \beta^2}t)$$

Suppose we assume that u vanishes on the top and bottom surfaces of the cylinder, ie, at $z = 0, L$. Then, $A_1 = 0$ and $\alpha = r\pi/L$ where r is an integer. Suppose we also assume that u vanishes on the curved surface of the cylinder, ie at $\rho = R$. Then, $J_m(\beta R) = 0$ or $\beta = \beta_{mn}/R$ where $\beta_{mn}, n = 1, 2, 3, ...$ are the zeros of $J_m(x)$. In other words, every solution of the heat equation that vanishes on the surface of the cylinder (both the flat and the curved) can be expressed as

$$u(t, \rho, \phi, z) = \sum_{m,n,r} J_m(\beta_{mn}\rho/R)sin(r\pi z/L)(B_1(m, n, r)cos(m\phi)$$

$$+B_2(m, n, r)sin(m\phi))exp(-D((r\pi/L)^2 + \beta_{mn}^2/R^2)t)$$

where m assumes all non-negative integer values, and n and r assume positive integer values. Note that this solution could also have been obtained from the Fourier series expansion

$$u(t, \rho, \phi, z) = \sum_{m,r}(Amr(t, \rho)cos(m\phi) + B_{mr}(t, \rho)sin(m\phi))sin(r\pi z/L)$$

[257] Show that $\phi_T(\omega) = \frac{1}{2\pi}\int_{-T}^{T} exp(-j\omega t)dt$ converges to $\delta(\omega)$ as $T \to \infty$. The convergence is to be understood in the weak sense. Let $f(\omega)$ be infinitely differentiable and rapidly decreasing function. Then since

$$\phi_T(\omega) = sin(\omega T)/\pi\omega$$

it follows that

$$\int_{-\infty}^{\infty} f(\omega)\phi_T(\omega)d\omega = \int_{-\infty}^{\infty} f(\omega)sin(\omega T)d\omega/\pi\omega$$

$$= \int_{-\infty}^{\infty} f(x/T)sin(x)dx/\pi x$$

As $T \to \infty$, this converges to

$$f(0)\int_{-\infty}^{\infty} sin(x)dx/\pi x = f(0)$$

[258] Let

$$f(t) = \int_{-\infty}^{\infty} F(\omega)exp(j\omega t)d\omega$$

where f, F are both integrable. Then show that

$$F(\omega) = \frac{1}{2\pi}\int_{-\infty}^{\infty} f(t)exp(-j\omega t)dt$$

Under what conditions is

$$\frac{d^n f(t)}{dt^n} = \int_{-\infty}^{\infty} (j\omega)^n F(\omega)exp(j\omega t)d\omega$$

true.

[259] Is it possible to construct two finite square matrices A, B such that

$$AB - BA = I$$

where I is the identity matrix.

hint:Take the trace on both sides and arrive at a contradiction. In quantum mechanics, we construct infinite matrices satisfying the above commutation relation. For example, in the quantum mechanical harmonic oscillator, we construct the creation and annihilation operators a, a^* satisfying this commutation relation. Note that if $A = ((a_{ij})), B = ((b_{ij}))$ are infinite matrices, such that $\sum_{i,k} |a_{ik} b_{ki}| < \infty$, ie the infinite series with terms $a_{ik} b_{ki}$ taken in any one order is absolutely summable, then $Tr(AB) = Tr(BA)$. If however, absolute summability does not hold good, then the result is not true in general.

[260] Consider the dynamical system

$$x_i'(t) = f_i(x_1(t), ..., x_n(t)), i = 1, 2, ..., n$$

Let $x_i(t)$ be a solution and let $x_i(t) + \delta x_i(t)$ be another solution with $\delta x_i(t)$ of first order of smallness. Then upto first order of smallness, we have

$$\delta x_i'(t) = \sum_{j=1}^{n} f_{i,j}(x_1(t), ..., x_n(t)) \delta x_j(t), i = 1, 2, ..., n$$

In particular suppose $\mathbf{x} = (x_1, ..., x_n)$ is a fixed point, so that

$$f_i(\mathbf{x}) = 0, i = 1, 2, ..., n$$

Then if $\mathbf{x} + \delta x_i(t)$ is a solution with $\delta x_i(t)$ of the first order of smallness, then

$$\delta x_i'(t) = \sum_{j=1}^{n} f_{i,j}(\mathbf{x}) \delta x_j(t), i = 1, 2, ..., n$$

It follows that $\mathbf{x}$ is a stable fixed point of the dynamical system iff all the eigenvalues of the matrix $\mathbf{F} = ((f_{i,j}(\mathbf{x})))_{1 \leq i,j \leq n}$ have negative real parts.

[261] Suppose our computer consists of b bit registers for storing signal samples as well as multipliers. Then if x, y are two b bit numbers, let $[xy]$ denote the number obtained by rounding the $2b$ bit number xy to b bits. If ϵ is the corresponding rounding error, then $[xy] = xy + \epsilon$. Note that $\epsilon \in [-\Delta, \Delta]$ where $\Delta = 2^{-b-1}$. Thus if we implement the difference equation

$$y[n] + a_1 y[n-1] + ... + a_p y[n-p] = b_0 x[n] + b_1 x[n-1] + ... + b_q x[n-q]$$

on the computer, we would actually be implementing

$$y[n] + \sum_{i=1}^{n} a_i y[n-i] + \sum_{i=1}^{p} \epsilon_i[n] = \sum_{i=0}^{q} b_i x[n-i] + \sum_{i=0}^{q} \epsilon_i'[n]$$

where the rounding errors $\epsilon_i[n]$ and $\epsilon'_i[n]$ assume values in $[-\Delta, \Delta]$. We assume that $\{\epsilon_i[n]\}, \{\epsilon'_j[n]\}, i = 1, 2, ..., p, j = 1, 2, ..., q+1$ are $p+q+1$ independent white noise sequences with each sample uniformly distributed over $[-\Delta, \Delta]$. Then $w[n] = \sum_{i=0}^{q} \epsilon'_i[n] - \sum_{i=1}^{p} \epsilon_i[n]$ is a white noise sequence with mean zero and variance $(p+q+1)\Delta^2/12$ and it propagates to the output giving the output power spectral density as

$$S(\omega) = \frac{p+q+1)\Delta^2/12}{|1 + \sum_{i=1}^{p} a_i exp(-ji\omega)|^2}$$

[262] Consider a one dimensional dynamical system $x'(t) = f(x(t))$. Make the change of variable $x = g(y)$. Then,

$$g'(y(t))y'(t) = f(g(y(t)))$$

or

$$y'(t) = \frac{f(g(y(t)))}{g'(y(t))}$$

More generally, consider the n dimensional dynamical system

$$x'_i(t) = f_i(x_1(t), ..., x_n(t)), i = 1, 2, ..., n$$

Make the change of variables

$$x_i = g_i(y_1, ..., y_n), i = 1, 2, ..., n$$

Then,

$$\sum_{j=1}^{n} \frac{\partial g_i(\mathbf{y}(t))}{\partial y_j} y'_j(t) = f_i(g(\mathbf{y}(t))), i = 1, 2, ..., n$$

Define the matrix

$$J(\mathbf{y}) = ((\frac{\partial g_i(\mathbf{y}(t))}{\partial y_j}))_{1 \leq i,j \leq n}$$

Then, the transformed dynamical system is

$$y'_i(t) = \sum_{j=1}^{n} J^{-1}(\mathbf{y}(t))_{ij} f_j(g(\mathbf{y}(t))), i = 1, 2, ..., n$$

[263] Consider the n-dimensional dynamical system

$$x'_i(t) = f_i(x_1(t), ..., x_n(t)), i = 1, 2, ..., n$$

Define

$$h(x_1, ..., x_{2n}) = \sum_{i=1}^{2n} f_i(x_1, ..., x_n)x_{n+i}$$

Suppose $(x_{0,i}, i = 1, 2, ..., 2n)$ is a minimum of h. Then,

$$\frac{\partial h(x_{0,1}, ..., x_{0,2n})}{\partial x_i} = 0, i = 1, 2, ..., 2n$$

This gives

$$f_i(x_{0,1}, ..., x_{0,n}) = 0, \sum_{j=1}^{n} \frac{\partial f_j(x_{0,1}, ..., x_{0,n})}{\partial x_i} x_{0,n+j} = 0, i = 1, 2, ..., n$$

If we assume that the matrix $((\frac{\partial f_i(x_{0,1},...,x_{0,n})}{\partial x_j}))_{1 \le i,j \le n}$ is non-singular, then

$$f_i(x_{0,1}, ..., x_{0,n}) = 0, x_{0,n+i} = 0, i = 1, 2, ..., n$$

[263] Let $\xi(t)$ be an n dimensional diffusion process satisfying the stochastic differential equation

$$d\xi(t) = b(\xi(t))dt + \sigma(\xi(t))dB(t)$$

where, for each $\xi \in \mathbb{R}^n$, $b(\xi) \in \mathbb{R}^n$ and $\sigma(\xi) \in \mathbb{R}^{n \times d}$ and $B(t)$ is a d-dimensional Brownian motion process. Derive the approximate mean and covariance propagation equations.

hint:

$$d\mathbb{E}\xi(t) = \mathbb{E}b(\xi(t))dt$$

$$d(\xi_i(t)\xi_j(t)) = \xi_i(t)d\xi_j(t) + \xi_j(t)d\xi_i(t) + d\xi_i(t).d\xi_j(t)$$

so that

$$d\mathbb{E}(\xi_i(t)\xi_j(t)) = \mathbb{E}(\xi_i(t)b_j(\xi(t)))dt + \mathbb{E}(\xi_j(t)b_i(\xi(t)))dt + \sum_k \mathbb{E}(\sigma_{ik}(\xi(t))\sigma_{jk}(\xi(t)))dt$$

Now write

$$\xi(t) = \mathbb{E}\xi(t) + \delta\xi(t)$$

so that $\mathbb{E}(\delta\xi(t)) = 0$ and expand the functions b, σ around $\mathbb{E}\xi(t)$ in powers of $\delta\xi(t)$ retaining upto second degree terms and hence derive the approximate mean and covariance propagation equations.

[264] Let A, B be two matrices of size $N \times N$. Define the operator $L : \mathbb{C}^{N \times N} \to \mathbb{C}^{N \times N}$ so that

$$L(X) = AX + XB$$

Then show that if n is a positive integer,

$$L^n(X) = \sum_{k=0}^{n} \binom{n}{k} A^k X B^{n-k}$$

hint: Define the operators L_A, R_B on $\mathbb{C}^{N \times N}$ so that

$$L_A(X) = AX, R_B(X) = XB$$

Then, $[L_A, R_B] = 0$ and

$$L = L_A + R_B$$

[265] Let $X_1, ..., X_n$ be independent $N(0,1)$ random variables on a probability space $(\Omega, \mathcal{F}, P)$. Let $\mu_1, ..., \mu_n$ be real numbers and define a measure Q on $(\Omega, \mathcal{F})$ so that

$$dQ(\omega) = exp(\sum_{i=1}^{n} \mu_i X_i(\omega) - \frac{1}{2} \sum_{i=1}^{N} \mu_i^2) dP(\omega)$$

Show that Q is a probability measure on $(\Omega, \mathcal{F})$ and that

$$\mathbb{E}_Q(exp(\sum_{i=1}^{n} c_i(X_i - \mu_i))) = \mathbb{E}_P(exp((\mu_i + c_i)X_i - \sum_i (\mu_i^2/2 + \mu_i c_i)))$$

$$= exp(\sum_i (\mu_i + c_i)^2/2 - \sum_i (\mu_i^2/2 + \mu_i c_i)) = exp(\sum_i c_i^2/2)$$

proving that $X_i - \mu_i, i = 1, 2, ..., n$ is under the probability measure Q, an independent set of $N(0,1)$ random variables.

[266] Let $x(t)$ be a diffusion process in $\mathbb{R}^1$ satisfying the stochastic differential equation

$$dx(t) = \mu(x(t))dt + \sigma(x(t))dB(t)$$

where B is standard Brownian motion. Define

$$M(t) = exp(a(x(t) - \int_0^t \mu(x(s))ds) - \frac{a^2}{2} \int_0^t \sigma(x(s))^2 ds)$$

Equivalently,

$$M(t) = exp(\xi(t))$$

where

$$\xi(t) = a(x(t) - \int_0^t \mu(x(s))ds) - \frac{a^2}{2} \int_0^t \sigma(x(s))^2 ds$$

Then,

$$d\xi(t) = a(dx(t) - \mu(x(t))dt) - a^2 \sigma(x(t))^2 dt/2$$

$$= a\sigma(x(t))dB(t) - a^2 \sigma(x(t))^2 dt/2$$

$$(d\xi(t))^2 = a^2 \sigma(x(t))^2 dt$$

It follows from Ito's formula that

$$dM(t) = exp(\xi(t))(d\xi(t) + (d\xi(t))^2/2) = exp(\xi(t))\sigma(\xi(t))dB(t)$$

proving that $M(.)$ is a Martingale with respect to the filtration generated by $B(.)$.

[267] Let $x(t_1, ..., t_r)$ be a r-dimensional signal with Fourier transform

$$X(\omega_1, ..., \omega_r) = \int_{\mathbb{R}^r} x(t_1, ..., t_r) exp(-j(\omega_1 t_1 + ... + \omega_r t_r)) dt_1 ... dt_r$$

Suppose

$$X(\omega_1, ..., \omega_r) = 0, (\omega_1, ..., \omega_r) \notin \Pi_{k=1}^r [-\sigma_k, \sigma_k]$$

Then, show that if $T_i = \pi/\sigma_i, i = 1, 2, ..., r$, then

$$x(t_1, ..., t_r) = \sum_{n_1, ..., n_r = -\infty}^{\infty} x(n_1 T_1, ..., n_r T_r) \Pi_{k=1}^r \frac{sin(\sigma_k(t - n_k T_k))}{\sigma_k(t - n_k T_k)}$$

This is Shannon's sampling theorem for multidimensional signals.

[268] Let $-\infty < a < b < \infty$ and let $f : [a, b] \to \mathbb{R}$ be continuous. Then show that f is uniformly continuous.

hint: Let $\epsilon > 0$. Then, for each $t \in [a, b]$, there is a $\delta(t) > 0$ such that $|s - t| < \delta(t), s \in [a, b]$ implies $|f(s) - f(t)| < \epsilon$. Now $\{(t - \delta(t), t + \delta(t))\}_{t \in [a,b]}$ is an open covering for $[a, b]$ and hence by the Heine-Borel property, there is a finite set $\{t_1, ..., t_r\}$ in $[a, b]$ such that

$$[a, b] \subset \bigcup_{k=1}^r (t_k - \delta(t_k), t_k + \delta(t_k))$$

Using this, the proof of uniform continuity can be easily completed.

[269] Let $\mathbf{B}(t), t \geq 0$ be a standard n-dimensional Brownian motion process and let for each $t \geq 0$, $\mathbf{A}(t)$ be an $n \times n$ matrix. Define the process

$$\mathbf{X}(t) = \int_0^t \mathbf{A}(s) d\mathbf{B}(s)$$

Then $\mathbf{X}(t), t \geq 0$ is a Gaussian process with zero mean and autocorrelation

$$\mathbb{E}(\mathbf{X}(t)\mathbf{X}(s)^T) = \int_0^s \mathbf{A}(u)\mathbf{A}(u)^T du, s \leq t$$

[270] Let (X_1, X_2) be iid $N(0, 1)$ variables and $(m_1, m_2) \in \mathbb{R}^2$. Then

$$\mathbb{E}[((X_1 + m_1)^2 + (X_2 + m_2)^2)^{-1/2}]$$

$$= (2\pi)^{-1} \int_{\mathbb{R}^2} (y_1^2 + y_2^2)^{-1/2} exp(-\frac{1}{2}(y_1 - m_1)^2 + (y_2 - m_2)^2) dy_1 dy_2$$

$$= (2\pi)^{-1} \int_0^\infty \int_0^{2\pi} exp(-\frac{1}{2}(r^2 + m^2 - 2mr.cos(\theta))) dr d\theta$$

where

$$m^2 = m_1^2 + m_2^2$$

Denote the above integral by $F(m)$. We wish to find m such that $F(m)$ is a maximum. Setting $F'(m) = 0$ gives us

$$\int_{[0,\infty)\times[0,2\pi)} (r.cos(\theta) - m).exp(-\frac{1}{2}(r^2 + m^2 - 2mr.cos(\theta)))drd\theta = 0$$

Note that $0 < F(m) < \infty$ for all $m \in [0, \infty)$ while $F(m) \to 0$ as $m \to \infty$. It follows that $F(m)$ attains its maximum at some finite value of m and hence it follows that

$$sup_{m\geq0}F(m) < \infty$$

In other words, we have proved that

$$sup_{m_1,m_2\in\mathbb{R}}\mathbb{E}[((X_1 + m_1)^2 + (X_2 + m_2)^2)^{-1/2}] < \infty$$

[271] Solve the n dimensional wave equation using n dimensional Fourier transforms. Note that the n-dimensional wave equation is given by

$$u_{,t}(t, x_1, ..., x_n) = \sum_{i=1}^{n} u_{,x_ix_i}(t, x_1, ..., x_n)$$

hint: Let

$$\hat{u}(t, k_1, ..., k_n) = \int_{\mathbb{R}^n} exp(-i(k_1x_1 + ... + k_nx_n))u(t, x_1, ..., x_n)dx_1...dx_n$$

Then,

$$\hat{u}_{,t}(t, \mathbf{k}) = -k^2\hat{u}(t, \mathbf{k})$$

It follows that

$$\hat{u}(t, \mathbf{k}) = A(\mathbf{k})exp(ikt) + B(\mathbf{k})exp(-ikt)$$

where

$$k = |\mathbf{k}| = (\sum_{i=1}^{n} k_i^2)^{1/2}$$

Now take the inverse Fourier transform to obtain the general solution.

[272] Suppose $\xi(t)$ is a one dimensional diffusion process satisfying the sde

$$d\xi(t) = \mu(\xi(t))dt + \sigma(\xi(t))dB(t)$$

where $B(.)$ is a one dimensional Brownian motion process. Let $f(t, x)$ be twice continuously differentiable and set

$$M(t) = f(t, \xi(t)) - \int_0^t (\frac{\partial}{\partial s} + L_s)f(s, \xi(s))ds$$

where

$$L_t = \mu(x)\frac{\partial}{\partial x} + \frac{\sigma^2(x)}{2}\frac{\partial^2}{\partial x^2}$$

Then show that M is a martingale with respect to the Brownian filtration.

[273] Let $(\Omega, \mathcal{F}, P)$ be a probability space and let $\{\mathcal{F}_t : t \geq 0\}$ be a filtration on this space. For any stop-time τ with respect to this filtration, ie, $\{\tau \leq t\} \in \mathcal{F}_t, t \geq 0$ define $\mathcal{F}_\tau$ to be the set of all $E \in \mathcal{F}$ for which $E \cap \{\tau \leq t\} \in \mathcal{F}_t, t \geq 0$. Then show that $\mathcal{F}_\tau$ is a σ-field. Show that if τ, σ are two stop times such that $\tau \leq \sigma$, then $\mathcal{F}_\tau \subset \mathcal{F}_\sigma$.

[274] Let $\Omega = C[0, \infty)$ be the space of all continuous functions $f : [0, \infty) \to \mathbb{R}$. For $x, y \in \Omega$, define

$$d(x,y) = \sum_{n=1}^{\infty} 2^{-n} \frac{sup_{t \in [0,n]}|x(t) - y(t)|}{1 + sup_{t \in [0,n]}|x(t) - y(t)|}$$

Show that d is a metric on Ω. Show that if $x_n \in \Omega, n = 1, 2, \dots$ and $x \in \Omega$, then $d(x_n, x) \to 0$ iff x_n converges to x uniformly on every compact subset K of $[0, \infty)$.

[275] Lotka-Volterra system. $x(t)$ is the population of prey in a pond at time t and $y(t)$ is the population of predator in the pond. If $x(t)$ is large, then the rate of increase of prey population is large while if $x(t)$ is small, the rate of increase of predator population is small. Thus we have the equation

$$y'(t) = (-a + bx(t))y(t)$$

where $a, b > 0$. Note that in the absence of any prey, ie, $x(t) = 0$, we have $y'(t) = -ay(t)$ meaning that the predator population decreases exponentially with time. If the predator population is small, the prey population increases exponentially with a rate c while if the predator population is large, the rate of increase of prey population is small. Thus we have

$$x'(t) = (c - ky(t))x(t)$$

where $c, k > 0$. $(0,0)$ is a trivial equilibrium point. A non-trivial equilibrium point is $(x_0, y_0) = (a/b, c/k)$. We linearize the system of differential equations about this equilibrium point by setting

$$x(t) = x_0 + \delta x(t), y(t) = y_0 + \delta y(t)$$

Neglect terms quadratic in $(\delta x(t), \delta y(t))$ to get a linear system of differential equations for $\delta x(t), \delta y(t)$. Show that this linearized system is oscillatory and hence the equilibrium point is a point of critical stability.

[276] Homotopic perturbation method for solving differential equations. Let L, M be two differential operators, not necessarily linear. We are given a function f and wish to determine a function g so that $(L + M)(g) = f$, or equivalently, $L(g) + M(g) = f$. Let $p \in [0, 1]$ and consider the differential equation

$$H(g, p) = (1 - p)(L(g) - L(g_0)) + p(L(g) + M(g) - f) = 0$$

where g_0 is a known function. We have

$$H(g, 0) = L(g) - L(g_0)$$

so g_0 satisfies the equation $H(g, p) = 0$ when $p = 0$. Further,

$$H(g, 1) = L(g) + M(g) - f$$

which means that the solution g that we are looking for satisfies $H(g, 1) = 0$. We can write

$$H(g, p) = L(g) - L(g_0) + p(M(g) + L(g_0) - f)$$

We expand the solution g in powers of p, ie,

$$g = \sum_{m=0}^{\infty} p^m g_m$$

Equating coefficients of successive powers of p we can obtain a series solution to $H(g, p) = 0$ for all $p \in [0, 1]$.

Acknowledgement: This method of solving nonlinear problems was brought to the author's attention by his colleague Akash Rathi.

[277] Suppose $M(t), t \geq 0$ is a continuous martingale. Then by Ito's formula,

$$d|M(t)| = sgn(M(t))dM(t) + \delta(M(t))d < M, M > (t)$$

so that

$$|M(t)| - |M(s)| = \int_s^t sgn(M(u))dM(u) + \int_s^t \delta(M(u))d < M, M > (u)$$

This prompts us to define the local time process for the martingale M at 0 as

$$L_t(M) = \int_0^t \delta(M(u))d < M, M > (u)$$

More generally for $x \in \mathbb{R}$,

$$|M(t) - x| - |M(s) - x| = \int_s^t sgn(M(u) - x)dM(u) + \int_s^t \delta(M(u) - x)d < M, M > (u)$$

so that the local time of M at x is defined to be the process

$$L_t^x(M) = \int_0^t \delta(M(u) - x)d < M, M > (u)$$

Note that

$$\int_{-\infty}^{\infty} f(x)L_t^x(M)dx = \int_0^t f(M(u))d < M, M > (u)$$

[278] Let $x(t), t \geq 0$ be a one dimensional diffusion process with drift coefficient $\mu(x)$ and diffusion coefficient $\sigma(x)$. Thus for some Brownian motion process $B(t), t \geq 0$, $x(t)$ satisfies the sde

$$dx(t) = \mu(x(t))dt + \sigma(x(t))dB(t)$$

Let $g(x)$ be a twice continuously differentiable function on $\mathbb{R}$. Define the process

$$M(t) = exp(g(x(t)) - \int_0^t (\mu(x(s))g'(x(s)) + \frac{\sigma^2(x(s))}{2}g''(x(s)) + \frac{\sigma^2(x(s))}{2}g'(x(s))^2)ds)$$

Let

$$y(t) = g(x(t)) - \int_0^t (\mu(x(s))g'(x(s)) + \frac{\sigma^2(x(s))}{2}g''(x(s)))ds$$

Then,

$$dy(t) = g'(x(t))\sigma(x(t))dB(t)$$

so the process $y(t), t \geq 0$ is a Martingale. Its quadratic variation is given by

$$< y, y > (t) = \int_0^t (g'(x(s))\sigma(x(s))^2 ds$$

or equivalently,

$$(dy(t))^2 = \sigma^2(x(t))g'(x(t))^2$$

It thus follows that

$$M(t) = exp(y(t) - \frac{1}{2} < y, y > (t))$$

Hence, $M(t), t \geq 0$ is a Martingale. This can be generalized to multidimensional diffusion processes. Let $\mu : \mathbb{R} \times \mathbb{R}^n \to \mathbb{R}^n$ and $\sigma : \mathbb{R} \to \mathbb{R}^n \to \mathbb{R}^{n \times p}$ be smooth functions and let $x(t) \in \mathbb{R}^n$ satisfy the sde

$$dx(t) = \mu(t, x(t))dt + \sigma(t, x(t))dB(t)$$

where $B(t), t \geq 0$ is a p dimensional Brownian motion process. Let $g : \mathbb{R} \times \mathbb{R}^n \to \mathbb{R}$ be a smooth function and for $t \in \mathbb{R}, x \in \mathbb{R}^n$, define $a(t, x) = \sigma(t, x)\sigma^T(t, x) \in \mathbb{R}^{n \times n}$. By Ito' formula,

$$dg(t, x(t)) = (g_{,t}(t, x(t)) + (L_t g)(t, x(t)))dt + \sum_{i,j} g_{,x_i}(t, x(t))\sigma_{ij}(t, x(t))dB_j(t)$$

where

$$L_t f(t, x) = \mu(t, x)^T \nabla f(t, x) + \frac{1}{2}Trace(a(t, x)\nabla \nabla^T f(t, x))$$

L_t is the generator of the diffusion $x(t), t \geq 0$. Thus if we define the process

$$y(t) = g(t, x(t)) - \int_0^t (g_{,s}(s, x(s)) + L_s g(s, x(s)))ds$$

then,

$$dy(t) = (\nabla g(t, x(t)))^T \sigma(t, x(t))dB(t)$$

and hence, $y(t), t \geq 0$ is a Martingale. Further,

$$< y, y > (t) = \int_0^t (\nabla g(s, x(s)))^T a(s, x(s)) \nabla g(s, x(s)) ds$$

Hence

$$M(t) = exp(y(t) - \frac{1}{2} < y, y > (t)) = exp(g(t, x(t)) - \int_0^t (g_{,s}(s, x(s)) + L_s g(s, x(s)))$$

$$+ \frac{1}{2}(\nabla g(s, x(s)))^T a(s, x(s)) \nabla g(s, x(s))) ds)$$

is a Martingale.

[279] Let $z = x + iy$ with x, y real. Then $z^2 = u + iv$ where $u = x^2 - y^2$ and $v = 2xy$. Since z^2 is analytic, both u, v satisfy Laplace's equation ie, $u_{,xx} + u_{,yy} = 0, v_{,xx} + v_{,yy} = 0$. This can also be verified by direct computation.

[280] Integrating factor method for solving first order differential equations. We want to solve the differential equation

$$\phi(x, y)dx + \psi(x, y)dy = 0$$

Multiply both sides by a function $M(x, y)$ so that

$$M\phi.dx + M\psi.dy = 0$$

M is to be selected so that the left side becomes a perfect differential, ie,

$$dS = M\phi.dx + M\psi.dy$$

for some function $S(x, y)$. This means that

$$S_{,x} = M\phi, S_{,y} = M\psi$$

and the condition for such an S to exist is that

$$(M\phi)_{,y} = (M\psi)_{,x}$$

The general solution is then given by

$$S(x, y) = c$$

where c is any constant. The problem of solving a nonlinear first order differential equation is thereby transformed to solving a linear partial differential equation for M.

[281] Suppose X is a $N(0, 1)$ random variable (Gaussian with mean zero and unit variance). Let $Y = f(X)$ where f has a Taylor series

$$f(x) = \sum_{n=0}^{\infty} f^{(n)}(0)x^n/n!$$

Then determine a series expansion for the moment generating function $M_Y(t) = \mathbb{E}(exp(tY))$ of Y.

hint. First work out this problem when f is a Polynomial. In that case setting

$$f(x) = a_0 + a_1 x + a_2 x^2 + ... + a_p x^p$$

we have

$$exp(tf(X)) = \sum_{n=0}^{\infty} t^n f(X)^n / n!$$

where

$$f(X)^n = (a_0 + a_1 X + ... + a_p X^p)^n / n! = \sum_{r_0+...+r_p=n} \frac{1}{r_0!...r_p!} a_0^{r_0} a_1^{r_1} ... a_p^{r_p} X^{r_1+2r_2+...+pr_p}$$

[282] Nonlinear state variable equations. Consider first a simple first order ordinary differential equation for a scalar state variable $x(t)$.

$$x'(t) = f(x(t)), t \geq 0$$

We write

$$f(x) = ax + g(x)$$

where

$$a = f(0), g(x) = \sum_{m=1}^{\infty} f^{(m)}(0) x^m / m!$$

Let $p \in [0, 1]$ and consider the differential equation

$$(1 - p)(x'(t) - ax(t) - x_0'(t) + ax_0(t)) + p(x'(t) - ax(t) - g(x(t))) = 0$$

When $p = 0$, a solution is

$$x(t) = x_0(t)$$

When $p = 1$, it reduces to the given differential equation. For general p, this equation expands to give

$$x'(t) - ax(t) - x_0'(t) + ax_0(t) + p(x_0'(t) - ax_0(t) - g(x(t))) = 0$$

Write

$$x(t) = \sum_{m=0}^{\infty} p^m x_m(t)$$

Substituting this into the above equation and equating coefficients of p^0 gives the trivial equation

$$x_0'(t) - ax_0(t) - x_0'(t) + ax_0(t) = 0$$

Equating coefficients of p gives

$$x_1'(t) - ax_1(t) + x_0'(t) - ax_0(t) - g(x_0(t)) = 0$$

This is a linear differential equation for x_1 and is easily solved using the integrating factor $exp(-at)$. Equating coefficients of p^2 gives

$$x_2'(t) - ax_2(t) - g'(x_0(t))x_1(t) = 0$$

This is again a linear differential equation for x_2 and is easily solved.

[283] Simulational problems in Mathematical Physics.

[1] Numerically solve the time dependent Schrodinger equation in one dimension for the potentials (a) $V(x) = Kx^2/2$, (b) $V(x) = Kx^2/2 + \epsilon.x^3$.

hint: For the potential $V(x)$, the time dependent Schrodinger equation is

$$\psi_{,t}(t,x) = (ih/4\pi m)\psi_{,xx}(t,x) - (2\pi i/h)V(x)\psi(t,x)$$

With a time discretization step of size τ and a spatial discretization step of size Δ, the Schrodinger equation approximates to the following partial difference equation.

$$\tau^{-1}(\psi[n+1,k] - \psi[n,k]) = (ih/4\pi m\Delta^2)(\psi[n,k+1] - 2\psi[n,k] + \psi[n,k-1]) - (2\pi i/h)V[k]\psi[n,k]$$

where

$$\psi[n,k] = \psi(n\tau, k\Delta)$$

[2] Simulate the motion of a particle in the periodic gravitational field generated by two masses M_1, M_2 moving in a circular orbit with uniform angular velocity ω about their centre of mass.

[3] Plot the trajectory traced by a point on the circumference of a disc of radius R as the disc rolls on a line with uniform translational velocity $R\omega$. Show that the trajectory is a cycloid.

[4] Plot the electric field lines when the field is produced by two charges $Q, -Q$ separated by a distance d.

hint: If the electric field is $E_x(x,y)\hat{x} + E_y(x,y)\hat{y}$, then the field lines of force satisfy are given by the solutions to the differential equation

$$\frac{dx}{E_x(x,y)} = \frac{dy}{E_y(x,y)}$$

or equivalently in parametric form by

$$\frac{dx}{d\lambda} = E_x(x,y), \frac{dy}{d\lambda} = E_y(x,y)$$

These equations are solved numerically by the discretization scheme

$$x[n+1] = x[n] + \Delta.E_x(x[n], y[n]), y[n+1] = y[n] + \Delta.E_y(x[n], y[n])$$

[5] Using variational calculus, determine the equation of a chain suspended between two ends in a uniform gravitational field.

hint: Let $y = f(x)$ be the equation of a chain of length L where the two endpoints are $(a, h_1), (b, h_2)$. The potential energy of the chain is then given by

$$U = \int_a^b mg\sqrt{1 + f'^2(x)}dx$$

This is to be minimized subject to the length constraint, ie,

$$\int_a^b \sqrt{1 + f'^2(x)}dx = L$$

Using Lagrange multipliers, formulate this constrained optimization problem as a problem in the calculus of variations, write down the Euler-Lagrange equations and solve it.

[6] Numerically solve the Fokker-Planck equation for the probability density of a one dimensional diffusion process with specified drift and diffusion coefficients. Use the method of moments.

hint: Let $\mu(x)$ be the drift coefficient and $\sigma(x)$ the diffusion coefficient. Let $p(t, x)$ be the probability density of the process at time t. Then p satisfies the Fokker-Planck equation

$$p_{,t} = -(\mu.p)_{,x} + \frac{1}{2}(a.p)_{,xx}$$

where

$$a(x) = \sigma^2(x)$$

Expand the solution $p(t, x)$ as a linear combination of test functions $\phi_i(x), i = 1, 2, ..., n$, ie,

$$p(t, x) = \sum_{i=1}^n c_i(t)\phi_i(x)$$

Substituting this into the Fokker-Planck equation gives

$$\sum_i c_i'(t)\phi_i(x) = -\sum_i c_i(t)(\mu.\phi_i)_{,x} + \frac{1}{2}\sum_i c_i(t)(a\phi_i)_{,xx}$$

Take the inner product on both sides with $\phi_j(x)$ to get

$$\sum_i c_i'(t) < \phi_i, \phi_j >= -\sum_i c_i(t) < (\mu\phi_i)_{,x}, \phi_j > +\frac{1}{2}\sum_i c_i(t) < (a\phi_i)_{,xx}, \phi_j >$$

Define

$$q_{ji} =< \phi_j, \phi_i >, s_{ji} = - < \phi_j, (\mu\phi_i)_{,x} > +\frac{1}{2} < \phi_j, (a\phi_i)_{,xx} >$$

Then the above system of ordinary differential equations can be expressed as

$$\sum_i q_{ji} c_i'(t) = \sum_i s_{ji} c_i(t)$$

or in matrix notation,

$$\mathbf{c}'(t) = \mathbf{Q}^{-1}\mathbf{S}\mathbf{c}(t)$$

where

$$\mathbf{Q} = ((g_{ij})), \mathbf{S} = ((s_{ij}))$$

The solution is

$$\mathbf{c}(t) = exp(t\mathbf{Q}^{-1}\mathbf{S})\mathbf{c}(0)$$

[284] Introduction to the thesis on parameter estimation in linear and nonlinear systems using wavelet based signal processing. One of the fundamental problems in signal processing is the identification of a system, linear or nonlinear based on input-output data. It usually happens that the system is known except for a finite set of unknown parameters and then the problem is to estimate this finite parameter set. Statistical methods for parameter estimation include Bayesian methods which can be cast in the following form. $\mathbf{X}$ is a random vector whose probability density $p(\mathbf{X}|\theta)$ depends on a random parameter vector θ having probability density $p(\theta)$. We then construct an estimate $\hat{\theta}(\mathbf{X})$ of θ as a Borel measurable function of the random vector $\mathbf{X}$ such that the average cost $\int C(\hat{\theta}(\mathbf{X}),\theta)p(\mathbf{X}|\theta)p(\theta)d\mathbf{X}d\theta$ is a minimum. Here, $C(\hat{\theta},\theta)$ increases as the discrepancy between the true parameter value θ and its estimate $\hat{\theta}$ increases. Different kinds of cost functions $C(\hat{\theta},\theta)$ yield different estimators. For example, if we take C as the square of the Euclidean distance between the true parameter and its estimate, then the optimal estimator $\hat{\theta}$ becomes the conditional expectation of the parameter given the observation, ie, $\hat{\theta}(\mathbf{X}) = \mathbb{E}(\theta|\mathbf{X})$. This estimator is called the minimum mean square estimator. Another important parameter estimator is the maximum aposteriori (MAP) estimator defined as that value of θ for which

$$p(\theta|\mathbf{X}) = p(\mathbf{X}|\theta).p(\theta)/p(\mathbf{X})$$

is a maximum or equivalently the maximum of the function $\theta \to p(\mathbf{X}|\theta)p(\theta)$ over all θ. In other words, for a given observation $\mathbf{X}$, we select the most probable value of θ. When the apriori density $p(\theta)$ of θ is not known, or when θ is modeled as an unknown non-random parameter, then we can use the maximum likelihood (ML) estimator as that value of θ for which $p(\mathbf{X}|\theta)$ is a maximum, ie, that value of the parameter for which the given observation is most likely to occur. When however we have no knowledge of the statistics of the observation, but have some idea about the nature of the system that generates the given data, we use the least squares method and its variants. For example, suppose that the model for data generation is given by

$$y(n) = F(y(n-1), ..., y(n-p), x(n), x(n-1), ..., x(n-q), \theta_1, ..., \theta_r)$$

with $x(.)$ as the known input process and $y(.)$ as the measured output process, then the unknown parameters θ_i may be estimated by minimizing

$$\sum_{n \in D} (y(n) - F(y(n-1), ..., y(n-p), x(n), x(n-1), ..., x(n-q), \theta_1, ..., \theta_r))^2$$

where D is the time index set for available input-output data. Here the function F is known but the parameters θ_i that are arguments of this function are unknown.

The present thesis deals with estimating parameters of linear and nonlinear differential equations arising in circuit theory. The Kirchhoff equations of the circuit lead to a parametric model for the system with the parameter/parameters giving important information about the system. This work is centred around problems involving parameter estimation in linear and nonlinear systems. To start with, we have considered a second order linear differential equation as the defining input-output equation for an LTI system and the problem of estimating the coefficients in this equation. There are many suboptimal ways to estimate these coefficients. One of them is to write down the exact solution of this differential equation in the absence of any input in terms of the natural frequency and damping factor which can be expressed as functions of the coefficients of the differential equation. Let a, b be the coefficients of this differential equation and ω and γ the natural frequency and damping factor. Then, the solution for given nonzero initial conditions has the form $F(t, \omega, \gamma)$ where F is a damped sinusoid. The parameters ω, γ can be estimated by measuring the actual response $y(t)$ and minimizing the distance between F and y over a given time duration $[0, T]$. The distance may be defined as an integral of the error square if continuous measurements are available or as a sum of the error square over discrete samples if the measurements are discrete. Further the algorithm can be made adaptive by using the stochastic gradient algorithm in which if at a given time instant t, $(y(t) - F(t, \hat{\omega}(t), \hat{\gamma}(t)))^2$ is large, then alter the instantaneous estimates $\hat{\omega}(t)$ and $\hat{\gamma}(t)$ at $t + \delta t$ so that this instantaneous error square decreases. Once estimates ω, γ have been estimated this way, we invert the equations defining ω, γ in terms of a, b to get estimates of a, b. A more direct method of estimating a, b from measurements is to consider the differential equation $y''(t) + ay'(t) + by(t) \approx 0$ and estimate a, b by minimizing $\int_0^T (y''(t) + ay'(t) + by(t))^2 dt$ with respect to a, b if continuous measurements are available or by minimizing $\sum_k (y''(t_k) + ay'(t_k) + by(t_k))^2$ if the measurements are taken at the discrete times t_k and $y'(t_k)$ and $y''(t_k)$ are approximated by finite differences. The problem with the least squares based techniques is that a very large amount of data storage is needed. We have to store the entire measured data and then perform the estimation. Thus, it is natural to look for a suboptimal method that involves lesser data storage. This can be done using wavelet based signal representation. Wavelets naturally take care of time-scale properties of the signal so that if the resolution is specified, then with very few coefficients, we can represent the signal to a very high degree of accuracy. To see this more clearly, we consider a signal $x(t)$ in $L^2[0, T]$. Let $e_i, i = 1, 2, \ldots$ be an orthonormal basis for $L^2[0, T]$. Then $x(t) = \sum_{i=1}^{\infty} < x, e_i > e_i(t)$. Truncating this expansion to n terms, we have the approximation $x(t) \approx \sum_{i=1}^{n} < x, e_i > e_i(t)$. Suppose on the other hand, we try to approximate the signal $x(t)$ by its samples spaced at Δ. If the signal has rapid fluctuations at some points, then the latter representation may not be accurate unless the sample spacing is very small. The representation of the signal using the basis $e_i, i = 1, 2, \ldots$ may be very good if the basis signals e_i are chosen appropriately. With this motivation, we look for a wavelet based method for parameter estimation. The idea is to store a finite set of the wavelet coefficients $c(n, k) = \int_{-\infty}^{\infty} y(t)\psi_{n,k}(t)dt, n, k \in \mathbb{Z}$ rather than the signal samples and base our parameter estimation algorithm on this storage. One way to do this for the above second order differential equation would be to express the output when the input is zero as $y(t) = F(t, \omega, \gamma)$ where F is a damped sinusoid. Then we discretize the possible range of the parameter vector (ω, γ) and for each (ω, γ) in this finite set D, we compute the wavelet coefficients

$$c(n, k|\omega, \gamma) = \int_{-\infty}^{\infty} F(t, \omega, \gamma)\psi_{n,k}(t)dt$$

Now, given measurements $y(t)$ of the output, we compute its wavelet coefficients

$$c(n, k) = \int_{-\infty}^{\infty} y(t)\psi_{n,k}(t)dt$$

and store these coefficients. The range R of the index pair (n, k) will have to be chosen properly depending on the duration of the signal $y(t)$ and its maximum frequency. If this selection is made appropriately, then very few coefficients would be required. Having done so, we select that value of (ω, γ) in the finite set D, so that

$$\sum_{(n,k)\in R} (c(n, k) - c(n, k|\omega, \gamma))^2$$

is a minimum.

From the above discussion, it is clear that an important step in system identification is to derive a parametric model for the system. Usually, the system is described by a system of differential equations on solving which, we obtain an input output equation. If the system of differential equations is nonlinear, it may be difficult and even impossible to obtain a closed form input-output equation. In this case, we look for an approximate input-output relation. One very powerful method for deriving this is perturbation theory. Herein we separate out the nonlinear parts in the differential equations from the linear part and introduce a fictitious parameter ϵ into the nonlinear part and assume that ϵ is small (signifying a weak nonlinearity). We then expand the output as a Taylor series in the parameter ϵ, subsitute this into the given differential equations and equate equal powers of ϵ on both sides. Doing so results in a sequence of linear differential equations, one for each coefficient of ϵ in the perturbation series for the output. This is a linear differential equation for the given coefficient but it will in general contain nonlinear functions of the lower order coefficients. We can regard this as an auxiliary linear system for the given coefficient with the lower order coefficients constituting the driving forces. Thus, it is possible to sequentially solve for each coefficient (also called perturbation term) and by truncating the perturbation series to a high enough order, we can arrive at a good approximate input output equation on which we can base our parameter estimation algorithms. The idea of using perturbation theory to obtain approximate solutions to nonlinear differential equations goes back to quantum mechanics where perturbation theory is used to obtain the approximate state evolution equation when the Hamiltonian is perturbed by a small time dependent term. The approximate state evolution equation so obtained can then be used to compute transition probabilities for the system from one stationary state to another of the unperturbed system after the perturbation has been switched on. To summarize this method in the case of nonlinear differential equations, we consider a system with states $x_i(t), i = 1, 2, ..., n$ satisfying the differential equations

$$\frac{dx_i(t)}{dt} = f_i(x_1(t), ..., x_n(t)) + u_i(t), i = 1, 2, ..., n$$

$u_i's$ are given inputs and $f_i : \mathbb{R}^n \to \mathbb{R}$ is a smooth nonlinear function. We assume that f_i can be decomposed as

$$f_i(x_1, ..., x_n) = \sum_{i=1}^{n} a_{ij}x_j + \epsilon.g_i(x_1, ..., x_n), i = 1, 2, ..., n$$

where g_i is a nonlinear function and ϵ is a small parameter. a_{ij} are constants. We try a solution

$$x_i(t) = x_{i0}(t) + \epsilon.x_{i1}(t) + O(\epsilon^2)$$

Substituting this into the above differential equations and equating coefficients of $\epsilon^i, i = 0, 1$ respectively gives us

$$\frac{dx_{i0}}{dt} = \sum_{j=1}^{n} a_{ij}x_{j0}(t) + u_i(t), i = 1, 2, ..., n,$$

$$\frac{dx_{i1}}{dt} = \sum_{j=1}^{n} a_{ij}x_{j1}(t) + g_i(x_{10}(t), ..., x_{1n}(t)), i = 1, 2, ..., n$$

These two equations can be expressed in vector notation as

$$\frac{d\mathbf{x}_0(t)}{dt} = \mathbf{A}\mathbf{x}_0(t) + \mathbf{u}(t),$$

$$\frac{d\mathbf{x}_1(t)}{dt} = \mathbf{A}\mathbf{x}_1(t) + \mathbf{g}(\mathbf{x}_0(t))$$

where

$$\mathbf{x}_0(t) = [x_{10}(t), x_{20}(t), ..., x_{n0}(t)]^T, \mathbf{x}_1(t) = [x_{11}(t), x_{21}(t), ..., x_{n1}(t)]^T$$

and the solutions for zero initial state can be obtained as

$$\mathbf{x}_0(t) = \int_0^t exp((t - \tau)\mathbf{A})\mathbf{u}(\tau)d\tau,$$

$$\mathbf{x}_1(t) = \int_0^t exp((t - \tau)\mathbf{A})\mathbf{g}(\mathbf{x}_0(\tau))d\tau$$

$$= \int_0^t exp((t - \tau)\mathbf{A})\mathbf{g}(\int_0^\tau exp((\tau - \alpha)\mathbf{A})\mathbf{u}(\alpha)d\alpha)d\tau$$

In the second part of this thesis, we consider a transistor amplifier circuit and attempt to estimate the cutin voltage V_T of the transistor from input output data. The first step in doing this is to derive using perturbation theory for differential equations, an approximate Volterra model for the system with the first and second order Volterra kernels being obtained as explicit functions of the transistor parameter V_T. The Volterra model has the form

$$y(t) = \int h(\tau, V_T)x(t - \tau)d\tau + \int g(\tau_1, \tau_2, V_T)x(t - \tau_1)x(t - \tau_2)d\tau_1 d\tau_2$$

where V_T is the transistor cutin voltage, ie, the parameter to be estimated. After discretizing this input-output equation, we get

$$y(n\Delta) = \sum_k h(k\Delta, V_T)x((n-k)\Delta)\Delta + \sum_{k,m} g(k\Delta, m\Delta, V_T)x((n-k)\Delta)x((n-m)\Delta)\Delta^2$$

and V_T can be estimated using the LMS algorithm

$$\hat{V}_T[n+1] = \hat{V}_T(n) - \mu.\frac{\partial}{\partial \hat{V}_T[n]}(y(n\Delta) - \sum_k h(k\Delta, \hat{V}_T[n])x((n-k)\Delta)\Delta$$

$$- \sum_{k,m} g(k\Delta, m\Delta, \hat{V}_T[n])x((n-k)\Delta)x((n-m)\Delta)\Delta^2)^2$$

with μ as the adaptation constant. Once again, as in the previous case, this estimation can be carried out with lesser data storage by passing over to the wavelet domain. This means that we represent the two Volterra kernels h, g as

$$h(\tau, V_T) = \sum_{n,k} c(n, k, V_T)\psi_{n,k}(\tau), g(\tau_1, \tau_2, V_T) = \int_{n,k,r,s} d(n, k, r, s)\psi_{n,k}(\tau_1)\psi_{r,s}(\tau_2)d\tau_1 d\tau_2$$

The wavelet coefficients c, d can be computed using

$$c(n, k, V_T) = \int h(\tau, V_T)\psi_{n,k}(\tau)d\tau, d(n, k, r, s, V_T) = \int g(\tau_1, \tau_2, V_T)\psi_{n,k}(\tau_1)\psi_{r,s}(\tau_2)d\tau_1 d\tau_2$$

To compute these integrals, we must discretize them to give

$$c(n, k, V_T) \approx \sum_n h(m\Delta, V_T)\psi_{n,k}(m\Delta).\Delta, d(n, k, r, s, V_T) \approx \sum_{m_1, m_2} g(m_1\Delta, m_2\Delta, V_T)\psi_{n,k}(m_1\Delta)\psi_{r,s}(m_2\Delta)\Delta^2$$

Substituting the wavelet representation of the Volterra kernels into the input-output equation gives us

$$y(t) = \sum_{n,k} c(n, k, V_T)\int x(t-\tau)\psi_{n,k}(\tau)d\tau + \sum_{n,k,r,s} d(n, k, r, s, V_T)\int x(t-\tau_1)x(t-\tau_2)\psi_{n,k}(\tau_1)\psi_{r,s}(\tau_2)d\tau_1 d\tau_2$$

Taking the wavelet transform once again on both sides gives

$$f(p, q) = \int y(t)\psi_{p,q}(t)dt = \sum_{n,k} c(n, k, V_T)\int x(t-\tau)\psi_{n,k}(\tau)\psi_{p,q}(t)dtd\tau$$

$$+ \sum_{n,k,r,s} d(n, k, r, s, V_T)\int x(t-\tau_1)x(t-\tau_2)\psi_{n,k}(\tau_1)\psi_{r,s}(\tau_2)\psi_{p,q}(t)dtd\tau_1 d\tau_2$$

This equation represents the Volterra system in the wavelet domain and the corresponding data storage to implement the parameter estimation in the wavelet domain is much less than that compared to the direct time domain least squares. The practical implementation of this algorithm is as follows. We discretize the range of the parameter V_T and denote this discretized range by Ω. For each $V_T \in \Omega$, we compute

$c(n, k, r, s, V_T), d(n, k, r, s, V_T)$ as given above and store these values. Now suppose the same circuit with V_T unknown is given. We take measurements on the input $x(t)$ and output $y(t)$ and compute

$$A(p, q|n, k) = \int x(t - \tau)\psi_{n,k}(\tau)\psi_{p,q}(t)dtd\tau,$$

$$B(p, q|n, k, r, s) = \int x(t - \tau_1)x(t - \tau_2)\psi_{n,k}(\tau_1)\psi_{r,s}(\tau_2)\psi_{p,q}(t)dtd\tau_1 d\tau_2$$

and

$$f(p, q) = \int y(t)\psi_{p,q}(t)dt$$

for $(n, k), (r, s) \in D$ and $(p, q) \in E$, with the number of index pairs $\mu(E)$ in E being at least $\mu(D)^2$ where $\mu(D)$ is the number of index pairs in D. We then obtain estimates for $c(n, k), d(n, k, r, s)$ using the wavelet domain input output equation

$$f(p, q) = \sum_{(n,k)\in D} A(p, q|n, k)c(n, k) + \sum_{(n,k),(r,s)\in D} B(p, q|n, k, r, s)d(n, k, r, s), (p, q) \in E$$

We then match the quantities $c(n, k), d(n, k, r, s)$ respectively with $c(n, k, V_T)$ and $d(n, k, r, s, V_T)$ using either the least squares method combined with a search algorithm or by a gradient based scheme to obtain an optimal estimate for V_T.

This method can be generalized to higher order Volterra systems. Suppose for example, we wish to estimate the parameter θ in a system defined by the input-output equation

$$y(t) = \sum_{m=1}^{p} \int h_m(\tau_1, ..., \tau_m, \theta)x(t - \tau_1)...x(t - \tau_m)d\tau_1...d\tau_m$$

we set

$$c_m(n_1, k_1, ..., n_m, k_m, \theta) = \int h_m(\tau_1, ..., \tau_m, \theta)\psi_{n_1,k_1}(\tau_1)...\psi_{n_m,k_m}(\tau_m)d\tau_1...d\tau_m$$

and

$$f(p, q) = \int y(t)\psi_{p,q}(t)dt,$$

$$A_m(p, q|n_1, k_1, ..., n_m, k_m) = \int x(t - \tau_1)...x(t - \tau_m)\psi_{n_1,k_1}(\tau_1)...\psi_{n_m,k_m}(\tau_m)\psi_{p,q}(t)dtd\tau_1...d\tau_m$$

Then, the input output equation in the wavelet domain has the form

$$f(p, q) = \sum_{m,n_1,k_1,...,n_m,k_m} A_m(p, q|n_1, k_1, ..., n_m, k_m)c_m(n_1, k_1, ..., n_m, k_m, \theta)$$

and this equation can be used as the starting point for developing algorithms to estimate θ.

[285] Conclusions of the thesis. This thesis has developed a wavelet based method for estimating the parameter V_T in a transistor amplifier circuit. More generally, the method can be extended for estimating the other transistor parameters appearing the the Ebers-Moll model, ie, $I_{C0}, I_{EO}, \alpha_C, \alpha_E$. The method

can also be used for estimating parameters of linear and nonlinear systems occurring in problems of mechanics, fluid dynamics and quantum mechanics. In quantum mechanics for example, the Hamiltonian operator will generally have the form $H = H_0 + V(t, \theta)$ where H_0 is the Hamiltonian of the unperturbed system and $V(t, \theta)$ is the perturbation depending on an unknown parameter θ. The problem is to estimate the parameter θ by taking measurements on the observables of the quantum mechanical system. For example, we can use time dependent perturbation theory to obtain an approximate evolution of the wave function in time and then calculate the approximate transition probabilities between stationary states of the unperturbed system during the time duration $[0, t]$. These expressions will be functions of θ. A wavelet based method along the lines developed in the thesis will then enable us to obtain estimates of the parameter θ. To be more precise, let the wave functions be expanded upto $O(\epsilon)$ so that

$$\psi(t) = \psi^{(0)}(t) + \epsilon.\psi^{(1)}(t) + O(\epsilon^2)$$

Plugging this into the Schrodinger evolution equation

$$\psi'(t) = (-2\pi i/h)(H_0 + \epsilon V(t, \theta))\psi(t)$$

gives us upto $O(\epsilon)$,

$$\psi^{(0)'}(t) = (-2\pi i/h)H_0\psi^{(0)}(t), \psi^{(1)'}(t) = (-2\pi i/h)(H_0\psi^{(1)}(t) + V(t, \theta)\psi^{(0)}(t))$$

Assume that the stationary states of the unperturbed Hamiltonian H_0 are $|n>, n = 1, 2, ...$ with corresponding energy levels $E_n, n = 1, 2, ...$ Assume that at $t = 0$, the state of the system is $|n>$. Then,

$$\psi^{(0)}(t) = exp(-2\pi i E_n t/h)|n>$$

$$\psi^{(1)}(t) = (-2\pi i/h) \int_0^t exp(-2\pi i H_0(t - s)/h))V(s, \theta)\psi^{(0)}(s)ds$$

It follows that upto $O(\epsilon^2)$ the probability for the system to make a transition from the state $|n>$ to the state $|m>$ with $m \neq n$ under the influence of the perturbation in the duration $[0, t]$ is given by

$$P_{n \to m}(t) = |<m|\epsilon\psi^{(1)}(t)>|^2 = (2\pi\epsilon/h)^2|\int_0^t <m|V(s, \theta)|n> exp(2\pi i(E_m - E_n)s)ds|^2$$

By measuring these transition probabilities as a function of time t, we can using the wavelet method estimate the parameter θ Suppose X is an observable on the quantum system. Then the change upto $O(\epsilon)$ in its average at time t given that the system started in the state $|n>$ is given by

$$<X>(t) = \epsilon(<\psi^{(0)}(t)|X|\psi^{(1)}(t)> + <\psi^{(1)}(t)|X|\psi^{(0)}(t)>= 2\epsilon.Re(<\psi^{(0)}(t)|X|\psi^{(1)}(t)>$$

$$= 2\epsilon.Re((-2\pi i/h) \int_0^t <\psi^{(0)}(t)|X.exp(-2\pi i H_0(t - s)/h)V(s, \theta)|\psi^{(0)}(s)> ds)$$

$$= 2\epsilon.Re((-2\pi i/h)exp(2\pi i E_n t/h). \int_0^t <n|X.exp(-2\pi i H_0(t - s)/h)V(s, \theta)|n> exp(-2\pi i E_n s/h)ds)$$

$$= \sum_k 2\epsilon.Re((-2\pi i/h)exp(2\pi i(E_n - E_k)t/h) \int_0^t <n|X|k><k|V(s, \theta)|n> exp(2\pi i(E_k - E_n)s/h)ds)$$

Using this equation and measurements on $< X > (t)$ at different times t, algorithms for estimating the parameter θ can be developed. Suppose

$$V(t, \theta) = \phi(t)V_0(\theta)$$

where $V_0(\theta)$ is a constant operator depending on the parameter θ and $\phi(t)$ is a scalar function of t. Then from the above formula,

$$< X > (t) = \int_0^t K(t, s, \theta)\phi(s)ds$$

where

$$K(t, s, \theta) = \sum_k [2\epsilon.Re((-2\pi i/h)exp(2\pi i(E_n - E_k)(t - s)/h) < n|X|k >< k|V_0(\theta)|n >]$$

The problem thus reduces to the problem of estimating the parameters of an LTI system. By taking into consideration $O(\epsilon^2)$ terms, it can be shown that $< X > (t)$ is related to $\phi(t)$ by a second order Volterra system and hence, the wavelet method for estimating the parameters of such systems developed in this thesis can be applied to the problem of estimating the parameter θ on which the perturbing potential depends.

The methods developed in this thesis can in principle be applied to amplifier circuits involving the cascade connection of several stages. In each stage, there is one transistor and the problem is to estimate the V_T for each transistor. The KCL and KVL for the circuit is written down taking into account the Ebers-Moll model for each transistor having a different V_T. Then by expressing the Ebers-Moll characteristic for each transistor as the sum of a linear and nonlinear component, the usual method of perturbation theory is developed to express the output in the form of a second order Volterra system

$$y(t) = \int_0^t h_1(\tau, V_{T1}, ..., V_{Tk})x(t - \tau)d\tau + \int_0^t \int_0^t h_2(\tau_1, \tau_2, V_{T1}, ..., V_{Tk})x(t - \tau_1)x(t - \tau_2)d\tau_1 d\tau_2$$

This equation is transformed into a second order equation for the wavelet coefficients of h_1, h_2 and the familiar gradient method is applied to minimize the sum of error squared with respect to $V_{T1}, ..., V_{Tk}$. This can be further generalized by taking into account higher order perturbation terms leading to a Volterra model of order r where $r = 2, 3, ...$ Such a model has the form

$$y(t) = \sum_{m=1}^{r} \int_{[0,t]^m} h_m(\tau_1, ..., \tau_m, V_{T1}, ..., V_{Tk})x(t - \tau_1)...x(t - \tau_m)d\tau_1...d\tau_m$$

This setup can be further generalized to multichannel Volterra systems obtained by applying perturbation theory to circuits involving several inputs and several outputs. For example, if the inputs are $x_i(t), i = 1, 2, ...M$ and the outputs are $y_i(t), i = 1, 2, ..., N$, then a second order Volterra model will have the form

$$y_i(t) = \sum_{j=1}^{M} \int_0^t h_{1,i,j}(\tau, \theta)x_j(t - \tau)d\tau + \sum_{j,m=1}^{M} \int_{[0,t]^2} h_{2,i,j,m}(\tau_1, \tau_2, \theta)x_j(t - \tau_1)x_m(t - \tau_2)d\tau_1 d\tau_2.i = 1, 2, ..., N$$

where θ is the unknown parameter to be estimated. The wavelet coefficients of the first and second order filter coefficients are

$$c_{ij}(n,k,\theta) = \int h_{1,i,j}(\tau,\theta)\psi_{nk}(\tau)d\tau,$$

$$d_{ijm}(n,k,p,q,\theta) = \int h_{2,i,j,m}(\tau_1,\tau_2,\theta)\psi_{nk}(\tau_1)\psi_{pq}(\tau_2)d\tau_1 d\tau_2$$

or equivalently,

$$h_{1,i,j}(\tau,\theta) = \sum_{n,k} c_{ij}(n,k,\theta)\psi_{nk}(\tau),\, h_{2,i,j,m}(\tau_1,\tau_2,\theta) = \sum_{n,k,p,q} d_{ijm}(n,k,p,q,\theta)\psi_{nk}(\tau_1)\psi_{pq}(\tau_2)$$

By plugging these expressions into the input-output equation, and then taking the wavelet transform of the output, we can derive a wavelet domain representation of this MIMO (Multi-input multi-output) system and use this representation to estimate the parameter θ.

[286] Frequency modulation. The message signal is $m(t)$ and ω_c is the carrier frequency. The FM signal is

$$x(t) = A.cos\left(\omega_c t + K\int_0^t m(s)ds\right) = A.Re\left(exp(j\omega_c t).exp\left(jK\int_0^t m(s)ds\right)\right)$$

Now,

$$exp\left(jK\int_0^t m(s)ds\right) = 1 + \sum_{n=1}^{\infty}(jK)^n \int_{0<s_1<...<s_n<t} m(s_n)m(s_{n-1})...m(s_1)ds_1 ds_2...ds_n$$

If $m(t)$ is white noise, then $\int_0^t m(s)ds$ is Brownian motion $B(t)$ and

$$x(t) = A.cos(\omega_c t + K.B(t))$$

Let

$$y(t) = A.sin(\omega_c t + K.B(t))$$

Then,

$$dx(t) = -y(t)(\omega_c dt + K.dB(t)) - K^2 x(t)dt/2,$$

$$dy(t) = x(t)(\omega_c dt + K dB(t)) - K^2 y(t)dt/2$$

Equivalently,

$$dx(t) = -(\omega_c y(t) + K^2 x(t)/2)dt - K y(t)dB(t),$$

$$dy(t) = \omega_c x(t) - K^2 y(t)/2)dt + K x(t)dB(t)$$

This system of stochastic differential equations can be solved numerically.

Chapter 2

Mechanics

Introduction: This chapter discusses problems in classical mechanics including motion of particles and rigid bodies subject to forces, differentiable manifolds, symplectic geometry, Hamiltonian and Lagrangian formulations of mechanics, mechanical systems with noisy perturbations and simulation of mechanical systems using the discretization scheme.

[1].Motion on an inclined plane: Assume that the inclined plane has an angle α and that its mass is M. Assume that a smooth ball of mass m rolls on the plane and let I denote the moment of inertia of the ball. After time t the ball has slid a distance $x(t)$ down the plane and the plane has moved a distance $y(t)$ horizontally on the smooth table. Let L denote the length of the inclined portion of the plane. Then the position of the ball after time t is given by

$$(y(t) + x(t).cos(\alpha), x(t).sin(\alpha))$$

and hence its translational kinetic energy is given by

$$T_{ball} = \frac{m}{2}(x'^2 + y'^2 + 2x'y'cos(\alpha))$$

The rotational kinetic energy of the ball is given by $IR^2x'^2/2$ since its angular velocity is x'/R. The Lagrangian of the system comprising the plane and the ball is therefore given by

$$L(x, y, x', y') = \frac{m}{2}(x'2 + y'^2 + 2x'y'.cos(\alpha)) + Mx'^2/2 - mgx.sin(\alpha)$$

The Euler-Lagrange equations of motion can now be immediately written down from this expression. In the presence of random forces like friction, the equations of motion assume the form

$$m(x'' + y''.cos(\alpha)) + Mx'' + mg.sin(\alpha) = f_2(t),$$

$$m(y'' + x''.cos(\alpha)) = f_2(t)$$

[2].Define the notions of static friction and kinetic friction by means of an example.

ans:Suppose a block of mass M is at rest on a table. The normal reaction is $N = Mg$. In order to make the block start moving we need to apply a horizontal force given by $\mu_s N$ where μ_s is the coefficient of static friction. This friction arises due to the fact that the block and the table when looked at through a microscope have teethed surfaces and hence they interlock. When the body is at rest, this interlock is much more deep that what it would be if the block were in motion. When the block is in motion, the force needed to continue keeping it in motion is $\mu_k N$. In view of what we just said, we have that $\mu_k < \mu_s$.

[3] Problems in classical mechanics.

1.Define a differentiable manifold, chart, atlas.

2.Define a curve on a differentiable manifold.

3.Define the differential of a curve on a manifold and use this to define a tangent vector to the manifold at a given point as a differential operator acting on differentiable functions.

4.What do you understand by a differential form on a manifold. Give an example taking the manifold as $\mathbb{R}^n$.

5.Define the tangent manifold.

6.Define a vector bundle and explain how the tangent bundle is an example of a vector bundle.

7. What do you understand by a submanifold of a manifold.

8. What do you understand by an imbedding of one differentiable manifold in another.

9. What do you understand by the cartesian product of two differentiable manifolds. How do you construct the tangent space to a point in the cartesian product of two differentiable manifolds.

10. How do you define the differential of an exterior form on a manifold.

ans: Suppose

$$\omega = \sum_{i_1,\ldots,i_k} \omega_{i_1,\ldots,i_k}(x) dx^{i_1} \wedge \ldots \wedge dx^{i_k}$$

where $\omega_{i_1,\ldots,i_k}(x)$ are differentiable functions. Then, the differential of ω is defined by

$$d\omega = \sum_{i_1,\ldots,i_k,j} \left(\frac{\partial \omega_{i_1,\ldots,i_k}}{\partial x^j}\right) dx^j \wedge dx^{i_1} \wedge \ldots \wedge dx^{i_k}$$

11. What do you understand by a derivation of an algebra. Show that a vector field on a manifold is a derivation.

hint: Suppose $t \to c(t)$ is a curve on a differentiable manifold M. Let $f : M \to \mathbb{R}$ be a differentiable function. Then we write $X(c(t)) = c'(t)$ if $\frac{d}{dt}f(c(t)) = X(c(t))(f)$. Hence $X(c(t))$ satisfies

$$X(c(t))(fg) = f(c(t))X(c(t))g + g(c(t))X(c(t))f$$

12. What do you understand by a closed form and an exact form on a manifold. Give an example taking the manifold as $\mathbb{R}^3$. What is Darboux's theorem on closed and exact forms.

hint: A differentiable form ω is said to be closed if $d\omega = 0$. The form ω is said to be exact if there exists another form α such that $\omega = d\alpha$.

13. Define a connection on a differentiable manifold.

hint: Let M be a differentiable manifold and let $\chi(M)$ be the space of all vector fields on M. A connection is a map $\nabla : \chi(M) \times \chi(M) \rightarrow \chi(M)$ such that $(X, Y) \rightarrow \nabla_X Y$ satisfying (a) $\nabla_{fX}(Y) = f\nabla_X(Y)$ and (b) $\nabla_X(fY) = X(f)Y + f\nabla_X(Y)$ for all differentiable functions f on M.

14. Suppose G is a matrix Lie group. Show that the one form $\omega = g^{-1}dg$ is invariant under left translations.

ans: We have for $h \in G$, $(hg)^{-1}d(hg) = g^{-1}h^{-1}hdg = g^{-1}dg$.

15. Show that on a matrix Lie group G, the one form $dg.g^{-1}$ is right invariant.

ans:

$$d(gh).(gh)^{-1} = dg.h.h^{-1}g^{-1} = dg.g^{-1}$$

16. Let M be a symplectic manifold with local coordinates $(q_1, ..., q_n, p_1, ..., p_n)$ and symplectic form

$$\omega = \sum_{i=1}^{n} dq_i \wedge dp_i$$

Let $H : M \rightarrow \mathbb{R}$ be a smooth function. Determine the components of the vector field X_H defined by the equation

$$dH(Y) = \omega(X_H, Y)$$

ans: Let

$$Y = \sum_i Y^i \frac{\partial}{\partial q_i} + \sum_i Z^i \frac{\partial}{\partial p_i}$$

and also let

$$X_H = \sum_i U^i \frac{\partial}{\partial q_i} + \sum_i V^i \frac{\partial}{\partial p_i}$$

where Y^i, Z^i, U^i, V^i are functions of $q_1, ..., q_n, p_1, ..., p_n$. Then,

$$\omega(X_H, Y) = \sum_i (U^i Z^i - V^i Y^i)$$

on the one hand and on the other,

$$dH(Y) = \sum_i \left(\frac{\partial H}{\partial q_i} Y^i + \frac{\partial H}{\partial p_i} Z^i \right)$$

Equating these two expressions and assuming Y^i, Z^i are arbitrary functions on M, it follows that

$$V^i = -\frac{\partial H}{\partial q_i}, U^i = \frac{\partial H}{\partial p_i}$$

16.Suppose M is a two dimensional manifold with local coordinates (x^1, x^2). Consider the one form ω on M defined by the equation

$$\omega = \omega_1 dx^1 + \omega_2 dx^2$$

State Stokes theorem for this form.

hint:

$$d\omega = (\omega_{2,1} - \omega_{1,2})dx^1 \wedge dx^2$$

so that

$$\int_M d\omega = \int (\omega_{2,1} - \omega_{1,2})dx^1 dx^2$$

on the one hand and on the other,

$$\int_{\partial M} \omega = \int_\Gamma (\omega_1 dx^1 + \omega_2 dx^2)$$

where $\Gamma = \partial M$ is the boundary of M.

Ref:Abraham and Marsden, Foundations of mechanics.

[4].Suppose a dynamical system is defined by the n coupled first order ordinary differential equations

$$\frac{dx_i}{dt} = f_i(x_1, ..., x_n, t), i = 1, 2, ..., n$$

An observable $\phi(x_1, ..., x_n, t)$ is given and we wish to compute its rate of change along the trajectory of the dynamical system. Show that

$$\frac{d\phi}{dt} = \sum_i \phi_{,i} x'_i + \phi_{,t} = \sum_i \phi_{,i} f_i + \phi_{,t}$$

where the notations

$$\phi_{,i} = \frac{\partial \phi}{\partial x_i}, \phi_{,t} = \frac{\partial \phi}{\partial t}$$

for the partial derivatives are used. Show further that

$$\frac{d^2 \phi}{dt^2} = \phi_{,ij} f_i f_j + \phi_{,i}(f_{,ij} f_j + f_{i,t}) + 2\phi_{,it} f_i + \phi_{,tt}$$

with summation over the repeated indices i, j being implied.

[5].Write down in component form the equations of motion for an n body system moving under the inverse square law of gravitational forces between the bodies.

ans:

$$x_i'' = \sum_{j:j\neq i} m_j(x_j - x_i)/r_{ij}^3, \quad y_i'' = \sum_{j:j\neq i} m_j(y_j - y_i)/r_{ij}^3, \quad z_i'' = \sum_{j:j\neq i} m_j(z_j - z_i)/r_{ij}^3$$

[6]. Consider a mechanical system with V a potential function and the equations of motion

$$x_i'' = -V_{,i}, i = 1, 2, ..., n$$

where $V = V(x_1, ..., x_n) = V(x)$. Define

$$I = \sum_{i=1}^{n} x_{i'2}$$

Compute I' and I''. Evaluate it for a system of n particles moving under the inverse square law of gravitation of each other.

hint:

$$I' = 2\sum_i x_i' x_i'' = -2\sum_i x_i' V_{,i}$$

$$I'' = -2(\sum_i x_i'' V_{,i} + \sum_{i,j} x_i' x_j' V_{,ij}) = 2\sum_i V_{,i}^2 - 2\sum_{i,j} V_{,ij} x_i' x_j'$$

For the inverse square law, we have the equations of motion

$$x_i'' = \sum_{j\neq i} m_j(x_j - x_i)/r_{ij}^3, \quad y_i'' = \sum_{j\neq i} m_j(y_j - y_i)/r_{ij}^3, \quad z_i'' = \sum_{j\neq i} m_j(z_j - z_i)/r_{ij}^3$$

In this case, we set

$$I = \sum_i m_i(x_i^2 + y_i^2 + z_i^2)$$

and find that

$$I' = 2\sum_i m_i(x_i x_i' + y_i y_i' + z_i z_i')$$

$$I'' = 2\sum_i m_i(x_i'^2 + y_i'^2 + z_i'^2) + 2\sum_i m_i(x_i x_i'' + y_i y_i'' + z_i z_i'')$$

$$= 4T + 2\sum_i (x_i f_{ix} + y_i f_{iy} + z_i f_{iz})$$

where

$$f_{ix} = \sum_{j\neq i} m_j(x_j - x_i)/r_{ij}^3, \quad f_{iy} = \sum_{j\neq i} m_j(y_j - y_i)/r_{ij}^3, \quad f_{iz} = \sum_{j\neq i} m_j(z_j - z_i)/r_{ij}^3$$

and

$$r_{ij} = ((x_i - x_j)^2 + (y_i - y_j)^2 + (z_i - z_j)^2)^{1/2}$$

Thus,

$$\sum_i x_i f_{ix} = \sum_{i \neq j} m_j x_i (x_j - x_i)/r_{ij}^3, \sum_i y_i f_{iy} = \sum_{i \neq j} m_j y_i (y_j - y_i)/r_{ij}^3, \sum_i z_i f_{ij} = \sum_{i \neq j} m_j z_i (z_j - z_i)/r_{ij}^3$$

[7]. A particle moves in accord with damping and fluctuating forces. The equation of motion is

$$dx(t) = v(t)dt, dv(t) = -\gamma.v(t)dt - U'(x(t)) + \sigma.dB(t)$$

where $U(.)$ is a potential and $B(.)$ is standard Brownian motion. Show that the probability density $p(t, x, v)$ of the position-velocity pair $(x(t), v(t))$ satisfies the Fokker-Planck equation

$$p_{,t} = -vp_{,x} + \gamma(vp)_{,v} + U'(x)p_{,v} + (\sigma^2/2)p_{,vv}$$

Solve this equation when $U(x) = -Kx$.

[8]. Suppose the potential function $U(x_1, ..., x_n)$ is a homogeneous function of the $x_i's$ of degree d. Then show that

$$\sum_i x_i U_{,i} = d.U$$

Deduce the virial theorem from this when the equations of motion are given by

$$x_i'' = -U_{,i}, i = 1, 2, ..., n$$

hint:

$$U(ax_1, ..., ax_n) = a^d U(x_1, ..., x_n)$$

so that we get on differentiating both sides with respect to a at $a = 1$,

$$\sum_i x_i U_{,i}(x) = d.U(x)$$

Now the moment of inertia of the system is defined as

$$I = \sum_i x_i^2$$

so that

$$I' = 2\sum x_i x_i', I'' = 2\sum x_i'^2 + \sum x_i x_i'' = 4T - \sum_i x_i U_{,i} = 4T - d.U$$

where the kinetic energy T is defined by

$$T = \sum_i x_i'^2/2$$

It follows that if motion is bounded, ie, positions x_i and velocities x_i' are bounded, then

$$4 <T> -d<U> = 0$$

where $< \xi >$ denotes the time average of the quantity ξ:

$$< \xi >= lim_{T \to \infty} \frac{1}{T} \int_0^T \xi(t) dt$$

Note that if $H = T + U$ is the total energy, the above equation can be expressed as

$$I'' = 4T - d.(H - T) = (d + 4)T - d.H = 4(H - U) - d.U = 4.H - (d + 4).U$$

Since H is a constant along the system trajectory, it follows that if the positions and velocities remain bounded, then

$$< T >= \frac{d}{d + 4} H, < U >= \frac{4}{d + 4} H$$

These identities are very important in the study of the n-body problem.

[9].Solve using perturbation theory for differential equations

$$x'' + \omega^2 x + \epsilon f(x, x') = 0$$

Your answer must be expressed as a power series in the parameter ϵ. To get the solution upto $O(\epsilon)$, we consider

$$x(t) = x_0(t) + \epsilon.x_1(t)$$

and obtain on equating coefficients of ϵ^0 and ϵ^1,

$$x_0'' + \omega^2 x_0 = 0, x_1'' + \omega^2 x_1 + f(x_0, x_0') = 0$$

[10].Project on applications of differential geometry and Lie group theory to dynamical systems. (Reference:Helgason, Differential geometry, Lie groups and symmetric spaces).

(a) Given a vector field on a differentiable manifold $\mathcal{M}$ in component form, determine the corresponding integral curves.

(b) Given a dynamical system in terms of the Hamiltonian $H(\mathbf{q}, \mathbf{p})$, determine whether or not a function $F(\mathbf{q}, \mathbf{p})$ is a constant of the motion using the Poisson bracket formalism. Specifically, we must check whether the quantity

$$\{H, F\} = \sum_i (H_{,q_i} F_{,p_i} - H_{,p_i} F_{,q_i})$$

is close to zero or not. The Poisson bracket $\{H, F\}$ is evaluated using a numerical scheme of discretizing the position and momentum variables and approximating the partial derivatives of H and F with respect to these variables using finite differences.

Suppose that the Lagrangian of a system $L(q, q')$ is invariant under a one parameter group of transformations. Then, there is associated with this group a conserved quantity called the Noether current. Apply this principle to different dynamical systems. Suppose $\mathcal{M}$ is a differentiable manifold and

$\phi : \mathcal{M} \to \mathcal{M}$ is a diffeomorphism. The derive the transformation law of a vector field and of a one form in $\mathcal{M}$ under ϕ and use this in the analysis of mechanical systems defined by a Hamiltonian.

[11].Using MATLAB write an optimization routine to compute the trajectories of a Lagrangian dynamical system. Suppose $L(q, q')$ is the Lagrangian. Then let $q[n] = q(n\Delta)$ denote the time discretized coordinate. The continuous time trajectory $q(t)$ is obtained by extremizing $\int_0^T L(q(t), q'(t))dt$ and the corresponding discretized trajectory is obtained by setting $T = N\Delta$ and extremizing $\sum_{n=0}^{N-1} L(q[n], (q[n+1] - q[n])/\Delta)$ with respect to $q[1], q[2], ..., q[N-1]$ keeping $q[0]$ and $q[N]$ fixed. Simulate various kinds of Lagrangian dynamical systems using such a discretization and optimization procedure.

[12]. Develop numerical algorithms for simulating the motion of a spinning top, first in classical mechanics and then in quantum mechanics.

hint: Suppose I_1, I_2, I_3 are the moments of inertia of the top along principal axes. Suppose first that no external torque acts on the top. Let $\omega_1, \omega_2, \omega_3$ denote the angular velocity components of the top along the principal axes. Then the angular momenta components along the principal axes are $\mathbf{L} = (I_1\omega_1, I_2\omega_2, I_3\omega_3)$ and hence, the equations of motion

$$\frac{\partial \mathbf{L}}{\partial t} + \omega \times \mathbf{L} = \mathbf{0}$$

gives us

$$I_1\omega_1' + \omega_2\omega_3(I_3 - I_2) = 0, I_2\omega_2' + \omega_3\omega_1(I_1 - I_3) = 0, I_3\omega_3' + \omega_1\omega_2(I_2 - I_1) = 0$$

These are called the Euler equations of motion and they form a set of three coupled first order nonlinear differential equations. To simulate the motion, we discretize these differential equations to get

$$\omega_1[n+1] - \omega_1[n] = (\Delta/I_1)(I_2 - I_3)\omega_2[n]\omega_3[n],$$

$$\omega_2[n+1] - \omega_2[n] = (\Delta/I_2)(I_3 - I_1)\omega_3[n]\omega_1[n],$$

$$\omega_3[n+1] - \omega_3[n] = (\Delta/I_3)(I_1 - I_2)\omega_1[n]\omega_2[n]$$

where Δ is the discretization step size. In the presence of an external torque like the torque due to the gravitational field, we must use the Euler angles to simulate the motion. Let ϕ, θ, ψ be the Euler angles. Then any rotation can be expressed as $R(\phi, \theta, \psi) = R_z(\phi)R_x(\theta)R_z(\psi)$. The components of the angular velocity relative to the principal axes of the top are computed by using the fact that if $\mathbf{r}_0$ is the position vector of a point in the top at time $t = 0$, then after time t, it becomes

$$\mathbf{r}_t = R(\phi(t), \theta(t), \psi(t))\mathbf{r}_0$$

and is rate of change can be computed in terms of ϕ', θ', ψ'. Specifically,

$$d\mathbf{r}_t/dt = (\phi' R_{,\phi} + \theta' R_{,\theta} + \psi' R_{,\psi})\mathbf{r}_0$$

$$= (\phi' R_{,\phi} + \theta' R_{,\theta} + \psi' R_{,\psi})R^{-1}\mathbf{r}_t = \Omega_t\mathbf{r}_t$$

Ω_t is a 3×3 skew symmetric matrix and its nonzero entries give the angular velocity $\tilde{\omega}_t$ about the space set of axes. The angular velocity tensor with respect to the body set of axes is then given by $R^{-1}.\Omega_t.R$ and its nonzero components give the angular velocity components ω_t about the body set of axes. From these components, we can easily compute the kinetic energy of the spinning top using the moments of inertial about the body set of axes. Then the Lagrangian can be computed and the Euler-Lagrange equations of motion can be formulated. Once we have the Lagrangian of the top in the form $\mathcal{L}(\phi, \theta, \psi, \phi', \theta', \psi')$, it is straightforward to obtain the Hamiltonian via the Legendre transformation. Specifically, we compute the canonical momenta

$$p_\phi = \frac{\partial \mathcal{L}}{\partial \phi'}, p_\theta = \frac{\partial \mathcal{L}}{\partial \theta'}, p_\psi = \frac{\partial \mathcal{L}}{\partial \psi'}$$

solve for ϕ', θ', ψ' in terms of $(\phi, \theta, \psi, p_\phi, p_\theta, p_\psi)$ and then form the Hamiltonian

$$H(\phi, \theta, \psi, p_\phi, p_\theta, p_\psi) = p_\phi \phi' + p_\theta \theta' + p_\psi \psi' - \mathcal{L}$$

and then, apply the canonical quantization procedure to formulate the Schrodinger wave equation for the spinning top:

$$H(\phi, \theta, \psi, -ih\frac{\partial}{\partial \phi}, -ih\frac{\partial}{\partial \theta}, -ih\frac{\partial}{\partial \psi})\Psi(\phi, \theta, \psi)$$

$$= E\Psi(\phi, \theta, \psi)$$

[13]. Assume that a particle moves along a trajectory $t \to \mathbf{R}(t)$. Determine the curvature of this trajectory at any given point.

hint:Let ds denote the infinitesimal distance measure along the trajectory. Then $ds(t) = |d\mathbf{R}(t)| = |\mathbf{R}'(t)|dt$. Let $d\theta$ denote the angle between two inifinitesimally separated tangent vector on the trajectory. Then

$$d\theta(t) = sin(d\theta(t)) =$$

$$|\mathbf{R}'(t) \times (\mathbf{R}'(t) + \mathbf{R}''(t)dt)/|\mathbf{R}'(t)|^2 = |\mathbf{R}'(t) \times \mathbf{R}''(t)|dt/|\mathbf{R}'(t)|^2$$

The curvature of the trajectory at a given point on it is defined by

$$K = d\theta/ds$$

Thus,

$$K(t) = |\mathbf{R}'(t) \times \mathbf{R}''(t)|/|\mathbf{R}'(t)|^3$$

[14] Projects on dynamical systems.

[1] Give examples of C^∞-manifolds along with local charts. What is the tangent plane TM_x to a manifold M at a given point $x \in M$. How do you define tangent vectors in terms of equivalence classes of curves. Show that this definition of tangent vectors is independent of the choice of the local chart.

[2] Give examples of fibre bundles. Show that the tangent bundle associated to a differentiable manifold is a fibre bundle.

[3] Suppose ω is a symplectic form on a differentiable manifold M. Let H be a smooth function on M and define a vector field X_H on M so that

$$dH(x)(Y) = \omega(X_H, Y)(x)$$

for all vector fields Y. Show how this definition of the Hamiltonian vector field X_H can be used to define the Hamiltonian equations of motion using coordinates.

[15] Determine the general solution for the displacement of the forced damped harmonic oscillator:

$$x''(t) + \gamma x'(t) + \omega_0^2 x(t) = f(t)$$

Separate out the natural response part and the forced response part. Note that the natural response is the general solution to the associated homogeneous equation while the forced response is a particular solution to the forced equation, namely that solution which is expressed as a linear functional of $f(.)$.

[16] Explain how the dissipative function can be introduced into the Euler-Lagrange equations of motion to take into account frictional damping.

hint:In the one dimensional case, consider

$$L(x, x') = mx'^2/2 - kx^2/2,\, F(x') = -\gamma x'^2/2$$

Then the equation of motion

$$\frac{d}{dt}\frac{\partial L}{\partial x'} - \frac{\partial L}{\partial x} = \frac{\partial F}{\partial x'}$$

gives

$$mx'' + \gamma x' + kx = 0$$

Generalize this to the multivariate case with

$$L(x_i, x_i', i = 1, 2, ..., n) = \sum_{i,j=1}^{n} K_{ij} x_i' x_j'/2 - \sum_{i,j=1}^{n} M_{ij} x_i x_j/2,$$

$$F(x_i', i = 1, 2, ..., n) = - \sum_{i,j=1}^{n} \gamma_{ij} x_i' x_j'/2$$

[17] Show that the motion of a particle in a central potential $U(r)$ gives two integrals of motion, namely the energy and angular momentum (first prove that motion takes place in a plane).

$$\frac{m}{2}(r'^2 + r^2\phi'^2) + U(r) = E,\, r^2\phi' = \beta/m$$

where E, β are constants. E is the energy while β is the angular momentum. Using these equations with $U(r) = -K/r$, obtain Rutherford's formula for the scattering cross section.

hint: Assume that the particle having mass m initially moves with a speed v and along a line parallel to the x axis at a distance s from the centre. s is the length of the perpendicular drawn from the origin to the initial line of motion. Then, $E = mv^2/2 - K/s$ while $\beta = msv$. Determine the final angle θ made by the particle's direction of motion at ∞ with the x axis. The scattering cross section is then given by $\sigma(\theta) = s.\frac{ds}{d\theta}$. Suppose that the initial flux of particles is I. Then the initial number of particles that flow across the cross section ds per unit time is $2\pi I s ds$ and hence after scattering, the number of particles that get scattered inside the solid angle $d\Omega = 2\pi sin(\theta)d\theta$ is given by $I(sin(\theta))^{-1}s(\theta)\frac{ds(\theta)}{d\theta}$.

[18] Derive an approximate discrete time algorithm for simulating the motion of two particles having masses m_1, m_2 moving in the potential field $U(\mathbf{r}_1, \mathbf{r}_2)$.

hint: The equations of motion are

$$\mathbf{r}_1' = \mathbf{v}_1, \mathbf{r}_2' = \mathbf{v}_2, m_1\mathbf{v}_1' = -\nabla_{\mathbf{r}_1}U(\mathbf{r}_1, \mathbf{r}_2), m_2\mathbf{v}_2' = -\nabla_{\mathbf{r}_2}U(\mathbf{r}_1, \mathbf{r}_2)$$

We define the forces acting on the particles

$$\mathbf{F}_1(\mathbf{r}_1, \mathbf{r}_2) = -\nabla_{\mathbf{r}_1}U(\mathbf{r}_1, \mathbf{r}_2), \mathbf{F}_2(\mathbf{r}_1, \mathbf{r}_2) = -\nabla_{\mathbf{r}_2}U(\mathbf{r}_1, \mathbf{r}_2)$$

so that the equations of motion after discretization yield

$$\mathbf{r}_1[k+1] = \mathbf{r}_1[k] + \Delta.\mathbf{v}_1[k], \mathbf{r}_2[k+1] = \mathbf{r}_2[k] + \Delta.\mathbf{v}_2[k],$$

$$\mathbf{v}_1[k+1] = \mathbf{v}_1[k] + \Delta.\mathbf{F}_1(\mathbf{r}_1[k], \mathbf{r}_2[k])/m_1,$$

$$\mathbf{v}_2[k+1] = \mathbf{v}_2[k] + \Delta.\mathbf{F}_2(\mathbf{r}_1[k], \mathbf{r}_2[k])/m_2$$

[19] Elastic Collision of two particles. Let $\mathbf{v}_1, \mathbf{v}_2$ denote the velocities of the two particles before the collision and $\mathbf{v}_1', \mathbf{v}_2'$, the same after the collision. Momentum conservation and energy conservation give

$$m_1\mathbf{v}_1 + m_2\mathbf{v}_2 = m_1\mathbf{v}_1' + m_2\mathbf{v}_2'$$

$$m_1|\mathbf{v}_1|^2 + m_2|\mathbf{v}_2|^2 = m_1|\mathbf{v}_1'|^2 + m_2|\mathbf{v}_2'|^2$$

The velocity of the centre of mass before the collision is given by

$$\mathbf{V} = \frac{m_1\mathbf{v}_1 + m_2\mathbf{v}_2}{m_1 + m_2}$$

and the velocity of the centre of mass after the collision is

$$\mathbf{V}' = \frac{m_1\mathbf{v}_1' + m_2\mathbf{v}_2'}{m_1 + m_2}$$

In view of the momentum conservation equation, we see that

$$\mathbf{V} = \mathbf{V}'$$

The velocities of the particles before the collision relative to the centre of mass frame are

$$\mathbf{u}_1 = \mathbf{v}_1 - \mathbf{V}, \mathbf{u}_2 = \mathbf{v}_2 - \mathbf{V}$$

and the velocities of the particles after the collision relative to the centre of mass frame are

$$\mathbf{u}_1' = \mathbf{v}_1' - \mathbf{V}, \mathbf{u}_2' = \mathbf{v}_2' - \mathbf{V}$$

The momentum conservation equations can thus be expressed as

$$m_1\mathbf{u}_1 + m_2\mathbf{u}_2 = \mathbf{0} = m_1\mathbf{u}_1' + m_2\mathbf{u}_2'$$

It follows that

$$\mathbf{u}_2 = -\frac{m_1\mathbf{u}_1}{m_2}, \mathbf{u}_2' = -\frac{m_1\mathbf{u}_1'}{m_2}$$

The relative velocity between the particles prior to the collision is

$$\mathbf{v} = \mathbf{v}_1 - \mathbf{v}_2 = \mathbf{u}_1 + \mathbf{V} - (\mathbf{u}_2 + \mathbf{V}) = \mathbf{u}_1 - \mathbf{u}_2 = (1 + m_1/m_2)\mathbf{u}_1,$$

and the relative velocity between the particles after the collision is

$$\mathbf{v}' = \mathbf{v}_1' - \mathbf{v}_2' = \mathbf{u}_1' - \mathbf{u}_2' = (1 + m_1/m_2)\mathbf{u}_1'$$

The energy conservation equation can be expressed as

$$m_1|\mathbf{V} + \mathbf{u}_1|^2 + m_2|\mathbf{V} + \mathbf{u}_2|^2 = m_1|\mathbf{V} + \mathbf{u}_1'|^2 + m_2|\mathbf{V} + \mathbf{u}_2'|^2$$

In view of the relation $m_1\mathbf{u}_1 + m_2\mathbf{u}_2 = \mathbf{0} = m_1\mathbf{u}_1' + m_2\mathbf{u}_2'$, this equation simplifies to

$$m_1|\mathbf{u}_1|^2 + m_2|\mathbf{u}_2|^2 = m_1|\mathbf{u}_1'|^2 + m_2|\mathbf{u}_2'|^2$$

Substituting for $\mathbf{u}_2$ and $\mathbf{u}_2'$ the expressions in terms of $\mathbf{u}_1$ and $\mathbf{u}_2$ respectively, we thus get

$$|\mathbf{u}_1|^2 = |\mathbf{u}_1'|^2$$

and therefore

$$|\mathbf{v}| = |\mathbf{v}'|$$

[20] Discretize the equations of motion of the restricted three body problem and implement this system using MATLAB.

hint: There are two centres of forces. The first is located at $(-\mu.cos(\omega t), -\mu.sin(\omega t))$ and the second at $((1-\mu).cos(\omega t), (1-\mu).sin(\omega t))$. Denote the former by $\rho(t)$ and the latter by $\sigma(t)$. Then the equations of motion are

$$\mathbf{r}''(t) = -\frac{K_1(\mathbf{r}(t) - \rho(t))}{|\mathbf{r}(t) - \rho(t)|^3} - \frac{K_2(\mathbf{r}(t) - \sigma(t))}{|\mathbf{r}(t) - \sigma(t)|^3}$$

This is a time varying nonlinear system.

[21] Consider the continuous time dynamical system: $\frac{dx(t)}{dt} = \mathbf{f}(\mathbf{x}(t))$. Let $\mathbf{x}_0(t)$ be a solution and consider a small perturbation around this solution $\mathbf{x}_0(t) + \delta\mathbf{x}(t)$. If this perturbation is to be a solution, then we must have approximately for $\delta\mathbf{x}(t)$ small,

$$\frac{d\mathbf{x}(t)}{dt} = \mathbf{A}(t)\delta\mathbf{x}(t)$$

where

$$\mathbf{A}(t) = ((a_{ij}(t))), a_{ij}(t) = \frac{\partial f_i(\mathbf{x}_0(t))}{\partial x_j}$$

This is called the linearized system. Give an iterative method of solving the linearized equation.

[22] Derive an approximate numerical method for solving the geodesic equation on a Riemannian manifold.

hint: Let $g_{ij}(x)$ be the metric,

$$\Gamma_{ijk}(x) = \frac{1}{2}(g_{ij,k} + g_{ik,j} - g_{jk,i})$$

the Christoffel symbol of the first kind and

$$\Gamma^i_{jk} = g^{im}\Gamma_{mjk}, ((g^{ij})) = ((g_{ij}))^{-1}$$

the Chistoffel symbol of the second kind. Then the geodesic equation is given by

$$\frac{d^2x^i(\tau)}{d\tau^2} + \Gamma^i_{jk}(x(\tau))\frac{dx^j(\tau)}{d\tau}\frac{dx^k(\tau)}{d\tau} = 0$$

Here τ is the proper time along the trajectory defined by the equation

$$d\tau^2 = g_{ij}(x(\tau))dx^i(\tau)dx^j(\tau)$$

The discretized equation is obtained by replacing τ with $n\Delta$, $x^i(\tau)$ with $x^i[n] = x^i(n\Delta)$, $\frac{dx^i(\tau)}{d\tau}$ with $(x^i[n+1] - x^i[n])/\Delta$ and $\frac{d^2x^i(\tau)}{d\tau^2}$ with $(x^i[n+2] - 2x^i[n+1] + x^i[n])/\Delta^2$.

[23] The equations of motion of a particle moving in the gravitational field of a spherical body are

$$\frac{d^2x_i}{dt^2} = -\frac{GMx_i}{(x_1^2 + x_2^2 + x_3^2)^{3/2}}, i = 1, 2, 3$$

Discretize this equation with respect to the time variable and simulate it using MATLAB.

[24] What is a maximal Abelian subspace with reference to a Lie algebra. Given a Cartan decomposition $\} = \approx \oplus \sqrt{}$ of a semisimple Lie algebra, let $\langle$ be a Cartan subalgebra of $\}$. Let θ be the automorphism of $\}$ such that $\theta(X + Y) = X - Y$ for $X \in \approx$ and $Y \in \sqrt{}$. Suppose $\langle$ is invariant under

θ. Then show that $\langle = \langle_t \oplus \langle_p$, where $\langle_t = \langle \cap \approx$ and $\langle_p = \langle \cap \sqrt{}$. Prove that $\langle_p$ is a maximal Abelian subspace of $\sqrt{}$.

[25] Derive Euler's equations for a spinning top using Eulerian angles and the Lagrangian approach, discretize these equations and implement them using MATLAB.

[26] Perturb the equations of motion of a spinning top by white noise and damping and express the resulting dynamical equations as a system of stochastic differential equations of the Ito type and simulte these using MATLAB.

[27] Optimal control theory: The state vector $\mathbf{x}(t)$ of a system satisfies the differential equation

$$\frac{d\mathbf{x}(t)}{dt} = \mathbf{f}(\mathbf{x}(t), \mathbf{u}(t), t)$$

where $\mathbf{u}(t)$ is the input process. We assume that $\mathbf{u}(t) \in \mathbb{R}^p$ and $\mathbf{x}(t) \in \mathbb{R}^n$ so that $\mathbf{a} : \mathbb{R}^n \times \mathbb{R}^p \times \mathbb{R}_+ \to \mathbb{R}^n$. Let $T > 0$ be fixed. The optimal control problem is to determine the input $\mathbf{u}(t), 0 \leq t \leq T$ such that

$$\int_0^T h(\mathbf{x}(t), \mathbf{u}(t), t)dt$$

is a minimum subject to the constraints (1) $\mathbf{x}(t)$ satisfies the above state equations with input $\mathbf{u}(t)$ over the interval $[0, T]$ and (2) $\psi(\mathbf{x}(T)) = 0$ where $\psi : \mathbb{R}^n \to \mathbb{R}^q$ is a given function. Using Lagrange multipliers, this constrained optimization problem amounts to extremizing

$$S[\mathbf{x}, \mathbf{u}, \lambda, \mu] = \int_0^T h(\mathbf{x}(t), \mathbf{u}(t), t)dt + \int_0^T \lambda^T(t)(\mathbf{x}'(t) - \mathbf{f}(\mathbf{x}(t), \mathbf{u}(t), t))dt$$

$$+\mu^T \psi(\mathbf{x}(T))$$

Here, $\lambda(t) \in \mathbb{R}^n$ and $\mu \in \mathbb{R}^q$.

[28] Linear-quadratic optimal control. The state equations are linear, ie,

$$\mathbf{x}'(t) = \mathbf{A}(t)\mathbf{x}(t) + \mathbf{G}(t)\mathbf{u}(t), t \geq 0$$

and the problem is to design the input $\mathbf{u}(t), 0 \leq t \leq T$ so that

$$\frac{1}{2}\int_0^T [\mathbf{x}^T(t)\mathbf{Q}(t)\mathbf{x}(t) + \mathbf{u}^T(t)\mathbf{R}(t)\mathbf{u}(t)]dt$$

is minimized where $\mathbf{Q}(t)$ and $\mathbf{R}(t)$ are positive definite matrices of sizes $n \times n$ and $p \times p$ respectively.

[29] Let $\mathcal{M}$ and $\mathcal{N}$ be two differentiable manifolds and let $\phi : \mathcal{M} \to \mathcal{N}$ be a differentiable mapping. Let $v \in T\mathcal{M}_x$ and $w \in T\mathcal{N}_{\phi(x)}$ be such that $w = d\phi(x)(v)$. Express the components of w in terms of the

components of v using appropriate local coordinates on the two manifolds. Suppose $\omega \in T\mathcal{N}^*_{\phi(x)}$. Then express the components of $(\phi^*\omega)(x) \in T\mathcal{M}^*_x$ in terms of the components of ω using local coordinates.

hint: Suppose $t \to c(t)$ is a curve in $\mathcal{M}$ such that $c(0) = x$ and $c'(0) = v$. Then for any function f on $\mathcal{M}$ that is differentiable in a neighbourhood of x, we have in terms of local coordinates on $\mathcal{M}$ and $\mathcal{N}$

$$d\phi(x)(v)(f) = \frac{d}{dt} f(\phi(c(t)))|_{t=0} = \sum_{i,j} \frac{\partial f(y)}{\partial y^i}|_{y=\phi(x)} \frac{\partial \phi^i(x)}{\partial x^j} v^j$$

from which we deduce that

$$(d\phi(x)(v))^i = \sum_j \frac{\partial \phi^i(x)}{\partial x^j} v^j$$

Likewise,

$$(\phi^*\omega)(x)(v) = \omega(d\phi(x)(v))$$

which reads in terms of components,

$$(\phi^*\omega)_i(x)v^i = \omega_j(x)\phi^j_{,i}(x)v^i$$

so that

$$(\phi^*\omega)_i(x) = \omega_j(x)\phi^j_{,i}(x)$$

[30] Let $\mathcal{L}_1$ and $\mathcal{L}_2$ be two Lagrangians such that $\mathcal{L}_1 - \mathcal{L}_2$ is a total time derivative. Then prove that the equations of motion corresponding to the two Lagrangians are the same.

[31] Write down the Lagrangian density for the one dimensional wave equation.

[32] Let $U(x_1, ..., x_n)$ be the potential field in which n particles are moving. The equations of motion are

$$x''_i = -U_{,i}(x), i = 1, 2, ..., n$$

or equivalently,

$$x'' = -\nabla U(x)$$

Assume that

$$U(x) = \frac{1}{2}x^T Cx + \epsilon.F(x)$$

where ϵ is a small parameter and C is a positive definite matrix. Solve for $x(t)$ upto $O(\epsilon)$.

hint: The equations of motion are

$$x'' = -Cx - \epsilon\nabla F(x)$$

Writing

$$x(t) = y(t) + \epsilon.z(t)$$

and equating terms upto $O(\epsilon)$ gives us

$$y''(t) = -Cy(t), z''(t) = -Cz(t) - \nabla F(y(t))$$

These equations can be solved easily using the Laplace transform. The first gives on Laplace transformation,

$$s^2 Y(s) - sy(0) - y'(0) = -CY(s)$$

so that

$$Y(s) = (s^2 I + C)^{-1}(sy(0) + y'(0))$$

and setting $g(t) = -\nabla F(y(t))$, the second gives

$$s^2 Z(s) - sz(0) - z'(0) = -CZ(s) + G(s)$$

so that

$$Z(s) = (s^2 I + C)^{-1}(sz(0) + z'(0)) + (s^2 I + C)^{-1} G(s)$$

These Laplace transforms can be easily inverted using the fact that the inverse Laplace transforms of $(s^2 I + C)^{-1}$ and $s(s^2 I + C)^{-1}$ are respectively $C^{-1/2}.sin(t.C^{1/2})$ and $cos(t.C^{1/2})$.

[33] Let X, Y be two smooth vector fields on a differentiable manifold $\mathcal{M}$ of dimension n. Let $t \to \phi_t(x)$ be the flow on $\mathcal{M}$ generated by X such that $\phi_0(x) = x$. Compute

$$Z(x) = lim_{t \to 0}((d\phi_t(Y)(x) - Y(x))/t$$

where by $Y(x)$ we mean the column vector representation of $Y(x)$ relative to a local coordinate chart. Note that $(d\phi_t(Y))(x) = (d\phi_t)_y(Y(y))$ where $y = \phi_t^{-1}(x)$.

[34] Assume that a particle moves in accord with the Hamilton equations of motion in conjunction with a random force, ie,

$$dq(t) = H_{,p}(q(t), p(t))dt, dp(t) = -H_{,q}(q(t), p(t))dt + \sigma.dB(t)$$

where B is a Brownian motion process. Show that the probability density $f(q, p, t)$ of $(q(t), p(t))$ satisfies the Fokker-Planck equation

$$f_{,t} = -(H_{,p}f)_{,q} + (H_{,q}f)_{,p} + \frac{\sigma^2}{2}f_{,pp}$$

$$= H_{,q}f_{,p} - H_{,p}f_{,q} + \frac{\sigma^2}{2}f_{,pp}$$

Derive a numerical algorithm for approximating this pde with a difference equation and explain how you can simulate this system on a computer.

[35] Explain how the Legendre transformation can be used to pass over from the Lagrangian to the Hamiltonian formulation of classical mechanics. Explain this using the notion of fibre derivative. Note that if (q, q') is in $T\mathcal{M}$ and (q, p) is in $T\mathcal{M}^*$, and the Lagrangian $L : T\mathcal{M} \to \mathbb{R}$ is given, then

the Fiber derivative of L is the map $FL : TM \to TM^*$ In terms of coordinates, $FL(q, q')$ is given by $\frac{\partial L(q,q')}{\partial q'_i} dq_i = p_i dq_i$, summation over the repeated index i being implied. The canonical momenta are $p_i = \frac{\partial L(q,q')}{\partial q'_i}$. The fiber derivative FL of the Lagrangian maps the point (q, q') in TM to the point $(q, p) \in TM^*$. Consider the canonical one form $\theta = p_i dq_i$ on TM^*. Note that $FL : TM \to TM^*$ and hence $FL^* : T^*TM^* \to T^*TM$ Then,

$$FL^*\theta = p_i o FL d(q_i o FL) = L_{,q'_i} dq_i$$

We also note that the canonical symplectic form is given by

$$\omega = dq_i \wedge dp_i = -d\theta, \theta = p_i dq_i$$

Then,

$$-FL^*\omega = d(p_i o FL) \wedge d(q_i o FL) = dL_{,q'_i} \wedge dq_i$$

$$= L_{,q'_i q_j} dq_j \wedge dq_i + L_{,q'_i q'_j} dq'_j \wedge dq_i$$

Equivalently,

$$-FL^*\omega = FL^* d\theta = dFL^*\theta = d(L_{,q'_i} dq_i) = L_{,q'_i q_j} dq_j \wedge dq_i + L_{,q'_i q'_j} dq'_j \wedge dq_i$$

[36] Discretize the equations of motion of a simple pendulum with external forcing and simulate this motion using MATLAB.

hint: In continuous time,

$$\theta''(t) = -a.sin(\theta(t)) + f(t)$$

which gives after discretization,

$$\theta[n+2] - 2\theta[n+1] + \theta[n] = -a\Delta^2.sin(\theta[n]) + \Delta^2.f[n]$$

where Δ is the time discretization step size and $\theta[n] = \theta(n\Delta), f[n] = f(n\Delta)$.

[37] Consider a Lagrangian $L(q_1, ..., q_n, q'_1, ..., q'_n, t)$. The associated Hamiltonian is

$$H(\mathbf{q}, \mathbf{p}) = \sum_{i=1}^{n} p_i q'_i - L,$$

where

$$p_i = \frac{\partial L}{\partial q'_i}, i = 1, 2, ..., n$$

Thus,

$$L = \sum_i p_i q'_i - H$$

Suppose $\mathbf{Q}, \mathbf{P}$ is a new set of canonical coordinates and momenta and K is the Hamiltonian function for these new canonical variables. Then if the equations of motion for $(\mathbf{q}, \mathbf{p}, H)$ correspond to the equations of motion of $(\mathbf{Q}, \mathbf{P}, K)$, the corresponding Lagrangians must differ by a total time derivative, ie,

$$\sum_i p_i q_i' - H = \sum_i P_i Q_i' - K + dF/dt$$

or in differential form

$$\sum_i p_i dq_i - H dt = \sum_i P_i dQ_i - K dt + dF$$

Assuming that F is expressible as a function of $\mathbf{q}, \mathbf{Q}, t$, we then get

$$\sum_i p_i dq_i - H = \sum_i P_i dQ_i - K + F_{,q_i} dq_i + F_{,Q_i} dQ_i + F_{,t} dt$$

so that

$$p_i = F_{,q_i}, P_i = -F_{,Q_i}, K = H + F_{,t}$$

This is defines a canonical transformation corresponding to the generator $F(\mathbf{q}, \mathbf{Q}, t)$. Now

$$\sum_i p_i dq_i - H dt = \sum_i (d(P_i Q_i) - Q_i dP_i) - K dt + dF = -\sum_i Q_i dP_i - K dt + d(F + \sum_i Q_i P_i)$$

Let

$$G(\mathbf{q}, \mathbf{P}, t) = \sum_i Q_i P_i + F$$

Then we get

$$p_i = G_{,q_i}, Q_i, G_{,P_i}, K = H + G_{,t}$$

This is another variant of a canonical transformation.

[38] Write a MATLAB programme for solving the gravitational n-body problem.

hint: The equations of motion are

$$x_i'' = K \sum_{j \neq i} m_j \frac{(x_j - x_i)}{((x_j - x_i)^2 + (y_j - y_i)^2 + (z_j - z_i)^2)^{3/2}},$$

$$y_i'' = K \sum_{j \neq i} m_j \frac{(y_j - y_i)}{((x_j - x_i)^2 + (y_j - y_i)^2 + (z_j - z_i)^2)^{3/2}},$$

$$z_i'' = K \sum_{j \neq i} m_j \frac{(z_j - z_i)}{((x_j - x_i)^2 + (y_j - y_i)^2 + (z_j - z_i)^2)^{3/2}},$$

where i=1,2,..., n. Transform these $3n$ second order differential equations into $6n$ first order differential equations and discretize these.

[39] Write a MATLAB programme for solving the Fokker-Planck equation for the gravitational n-body problem when there are white noise forcing terms which perturb the motion.

hint:

$$dx_i = v_{xi}dt,$$

$$dv_{xi} = K \sum_{j \neq i} m_j \frac{(x_j - x_i)}{((x_j - x_i)^2 + (y_j - y_i)^2 + (z_j - z_i)^2)^{3/2}} dt + \sigma_{xi} dB_{xi}(t)$$

$$dy_i = v_{yi}dt,$$

$$dv_{yi} = K \sum_{j \neq i} m_j \frac{(y_j - y_i)}{((x_j - x_i)^2 + (y_j - y_i)^2 + (z_j - z_i)^2)^{3/2}} dt + \sigma_{yi} dB_{yi}(t)$$

$$dz_i = v_{zi}dt,$$

$$dv_{zi} = K \sum_{j \neq i} m_j \frac{(z_j - z_i)}{((x_j - x_i)^2 + (y_j - y_i)^2 + (z_j - z_i)^2)^{3/2}} dt + \sigma_{zi} dB_{zi}(t)$$

where $i = 1, 2, ..., n$ and $B_{xi}, B_{yi}, B_{zi}, i = 1, 2, ..., n$ are $3n$ independent standard Brownian motion processes. Combine this with the fact that if $\xi_i, i = 1, 2, ..., n$ satisfy the system of stochastic differential equations

$$d\xi_i = f_i(\xi_1, ..., \xi_n)dt + \sum_{j=1}^{p} \sigma_{ij}(\xi_1, ..., \xi_n)dB_j(t), i = 1, 2, ..., n$$

where $B_1, ..., B_p$ are standard independent Brownian motion processes, then the joint probability density $p_t(\xi_1, ..., \xi_n)$ of $\xi_1(t), ..., \xi_n(t)$ satisfies the Fokker-Planck equation

$$\frac{\partial p_t}{\partial t} = -\sum_{j=1}^{n} \frac{\partial (f_j p)}{\partial \xi_j} + \frac{1}{2} \sum \frac{\partial^2 (a_{ij} p)}{\partial \xi_i \partial \xi_j}$$

where

$$a_{ij} = \sum_k \sigma_{ik} \sigma_{jk}$$

[40] Consider the Hamiltonian

$$H(q_1, ..., q_n, p_1, ..., p_n) = \sum_{i=1}^{n} p_i^2 / 2m_i + U(q_1, ..., q_n)$$

Define

$$J = \sum_{i=1}^{n} p_i q_i$$

Then using the Hamilton equations

$$q_i' = H_{,p_i} = p_i/m_i, p_i' = -H_{,q_i} = -U_{,q_i}$$

we get

$$J' = \sum_i (p_i'q_i + p_iq_i') = -\sum_i q_iU_{,q_i} + \sum_i p_i^2/m_i$$

Suppose U is a homogeneous function of $q_1, ..., q_n$ of degree λ. Then

$$\sum_i q_iU_{,q_i} = \lambda U$$

and we get the Virial theorem

$$J' = -\lambda.U + 2T$$

where

$$T = \sum_i p_i^2/2m_i$$

is the kinetic energy of the system. Suppose the motion is bounded so that $p_i, q_i, i = 1, 2, ..., n$ are bounded functions of time. Then, J is also a bounded function of time and on taking time averages in the Virial theorem, we get

$$\lambda < U >= 2 < T >$$

with $< X >$ denoting time average of X, ie,

$$< X >= lim_{T \to \infty} \frac{1}{T} \int_0^T X(t)dt$$

[41] Write down the angular momentum integrals for the n-body problem.

[42] Solve the classical harmonic oscillator problem using the Hamilton-Jacobi method.

hint: The Hamiltonian for the harmonic oscillator is

$$H(q,p) = p^2/2m + m\omega^2q^2/2$$

The Hamilton Jacobi equation for the action $S(q,t)$ is thus

$$S_{,t} + \frac{1}{2m}S_{,q}^2 + m\omega^2q^2/2 = 0$$

Assume S to be of the form

$$S(q,t) = W(q) - \alpha t$$

where α is a constant. Then W satisfies

$$\alpha = \frac{1}{2m}(W'(q))^2 + m\omega^2q^2/2$$

and this can be easily integrated.

[43] The equations of motion of a planar harmonic oscillator. The Lagrangian is given by

$$L(x, y, x', y') = \frac{m}{2}(x'^2 + y'^2) - \frac{m\omega^2}{2}(x^2 + y^2)$$

The equations of motion are

$$x'' = -\omega^2 x, y'' = -\omega^2 y$$

and the general solution is given by

$$x(t) = A_1.cos(\omega t + \phi_1), y(t) = A_2.cos(\omega t + \phi_2)$$

Determine the equation of the curve traced by the oscillator in the $x - y$ plane. The equation must be of the form $f(x, y) = 0$. This is done by eliminating t between the two equations.

[44] Let two forces $\mathbf{F}_1$ and $\mathbf{F}_2$ be applied at a point. Then prove using Cartesian coordinates that the magnitude of the net force on the point is given by

$$|\mathbf{F}_1 + \mathbf{F}_2| = \sqrt{F_1^2 + F_2^2 + 2F_1 F_2 cos(\theta)}$$

where $F_i = |\mathbf{F}_i|$ and θ is the angle between the two force vectors.

[45] Show that if $g_{\mu\nu}$ is the metric of space-time, then a system of coordinates can always be chosen so that $g = -1$ everywhere where $g = det((g_{\mu\nu}))$. In such a system, simplify the expression for the Ricci tensor and derive the Einstein field equations in free space from an action principle.

[46] Let K be a Cartesian system and K' another Cartesian system whose origin coincides with that of K and that rotates with respect to K with an angular velocity ω. Show that if $\mathbf{P}$ is a vector whose rates of change with respect to K and K' are respectively given by $(d\mathbf{P}/dt)_K$ and $(d\mathbf{P}/dt)_{K'}$, then

$$(d\mathbf{P}/dt)_K = (d\mathbf{P}/dt)_{K'} + \omega \times \mathbf{P}$$

hint: Write down $\mathbf{P}$ as a linear combination of the unit coordinate vectors of K', differentiate with respect to time using the fact that if $\mathbf{u}$ is a unit vector fixed in K', then its rate of change relative to K is given by $(d\mathbf{u}/dt)_{K'} = \omega \times \mathbf{u}$.

[47] Let M be a Riemannian manifold and let $(X, Y) \to \nabla_X Y$ be a connection in M. X, Y are smooth vector fields on M. Let $t \to \gamma(t)$ be a curve in M and relative to local coordinates (x^i), let $\gamma'(t) = \gamma^{i'}(t)\partial_i$ be the tangent vector to this curve at $\gamma(t)$. Here, $\partial_i = \partial/\partial x^i$ and $\gamma^i = x^i(\gamma)$. Define the functions Γ^i_{jk} locally by

$$\nabla_{\partial_j}\partial_k = \Gamma^i_{jk}\partial_i$$

Show that the geodesic equation

$$\nabla_{\gamma'}\gamma' = 0$$

can be expressed in local coordinates as

$$\gamma^{i''}(t) + \Gamma^i_{jk}(\gamma(t))\gamma^{j'}(t)\gamma^{k'}(t) = 0$$

[48] Consider a system of N point particles of masses $m_1, ..., m_N$ having positions given by

$$\mathbf{r}_i(t, q_1, ..., q_k), i = 1, 2, ..., N$$

Here, $N \geq k$ and the $q_i's$ are generalized coordinates, ie, they can vary freely. The equations giving the $\mathbf{r}_i's$ in terms of the $q_j's$ define the constraint equations. The velocity of the ith particle is

$$\frac{d\mathbf{r}_i}{dt} = \frac{\partial \mathbf{r}_i}{\partial t} + \sum_{j=1}^{k} \frac{\partial \mathbf{r}_i}{\partial q_j} q'_j$$

where

$$q'_j = \frac{dq_j}{dt}$$

Hence the kinetic energy of the system is

$$T(q_1, ..., q_N, q'_1, ..., q'_N, t) = \sum_{i=1}^{N} |\frac{m_i}{2}\frac{d\mathbf{r}_i}{dt}|^2 = \sum_{i=1}^{N} \frac{m_i}{2} |\frac{\partial \mathbf{r}_i}{\partial t}|^2$$

$$+ \sum_{i,j,r} \frac{m_i}{2}(\frac{\partial \mathbf{r}_i}{\partial q_j}, \frac{\partial \mathbf{r}_i}{\partial q_r})q'_j q'_r$$

$$+ \sum_{i,j} m_i(\frac{\partial \mathbf{r}_i}{\partial t}, \frac{\partial \mathbf{r}_i}{\partial q_j})q'_j$$

The kinetic energy can therefore be put in the form

$$T = \frac{1}{2}\sum_{i,j} a_{ij}(t, q_1, ..., q_N)q'_i q'_j + \sum_i b_i(t, q_1, ..., q_N)q'_i + c(t, q_1, ..., q_N)$$

Identify the functions a_{ij}, b_i, c.

[49] Write a MATLAB programme to simulate the restricted three body problem in which two masses m_1, m_2 are moving in circular orbits around their centre of mass with an angular velocity ω and the third body of mass m_3 is moving in the periodic gravitational field produced by these two bodies.

[50] The equations of motion of a spinning top when no external torque acts on it are

$$d\mathbf{L}/dt + \omega \times \mathbf{L} = \mathbf{0}$$

where ω is the angular velocity and $\mathbf{L}$ is the angular momentum. Here, d/dt is taken relative to a frame fixed to the top. I_1, I_2, I_3 denote the moments of inertia of the top about its principal axes and $\omega_1, \omega_2, \omega_3$ are the components of the angular velocity of the top about these axes. Then relative to this frame, $\mathbf{L} = (I_1\omega_1, I_2\omega_2, I_3\omega_3)$ and the above equation can be expressed as

$$I_1\omega_1' + (I_2 - I_3)\omega_2\omega_3 = 0, \quad I_2\omega_2' + (I_3 - I_1)\omega_3\omega_1 = 0$$

$$I_3\omega_3' + (I_1 - I_2)\omega_1\omega_2 = 0$$

These equations can be solved numerically. Note that these three equations admit an energy integral

$$I_1\omega_1^2 + I_2\omega_2^2 + I_3\omega_3^2 = 2E$$

where E is a constant equal to the energy of the top.

Chapter 3

Fluid Dynamics

Introduction: This chapter discusses basic problems in fluid dynamics including the Navier-Stokes equation, viscous forces in a fluid, vorticity, velocity potential and streamlines, numerical simulation of the fluid dynamical equations, equation of continuity or mass conservation, energy equation in a fluid and motion of an adiabatic fluid.

Problems in fluid dynamics:

[1] Write down an expression for the amount of fluid mass flowing out of a closed volume V per unit time. Assume that S is the bounding surface for V and that $\mathbf{v}$ is the velocity field while ρ is the density field.

ans:

$$Q = \int_S \rho \mathbf{v}.\mathbf{n}dS = \int_V \nabla.(\rho\mathbf{v})d^3x$$

[2] Write down the expression for the rate of change of a scalar field quantity along the path followed by a fluid particle. Assume that $\mathbf{v}$ is the velocity field.

ans:

$$\frac{DQ}{Dt} = \frac{\partial Q}{\partial t} + \mathbf{v}.\nabla Q$$

[3] Suppose ρ is the mass density and v_i are the components of the velocity in a cartesian system. Let p be the pressure field. The energy momentum tensor is defined by

$$\Pi_{ij} = \rho v_i v_j + p\delta_{ij}$$

Show that the mass conservation equation $(\rho v_i)_{,i} + \rho_{,t} = 0$ in conjunction with the momentum conservation equation

$$(\rho v_i)_{,t} + \Pi_{ij,j} = 0$$

leads to the Navier-Stokes equation

$$\rho(v_{i,t} + v_j v_{i,j}) = -p_{,i}$$

In these formulas, summation over repeated indices is implied.

[4] Write down the mass conservation equation in the cylindrical coordinate system

hint:The divergence of a vector field $\mathbf{f}$ in the cylindrical system is given by

$$divf = \frac{1}{\rho}\frac{\partial}{\partial\rho}(\rho f_\rho) + \frac{1}{\rho}\frac{\partial f_\phi}{\partial\phi} + \frac{\partial f_z}{\partial z}$$

[5] What do you understand by the strain rate in a fluid. Define the viscous stress tensor in terms of the strain rate. How do you compute the viscous force per unit volume due to viscosity.

hint:

$$S_{ij} = \eta_1(v_{i,j} + v_{j,i}) + \eta_2(\nabla.v)\delta_{ij}$$

is the viscous stress tensor. The force per unit volume due to viscosity is then given by

$$f_i = S_{ij,j} = \eta_1(v_{i,jj} + v_{j,ij}) + \eta_2(\nabla.v)_{,i}$$

$$= \eta_1\nabla^2 v_i + (\eta_1 + \eta_2)(\nabla.v)_{,i}$$

[6] Suppose that the two dimensional velocity field is given in plane polar coordinates by the formula

$$\mathbf{v} = v_\rho(\rho, \phi)\hat{\rho} + v_\phi(\rho, \phi)\hat{\phi}$$

where

$$\hat{\rho} = \hat{x}.cos(\phi) + \hat{y}.sin(\phi), \hat{\phi} = -\hat{x}.sin(\phi) + \hat{y}.cos(\phi)$$

Then derive the formula for the acceleration

$$(\mathbf{v}, \nabla)\mathbf{v}$$

in the polar system.

hint:

$$\frac{\partial\hat{\rho}}{\partial\rho} = 0, \frac{\partial\hat{\rho}}{\partial\phi} = \hat{\phi},$$

$$\frac{\partial\hat{\phi}}{\partial\rho} = 0, \frac{\partial\hat{\phi}}{\partial\phi} = -\hat{\rho}$$

Use these formulae in conjunction with

$$\nabla = \hat{\rho}.\frac{\partial}{\partial\rho} + \frac{\hat{\phi}}{\rho}\frac{\partial}{\partial\phi}$$

[7] Derive the expressions for the strain rate tensor of a fluid in cylindrical and spherical polar coordinates.

hint: In cartesian coordinates, the strain rate tensor components are given by

$$S_{xx} = v_{x,x}, S_{yy} = v_{y,y}, S_{zz} = v_{z,z}, S_{xy} = \frac{1}{2}(v_{x,y} + v_{y,x}) = S_{yx},$$

$$S_{yz} = S_{zy} = \frac{1}{2}(v_{y,z} + v_{z,y}), S_{zx} = S_{xz} = \frac{1}{2}(v_{x,z} + v_{z,x})$$

The corresponding components in cylindrical coordinates are given by

$$S_{\rho\rho} = \hat{\rho}.S.\hat{\rho}, S_{\phi\phi} = \hat{\phi}.S.\hat{\phi}, S_{zz} = \hat{z}.S.\hat{z},$$

$$S_{\rho\phi} = \hat{\rho}.S.\hat{\phi}, S_{\phi z} = \hat{\phi}.S.\hat{z}, S_{z\rho} = \hat{z}.S.\hat{\rho}$$

[8] Define vorticity of a fluid in terms of its velocity field and derive the differential equation satisfied by the vorticity starting with the Navier-Stokes equation.

hint:Vorticity is defined as the curl of the velocity field:$\Omega = \nabla \times v$. Consider the Navier-Stokes equation for an incompressible fluid:

$$(v_{,t} + v.\nabla v) = -\nabla p/\rho + \nu\nabla^2 v$$

Using the identity

$$(\nabla \times v) \times v = v.\nabla v - \frac{1}{2}\nabla v^2$$

we get

$$(\Omega_{,t} + \nabla \times (\Omega \times v)) = \nu\nabla^2\Omega$$

[9] Derive the components of the vorticity of a fluid in the spherical polar coordinate system.

hint: Use the expression for curl of a vector field in the spherical polar coordinate system.

[10] Derive the expression for the components of the vorticity in an arbitrary orthogonal coordinate system. Note that a system of coordinates (q_1, q_2, q_3) is an orthogonal system of coordinates if the conditions $(\nabla q_i, \nabla q_j) = a_i(q)\delta_{ij}$ are satisfied where the $a_i's$ are some functions of q. Alternatively, a system (q_1, q_2, q_3) is orthogonal if when the position vector $\mathbf{r}$ is expressed as a functions of the $q's$, then

$$(\frac{\partial \mathbf{r}}{\partial q_i}, \frac{\partial \mathbf{r}}{\partial q_j}) = b_i(q)\delta_{ij}$$

are satisfied where $b_i's$ are some functions.

[11] Assume that a gas is given whose equation of state $p = p(\rho)$, ie, pressure as a function of the density is known except for some unknown parameters $\theta_1, ..., \theta_q$, ie the equation of state is

$$p = F(\rho, \theta_1, ..., \theta_q)$$

where the function F is known but the parameters $\theta_1, ..., \theta_q$ are unknown. The objective is to estimate the parameters $\theta_1, ..., \theta_n$ by taking measurements on the velocity field of the gas at different points. The dynamics of the gas is governed by the mass conservation equation, ie, the equation of continuity and the Navier-Stokes equation.

$$div(\rho\mathbf{v}) + \rho_{,t} = 0,$$

$$\rho(\mathbf{v}.\nabla\mathbf{v} + \mathbf{v}_{,t}) = -\nabla p + \eta\nabla^2\mathbf{v}$$

where η is the viscosity of the gas. We may also treat the viscosity η as an unknown parameter to be estimated. Note that if $\Omega = \nabla \times \mathbf{v}$, ie Ω is the vorticity, then

$$\Omega \times \mathbf{v} = \mathbf{v}.\nabla \mathbf{v} - \frac{1}{2}\nabla v^2$$

If ρ were a constant as for a liquid, then we could infer the vorticity equation by taking the curl of the Navier-Stokes equation

$$\Omega_{,t} + \nabla \times (\Omega \times \mathbf{v}) = \eta \nabla^2 \Omega$$

However for a gas, ρ is not a constant and we cannot infer this equation directly. However, if $\mu = \eta/\rho$ were a constant then by defining the function $h(\rho) = \int_{\rho_0}^{\rho} dp(\rho)/\rho$, we get $\nabla h = (p'(\rho)/\rho)\nabla \rho = \nabla p/\rho$, whence the Navier-Stokes equation gives

$$\Omega_{,t} + \nabla \times (\Omega \times \mathbf{v}) = \mu \nabla^2 \Omega$$

Suppose we assume that the flow of the gas takes place inside a cubical box of side length L parallel to the x, y, z axes respectively. Then we can discretize the mass conservation equation as well as the Navier-Stokes equation and apply the least squares algorithm to estimate the parameters $\theta_1, ..., \theta_q$. The discretized mass conservation equation reads

$$\Delta^{-1}[\rho[n+1,k,m,r]v_x[n+1,k,m,r] + \rho[n,k+1,m,r]v_y[n,k+1,m,r]$$

$$+\rho[n,k,m+1,r]v_z[n,k,m+1,r] - \rho[n,k,m,r](v_x[n,k,m,r] + v_y[n,k,m,r] + v_z[n,k,m,r])]$$

$$+\tau^{-1}(\rho[n,k,m,r+1] - \rho[n,k,m,r]) = 0$$

The discretized Navier-Stokes equations are obtained as follows. Set

$$F(\rho) = p'(\rho)/\rho$$

Note that since p is a function of the unknown parameters $\theta_1, ..., \theta_q$, F also is a function of these unknown parameters, ie, $F(\rho, \theta_1, ..., \theta_q)$. Then

$$\tau^{-1}(v_x[n,k,m,r+1] - v_x[n,k,m,r]) + \Delta^{-1}v_x[n,k,m,r](v_x[n+1,k,m,r] - v_x[n,k,m,r])$$

$$+\Delta^{-1}v_y[n,k,m,r](v_x[n,k+1,m,r] - v_x[n,k,m,r]) + \Delta^{-1}v_z[n,k,m,r](v_x[n,k,m+1,r] - v_x[n,k,m,r])$$

$$= -\Delta^{-1}F(\rho[n,k,m,r], \theta_1, ..., \theta_q)(\rho[n+1,k,m,r] - \rho[n,k,m,r])$$

$$+\Delta^{-2}\frac{\eta}{\rho[n,k,m,r]}(v_x[n+1,k,m,r] + v_x[n-1,k,m,r]$$

$$+v_x[n,k+1,m,r] + v_x[n,k-1,m,r] + v_x[n,k,m+1,r] + v_x[n,k,m-1,r] - 6v_x[n,k,m,r])$$

and similarly for the y and z components. An adaptive algorithm can be applied to these equations to esimate the parameters $\theta_1, ..., \theta_q$ from the Navier-Stokes equations provided that the density $\rho[n,k,m,r]$ is known. If it is unknown, then additional information furnished by the mass conservation equation must be used. An example is obtained by setting $p(\rho, C, \gamma) = C\rho^{\gamma}$, where C, γ are the unknown parameters, ie this corresponds to an adiabatic gas with γ as the ratio of specific heats at constant pressure C_p and at constant volume C_v. Suppose that the velocity field $\mathbf{v}(\mathbf{r}, t)$ is known and that the density at time $t = 0$,

$\rho(\mathbf{r}, 0)$ is also known. Then the density at all times $\rho(\mathbf{r}, t)$ can be determined from the mass conservation equation. This enables us to incorporate the mass conservation equation into the adaptive algorithm for estimating the parameters $\theta_1, ..., \theta_q$. Specifically, from knowledge of $\mathbf{v}[n, k, m, r]$, $\rho[n, k, m, r]$ for a fixed r and all n, k, m, $\rho[n, k, m, r + 1]$ can be determined from the discretized mass conservation equation.

[12] From the mass conservation equation and Navier-Stokes equation for an incompressible fluid, we have

$$div\mathbf{v} = 0,$$

$$\Omega_{,t} + \nabla \times (\Omega \times \mathbf{v}) = \nu\nabla^2\Omega$$

where ν is the ratio of the viscosity to the density and $\Omega = \nabla \times \mathbf{v}$ is the vorticity. The equation $\nabla.v = 0$ implies $v = \nabla \times A$ where by replacing A by A plus the gradient of a scalar field, we can assume $\nabla.A = 0$. Then,

$$\Omega = \nabla(\times\nabla \times A) = -\nabla^2 A$$

Thus, the vorticity equation gives

$$-\nabla^2 A_{,t} + \nabla \times (-(\nabla^2 A) \times (\nabla \times A)) = -\nu\nabla^2\nabla^2 A$$

or

$$\nabla^2(A_{,t} - \nu\nabla^2 A) = -\nabla \times (\nabla^2 A \times (\nabla \times A))$$

Defining $K(x, x') = \frac{1}{4\pi|x-x'|}$, it follows that

$$A_{,t}(t, x) = \nu\nabla^2 A(t, x) + \int_{\mathbb{R}^3} K(x, x')\nabla \times (\nabla^2 A \times (\nabla \times A(t, x')))d^3x'$$

This equation is first order in time and can be used to simulate the fluid in any region of space by adopting a discretization scheme in time. Note that A can be determined from v since

$$\nabla^2 A = -\Omega$$

implies

$$A(x) = \frac{1}{4\pi} \int_{\mathbb{R}^3} K(x, x')\Omega(x')d^3x'$$

[13] Parameter estimation problems in fluid dynamics.

[1] Estimate the viscosity of an incompressible liquid from velocity measurements. Suppose the flow takes place in two dimensions. Then

$$\mathbf{v} = v_x(x, y)\hat{x} + v_y(x, y)\hat{y}$$

The equation of continuity (mass conservation) for the incompressible fluid is

$$div\mathbf{v} = 0$$

or in terms of components,

$$v_{x,x} + v_{y,y} = 0$$

Thus there exists a function $\psi(x,y)$ such that

$$v_x = \psi_{,y},\ v_y = -\psi_{,x}$$

The vorticity $curl\,\mathbf{v}$ is $\Omega_z(x,y)\hat{z}$ where

$$\Omega_z = v_{y,x} - v_{x,y} = -\nabla^2\psi = -(\psi_{,xx} + \psi_{,yy})$$

The Navier-Stokes equation is

$$\mathbf{v}_{,t} + \Omega \times \mathbf{v} + \frac{1}{2}\nabla(v^2) = -\nabla p/\rho + \nu\nabla^2\mathbf{v}$$

Taking curl we get

$$\Omega_{,t}\hat{z} + curl(\Omega \times \mathbf{v}) = \nu\nabla^2\Omega$$

Now,

$$\Omega \times \mathbf{v} = \Omega_z\hat{z} \times (v_x\hat{x} + v_y\hat{y}) = \Omega_z\psi_{,y}\hat{y} + \Omega_z\psi_{,x}\hat{x}$$

and hence,

$$curl(\Omega \times \mathbf{v}) = \hat{z}((\Omega_z\psi_{,y})_{,x} - (\Omega_z\psi_{,x})_{,y})$$
$$= \hat{z}(\Omega_{z,x}\psi_{,y} - \Omega_{z,y}\psi_{,x})$$

Substituting all these expressions, we get

$$-\nabla^2\psi_{,t} - (\nabla^2\psi_{,x})\psi_{,y} + (\nabla^2\psi_{,y})\psi_{,x} = -\nu\nabla^2\nabla^2\psi$$

Suppose we take measurements on the velocity field in a domain D. Then, we can determine the stream function ψ in this region. We then wish to use this data to estimate ν on a real time basis by applying the gradient algorithm to the above equation in the form

$$\hat{\nu}(t + \delta t) = \hat{n}u(t) - \mu.\frac{\partial}{\partial\hat{\nu}(t)}\int_D (\nabla^2\psi_{,t} + (\nabla^2\psi_{,x})\psi_{,y} - (\nabla^2\psi_{,y})\psi_{,x}$$

$$-\hat{\nu}(t)\nabla^2\nabla^2\psi)^2\,dxdy$$

with μ as the adaptation step size.

[14] Write down the Navier-Stokes equation for fluid flow in cylindrical and spherical polar coordinates.

hint: If $\mathbf{v}$ is the velocity field expressed using an orthogonal system of curvilinear coordinates and p is the pressure, evaluate in this coordinate system the quanatities $\mathbf{v}.\nabla\mathbf{v}$, ∇p and $\nabla^2\mathbf{v}$.

[15] Development of signal processing algorithms for estimating fluid parameters like density and viscosity. The equations of motion of the fluid are the Navier-Stokes equations and the mass conservation

equation. The former is simply Newton's second law of motion for the fluid. The idea is to discretize these partial differential equations by introducing a time step size and a spatial pixel size and then minimize the sum of squares of the difference between the left and right sides over all spatio-temporal pixels with respect to the fluid parameters viscosity and density. In case the fluid is a gas, we use the equation of state $p = C\rho^\gamma$ where p is the pressure, ρ is the density and C, γ are two constants. These constants are also estimated by minimizing the sum of error squared. For example, if the motion of the gas takes place along only one dimension, the equations for the density $\rho(t, x)$ and velocity $v(t, x)$ are

$$\rho_{,t} + (\rho.v)_{,x} = 0,$$

$$\rho(v_{,t} + vv_{,x}) = -p_{,x} + \eta.v_{,xx} + f$$

where f is the external force and for any dynamical variable $g(t, x)$, $g_{,t} = \frac{\partial g}{\partial t}$ and $g_{,x} = \frac{\partial g}{\partial x}$. Discretization of these equations with time step size τ, spatial step size Δ, $\rho[n, k] = \rho(n\tau, k\Delta), v[n, k] = v(n\tau, k\Delta)$, gives

$$\tau^{-1}(\rho[n + 1, k] - \rho[n, k]) + \Delta^{-1}(\rho[n, k + 1]v[n, k + 1] - \rho[n, k]v[n, k]) = 0$$

$$\rho[n, k](\tau^{-1}(v[n + 1, k] - v[n, k]) + \Delta^{-1}v[n, k](v[n, k + 1] - v[n, k])) = -\Delta^{-1}(p[n, k + 1] - p[n, k])$$

$$+\Delta^{-2}\eta(v[n, k + 1] - 2v[n, k] + v[n, k - 1]) + f[n, k]$$

where $p[n, k] = C.\rho[n, k]^\gamma$, These equations can be easily simulated and the least squares algorithm applied for estimating η, C, γ.

[16] Energy equation for an adiabatic fluid. Let $\mathbf{v}$ denotet the velocity field, ρ the density field, ϵ the internal energy per unit mass of the fluid and p the pressure. $1/\rho$ is then the volume per unit mass and $w = \epsilon + p/\rho$ is the enthalpy per unit mass. The Navier-Stokes equation is

$$\rho(\mathbf{v}.\nabla\mathbf{v} + \mathbf{v}_{,t}) = -\nabla p$$

and the mass conservation equation is

$$\rho_{,t} + div(\rho\mathbf{v}) = 0$$

Let s denote the entropy per unit mass. Then the first law of thermodynamics applied to a unit mass of the fluid is given by

$$Tds = d\epsilon + p.d(1/\rho) = d\epsilon - pd\rho/\rho^2$$

We assume that the fluid is adiabatic, so that

$$0 = Ds/Dt = \mathbf{v}.\nabla\mathbf{s} + s_{,t}$$

The macroscopic kinetic energy of the fluid per unit mass is $v^2/2$ and hence the total energy per unit mass is given by $\epsilon + v^2/2$. The total energy per unit volume is $\rho(\epsilon + v^2/2)$. Its rate of increase is given by

$$(\rho(\epsilon + v^2/2))_{,t} = \rho_{,t}(\epsilon + v^2/2) + \rho(\epsilon_{,t} + \mathbf{v}.\mathbf{v}_{,t})$$

Now, by mass conservation,

$$\rho_{,t} = -div(\rho\mathbf{v}),$$

by the first law of thermodynamics,

$$\epsilon_{,t} = Ts_{,t} + p\rho_{,t}/\rho^2$$

and by the Navier-Stokes equation,

$$\mathbf{v}_{,t} = -\mathbf{v}.\nabla\mathbf{v} - \nabla p/\rho = -\Omega \times \mathbf{v} - \nabla v^2/2 - \nabla p/\rho$$

where

$$\Omega = \nabla \times \mathbf{v}$$

is the vorticity. Thus,

$$(\rho(\epsilon + v^2/2))_{,t} = -div(\rho\mathbf{v})(\epsilon + v^2/2) + \rho(Ts_{,t} - p.div(\rho\mathbf{v})/\rho^2) - \rho\mathbf{v}.\nabla(v^2/2) - \mathbf{v}.\nabla p$$

$$= -div(\rho\mathbf{v})(\epsilon + v^2/2 + p/\rho) - \rho\mathbf{v}(\nabla(v^2/2)) - \rho T\mathbf{v}.\nabla\mathbf{s} - \mathbf{v}.\nabla p$$

$$= -div(\rho\mathbf{v})(\epsilon + v^2/2 + p/\rho) - \rho\mathbf{v}.(\nabla(v^2/2)) - \mathbf{v}.(\rho T\nabla s + \nabla p)$$

Since $w = \epsilon + p/\rho$, it follows that

$$dw = d\epsilon + dp/\rho - pd\rho/\rho^2 = Tds + dp/\rho$$

so that

$$\rho T\nabla s + \nabla p = \rho\nabla w$$

This gives

$$(\rho(\epsilon + v^2/2))_{,t} = -div(\rho\mathbf{v}(w + v^2/2)) - \rho\mathbf{v}.\nabla(v^2/2 + w)$$

$$= -div(\rho\mathbf{v}(w + v^2/2))$$

[17]

The Navier-Stokes equation for a two dimensional viscous fluid is written down. Using the incompressibility condition, we then write down the mass conservation equation which immediately leads to the existence of a stream function from which the velocity field can be derived. By combining with the curl of the Navier-Stokes equation, we eliminate the pressure leading to a nonlinear partial differential equation for the stream function. By using the Green's function for the two dimensional Laplacian operator, we cast this pde in a form that is suitable for MATLAB simulation, ie, in a form where the stream function can be computed iteratively in the time variable.

Problem formulation

A viscous incompressible fluid in two dimensions is completely characterized by its two dimensional velocity field

$$\mathbf{v}(t, x, y) = v_x(t, x, y)\hat{x} + v_y(t, x, y)\hat{y}$$

and its pressure $p(t, x, y)$. The density ρ is a constant in view of incompressibility. The Navier-Stokes equation for such a fluid is given by

$$\rho(\mathbf{v}.\nabla\mathbf{v} + \mathbf{v}_{,t}) = -\nabla p + \eta\nabla^2\mathbf{v}$$

which in terms of components reads

$$v_x v_{x,x} + v_y v_{x,y} + v_{x,t} = -p_{,x}/\rho + \nu(v_{x,xx} + v_{x,yy})$$

$$v_x v_{y,x} + v_y v_{y,y} + v_{y,t} = -p_{,y}/\rho + \nu(v_{y,xx} + v_{y,yy})$$

where η is the viscosity and $\nu = \eta/\rho$. The notations $X_{,t}, X_{,x}, X_{,y}$ where $X(t,x,y)$ is a function of (t,x,y) denote $\frac{\partial X}{\partial t}, \frac{\partial X}{\partial x}$ and $\frac{\partial X}{\partial y}$ respectively. This equation along with the mass conservation equation

$$v_{x,x} + v_{y,y} = 0$$

for an incompressible fluid defines a set of three equations for the three variables v_x, v_y, p. The incompressibility equation implies the existence of a stream function $\psi(t,x,y)$ such that

$$v_x = \psi_{,y}, v_y = -\psi_{,x}$$

substituting this into the two Navier-Stokes equations gives us

$$\psi_{,y}\psi_{,xy} - \psi_{,x}\psi_{,yy} + \psi_{,ty} = -p_{,x}/\rho + \nu(\psi_{,xxy} + \psi_{,yyy})$$

$$-\psi_{,y}\psi_{,xx} + \psi_{,x}\psi_{,xy} - \psi_{,tx} = -p_{,y}/\rho - \nu(\psi_{,xxx} + \psi_{,xyy})$$

This is a system of two partial differential equations for the two variables ψ, p. To eliminate p and thereby obtain an equation involving only ψ, we take $\frac{\partial}{\partial y}$ of the first equation and subtract from it $\frac{\partial}{\partial x}$ of the second equation to get

$$(\psi_{,y}\psi_{,xy})_{,y} + (\psi_{,y}\psi_{,xx})_{,x} - (\psi_{,x}\psi_{,yy})_{,y} - (\psi_{,x}\psi_{,xy})_{,x} + \psi_{,tyy} + \psi_{,txx}$$

$$= \nu(\psi_{,xxy} + \psi_{,yyy})_{,y} + \nu(\psi_{,xxx} + \psi_{,xyy})_{,x}$$

which can be simplified to

$$\nabla^2\psi_{,t} + \psi_{,y}\psi_{,xyy} + \psi_{,y}\psi_{,xxx} - \psi_{,x}\psi_{,yyy} - \psi_{,x}\psi_{,xxy}$$

$$= \nu\nabla^2\nabla^2\psi$$

or

$$\nabla^2\psi_{,t} + \psi_{,y}\nabla^2\psi_{,x} - \psi_{,x}\nabla^2\psi_{,y} = \nu\nabla^2\nabla^2\psi$$

where

$$\nabla^2 = \frac{\partial^2}{\partial x^2} + \frac{\partial^2}{\partial y^2}$$

is the two dimensional Laplacian operator. From the above pde satisfied by ψ, it follows that

$$\psi_{,t}(t,x,y) = -\int G(x,y|x',y')F(\psi)(t,x',y')dx'dy'$$

where G is the Green's function for the two dimensional Laplace operator and the integration is carried out over the entire plane. Also,

$$F(\psi) = -\psi_{,y}\nabla^2\psi_{,x} + \psi_{,x}\nabla^2\psi_{,y} + \nu\nabla^2\nabla^2\psi$$

This gives an integro partial differential equation satisfied by $\psi(t, x, y)$ and is particularly suitable for simulation. Suppose t, x, y are discretized with time step τ and spatial steps Δ. The discretized stream function is then

$$\psi[n, k, m] = \psi(n\tau, k\Delta, m\Delta), n, k, m = 0, 1, 2, \ldots$$

and the equation satisfied by ψ after discretization approximates to

$$\psi[n + 1, k, m] = \psi[n, k, m] + \tau\Delta^2 \sum_{p,q} G[k, m|p, q] F(\psi)[n, p, q]$$

$F(\psi)[n, p, q]$ are finite difference approximates to $F(\psi)(t, x, y)$. For example, the term $\psi_{,y}$ is approximated as

$$\Delta^{-1}(\psi[n, p, q + 1] - \psi[n, p, q])$$

The term $\psi_{,x}$ is approximated as

$$\Delta^{-1}(\psi[n, p + 1, q] - \psi[n, p, q])$$

The term $\nabla^2\psi$ is approximated as

$$\Delta^{-2}(\psi[n, p + 1, q] + \psi[n, p - 1, q] + \psi[n, p, q + 1] + \psi[n, p, q - 1] - 4\psi[n, p, q])$$

and likewise for the other terms.

Chapter 4

Electromagnetics

4. Quantum mechanics

Introduction: This chapter discusses problems of quantum scattering theory, Feynman path integrals, quantum field theory based on creation and annihilation operator fields, The WKB method for approximate determination of wave functions of a quantum system, The Dirac equation for the electron, time dependent and time independent perturbation theory and notions of distance between two quantum states.

[1].Express in matrix notation, the Lorentz transformation formulae.

ans: Suppose K is a reference frame and K' is another frame that moves relative to K with a uniform velocity $\mathbf{v}$. Then the spatial coordinates in the two frames are connected via the formulae

$$\mathbf{x}'_{\parallel} = \gamma(\mathbf{x}_{\parallel} - \mathbf{v}t),$$

$$\mathbf{x}'_{\perp} = \mathbf{x}_{\perp}$$

where $\parallel$ denotes the component parallel to $\mathbf{v}$ and $\perp$ denotes the component perpendicular to $\mathbf{v}$. These can be expressed in a single equation as

$$\mathbf{x}' = \mathbf{x}'_{\parallel} + \mathbf{x}'_{\perp} = \gamma(\mathbf{x}_{\parallel} - \mathbf{v}t) + \mathbf{x}_{\perp}$$

$$= (\gamma - 1)(\mathbf{x}.\mathbf{v}\mathbf{v}/v^2 - \mathbf{v}t) + \mathbf{x}$$

This combined with the time equation

$$t' = \gamma(t - \mathbf{v}.\mathbf{x}/c^2)$$

furnishes the complete set of Lorentz transformation equations

[2].Express the Lorentz boost along the z axis in matrix notation.

hint:

$$t' = \gamma(t - vz), x' = x, y' = y, z' = \gamma(z - vt)$$

The spatio-temporal vector to be transformed is (t, x, y, z) and the matrix can now be formed.

[3]. Derive the formula for Lorentz transformation of single particle states.

ans:

Let $\psi_{k,\sigma}$ denote the state corresponding to particle four momentum k and spin component σ along the direction of k. Let Λ be a Lorentz transformation and $\Lambda \to U(\Lambda)$ a representation of the Lorentz group acting on single particle states. Then Let $L(p)$ denote a Lorentz transformation that carries the four vector k to the four vector p. We have

$$\psi_{p,\sigma} = N(p)U(L(p))\psi_{k,\sigma}$$

so that if Λ is an arbitrary Lorentz transformation, then

$$U(\Lambda)\psi_{p,\sigma} = N(p)U(\Lambda.L(p))\psi_{k,\sigma}$$

$$= N(p)U(L(\Lambda p))U(L(\Lambda p)^{-1}\Lambda L(p))\psi_{k,\sigma}$$

Now $U(\Lambda p)^{-1}\Lambda L(p)$ carries k to p, then to Λp and finally back to k. It follows that if $W(\Lambda, p) = U(L(\Lambda p)^{-1}\Lambda p)$ fixes k and hence we can write

$$U(W(\Lambda, p))\psi_{k,\sigma} = \sum_{\sigma'} D_{\sigma',\sigma}(W(\Lambda, p))\psi_{k,\sigma'}$$

Combining these two equations gives us

$$\psi_{p,\sigma} = (N(p)/N(\Lambda p)) \sum_{\sigma'} D_{\sigma',\sigma}(W(\Lambda, p))\psi_{\Lambda p,\sigma'}$$

Here, $N(p)$ is a normalization factor and D is a unitary representation of the Lorentz group that fixes the four vector k. The normalization factor $N(p)$ has been computed by Weinberg in his book "The Quantum theory of fields, Vol.I" by considering invariant four dimensional integrals.

[4]. Define (a) in and out states and (b) The scattering matrix in quantum scattering theory.

hints: Let H_0 be the Hamiltonian of the unperturbed system and V the perturbation. The total Hamiltonian is $H = H_0 + V$. Let Φ_α be the energy eigenstates of the unperturbed system so that

$$H_0\Phi_\alpha = E_\alpha\Phi_\alpha,$$

Let $\psi_{\alpha\pm}$ be states defined by

$$\psi_{\alpha\pm} = \Phi_\alpha + (E_\alpha - H_0 \pm i\epsilon)^{-1}V\psi_{\alpha\pm}$$

where ϵ is a small positive number. These are called the Lippman-Schwinger equations. Then we have the equation

$$(E_\alpha - H_0 - V)\psi_{\alpha\pm} = 0$$

or

$$(E_\alpha - H)\psi_{\alpha\pm} = 0$$

This means that $\psi_{\alpha\pm}$ are eigenstates of the perturbed Hamiltonian corresponding to the same energy E_α. We wish to explore the relationship between these perturbed states and the corresponding unperturbed states Φ_α. Let $g(\alpha)$ be a well behaved function of the continuum index α. Then consider the state

$$\Phi_g(t) = \int exp(-iE_\alpha t)g(\alpha)\Phi_\alpha d\alpha$$

and

$$\psi_{g\pm}(t) = \int exp(-iE_\alpha t)g(\alpha)\psi_{\alpha\pm}d\alpha$$

We want $\psi_{g+}(t)$ to represent a state evolving in accordance with the perturbed Hamiltonian H such that in the remote past, ie, as $t \to -\infty$, this state converges to $\Phi_g(t)$, the state evolving in accord with the unperturbed Hamiltonian H_0. First observe that

$$\Phi_g(t) = exp(-itH_0)\Phi_g, \psi_{g\pm}(t) = exp(-itH)\psi_{g,\pm}$$

where

$$\Phi_g = \int g(\alpha)\Phi_\alpha d\alpha, \psi_g = \int g(\alpha)\psi_{\alpha\pm}d\alpha$$

Then the equation

$$lim_{t\to-\infty}(\psi_{g,+}(t) - \Phi_g(t)) = 0$$

holds iff

$$lim_{t\to-\infty}(exp(-itH)\psi_{g+} - exp(-itH_0)\Phi_g) = 0$$

iff

$$\psi_{g,+} = \Omega(+)\Phi_g$$

where

$$\Omega(+) = lim_{t\to-\infty}exp(itH).exp(-itH_0)$$

Likewise

$$\psi_{g-} = \Omega(-)\Phi_g$$

where

$$\Omega(-) = lim_{t\to\infty}exp(itH)exp(-itH_0)$$

The scattering matrix elements are defined by

$$S_{\beta,\alpha} = (\psi_{\beta-}, \psi_{\alpha+})$$

We then have

$$S_{\beta,\alpha} = (\Omega(-)\Phi_\beta, \Omega(+)\Phi_\alpha) = (\Phi_\beta, \Omega(-)^*\Omega(+)\Phi_\alpha)$$

$$= (\Phi_\beta, S\Phi_\alpha)$$

where

$$S = \Omega(-)^*\Omega(+)$$

is the scattering matrix, or scattering operator. Note that the scattering matrix elements have been defined with respect to the free particle states. From the Lippman-Schwinger equations, we have

$$\psi_{g+}(t) = \int g(\alpha) exp(-iE_\alpha t)\psi_{\alpha+} d\alpha$$

$$= \int g(\alpha) exp(-iE_\alpha t)\Phi_\alpha d\alpha + g(\alpha) exp(-iE_\alpha t)\int (E_\alpha - H_0 + i\epsilon)V\psi_{\alpha+} d\alpha$$

$$= \Phi_g(t) + \int \Phi_\beta g(\alpha) exp(-iE_\alpha t)\frac{(\Phi_\beta, V\psi_{\alpha+})}{E_\alpha - E_\beta + i\epsilon} d\alpha d\beta$$

Now suppose $t \to -\infty$. Assuming that the $E'_\alpha s$ are all positive, it follows that the contour of integration over α can be completed using the infinite semicircle in the upper half complex plane since $Im(E) > 0$ and $t \to -\infty$ implies $exp(-iEt) \to 0$. This means that

$$lim_{t\to-\infty}(\psi_{g+}(t) - \Phi_g(t)) = 0$$

as desired. Let us look at the limit of $\psi_{g+}(t)$ as $t \to +\infty$. To this end note that

$$(\Phi_\beta, \psi_{g+}(t)) = \int g(\alpha) exp(-iE_\alpha t)\delta(E_\beta - E_\alpha) d\alpha - \int \frac{T_{\beta\alpha+}}{E_\alpha - E_\beta + i\epsilon} exp(-iE_\alpha t)g(\alpha) d\alpha$$

where

$$T_{\beta\alpha+} = (\Phi_\beta, V\psi_{\alpha+})$$

As $t \to \infty$, the contour integral in the second term can be taken over the lower half semicircle and noting that the pole at $E_\beta - i\epsilon$ falls in the lower half complex plane, we get for large positive t,

$$(\Phi_\beta, \psi_{g+}(t)) = g(\beta) exp(-iE_\beta t) + \int exp(-iE_\alpha t)T_{\beta\alpha+}\delta(E_\alpha - E_\beta)g(\alpha) d\alpha$$

On the other hand, we have

$$(\psi_{\beta-}, \psi_{\alpha+}) = S_{\beta\alpha} = (\Phi_\beta, S\Phi_\alpha)$$

so that

$$\psi_{\alpha+} = \int S_{\beta\alpha}\psi_{\beta-} d\beta$$

so that

$$(\Phi_\beta, \psi_{\alpha+}) = \int S_{\gamma\alpha}(Phi_\beta, \psi_{\gamma-}) d\gamma$$

$$= \int S_{\gamma\alpha}(\Phi_\beta, \Phi_\gamma + (E_\gamma - H_0 - i\epsilon)^{-1}V\psi_{\gamma-}) d\gamma$$

$$= S_{\beta\alpha} + \int d\gamma S_{\gamma\alpha}(\Phi_\beta, (E_\gamma - H_0 - i\epsilon)^{-1}V\psi_{\gamma-})$$

$$= S_{\beta\alpha} + \int d\gamma S_{\gamma\alpha}\frac{(\Phi_\beta, V\psi_{\gamma-})}{E_\gamma - E_\beta - i\epsilon}$$

Many more identities for the scattering matrix can be derived using this formalism. See "Quantum Field Theory" by Steven Weinberg. Now observe that

$$\psi_{g+}(t) = \int g(\alpha)\psi_{\alpha+}(t)d\alpha$$

$$= \int g(\alpha)S_{\beta\alpha}\psi_{\beta-}(t)d\beta d\alpha$$

Now

$$lim_{t\to\infty}(\psi_{\beta-}(t) - \Phi_{\beta}(t)) = 0$$

so that for large positive t, we have

$$\psi_{g+}(t) = \int g(\alpha)S_{\beta\alpha}\Phi_{\beta}(t)d\beta d\alpha$$

and hence we get for large positive t,

$$\int g(\alpha)S_{\beta\alpha}exp(-iE_{\beta}t)d\alpha$$

$$= g(\beta)exp(-iE_{\beta}t) + \int exp(-iE_{\alpha}t)T_{\beta\alpha+}\delta(E_{\alpha} - E_{\beta})g(\alpha)d\alpha$$

Noting that $S_{\beta\alpha}$ already contains a factor $\delta(E_{\alpha} - E_{\beta})$, it follows that

$$S_{\beta\alpha} = \delta(E_{\beta} - E_{\alpha}) + T_{\beta\alpha+}\delta(E_{\beta} - E_{\alpha})$$

and this is one very important formula for the scattering matrix. Some additional interesting formulae may be derived. For example

$$(\psi_{\beta-}, \psi_{\alpha+})$$

[5]. We consider a quantum field of the form

$$\psi_l(x) = \sum_{\sigma} \int (u_l(x,p,\sigma)a(p,\sigma) + v_l(x,p,\sigma)a^*(p,\sigma))d^3p$$

where $a(p,\sigma,n)$ is the operator that annihilates a particle having four momentum p and spin component σ while $a^*(p,\sigma)$ creates a particle having four momentum p and spin component σ. Let Λ denote a Lorentz transformation and $U(\Lambda)$ the representation at Λ that acts on states. For Lorentz covariance of the theory, we require that

$$U(\Lambda)\psi_l(x)U(\Lambda)^{-1} = \sum_{l'} D_{l,l'}(\Lambda^{-1})\psi_{l'}(\Lambda x)$$

Suppose Φ_0 is the vacuum. Then,

$$\Phi_{p,\sigma} = a^*(p,\sigma)\Phi_0$$

is a state in which there is one particle having four momentum p and spin component σ. The Lorentz transformation law of this state has the form

$$U(\Lambda)\Phi_{p,\sigma} = N(\Lambda,p)\sum_{\sigma'} D_{\sigma',\sigma}(W(\Lambda,p))\Phi_{\Lambda p,\sigma'}$$

Also,

$$U(\Lambda)\Phi_0 = \Phi_0$$

and hence we get

$$U(\Lambda)a^*(p,\sigma)U(\Lambda)^{-1} = N(\Lambda,p)\sum_{\sigma'} D_{\sigma',\sigma}(W(\Lambda,p))a^*(\Lambda p,\sigma')$$

Since $U(\Lambda)$ is unitary, it follows by taking adjoints that

$$U(\Lambda)a(p,\sigma)U(\Lambda)^{-1} = N(\Lambda,p)\sum_{\sigma'} D_{\sigma,\sigma'}(W(\Lambda,p)^{-1})a(\Lambda p,\sigma')$$

Now define

$$\psi_{l+}(x) = \sum_{\sigma}\int u_l(x,p,\sigma)a(p,\sigma)d^3p,$$

$$\psi_{l-}(x) = \sum_{\sigma}\int v_l(x,p,\sigma)a^*(p,\sigma)d^3p$$

Then, on the one hand,

$$U(\Lambda)\psi_{l+}(x)U(\Lambda)^{-1} = \sum_{l'} D_{l,l'}(\Lambda^{-1})\psi_{l'+}(\Lambda x)$$

$$= \sum_{l',\sigma} D_{l,l'}(\Lambda^{-1})\int u_{l'}(\Lambda x,p,\sigma)a(p,\sigma)d^3p$$

and on the other,

$$U(\Lambda)\psi_{l+}(x)U(\Lambda)^{-1} = \sum_{\sigma}\int u_l(x,p,\sigma)U(\Lambda)a(p,\sigma)U(\Lambda)^{-1}d^3p$$

$$= N(\Lambda,p)\sum_{\sigma,\sigma'}\int u_l(x,p,\sigma)D_{\sigma,\sigma'}(W(\Lambda,p)^{-1})a(\Lambda p,\sigma')d^3p$$

Note that in all these computations,

$$p^2 = p^{02} - \mathbf{p}^2 = m^2$$

or equivalently,

$$p^0 = E(\mathbf{p}) = \sqrt{m^2 + \mathbf{p}^2}$$

Comparing these two equations, we arrive at a constraint on the functions u_l. We leave it as an exercise to state these constraint equations.

[6]. Feynman's path integral approach to quantum mechanics: Let $|q, t> = exp(itH)|q>$ and $|p, t> = exp(itH)|p>$. Then

$$< q'', t + \tau | q', t > = < q'' | exp(-i\tau H) | q' >$$

We wish to evaluate this. Let $H = H(q, p)$ be such that in the expansion of H, all the terms involving p appear to the left of q. Then,

$$< q'' | exp(-i\tau H(q, p)) | q' > = \int < q'' | p' >' \, exp(-i\tau H(q, p')) < p' | q' > dp'$$

$$= \int exp(i(p'(q' - q'') - \tau H(q', p'))dp'$$

[7]. Explain the notion of path integrals for bosonic and fermionic fields. For bosonic fields, take the electromagnetic field as the prototype and calculate its propagator. For fermionic fields, take the Dirac field as the prototype and calculate its propagator.

hint: For a single bosonic field $\phi(x)$ corresponding to a particle of mass m, the Lagrangian density is given by

$$\mathcal{L}(\phi) = \frac{1}{2}\partial_\mu \phi \partial^\mu \phi - m^2 \phi^2 / 2$$

The path integral corresponding to this Lagrangian is therefore a Gaussian integral and it can be evaluated using the familiar formula

$$\int exp(-x^T A x / 2) exp(x^T u) dx = (2\pi)^{n/2} det(A)^{-1/2} exp(u^T A^{-1} u / 2)$$

[8]. Explain how you would implement the Dirac equation for a free particle on the DSP processor.

hint:

$$(\gamma^\mu \partial_\mu - m)\psi = 0$$

We are given $\psi(t, x, y, z)$ at $t = 0$ and wish to calculate $\psi(t, x, y, z)$ at $t > 0$. This can be done by discretizing the above equation:

$$\gamma^0 \partial_0 \psi(t, x)$$

is replaced by

$$\Delta^{-1}\gamma^0(\psi[n + 1, k, m, p] - \psi[n, k, m, p])$$

where Δ is the time discretization step size. Also

$$\gamma^1 \partial_x \psi(t, x, y, z)$$

is replaced by

$$\delta^{-1}\gamma^1(\psi[n, k + 1, m, p] - \psi[n, k, m, p])$$

where δ is the spatial discretization step in the x direction. Likewise,

$$\gamma^2 \partial_y \psi(t, x, y, z)$$

is replaced by

$$\delta^{-1}\gamma^2(\psi[n, k, m+1, p] - \psi[n, k, m, p])$$

[9]. Calculate the transition probability amplitudes for an atom from one stationary state to another under the influence of a sinusoidally varying electric field.

hint:Assume that the electrostatic potential is zero and evaluate the magnetic vector potential A in terms of the electric field and substitute in the formula $(p + eA)^2/2m$ where e is the electronic charge.

[10]. Derive discrete algorithms for computing the scattering matrix elements of a perturbed quantum system. Assume that the unperturbed Hamiltonian is H_0 and that the perturbed Hamiltonian is $H = H_0 + V$.

Hint: Let Let $\Phi(t)$ be a free state so that its evolution law has the form

$$\Phi(t) = exp(-itH_0)\Phi(0)$$

and $\psi + (t)$ be the scattered state into which this free state evolves, so that

$$\psi + (t) = exp(-itH)\psi(0)$$

As $t \to -\infty$, it follows that $\psi_+ t) \approx \Phi(t)$. Therefore,

$$\psi + (0) = lim_{t \to -\infty} exp(itH)exp(-itH_0)\Phi(0)$$

and hence, if we define the Moller operator

$$\Omega(+) = lim_{t \to -\infty} exp(itH)exp(-itH_0)$$

then we have

$$\psi_+(0) = \Omega(+)\Phi(0)$$

likewise

$$\psi_-(0) = \Omega(-)\Phi(0)$$

where

$$\Omega(-) = lim_{t \to \infty} exp(itH)exp(-itH_0)$$

The scattering operator is given by

$$S = \Omega(-)^*\Omega(+)$$

and to evaluate this, we must therefore develop algorithms for calculating the Moller operators $\Omega(\pm)$. Consider

$$U(t) = exp(itH)exp(-itH_0)$$

We have

$$U'(t) = iexp(itH).(H - H_0).exp(-itH_0) = i.exp(itH).V.exp(-itH_0)$$

$$= iexp(itH)exp(-itH_0)W(t) = iU(t)W(t)$$

where

$$W(t) = exp(itH_0).V.exp(-itH_0)$$

It follows that

$$U(t) = I + \sum_{n=1}^{\infty} i^n \int_{0<t_1<...<t_n<t} W(t_1)W(t_2)...W(t_n)dt_1 dt_2...dt_n$$

An integral of the form

$$\int_{t_1<...<t_n<t} W(t_1)...W(t_n)dt_1...dt_n$$

can be evaluated numerically as follows:Let Δ be the time discretization step size. Then for $t = N\Delta$, the above multiple integral is approximated by the discrete sum

$$\Delta^n \sum_{0 \leq r_1 \leq r_2 \leq ... \leq r_n \leq N} W[r_1]W[r_2]...W[r_n], W[j] = W(j\Delta)$$

[11]. Iterative evaluation of the scattering operator. From the Lippman-Schwinger equation

$$\psi_{\alpha+} = \Phi_\alpha + (E_\alpha - H_0 + i\epsilon)^{-1}V\psi_{\alpha+}$$

$$= \Phi_\alpha + \int \Phi_\beta \frac{<\Phi_\beta, V\psi_{\alpha+}>}{E_\alpha - E_\beta + i\epsilon} d\beta$$

Define the operator

$$T_\alpha = (E_\alpha - H_0 + i\epsilon)^{-1}V$$

Then,

$$\psi_{\alpha+} = \Phi_\alpha + T_\alpha\psi_{\alpha+}$$

$$= (I + \sum_{n=1}^{\infty} T_\alpha^n)\Phi_\alpha$$

We need a DSP method for evaluating the operators T_α^n. Note that

$$T_\alpha\psi = \int \Phi_\beta < \Phi_\beta, V\psi > d\beta/(E_\alpha - E_\beta + i\epsilon)$$

or equivalently,

$$T_\alpha = W_\alpha V$$

where

$$W_\alpha = (E_\alpha - H_0 + i\epsilon)^{-1} = \int \frac{|\Phi_\beta><\Phi_\beta|}{E_\alpha - E_\beta + i\epsilon} d\beta$$

Then,

$$T_\alpha^n = (W_\alpha V)^n = W_\alpha.V.W_\alpha.V...W_\alpha.V$$

where W_α and V both occur n times in this product. For example,

$$(W_\alpha V)^2 = W_\alpha V W_\alpha V$$

$$= \int \frac{\Phi_{\beta'}><\Phi_{\beta'}|V|\Phi_\beta><\Phi_\beta|V}{(E_\alpha - E_{\beta'} + i\epsilon)(E_\alpha - E_\beta + i\epsilon)} d\beta.d\beta'$$

[12]. Quantization of the motion of a simple pendulum:

$$L(\theta, \theta') = mL^2\theta'^2/2 + mgL.cos(\theta)$$

$$p = L_{,\theta'} = mL^2\theta'$$

so that

$$H(\theta, p) = p\theta' - L = p^2/2mL^2 - mgL.cos(\theta)$$

This is the Hamiltonian of the pendulum. The stationary state Schrodinger equation is

$$-\frac{h^2}{8\pi^2 mL^2}\psi''(\theta) - mgL.cos(\theta)\psi(\theta) = E\psi(\theta)$$

To determine the energy levels approximately, we set

$$cos(\theta) \approx 1 - \theta^2/2 + \theta^4/24$$

so that with neglect of a constant, the Hamiltonian is the sum of the Hamiltonian of a Harmonic oscillator plus and anharmonic term:

$$H(\theta, p) = H_{harm}(\theta, p) - mgL\theta^4/24$$

where

$$H_{harm}(\theta, p) = -\frac{h^2}{8\pi^2 mL^2}\frac{\partial^2}{\partial\theta^2} + mgL\theta^2/2$$

[13]. Quantization of the motion of a double pendulum. This is more tricky. First, we must write down the Lagrangian of the system, $L(\theta_1, \theta_2, \theta_1', \theta_2')$, then compute the canonical momenta

$$p_i = L_{,\theta_i'}, i = 1, 2$$

Then solve for $\theta_i', i = 1, 2$ in terms of $p_i, \theta_i, i = 1, 2$ and then apply the Legendre transformation to obtain the Hamiltonian $H(\theta_1, \theta_2, p_1, p_2)$. The final step is to replace p_a by $-ih\frac{\partial}{\partial\theta_a}, a = 1, 2$ and then make approximations to terms of the form $cos(\theta_1 - \theta_2)$ and $sin(\theta_1 - \theta_2)$. After this, the change in the energy levels may be determined using the standard tools of perturbation theory.

[14]. Born-Oppenheimer approximation to the computation of the energy levels of a crystal. The crystal is assumed to comprise of N nuclei and K electrons. Let $\mathbf{R}_i, i = 1, 2, ..., N$ denote the nuclear positions and $-ih\nabla_{R_i}$ the corresponding nuclear momentum operators. Let $\mathbf{r}_i, i = 1, 2, ..., K$ denote the electron positions and ∇_{r_i} the corresponding electron momentum operators. The nuclear kinetic energy is given by

$$T_N = -\frac{h^2}{2M} \sum_{i=1}^{N} \nabla^2_{R_i}$$

The electron kinetic energy is given by

$$T_e = -\frac{h^2}{2m} \sum_{i=1}^{K} \nabla^2_{r_i}$$

The electron-electron interaction potential energy is given by

$$V_e = \sum_{1 \leq i < j \leq K} \frac{e^2}{r_{ij}}, r_{ij} = |\mathbf{r}_i - \mathbf{r}_j|$$

The nucleus-nucleus interaction potential energy is given by

$$V_N = \sum_{1 \leq i < j \leq N} \frac{Z^2 e^2}{R_{ij}}, R_{ij} = |\mathbf{R}_i - \mathbf{R}_j|$$

The electron-nucleus interaction energy is given by

$$V_{eN} = - \sum_{1 \leq i \leq K, 1 \leq j \leq N} \frac{Z e^2}{|\mathbf{r}_i - \mathbf{R}_j|}$$

We note that T_N depends only on derivatives with respect to nuclear coordinates, T_e depends only on derivatives with respect to electron coordinates, V_e depends only on electron coordinates, V_N depends only on nuclear coordinates, and finally V_{eN} depends both on nuclear and electron coordinates. The total Hamiltonian of the system is given by

$$H = (T_N + T_e + V_e + V_N + V_{eN})$$

We set

$$H_e = T_e + V_e + V_{eN}$$

so that

$$H = H_e + H_N, H_N = T_N + V_N$$

Keeping the nuclear coordinates $\mathbf{R}_i$ as fixed, we determine the electron energy by solving the eigenvalue problem

$$H_e(\mathbf{r}, \mathbf{R})\phi_k(\mathbf{r}, \mathbf{R}) = E_{ek}(\mathbf{R})\phi_k(\mathbf{r}, \mathbf{R})$$

Here, the nuclear coordinates $\mathbf{R}$ appear as parameters in both the electronic wave function ϕ_k and in the electronic energy E_{ek}. We then set the total wave function

$$\psi_k(\mathbf{r},\mathbf{R}) = \phi_k(\mathbf{r},\mathbf{R})\chi_k(\mathbf{R})$$

Substitute this into the eigenvalue equation

$$H\psi_k = E_k\psi_k$$

to get

$$(H_e + H_N)\phi_k\chi_k = E_k\phi_k\chi_k$$

Now observe that since χ_k depends only on $\mathbf{R}$ and H_e does not contain derivatives with respect to $\mathbf{R}$, it follows that

$$H_e\phi_k\chi_k = \chi_k H_e\phi_k = \chi_k E_{ek}\phi_k = E_{ek}(\mathbf{R})\phi_k(\mathbf{r},\mathbf{R})\chi_k(\mathbf{R})$$

Hence the above equation reduces to

$$(H_N(\mathbf{r},\mathbf{R}) + E_{ek}(\mathbf{R}))\phi_k(\mathbf{r},\mathbf{R})\chi_k(\mathbf{R}) = E_k\phi_k(\mathbf{r},\mathbf{R})\chi_k(\mathbf{R})$$

In the zeroth order approximation, we assume that the nuclei are too heavy and hence their kinetic energy operator T_N has no action on the electron wave function $\phi_k(\mathbf{r},\mathbf{R})$ which are computed keeping the nuclear coordinates fixed. Hence, in this approximation, we get

$$H_N\phi_k\chi_k \approx \phi_k T_N\chi_k + V_N\phi_k\chi_k = \phi_k(T_N + V_N)\chi_k$$

Then the above equation gives

$$(T_N(\mathbf{R}) + V_N(\mathbf{R}) + E_{ek}(\mathbf{R}))\chi_k(\mathbf{R}) = E_k\chi_k(\mathbf{R})$$

This equation states that to complete the solution to the energy level determination problem, we assume that the nuclei are moving in the extra potential energy given by the electron energy levels.

[15]. Discuss one loop radiative corrections for the photon propagator in quantum electrodynamics: Consider an initial photon line labeled k_1,μ_1 corresponding to the momentum space wave function connected to the other end labeled k_1,μ_2. Another photon line labeled k_2,μ_3 is connected to the other end labeled k_2,μ_4. The photon lines are depicted using wavy lines. Now there are two electron paths between the points labeled μ_2,k_1 and μ_3,k_2. The first goes on top and is labeled p_1. It is directed from μ_2 to μ_3. Its two ends are labeled α_1 and α_2 respectively. The second line goes on bottom from 3 to 2 and its two ends are labeled β_1 and β_2 and the line itself is labeled p_2. Let $S_{\beta\alpha}(p)$ denote the electron propagator and $\Delta_{\beta\alpha}(k)$ the photon propagator. Then, the value of this Feynman diagram that represents a photon propagating from (k_1,μ_1) to (k_2,μ_4) is given by

$$\int \Delta_{\mu_2,\mu_1}(k_1)(\gamma^{\mu_1})_{\alpha_1,\beta_2} S_{\alpha_2,\alpha_1}(p_1) S_{\beta_2,\beta_1}(p_2)(\gamma^{\mu_3})_{\beta_1,\alpha_2}\Delta_{\mu_4,\mu_3}(k_2)$$

$$\delta(k_1 - p_1 + p_2)\delta(k_2 - p_1 + p_2)d^4p_1 d^4p_2$$

This evaluates to

$$\delta(k_1 - k_2) \int \Delta_{\mu_2,\mu_1}(k_1)(\gamma^{\mu_1})_{\alpha_1,\beta_2} S_{\alpha_2,\alpha_1}(p_1) S_{\beta_2,\beta_1}(p_1 - k_1)(\gamma^{\mu_3})_{\beta_1,\alpha_2} \Delta_{\mu_4,\mu_3}(k_1) d^4 p_1$$

$$= \delta(k_1 - k_2) \int \Delta_{\mu_2,\mu_1}(k_1) Tr(S(p_1)\gamma^{\mu_1} S(p_1 - k_1)\gamma^{\mu_3}) \Delta_{\mu_4,\mu_3}(k_1) d^4 p_1$$

Note that the photon propagator in the Feynman gauge is given by

$$\Delta_{\mu\nu}(k) = \frac{g_{\mu\nu}}{k^2 - i\epsilon}$$

[16]. What do you understand by a vector operator in quantum mechanics ?

Hint: Let $\mathcal{H}$ be the Hibert space of quantum states and let $R \to \pi(R)$ be a unitary representation of the rotation group $SO(3)$ in the Hilbert space $\mathcal{H}$. Write $R = exp(i(\theta_1 L_1 + \theta_2 L_2 + \theta_3 L_3))$ so that L_1, L_2, L_3 are the generators of $SO(3)$. Then,

$$\pi(R) = exp(i(\theta_1 J_1 + \theta_2 J_2 + \theta_3 J_3))$$

where

$$J_a = \tilde{\pi}(L_a), a = 1, 2, 3$$

Here, $\tilde{\pi}$ is the representation of the Lie algebra of $SO(3)$ in the Hilbert space $\mathcal{H}$. Specifically,

$$\pi(R) = \pi(exp(i\sum_{a=1}^{3}\theta_a L_a) exp(i(\sum_{a=1}^{3}\theta_a J_a)))$$

We know that the angular momenta $L_a, a = 1, 2, 3$ satisfy the commutation relations

$$[L_a, L_b] = i\epsilon_{abc} L_c$$

where summation over the repeated index c is implied. Here, $\epsilon_{123} = \epsilon_{231} = \epsilon_{312} = 1$ and $\epsilon_{213} = \epsilon_{132} = \epsilon_{321} = -1$. Since $\tilde{\pi}$ is a Lie algebra homomorphism, it follows that the generators J_a of the Lie algebra $\tilde{\pi}(so(3))$ satisfy the same commutation relations:

$$[J_a, J_b] = i\epsilon_{abc} J_c$$

Suppose S is a scalar operator in the Hilbert space $\mathcal{H}$. Then S is invariant under rotations, so

$$[J_a, S] = 0, a = 1, 2, 3$$

This implies that

$$\pi(R)S\pi(R)^{-1} = S, R \in SO(3)$$

Suppose $\psi \in \mathcal{H}$. Then under a rotation R, ψ transforms to $\pi(R)\psi$ and under the same rotation, $S\psi$ transforms to $\pi(R)S\psi$. So invariance of S under R means that for all $\psi \in \mathcal{H}$, we must have

$$S\pi(R)\psi = \pi(R)S\psi$$

or

$$S\pi(R) = \pi(R)S$$

or

$$S = \pi(R)S\pi(R)^{-1}$$

which is the correct invariance condition stated above. Suppose $V_a, a = 1, 2, 3$ is a vector operators. Then by definition, we must have

$$\pi(R)V_a\pi(R)^{-1} = \sum_b R_{ab}V_b$$

and taking R as infinitesimal, ie

$$R = I + i\epsilon_a L_a$$

gives

$$\pi(R) = I + i\epsilon_a J_a$$

and

$$(I + i\epsilon_b J_b)V_a(I - i\epsilon_b J_b) = V_a + i\epsilon_b(L_b)_{ac}V_c$$

from which we deduce that

$$[J_b, V_c] = i\epsilon_{bca}V_a, b, c = 1, 2, 3$$

This is the correct transformation law for a vector operator.

Matrix elements of a scalar operator: Let $|j, m >$ denote the common eigenstate of $J^2 = \sum_{a=1}^{3} J_a^2$ and J_3 corresponding respectively to the eigenvalues $j(j + 1)$ and m, ie,

$$J^2|j, m >= j(j + 1)|j, m >, J_3|j, m >= m|j, m >$$

Define the ladder operators

$$J_+ = J_1 + iJ_2, J_- = J_+^* = J_1 - iJ_2$$

Then we have

$$[J_+, J_3] = [J_1 + iJ_2, J_3] = [J_1, J_3] + i[J_2, J_3] = -iJ_2 - J_1 = -J_+$$

This equation can be expressed as

$$J_+ J_3 - J_3 J_+ = -J_+$$

It follows that

$$J_+ J_3|j, m > -J_3 J_+|j, m >= -J_+|j, m >$$

or

$$(m + 1)J_+|j, m >= J_3 J_+|j, m >$$

In other words, there exists a constant $c(j, m)$ such that

$$J_+|j, m >= c(j, m)|j, m + 1 >$$

Likewise,

$$J_-|j, m >= d(j, m)|j, m - 1 >$$

for some constant $d(j, m)$. Now consider the commutation relations

$$[J_+, V_1] = [J_1 + iJ_2, V_1] = -\epsilon_{213} V_3 = V_3$$

$$[J_+, V_2] = [J_1 + iJ_2, V_2] = i\epsilon_{123} V_3 = iV_3$$

$$[J_+, V_3] = [J_1 + iJ_2, V_3] = [J_1, V_3] + i[J_2, V_3] = i\epsilon_{132} V_2 - \epsilon_{231} V_1 = -iV_2 - V_1 = -V_+$$

Thus,

$$[J_+, V_+] = [J_+, V_1 + iV_2] = [J_+, V_1] + i[J_+, V_2] = V_3 - V_3 = 0$$

$$[J_+, V_-] = [J_+, V_1 - iV_2] = [J_+, V_1] - i[J_+, V_2] = V_3 + V_3 = 2V_3$$

Suppose S is a scalar operator. Then,

$$[J_+, S] = 0, [J_-, S] = 0.[J_3, S] = 0$$

and hence

$$0 = < j, m'|J_+S - SJ_+|j, m >= (\bar{d}(j, m') - c(j, m)) < j, m'|S|j, m >$$

Also,

$$0 = < j, m'|J_3 S - SJ_3|j, m >= (m' - m) < j, m'|S|j, m >$$

from which we deduce that if $m' \neq m$, then, $< j, m'|S|j, m >= 0$. This is an important selection rule. The selection rules for vector operators are obtained using the Wigner-Eckart theorem (Ref:Arno Bohm, "Quantum mechanics, foundations and applications").

[17]. ϕ^4 theory. We consider a scalar field $\phi(t, x)$ in space-time whose dynamics is described by the Lagrangian density

$$\mathcal{L}(\phi, \partial_\mu \phi) = \frac{1}{2} \partial_\mu \phi \partial^\mu \phi - \frac{1}{2} \phi^2 - V(\phi)$$

The action principle $\delta \int \mathcal{L} d^4x = 0$ gives us the field equations

$$\partial_\mu \partial^\mu \phi + \phi + V'(\phi) = 0$$

This is classical field theory. It can be quantized successfully using the Feynman path integral for fields. Note that the transition amplitudes can be approximated to any degree by using the following method: If $\mathcal{L} = \mathcal{L}_0 - V(\phi)$ and $S_0 = \int \mathcal{L}_0 d^4x$ and $S = \int \mathcal{L} d^4x = S_0 - \int V d^4x$, then

$$\int exp(iS)\mathcal{D}\phi = \int exp(iS_0)(\sum_{n \geq 0}(-i)^n(\int V(\phi)d^4x)^n/n!)\Pi_x d\phi(x)$$

$$= \sum_{n \geq 0}(-i)^n \int d^4x_1...d^4x_n \int exp(iS_0(\phi))V(\phi(x_1))V(\phi(x_2))...V(\phi(x_n))\Pi_x d\phi(x)$$

[18]. Syllabus for a course on quantum computation and quantum system simulation.

1.Definition of the state space of a finite state quantum system as a finite dimensional inner product space.

2.The notion of a quantum probability space $(\mathcal{H}, P(\mathcal{H}), \rho)$ where $\mathcal{H}$ is a Hilbert space, $P(\mathcal{H})$ is the lattice of orthogonal projections in $\mathcal{H}$ and ρ is a density operator, ie, a positive definite Hermitian operator in $\mathcal{H}$ having unit trace. An event is an element P in $P(\mathcal{H})$, the probability of which is given by $\mu(P) = Tr(\rho.P)$.

3.Gleason's theorem on the representation of probability measures on the projection lattice in terms of density matrices.

4.Wigner's theorem on automorphisms of $P(\mathcal{H})$. Every automorphism can be realized via a unitary or antiunitary operator. Specifically, if $\tau : P(\mathcal{H}) \to P(\mathcal{H})$ is an automorphism, then there exists a unitary or antinunitary operator U in $\mathcal{H}$ such that

$$\tau(P) = UPU^*, P \in P(\mathcal{H})$$

5.Hahn-Hellinger theorem on the general form of spectral measures on a separable Hilbert space.

6.System and bath dynamics. The Hamiltonian of the system and bath has the form

$$H = H_s + H_r + H_{sr}$$

where H_s is the system Hamiltonian, H_r the bath Hamiltonian and H_{sr} is the interaction Hamiltonian between the system and bath. H_s and H_r commute and we set $H_0 = H_s + H_r$. Let $\mathcal{H}_s$ denote the Hilbert space of the system variables, $\mathcal{H}_r$ the Hilbert space of the bath variables. Then the Hilbert space of the total system plus bath is given by $\mathcal{H}_s \otimes \mathcal{H}_r$. The evolution of the total state of the system plus bath is given by

$$\psi_t = exp(-it(H_0 + H_{sr}))\psi_0$$

We look at the interaction representation of this dynamics. Set

$$\psi_t = exp(-itH_0)\phi_t$$

Then,

$$id\phi_t/dt = \tilde{H}_{sr}(t)\phi_t$$

where

$$\tilde{H}_{sr}(t) = exp(itH_0)H_{sr}exp(-itH_0)$$

A given observable X in the interaction representation evolves as

$$X(t) = exp(itH_0)Xexp(-itH_0)$$

[19] Problems in quantum electrodynamics

[1] Derive the formula for the energy of an electromagnetic field in the Coulomb gauge in terms of the Fourier components of the magnetic vector potential. Using this, explain how an analogy can be made between the electromagnetic field and an ensemble of harmonic oscillators.

[2] Explain how using the previous problem, the electromagnetic field can be quantized and derive the commutation relations between the magnetic vector potential at two space-time points.

[3] Derive the equation for the evolution of a state in the interaction representation when the Hamiltonian has the form $H = H_0 + V(t)$ where H_0 is the unperturbed Hamiltonian and $V(t)$ is the interaction Hamiltonian. Hence arrive at the relationship between the interaction picture representation and the Heisenberg picture representation of states and observables.

[4] Prove that the four current density for the Dirac equation has the form $J^\mu = \bar{\psi}\gamma^\mu\psi$ where $\bar{\psi} = \psi^*\gamma^0$. Prove using the Dirac equation for the free particle that the four divergence of J^μ vanishes.

[5] If $H(t)$ is the Hamiltonian of a quantum mechanical system then show that $T(exp(\frac{-2\pi i}{h}\int_{t_1}^{t_2} H(t)dt))$ where T is the time ordering operator is the evolution operator that carries the state at time t_1 to the state at time t_2. How is this operator computed in the interaction representation for the interaction between electrons and the electromagnetic field. Note that the interaction Hamiltonian density has the form $J_\mu A^\mu$ where $J^\mu = \bar{\psi}\gamma^\mu\psi$.

[6] What do you understand by the photon propagator and the electron propagator. Explain how you would use these in the computation of scattering amplitudes.

[7] What is the meaning of the exact photon propagator. What are one loop radiative corrections.

[8] How do you calculate the transition probabilities for an atom between two stationary states in the presence of electromagnetic radiation.

[9] Derive the formula for the scattering amplitude of an electron by photon. The initial state is $|p, k >$ corresponding to the electron having momentum p and the photon having momentum k and the final state is $|p', k' >$ corresponding to the electron having momentum p' and the photon having momentum k'. The momentum conservation implies $p + k = p' + k'$ and this introduces a momentum conserving δ function $\delta(p + k - p' - k')$ into the scattering amplitude. The evolution operator in the interaction representation is $T(exp(-\frac{2\pi i}{h}\int J^\mu A_\mu d^4x))$ where T is the time ordering operator.

Reference:Landau and Lifshitz, Quantum Electrodynamics.

[20]. Develop a numerical package using MATLAB for Feynman's path integral approach to quantum mechanics. Suppose $L(q(t), q'(t), t)$ is the Lagrangian of a dynamical system. Then the quantum mechanical evolution kernel for the time duration $[0, T]$ with $q(0) = x, q(T) = y$ is given by

$$K_T(y, x) = \int_{0,x}^{T,y)} exp(\int_0^T L(q(t), q'(t), t)dt)\mathcal{D}q$$

Where $\mathcal{D}q$ denotes the standard measure on the path space over the duration $[0, T]$. Numerical implementation of this path integral is carried out by approximating $K_T(x, y)$ with

$$\tilde{K}_N(y, x) = \int_{\mathbb{R}^{N-1}} exp(\Delta . \sum_{n=0}^{N-1} L(q[n], (q[n+1] - q[n])/\Delta, n\Delta)\Pi_{j=1}^{N-1}dq[j]$$

where $q[0] = x, q[N] = y$. If $\psi_0(x)$ is the wave function at time 0, then the wave function at time T after appropriate normalization of the kernel K_T is given by

$$\psi_T(x) = \int K_T(y, x)\psi_0(x)dx$$

Example: For the forced harmonic oscillator, with forcing $f(t)$, the action has the form

$$S[x, T, f] = \frac{m}{2}\int_0^T x'^2(t)dt - \frac{m\omega^2}{2}\int_0^T x^2(t)dt - \int_0^T f(t)x(t)dt$$

and the path integral can be evaluted using formulae for integrating multivariate Gaussian densities. The discretized action can be expressed as

$$S_d[x, N, f] = \frac{m\Delta}{2}\sum_{n=0}^{N-1}(x[n+1] - x[n])^2/\Delta - \frac{m\omega^2\Delta}{2}\sum_{n=0}^{N-1}x^2[n] - \Delta.\sum_{n=0}^{N-1}f[n]x[n]$$

The path integral is calculated by setting $x[0] = a, x[N] = b$ and computing

$$K_N(b, a) = \int_{\mathbb{R}^{N-1}} exp(2\pi i S_d[x, N, f]/h)\Pi_{k=1}^{N-1}dx[k]$$

[21]. Develop a numerical package for quantum mechanical scattering theory. Specifically, suppose $U(r)$ is the scattering potential, ie, the interaction potential energy between the projectile and the target. Let E be the kinetic energy of the incident particles and the scatterer without interaction. Then by conservation of energy, the scattered wave function $u(r, \theta, \phi)$ satisfies the stationary Schrodinger equation

$$\nabla^2 u(r, \theta, \phi) + \frac{8\pi^2 m}{h^2}(E - U(r))u(r, \theta, \phi) = 0$$

This partial differential equation can be transformed into a radial ordinary differential equation by expanding u in terms of the spherical harmonic functions, ie,

$$u(r, \theta, \phi) = \sum_{l,m} R_{lm}(r)Y_{lm}(\theta, \phi)$$

To do so we note that

$$-\frac{h^2}{2m}\nabla^2 = -\frac{h^2}{2mr}\frac{\partial^2}{\partial r^2}r + L^2/2mr^2$$

where L^2 is the angular momentum squared operator. Then,

$$L^2 Y_{lm} = l(l+1)h^2 Y_{lm}$$

and hence R_{lm} satisfies the radial equation

$$-\frac{h^2}{2mr}(rR_{lm}(r))'' + l(l+1)h^2 R_{lm}(r)/2mr^2 + U(r)R_{lm}(r) = E.R_{lm}(r)$$

For each value of E in the continuous spectrum of the Schrodinger operator corresponding to the interacting projectile and target, we obtain a wave function $R_{lm}(r|E)$ and from this expression, we can determine the number of particles getting scattered into unit solid angle per unit time.

Reference:L.Schiff, Quantum Mechanics.

[22]. (Application of quantum mechanics to electronics). Develop a numerical package for determining the energy band structure of a crystal corresponding to the situation in which many electrons interact with many nuclei in such a way that the nuclear motions are small corresponding to the electron motion. This is the Born-Oppenheimer approximation. Let $\mathbf{R}_i, i = 1, 2, ..., N$ denote the nuclear positions and $\mathbf{r}_i, i = 1, 2, ..., NZ$ the electron positions. Then with M denoting the mass of each nucleus and m the mass of each electron, the total Hamiltonian of the crystal is given by

$$H(\mathbf{R}_1, ..., \mathbf{R}_N, \mathbf{r}_1, ..., \mathbf{r}_{NZ}) =$$

$$-\frac{h^2}{2M} \sum_{i=1}^{N} \nabla^2_{\mathbf{R}_i} - \frac{h^2}{2m} \sum_{i=1}^{NZ} \nabla^2_{\mathbf{r}_i}$$

$$+ \sum_{i \neq j, 1 \leq i,j \leq N} Z^2 e^2 / |\mathbf{R}_i - \mathbf{R}_j| + \sum_{i \neq j, 1 \leq i,j \leq NZ} e^2 / |\mathbf{r}_i - \mathbf{r}_j|$$

Approximate methods for determining the eigenvalues and eigenfunctions of H have been developed in the literature.

[23]. Quantum computation using MATLAB.

1. Define a single qubit state and give examples of a unitary gate acting on single qubit states.

2. Prove that any element of $SU(2)$ can be expressed as

$$A = exp(i(\phi_x \sigma_x + \phi_y \sigma_y + \phi_z \sigma_z))$$

where $\sigma_x, \sigma_y, \sigma_z$ are the Pauli spin matrices and ϕ_x, ϕ_y, ϕ_z are real numbers. What is the action of A on the eigenvectors of σ_z.

3.What do you understand by a generalized measurement on a quantum system. How does this notion lead to the wavefunction collapse postulate.

hint: Let $\mathcal{H}$ be the Hilbert space of a quantum system and let $M_m, m = 1, 2, ..., N$ be bounded linear operators in $\mathcal{H}$ satisfying $\sum_{m=1}^{N} M_m^* M_m = I$. Let ρ be a mixed state of the quantum system, ie, ρ is a positive definite operator in $\mathcal{H}$ satisfying $Tr(\rho) = 1$. When the measurement M_m is taken on the system, the outcome is m and this has a probability of $p(m) = Tr(\rho M_m^* M_m)$. After this measurement has been taken, the state of the system collapses to $\rho_m = M_m.\rho.M_m^*/p(m)$. When the measurement $\{M_m\}$ is made but no particular outcome is observed, the state of the system following the measurement is given by $\rho_M = \sum_{m=1}^{N} p(m)\rho_m = \sum_m M_m \rho M_m^*$. Note that this formula is valid since

$$Tr(\sum_m M_m \rho M_m^*) = Tr(\rho \sum_m M_m^* M_m) = Tr(\rho) = 1$$

4. Controlled Not (CNOT) gate. This is a two qubit gate. The Hilbert space is $\mathbb{C}^4$ and the standard basis for this space is given by $|x_1 x_2 >$ where $x_1, x_2 = 0, 1$. Let U denote the matrix of the gate relative to this basis. Then, the action of U is defined by

$$U|x_1, x_2 >= |x_1, x_1 \oplus x_2 >$$

In full expansion, this is

$$|00 >\to |00 >, |01 >\to |01 >, |10 >\to |11 >, |11 >\to |10 >$$

which corresponds to doing nothing on the total state if the first qubit is zero and flipping the second qubit if the first qubit is one.

5.Controlled U gate. Let U be a 2×2 unitary matrix, ie, U can be viewed as a single qubit quantum gate. The controlled U gate is a two qubit gate that does nothing if the first qubit is zero and if the first qubit is one, then nothing is done on the first qubit while the second qubit undergoes the transfromation U. In the standard basis $|x_1, x_2 >$ where $x_1, x_2 = 0, 1$, the action this gate V can be expressed as

$$V|x_1, x_2 >= |x_1 > U^{x_1}|x_2 >$$

or equivalently, it terms of components,

$$V|00 >= |00 >, V|01 >= |01 >, V|10 >= |1 > U|0 >, V|11 >= |1 > U|1 >$$

or equivalently,

$$V|0, x >= |0, x >, V|1, x >= |1 > U|x >$$

6.Notion of distance between two gates. Suppose U, V are two unitary operators in a Hilbert space $\mathcal{H}$. We can regard U, V as a pair of quantum gates. The distance between these two gates is defined by

$$E(U, V) = sup(\| (U - V)\psi \| : \psi \in \mathcal{H}, \| \psi \| = 1)$$

The following identity is obvious: If $U_i, V_i, i = 1, 2$ are four unitary operators in $\mathcal{H}$, then

$$\| (U_1 U_2 - V_1 V_2)\psi \| = \| (U_1 U_2 - U_1 V_2 + U_1 V_2 - V_1 V_2)\psi \|$$

$$\leq \| U_1((U_2 - V_2)\psi \| + \| (U_1 - V_1)V_2\psi \|$$

$$\leq E(U_2, V_2) + E(U_1, V_1)$$

By induction, it follows that if $U_i, V_i, i = 1, 2, ..., n$ are $2n$ unitary operators in $\mathcal{H}$, then

$$E(U_1 U_2 ... U_n, V_1 V_2 ... V_n) \leq \sum_{i=1}^{n} E(U_i, V_i)$$

Reference:Quantum Computation and Quantum Information, by Nielsen and Chuang.

7. Euler angle representation of 2×2 unitary gates. Let $\sigma_x, \sigma_y, \sigma_z$ be the usual Pauli spin matrices, ie,

$$\sigma_x = \begin{pmatrix} 0 & 1 \\ 1 & 0 \end{pmatrix}, \sigma_y = \begin{pmatrix} 0 & -i \\ i & 0 \end{pmatrix},$$

$$\sigma_z = \begin{pmatrix} 1 & 0 \\ 0 & -1 \end{pmatrix}$$

Then show that any 2×2 unitary matrix can be expressed as

$$U = exp(i\alpha) R_z(\phi) R_x(\theta) R_z(\psi)$$

where $\alpha, \phi, \theta, \psi$ are real numbers and

$$R_z(\phi) = exp(i\phi\sigma_z), R_x(\theta) = exp(i\theta\sigma_x)$$

Hint:Any element of $U(2)$ ($U(2)$ is the set of 2×2 unitary matrices) can be expressed as

$$U = exp(i\alpha)V$$

where $V \in SU(2)$ where α is real and $SU(2)$ is the set of 2×2 unitary matrices having determinant one. Further, any element V of $SU(2)$ can be expressed as

$$V = \begin{pmatrix} \alpha & \beta \\ -\bar{\beta} & \bar{\alpha} \end{pmatrix}$$

where α, β are arbitrary complex numbers satisfying

$$|\alpha|^2 + |\beta|^2 = 1$$

8.Suppose ρ is the state of a quantum system and $\{M_m\}$ a measurement. Compute the average value of an observable X after the measurement $\{M_m\}$ has been made. Assume that the outcome of the measurement is not known.

hint: After the measurement has been made, the state of the system becomes

$$\rho' = \sum_m .M_m^* . \rho M_m$$

and the avergate value of X in this state is

$$tr(\rho'X) = \sum_m tr(\rho M_m X M_m^*)$$

9.Suppose a quantum system evolves in accord with the Hamiltonian H_i with probability p_i. Then determine the state $\rho(T)$ after time T when the state at time 0 is given by ρ.

ans:

$$\rho(T) = \sum_i p_i U_i(T)\rho U_i^*(T), U_k(T) = exp(-iTH_k)$$

10.Suppose that a two state quantum system having spin operator

$$\mathbf{s} = \frac{1}{2}(\sigma_x \hat{x} + \sigma_y \hat{y} + \sigma_z \hat{z})$$

where $\sigma_x, \sigma_y, \sigma_z$ are the Pauli spin matrices, has a Hamiltonian determined by the spin interacting with a magnetic field

$$\mathbf{B}(t) = B_x cos(\omega t)\hat{x} + B_z \hat{z}$$

where B_x, B_z are constants. Then determine the state after time t.

hint: The spin magnetic moment of the system is proportional to $\mathbf{s}$. It is therefore given by

$$\mu = C\mathbf{s}$$

where C is a constant. The Hamiltonian is thus given by

$$H(t) = -(\mu, \mathbf{B}(t)) = -C(\mathbf{s}, \mathbf{B}(t)) = -(CB_x/2)\sigma_x.cos(\omega t) - (CB_z/2)\sigma_z$$

Noting that

$$\sigma_x = \begin{pmatrix} 0 & 1 \\ 1 & 0 \end{pmatrix}, \sigma_z = \begin{pmatrix} 1 & 0 \\ 0 & -1 \end{pmatrix}$$

we get

$$H(t) = -(C/2)\begin{pmatrix} -B_z & B_x cos(\omega t) \\ -B_x cos(\omega t) & B_z \end{pmatrix}$$

Schrodinger's equation for the state $(c_1(t), c_2(t))^T$ is

$$\begin{pmatrix} c_1'(t) \\ c_2'(t) \end{pmatrix} = -iH(t)\begin{pmatrix} c_1(t) \\ c_2(t) \end{pmatrix}$$

and this expands to give

$$c_1'(t) = -(iCB_z/2)c_1(t) + (iCB_x/2).cos(\omega t)c_2(t),$$

$$c_2'(t) = -(iCB_x/2)cos(\omega t)c_1(t) + (iCB_z/2)c_2(t)$$

This is a system of two coupled linear first order differential equations with time varying coefficients. It can be reduced to a single linear second order differential equation with time varying coefficients.

11. Nuclear magnetic resonance. Here we have an electron which can be in two spin states or a superposition of both. The spin of the electron is given by the matrix valued vector

$$\mathbf{s} = \frac{h}{4\pi}(\hat{x}\sigma_x + \hat{y}\sigma_y + \hat{z}\sigma_z)$$

This spin interacts with the magnetic field given by

$$\mathbf{B}(t) = B_0\hat{z} + B_1 cos(\omega t)\hat{x}$$

where $|B_1| << |B_0|$. The spin magnetic moment of the atom is the matrix valued vector

$$\mu = e\mathbf{s}/m = eh(\sigma_x\hat{x} + \sigma_y\hat{y} + \sigma_z\hat{z})/8\pi m$$

The interaction energy of the spin with the magnetic field is given by

$$H(t) = -(\mu, \mathbf{B}) = (eh/8\pi m)(B_0\sigma_z + B_1 cos(\omega t)\sigma_x)$$

$$= C\begin{pmatrix} B_0 & B_1 cos(\omega t) \\ B_1 cos(\omega t) & -B_0 \end{pmatrix}$$

where C is a constant. The Schrodinger equation for the this two state quantum system is then given by

$$\mathbf{c}' = -(2\pi i/h)H(t)\mathbf{c}(t)$$

or in terms of components,

$$c_1' = -(2\pi i C/h)(B_0 c_1(t) + B_1 cos(\omega t)c_2(t)), c_2' = -(2\pi i C/h)(-B_0 c_2(t) + B_1 cos(\omega t)c_1(t))$$

We solve this differential equation approximately using first order perturbation theory. Set

$$c_1(t) = c_1^{(0)}(t) + c_1^{(1)}(t), c_2(t) = c_2^{(0)}(t) + c_2^{(1)}(t)$$

where $c_1^{(0)}$ and $c_2^{(0)}$ are of the zeroth order of smallness and $c_1^{(1)}(t)$ and $c_2^{(1)}(t)$ are of the first order of smallness. B_0 is of zeroth order of smallness and B_1 is of the first order of smallness. Then,

$$c_1^{(0)'} = -(2\pi i B_0 C/h)c_1^{(0)}, c_2^{(0)'} = (2\pi i B_0 C/h)c_2^{(0)}$$

Assuming that before the perturbing magnetic field (the component along the x axis) is switched on, the system is in the upspin state, we have

$$c_1^{(0)} = exp(-2\pi i B_0 Ct/h), c_2^{(0)} = 0$$

The equations for the first order perturbed terms in the state are

$$c_1^{(1)'} = -(2\pi i B_0 C/h)c_1^{(1)} - (2\pi i C B_1/h)cos(\omega t)c_2^{(0)},$$

$$c_2^{(1)'} = (2\pi i B_0 C/h)c_2^{(1)} - (2\pi i C B_1/h)cos(\omega t)c_1^{(0)}$$

12.Project on quantum field theory:Implement using MATLAB an n-particle wave function for bosons. Specifically, suppose Φ_0 is the vacuum state in which there is no boson. Let $a(p), a^*(p)$ denote respectively the operator of annihilation of a boson of momentum p and the operator of creation of a boson of momentum p. Here $p = (p^0, \mathbf{p})$ is a four vector and $p^0 = \sqrt{m^2 + \mathbf{p}^2}$ is the energy of a boson having three momentum $\mathbf{p}$. Then the state in which there are n bosons having momenta $p_1, ..., p_n$ can be realized as

$$\Phi[p_1, ..., p_n] = a^*(p_1)...a^*(p_n)\Phi_0$$

The commutation relations satisfied by the creation and annihilation operators are

$$[a(p), a^*(p')] = \delta(\mathbf{p} - \mathbf{p}'), [a(p), a(p')] = 0, [a^*(p), a^*(p')] = 0$$

Suppose $\Phi[p_1, ..., p_n]$ is an n particle boson state. Then in view of the symmetry of the bosonic wave function, we must have

$$a^*(p)\Phi[p_1, ..., p_n] = \Phi[p, p_1, ..., p_n]$$

Note that since the operators $a^*(p), \mathbf{p} \in \mathbb{R}^3$, all commute, it follows that

$$\Phi[p_{\sigma 1}, ..., p_{\sigma n}] = \Phi[p_1, ..., p_n]$$

for all permutations σ of $\{1, 2, ..., n\}$.

$$a(p)\Phi[p_1, ..., p_n] = \sum_{i=1}^{n} \delta(\mathbf{p} - \mathbf{p}_i)\Phi[p_1, ..., \hat{p}_i, ..., p_n]$$

where $\hat{p}_i$ means omission of p_i. Note that this formula agrees with the fact that if p differs from all the $p_i's$, then $a(p)\Phi[p_1, ..., p_n] = 0$ since annihilation of a boson of momentum p from a state that does not have any boson of momentum p is zero.

[24].Implement using MATLAB the computation of the scattering matrix for a given unperturbed Hamiltonian and a perturbing Hamiltonian.

hint:Refer to Weinberg, Quantum field theory. Let H_0 be the unperturbed Hamiltonian and V the perturbation. The perturbed Hamiltonian is thus given by $H = H_0 + V$. Let Φ be a time independent state of the unperturbed system so that its evolution law is given by $\Phi(t) = exp(-itH_0)\Phi$. Let Ψ_+ be the corresponding time independent state of the perturbed system so that its evolution law is given by $\Psi(t) = exp(-itH)\Psi_+$. There was no interaction at $t \to -\infty$ and hence we must have

$$lim_{t \to -\infty} exp(-itH)\Psi_+ - exp(-itH_0)\Phi = 0$$

The state Φ evolves into Ψ_+. Ψ_+ is an instate. Likewise the corresponding outstate Ψ_- evolves into the unperturbed state Φ as $t \to \infty$. Thus,

$$lim_{t \to \infty} exp(-itH)\Psi_- exp(-itH_0)\Phi = 0$$

We thus define the Moller operators

$$\Omega(t) = exp(itH)exp(-itH_0)\Phi$$

and get

$$\Psi_+ = \Omega(-\infty)\Phi, \Psi_- = \Omega(\infty)\Phi$$

so that from unitarity of $\Omega(\pm\infty)$, we get

$$\Psi_+ = \Omega(-\infty)\Omega(\infty)^*\Psi_-, \Psi_- = \Omega(\infty)\Omega(-\infty)^*\Psi_+$$

We write this as

$$\Psi_- = S\Psi_+, S = \Omega(\infty)\Omega(-\infty)^*$$

S is called the scattering matrix.

[25]. Consider the four state quantum system corresponding to the spins of two independent particles. The Hamiltonian of this system is given by

$$H = a_1 Z_1 + a_2 Z_2 + a_{12} Z_1 Z_2$$

where Z_1, Z_2 are respectively the z components of the two spins. More precisely, if Z is the Pauli spin matrix σ_z, then

$$H = a_1 . Z \otimes I + a_2 I \otimes Z + a_{12} Z \otimes Z$$

Note that Z operates on $\mathbb{C}^2$ and hence H operates on $\mathbb{C}^2 \otimes \mathbb{C}^2 = \mathbb{C}^4$. Determine the unitary evolution operator $U(t) = exp(-itH)$.

hint:

$$Z = \begin{pmatrix} 1 & 0 \\ 0 & -1 \end{pmatrix}$$

so that

$$exp(-itZ) = \begin{pmatrix} exp(-it) & 0 \\ 0 & exp(it) \end{pmatrix}$$

Further,

$$exp(-itH) = (exp(-ia_1 tZ) \otimes exp(-ia_2 tZ)).exp(-ia_{12} tZ \otimes Z)$$

Also,

$$(Z \otimes Z)^{2n} = I, (Z \otimes Z)^{2n+1} = Z \otimes Z$$

so that

$$exp(-itZ \otimes Z) = cos(t)I - i.sin(t).Z \otimes Z$$

The previous formula can also be expressed in an alternate way. $Z^2 = I$ implies $Z^{2n} = I, Z^{2n+1} = Z$ and hence

$$exp(-itZ) = cos(t)I - i.sin(t)Z$$

so that

$$exp(-itH) = ((cos(a_1 t)I - i.sin(a_1 t)Z) \otimes (cos(a_2 t)I - i.sin(a_2 t)Z))(cos(a_{12} t)I - i.sin(a_{12} t)Z \otimes Z)$$

[26]. Let $\mathcal{H}_1$ be the Hilbert space of the system and $\mathcal{H}_2$ the Hilbert space of the environment. The Hilbert space of the system plus environment is then $\mathcal{H} = \mathcal{H}_1 \otimes \mathcal{H}_2$. Let $A_i, i = 1, 2, ..., N$ be operators in $\mathcal{H}_1$ and $B_1, ..., B_N$ operators in $\mathcal{H}_2$ so that

$$\rho = \sum_{i=1}^{N} A_i \otimes B_i$$

is the state of the system plus environment. The state of the system alone is given by

$$\rho_1 = Tr_2(\rho) = \sum_{i=1}^{N} Tr(B_i) A_i$$

and the state of the environment alone is given by

$$\rho_2 = Tr_1(\rho) = \sum_{i=1}^{N} Tr(A_i) B_i$$

[27]. Let $\mathcal{H}_1, \mathcal{H}_2$ be two Hilbert spaces and let U be a unitary operator in $\mathcal{H}_1 \otimes \mathcal{H}_2$. Choose a vector $|e_0> \in \mathcal{H}_2$ and an operator ρ in $\mathcal{H}_1$. Express $\tilde{\rho} = Tr_2(U(\rho \otimes |e_0> < e_0|)U^*)$ as $\sum_k E_k \rho E_k^*$, where the $E_k' s$ are operators in $\mathcal{H}_1$.

hint: We can write

$$U = \sum_i U_{1i} \otimes U_{2i}$$

where U_{1i} and U_{2i} are respectively operators in $\mathcal{H}_1$ and $\mathcal{H}_2$. Then,

$$U(\rho \otimes |e_0> < e_0|)U^* = \sum_{i,j}(U_{1i}\rho U_{1j}^*) \otimes (U_{2i}|e_0> < e_0|U_{2j}^*)$$

so that

$$\tilde{\rho} = Tr_2(U(\rho \otimes |e_0> < e_0|)U^*) = \sum_{i,j} a_{ij} U_{1i}\rho U_{1j}^*$$

where

$$a_{ij} = < e_0|U_{2j}^* U_{2i}|e_0 >$$

Note that

$$\bar{a}_{ij} = < e_0|U_{2i}^* U_{2j}|e_0 > = a_{ji}$$

and hence $((a_{ij}))$ is a Hermitian positive definite matrix. Hence it can be diagonalized giving

$$a_{ij} = \sum_k d_k v_{ik} \bar{v}_{jk}$$

It follows that

$$\tilde{\rho} = \sum_{k,i,j} d_k v_{ik} \bar{v}_{jk} U_{1i}\rho U_{1j}^*$$

Now define

$$E_k = \sqrt{d_k} \sum_i v_{ik} U_{1i}$$

so that

$$E_k^* = \sqrt{d_k} \sum_j \bar{v}_{jk} U_{1j}^*$$

and we have

$$\tilde{\rho} = \sum_k E_k \rho E_k^*$$

[28]. Define the Shannon entropy of a random vector and state the important inequalities for the shannon entropy.

hint:The inequalities are

$$H(X) + H(Y) \leq H(X,Y)$$

$$H(X|Y) \leq H(X)$$

$$H(X,Y,Z) + H(Y) \leq H(X,Y) + H(Y,Z)$$

The proofs are based on the inequality $log(x) \leq x - 1$ for all $x > 0$. This is proved by defining $f(x) = x - 1 - log(x)$ for $x > 0$ and noting that $f(1) = 0$ and $f'(x) = 1 - 1/x$ which is negative for $x < 1$ and positive for $x > 1$ implying that $f(x)$ is decreasing for $x < 1$ and increasing for $x > 1$ so that $f(x) \geq 0$ for all $x > 0$.

$$H(X) + H(Y) - H(X,Y) = \sum_{x,y} p(x,y) log(p(x)p(y)/p(x,y)) \leq \sum_{x,y} p(x,y)(p(x)p(y)/p(x,y) - 1) = 0$$

$$H(X|Y) - H(X) = \sum_{x,y} p(x,y) log(p(x)/p(x|y))$$

$$= \sum_{x,y} p(x,y) log(p(x)p(y)/p(x,y)) \leq \sum_{x,y} p(x,y)(p(x)p(y)/p(x,y) - 1) = 0$$

Finally,

$$H(X,Y,Z) + H(Y) - H(X,Y) - H(Y,Z) = \sum_{x,y,z} p(x,y,z) log(p(x,y)p(y,z)/p(x,y,z)p(y))$$

$$\leq \sum_{x,y,z} p(x,y,z)(p(x,y)p(y,z)/p(x,y,z)p(y) - 1) = \sum_{x,y,z} (p(x,y)p(y,z)/p(y) - p(x,y,z)) = 0$$

The first inequality is called the subadditivity property and the last inequality is called the strong subadditivity property of the entropy.

[29]. Let $p(x), q(x), x \in E$ be two probability distributions on a finite set E. The distance between these two probability distributions is defined as

$$D(p,q) = \frac{1}{2} \sum_{x \in E} |p(x) - q(x)|$$

Show that D is a metric on the space of probability distributions on E and that

$$D(p,q) = max_{S \subseteq E} \sum_{x \in S} (p(x) - q(x))$$

hint:Let F be the set of all $x \in E$ for which $p(x) \geq q(x)$. Thus $F^c = E - F$ is the set of all $x \in E$ for which $p(x) < q(x)$. Then,

$$\sum_{x \in E} |p(x) - q(x)| = \sum_{x \in F} (p(x) - q(x)) - \sum_{x \in F^c} (p(x) - q(x)) = 2 \sum_{x \in F} (p(x) - q(x))$$

since

$$\sum_{x \in F^c} p(x) = 1 - \sum_{x \in F} p(x), \sum_{x \in F^c} q(x) = 1 - \sum_{x \in F} q(x)$$

Further, if A is any subset of E, then

$$\sum_{x \in A} (p(x) - q(x)) = \sum_{x \in A \cap F} (p(x) - q(x)) + \sum_{x \in A \cap F^c} (p(x) - q(x)) \leq \sum_{x \in F} (p(x) - q(x))$$

since $p(x) - q(x) \geq 0$ for $x \in F$ while $p(x) - q(x) < 0$ for $x \in F^c$.

[30]. Let ρ, σ be two quantum states on the same Hilbert space $\mathcal{H}$. We define the trace distance between these two states as

$$D(\rho, \sigma) = \frac{1}{2} tr(|\rho - \sigma|)$$

where if A is any operator in $\mathcal{H}$, $|A| = (AA^*)^{1/2}$. Suppose that ρ and σ commute. Then, we can choose an orthonormal basis $\{|i >\}$ for the Hilbert space such that

$$\rho = \sum_i |i > p_i < i|, \sigma = \sum_i |i > q_i < i|$$

where $p_i, q_i \geq 0, \sum_i p_i = \sum_i q_i = 1$. Then,

$$D(\rho, \sigma) = \frac{1}{2} \sum_i |p_i - q_i| = D(p, q)$$

[31]. Polar decomposition. Suppose $\mathcal{H}$ is a finite dimensional Hilbert space and A is a linear operator in $\mathcal{H}$. Show that $A = |A|U$ where $|A|$ is positive semidefinite and U is unitary.

hint: Apply the spectral theorem to the positive semidefinite matrix $(AA^*)^{1/2}$.

[32]. For simulating a quantum system, the following weak version of the Campbell-Baker-Hausdorff formula is used. If A, B are operators in a Hilbert space, then

$$exp((A + B)\Delta t) = exp(A.\Delta t).exp(B.\Delta t).exp(-[A, B]\Delta t^2/2) + O(\Delta t^3)$$

This can be proved as follows:

$$exp((A+B)\Delta t) = I + (A+B)\Delta t + (A+B)^2 \Delta t^2/2 + O(\Delta t^3)$$

$$= I + (A+B)\Delta t + (A^2 + B^2 + AB + BA)\Delta t^2/2 + O(\Delta t^3)$$

on the one hand and on the other,

$$exp(A.\Delta t).exp(B.\Delta t).exp(-[A,B]\Delta t^2/2)$$

$$= (I + A\Delta t + A^2\Delta t^2/2)(I + B\Delta t + B^2\Delta t^2/2)(I - [A,B]\Delta t^2/2) + O(\Delta t^3)$$

$$= I + (A+B)\Delta t + \Delta t^2(A^2/2 + B^2/2 + AB - [A,B]/2) + O(\Delta t^3)$$

$$= I + (A+B)\Delta t + (\Delta t^2/2)(A^2 + B^2 + AB + BA) + O(\Delta t^3)$$

proving the claim.

For example, suppose the Hilbert space is $L^2(\mathbb{R})$. Take $A = p^2/2m$ and $B = V(q)$, where $p = -i\frac{\partial}{\partial q}$. Then,

$$[A,B] = [p^2/2m, V(q)] = (p/2m)[p, V(q)] + [p, V(q)](p/2m)$$

$$= (-/2m)V'(q) - (i/2m)V'(q)p = -\frac{i}{2m}(pV'(q) + V'(q)p)$$

The above formula can then be used to compute to a high degree of accuracy, the evolution of an initial state under Schrodinger's dynamics, ie,

$$exp(-i\Delta t(A+B)) = exp(-i\Delta t(p^2/2m + V(q)))$$

[33]. Distance measure in quantum mechanics. Suppose $|\psi >$ is the state of a quantum system with Hilbert space $\mathcal{H}$ and U, V are two unitary operators in the Hilbert space $\mathcal{H}$. Let M be a pov measurement so that the probability of M being true in the state $|\psi >$ is $< \psi|M|\psi >$. If the state evolves according to U, then, the probability of M being true after the evolution is $P_U =< \psi|U^*MU|\psi >$ while on the other hand, if the state evolves according to V, then the probability of M being true after the evolution is $P_V =< \psi|V^*MV|\psi >$. The distance between U, V relative to the measurement M, can thus be defined as

$$d_M(U,V) = sup_\psi|P_U - P_V| = sup_\psi| < \psi|U^*MU - V^*MV|\psi > |$$

Obviously,

$$d_M(U,V) \leq\| U^*MU - V^*MV \|$$

where $\| . \|$ is the spectral norm. Suppose H is a Hermitian operator. Then $| < \psi|H|\psi > |$ attains its maximum when $|\psi >$ is an eigenvector of H corresponding to the eigenvalue of H having maximum magnitude. It is clear then that

$$d_M(U,V) =\| U^*MU - V^*MV \|=\| M - UV^*MVU^* \|$$

[34].Variational method in quantum mechanics. Let H be the Hamiltonian operator of a quantum system and let its eigenfunctions be $\psi_n, n = 1, 2, ...$ with corresponding eigenvalues $E_n, n = 1, 2, ...$ Assume that $E_0 \leq E_1 \leq ... \leq E_n \leq E_{n+1} < ....$ Show that if we set $\psi = \sum_n c_n \psi_n$, with the normalization $< \psi, \psi >= \sum_n |c_n|^2$, then

$$< \psi|H|\psi >= \sum_n |c_n|^2 E_n \geq E_0$$

Hence deduce that the ground state wave function can be computed from the variational principle

$$\delta(< \psi|H|\psi > - \lambda < \psi, \psi >) = 0$$

Apply the variational principle to obtain the Hartree-Fock equations for the quantum many body problem.

hint: Consider for simplicity the two body problem with Hamiltonian

$$H = H_1 + H_2 + V_{12}$$

where H_1 depends only on the coordinates and momenta of the first particle, H_2 on only those of the second particle and V_{12} is the interaction potential dependent on the coordinates of both the particles. We try a wave function of the product type

$$\psi(r_1, r_2) = \psi_1(r_1)\psi_2(r_2)$$

where the functions ψ_1, ψ_2 are chosen so that $< \psi|H|\psi >$ is a minimum subject to the constraints that ψ_1, ψ_2 are normalized and orthogonal to each other. Derive the equations satisfied by ψ_1 and ψ_2. Now generalize this to the case of N particles with Hamiltonian given by

$$H = \sum_{i=1}^{N} H_i + \sum_{1 \leq i < j \leq N} V_{ij}$$

where H_i depends only on the coordinates and momenta of the ith particle while V_{ij} depends only on the coordinates and momenta of the ith and j^{th} particles.

[35]. Let A be a Hermitian matrix of size $n \times n$. Assume that its eigenvalues are distinct: $\lambda_1 > \lambda_2 > ... > \lambda_n$. Let $e_1, ..., e_n$ be the corresponding eigenvectors. Thus,

$$Ae_i = \lambda_i e_i, i = 1, 2, ..., n, < e_i, e_j >= \delta_{ij}$$

Let M be any subspace of $\mathbb{C}^n$ of dimension k where $1 \leq k \leq n$. Consider the $n - k + 1$ dimensional subspace $W_k = span\{e_k, e_{k+1}, ..., e_n\}$. Then $M \cap W_k$ is nonzero since

$$dim(M \cap W_k) = dim(M) + dim(W_k) - dim(M + W_k) \geq dim(M) + dim(W_k) - n = k + n - k + 1 - n = 1$$

Given any $x \in M \cap W_k$ such that $\| x \|= 1$, we have the expansion $x = \sum_{j=k}^{n} c_j e_j$ where $\sum_{j=k}^{n} |c_j|^2 = 1$. Then,

$$< x|A|x >= \sum_{j=k}^{n} |c_j|^2 \lambda_j \leq \lambda_k$$

Thus, there exists a vector $x \in M$ having unit norm such that $< x|A|x > \le \lambda_k$. In other words,

$$inf\{< x|A|x >: x \in M, \| x \| = 1\} \le \lambda_k$$

But choosing $M = span\{e_1, ..., e_k\}$, it follows that $e_k \in M$ and $< e_k|A|e_k >= \lambda_k$. Thus, we have proved that

$$sup_{dim M=k} inf\{< x|A|x >: x \in M, \| x \| = 1\} = \lambda_k$$

This formula provides us with a variational method for approximately determining the eigenvalues of A.

[36].WKB method for determining the solution to the time independent Schrodinger equation. The equation reads

$$\psi''(x) + \frac{2m}{h^2}(E - V(x))\psi(x) = 0$$

Define

$$\psi(x) = exp(iu(x))$$

Then,

$$\psi' = iu'.exp(iu), \psi'' = (iu'' - u'^2)exp(iu)$$

Substituting this into the differential equation gives

$$iu''(x) - u'^2(x) + \frac{2m}{h^2}(E - V(x)) = 0$$

Define

$$k(x)\sqrt{2m(E - V(x))/h^2}$$

Then,

$$iu'' - u'^2 + k^2(x) = 0$$

If $|u''| << |u'|^2$, then

$$u'(x) \approx \pm k(x)$$

The condition for this to happen is that $|k'| << |k|^2$. We denote this solution by u_0.

[37]. Explain how you would quantize the radiation field of photons.

hint:Start with the wave equation for the vector potential and impose the Coulomb gauge condition. It will follow that each Fourier component of the vector potential is a vector that is orthogonal to the direction of propagation of the corresponding plane wave. Then write down the expression for the total energy of the radiation field in terms of the vector potential. This expression will be a quadratic functional of the potential field. Application of Parseval's theorem then shows that the total energy can be expressed as an integral in the spatial frequency space, ie, wave-number space. This integral is a quadratic functional of the spatial Fourier components of the vector potential and this result prompts us to regard the radiation field of photons to be comprised of an ensemble of quantum simple harmonic oscillators based on which the quantum theory of radiation can be developed.

[38]. Let γ^μ be the usual Dirac matrices, ψ a four component wave function and A_μ a four component potential of the electromagnetic field. Write down the Lagrangian density as a function of $\psi, \bar\psi$ and A_μ such that setting the variation of the corresponding action to zero with respect to these variables results in (1) The Dirac equation and (2) the Maxwell equations with the four current density given by $J^\mu = -e\bar\psi.\gamma^\mu\psi$.

[39] Justify the assumption that the creation and annihilation operators for Fermions satisfy canonical anticommutation relations while the creation and annihilation operators for bosons satisfy canonical commutation relations.

[40] Suppose A_μ is the electromagnetic four potential and ψ the wave function of the matter part. Consider the gauge covariant derivative $D_\mu = \partial_\mu + ieA_\mu$. Assume that ψ transforms to $\psi' = exp(i\alpha)\psi$ where α is a function of the space-time coordinates. Determine the transformation law of A_μ, ie, $A_\mu \to A'_\mu$ so that $D'_\mu\psi' = exp(i\alpha)D_\mu\psi$ where $D'_\mu = \partial_\mu + ieA'_\mu$. Explain the significance of this with respect to the Schrodinger and Dirac wave equations for the electron moving in an electromagnetic field.

[41] DSP implementation of the Heisenberg equations of motion for a finite state quantum system. H is an $n \times n$ complex Hermitian matrix and $X(t)$ for each $t \geq 0$ is an $n \times n$ complex Hermitian matrix that satisfies the differential equation

$$X'(t) = i(HX(t) - X(t)H), t \geq 0$$

Solve this differential equation approximately by a discretization process and compare with the exact solution

$$X(t) = exp(itH).X(0).exp(-itH)$$

To compute $exp(itH)$, we can make use of Laplace transforms so that $exp(itH)$ is the inverse Laplace transform of $(sI - iH)^{-1}$.

[42] Define a vector operator in quantum mechanics in terms of the commutation relations between its components and the angular momentum operators. Give an example.

hint:Consider the position and momentum observables and write down their commutation relations with the components of the angular momenta.

[43] State and explain the Wigner-Eckart theorem in quantum mechanics regarding matrix elements of tensor operators.

[44] What are the parity and time reversal operators in quantum mechanics.

[45] Quantum mechanical theory of radiation.

[1] Quantization of the Maxwell field.

[2] Commutation relations between the field operators.

[3] Path integrals in quantum mechanics and quantum field theory.

[4] The theory of the quantized Klein-Gordon field.

[5] Relationship between Lagrangian and Hamiltonian field theories.

[6] Quantization of the Dirac field.

[7] Evaluation of the propagators for the Maxwell and Dirac fields from the commutation relations for the creation and annihilation operators.

[8] Boson and Fermion field quantization in terms of symmetric and antisymmetric tensor products.

[9] Evaluation of the propagators for the Maxwell and Dirac fields using the Feynman path integrals.

[10] Computing the scattering amplitudes of electrons, positrons and photons using the Feynman diagrams.

[11] Scattering matrix in the Lipmann-Schwinger formalism of in-states and out-states.

[98] Give a numerical method for simulating the time dependent Schrodinger equation with a cylindrically symmetric potential using the cylindrical coordinate system. The wave functions $\psi(t, \rho, \phi, z)$ satisfies

$$-\frac{h^2}{8\pi^2 m}(\rho^{-1}(\rho\psi_{,\rho})_{,\rho} + \rho^{-2}\psi_{,\phi\phi} + \psi_{,zz}) + V(\rho)\psi = \frac{ih}{2\pi}\psi_{,t}$$

[46] State and prove Liouville's theorem for the density operator in quantum mechanics.

hint: Let H be the Hamiltonian operator and consider the evolution of a pure state in accordance with the Schrodinger wave equation $i\frac{d}{dt}|\psi(t) >= H|\psi(t) >$. The state of a system $\rho(t)$ has the expansion $\rho(t) = \sum_i |\psi_i(t) > p_i < \psi_i(t)|$ where $p_i's$ are constants and $|\psi_i(t) >$ evolve in accord with the Schrodinger equation with Hamiltonian H. Deduce that

$$i\rho'(t) = H\rho(t) - \rho(t)H$$

and hence deduce that

$$\rho(t) = exp(-itH)\rho(0)exp(itH)$$

Show that if $|\phi >$ is an eigenstate of H, then $< \phi|\rho(t)|\phi >=< \phi|\rho(0)|\phi >$, ie, the average of $\rho(t)$ in an eigenstate is constant in time. More generally, if $|\phi_k >, k = 1, 2, ..., r$ are orthonormal eigenstates of H, and $p_1, ..., p_r$ positive constants adding up to unity and if $\sigma = \sum_k |\phi_k > p_k < \phi_k|$, then show that

$$Tr(\sigma\rho(t)) = Tr(\sigma\rho(0))$$

[47] Explain physically the meaning of the quantum mechanical equation

$$< \psi|\phi >= \sum_k < \psi|k >< k|\phi >$$

where $\{|k>, k = 1, 2, ...\}$ is an orthonormal set based on the interference of electrons passing through several slits on a screen and forming a pattern on a screen placed at a definite distance from the screen with slits. Reference(Feynman lectures on physics, vol.III). Generalize this to include several intermediate screens numbered $1, 2, ..., N$ with slits and hence explain the equation

$$< \psi | \phi > = \sum_{i_1, i_2, ..., i_N} < \psi | 1, i_1 >< 1, i_1 | 2, i_2 > ... < N - 1, i_{N-1} | N, i_N >< N, i_N | \phi >$$

[48] Give an elementary proof of the path integral based solution to the time dependent Schrodinger equation.

hint: Let $V(x)$ be the potential in which a quantum particle moves. Then define

$$K_T(b, a) = C \int_{(0,a)}^{(T,b)} exp(\int_0^T (\frac{m}{2} x'^2(t) - V(x(t))) dt) Dx$$

where C is a normalizing constant. Then the wave function at time T $\psi_T(x)$ is given by

$$\psi_T(b) = \int K_T(b, a) \psi_0(a) da$$

This is proved by computing ψ_{T+dT} in terms of ψ_T and comparing with the time dependent Schrodinger equation.

[49] Derive the second order ordinary differential equation satisfied by the radial component of the wave function of the hydrogen atom. Solve this equation by the series method.

[50] Explain the phenomenon of quantum mechanical tunneling by considering the stationary Schrodinger equation in the one dimensional potential $V(x) = 0$ for $x < 0$ and $x > L$ and $V(x) = V_0$ for $0 \leq x \leq L$ where $V_0 > 0$.

[51] Evaluate the path integral for the harmonic oscillator using Fourier series. The definition of the path integral is

$$\int_{(0,a)}^{(T,b)} exp(\frac{m}{2} \int_0^T x'^2(t) dt - \frac{m\omega^2}{2} \int_0^T x^2(t) dt) \Pi_{t \in [0,T]} dx(t)$$

[52] The Klein-Gordon field. The Lagrangian density of the field is

$$\mathcal{L}(\phi, \partial_\mu \phi) = \frac{1}{2} \partial_\mu \phi . \partial^\mu \phi - m^2 \phi^2 / 2$$

$$= \frac{1}{2} (\partial_0 \phi)^2 - \frac{1}{2} |\nabla \phi|^2 - m^2 \phi^2 / 2$$

The classical field equations are

$$\partial_\mu \frac{\partial \mathcal{L}}{\partial \partial_\mu \phi} = \frac{\partial \mathcal{L}}{\partial \phi}$$

We have

$$\frac{\partial \mathcal{L}}{\partial \partial_\mu \phi} = \partial^\mu \phi,$$

$$\frac{\partial \mathcal{L}}{\partial \phi} = -m^2 \phi$$

so that the field equations are

$$\partial_\mu \partial^\mu \phi = -m^2 \phi$$

or

$$\nabla^2 \phi - \partial_t^2 \phi - m^2 \phi = 0$$

which is the Klein-Gordon equation. The Hamiltonian density of the field is given by

$$\mathcal{H} = \pi \phi_{,t} - \mathcal{L}, \pi = \frac{\partial \mathcal{L}}{\partial \phi_{,t}}$$

A quick computation shows that

$$\mathcal{H} = \frac{1}{2}(\phi_{,t}^2 + |\nabla \phi|^2 + m^2 \phi^2)$$

The solution to the Klein-Gordon field equations can be expressed as

$$\phi(x) = \int \frac{1}{\sqrt{2p^0}}(a(\mathbf{p})exp(ip_\mu x^\mu) + a^*(\mathbf{p})exp(-ip_\mu x^\mu))d^3p$$

where

$$p^0 = \sqrt{m^2 + |\mathbf{p}|^2}$$

Here $x = (x^\mu)$. This expression is used in quantizing the Klein-Gordon field wherein $a(\mathbf{p})$ and $a^*(\mathbf{p})$ become operators obeying the commutation relations $[a(\mathbf{p}), a^*(\mathbf{p}')] = \delta(\mathbf{p} - \mathbf{p}')$. This leads immediately to the canonical commutation relations $[\phi(\mathbf{p}), \pi(\mathbf{p}')] = i\delta(\mathbf{p} - \mathbf{p}')$. The Hamiltonian $H = \int \mathcal{H}d^3x$ is expressed as $C. \int p^0(a(\mathbf{p})a^*(\mathbf{p}') + a^*(\mathbf{p}')a(\mathbf{p}))d^3p$ where C is a constant. This leads at once to the picture of the Klein-Gordon field as a continuous ensemble of harmonic oscillators.

[53] Show that the discretization of the stationary state Schrodinger wave equation in one dimension leads to an eigenvalue problem for a tridiagonal matrix, provided that the second order differential operator d^2/dx^2 is replaced by a second order finite difference, ie,

$$f''(x) \to \frac{f((k+1)\Delta) - 2f(k\Delta) + f((k-1)\Delta)}{\Delta^2}$$

[54] Topics in quantum field theory.

[1] The quantum mechanical harmonic oscillator, eigenvalues and eigenstates, the creation and annihilation operators.

[2] Bosons and Fermions, wavefunctions for these particles in the form of symmetric and antisymmetric states.

[3] Creation and annihilation operators for Bosons and Fermions.

[4] Quantization of the Klein-Gordon field using canonical commutation relations for the position and momentum operators of the field derived from the Lagrangian density.

[5] Quantization of the electromagnetic field using canonical commutation relations.

[6] Feynman path integrals in non-relativistic quantum mechanics.

[7] Quantization of the Klein-Gordon and electromagnetic fields using path integrals.

[8] The Dirac field of electrons and positrons.

[9] Computing the eigenfunctions of the Dirac operator in a Coulomb potential (Relativistic analysis of the Quantum Coulomb problem)

[10] Dyson series expansions for the quantum evolution operator when the Hamiltonian is time dependent. The Chronological product and the time ordering operator.

[55] Problems in quantum computation.

[1]. The density matrix of a quantum system is ρ. ρ is positive semidefinite with $tr(\rho) = 1$. The underlying Hilbert space is assumed to have dimension m. ρ therefore has a spectral representation of the form

$$\rho = \sum_{m=1}^{n} p_m |m><m|$$

where $< m|r >= \delta_{m,r}$ and $p_m \geq 0, \sum_m p_m = 1$. ρ has the following intepretation. If the system is in the state ρ, then with probability p_m the system is in the state $|m>$. Let X be an observable on the underlying Hilbert space. Then the average value of X in the state ρ is given by

$$< X >_\rho = \sum_m p_m < m|X|m >= tr(\rho.X)$$

Note that $< m|X|m >$ is the average of X in the state $|m>$. Suppose X has the spectral representation

$$X = \sum_m c_m |e_m >< e_m|$$

then

$$< X >_\rho = \sum_{m,k} p_m c_k | < e_k|m > |^2 = \sum_k c_k p(e_k)$$

where

$$p(e_k) = \sum_m p(e_k|m) p_m, p(e_k|m) = | < e_k|m > |^2$$

[2] Let $\mathcal{H}_i, i = 1, 2$ be two finite dimensional Hilbert spaces and let $\mathcal{H} = \mathcal{H}_1 \otimes \mathcal{H}_2$ be their tensor product. Let $e_i, i = 1, 2, ... n_1$ be an orthonormal basis for $\mathcal{H}_1$ and $f_i, i = 1, 2, ..., n_2$ an orthonormal basis for $\mathcal{H}_2$. Let $\psi \in \mathcal{H}$ be arbitrary. Then since $e_i \otimes f_j, i = 1, 2, ..., n_1, j = 1, 2, ..., n_2$ is an orthonormal basis for $\mathcal{H}$, it follows that there exist $n_1 n_2$ complex numbers c_{ij} such that

$$\psi = \sum_{i,j} c_{ij} e_i \otimes f_j$$

The singular value decomposition applied to the matrix $((c_{ij}))$ gives

$$c_{ij} = \sum_r \lambda_r u_{ir} \bar{v}_{jr}$$

where r runs from 1 to $min(n_1, n_2)$ and $((u_{im}))$ and $((v_{im}))$ are respectively $n_1 \times n_1$ and $n_2 \times n_2$ unitary matrices. Define

$$e'_r = \sum_i u_{ir} e_i, \ f'_r = \sum_j \bar{v}_{jr} f_j$$

Then it is clear that $e'_r, r = 1, 2, ..., min(n_1, n_2)$ and $f'_r = \sum_j \bar{v}_{jr} f_j, r = 1, 2, ..., min(n_1, n_2)$ are respectively orthonormal sets in $\mathcal{H}_1$ and $\mathcal{H}_2$ and

$$\psi = \sum_r \lambda_r e'_r \otimes f'_r$$

This is called the Schmidt decomposition.

[3] Let $\mathcal{H}_i, i = 1, 2$ be two finite dimensional Hilbert spaces and let $|\psi>$ be any unit vector in $\mathcal{H}_1 \otimes \mathcal{H}_2$. Then as seen above, there exist orthogonal sets $|e_r>, r = 1, 2, ..., m$ and $|f_r>, r = 1, 2, ..., m$ in $\mathcal{H}_1$ and $\mathcal{H}_2$ respectively such that

$$|\psi> = \sum_r |e_r> |f_r>$$

Note that $|e_r> |f_r>$ is the same as $e_r \otimes f_r$. It follows that

$$\rho_1 = Tr_2(|\psi><\psi|) = \sum_r <f_r|f_r> |e_r><e_r|, \rho_2 = Tr_1(|\psi><\psi|) = \sum_r <e_r|e_r> |f_r><f_r|$$

and thus ρ_1 and ρ_2 have the same set of non-zero eigenvalues, namely $\{<e_r|e_r><f_r|f_r>, r = 1, 2, ..., m\}$. In particular, the quantum entropies of ρ_1 and ρ_2 are the same.

[4] Let $\mathcal{H}_i, i = 1, 2$ be two finite dimensional Hilbert spaces and let A be a linear operator in $\mathcal{H}_1 \otimes \mathcal{H}_2$. Suppose $A = \sum_i B_i \otimes C_i$ where for each i, B_i is a linear operator in $\mathcal{H}_1$ and C_i is a linear operator in $\mathcal{H}_2$. Then $tr_1(A)$ is the linear operator in $\mathcal{H}_2$ given by

$$tr_1(A) = \sum_i tr(B_i) C_i$$

and $tr_2(A)$ is the linear operator in $\mathcal{H}_1$ given by

$$tr_2(A) = \sum_i tr(C_i)B_i$$

Suppose $|e_0>$ is a unit vector in $\mathcal{H}_2$ and U is a unitary operator in $\mathcal{H}_1 \otimes \mathcal{H}_2$. Let ρ be a density operator (state) in $\mathcal{H}_1$. We wish to compute $tr_2(U(\rho \otimes |e_0><e_0|)U^*)$. To do so, write

$$U = \sum_i A_i \otimes B_i$$

where for each i, A_i is a linear operator in $\mathcal{H}_1$ and B_i is a linear operator in $\mathcal{H}_2$. Then,

$$U(\rho \otimes |e_0><e_0|)U^* = \sum_{i,j}(A_i \otimes B_i)(\rho \otimes |e_0><e_0|)(A_j^* \otimes B_j^*)$$

$$= \sum_{i,j}(A_i\rho A_j^*) \otimes (B_i|e_0><e_0|B_j^*)$$

and hence

$$tr_2(U(\rho \otimes |e_0><e_0|)U^*) = \sum_{i,j} tr(B_i|e_0><e_0|B_j^*)A_i\rho A_j^*$$

$$= \sum_{i,j} <e_0|B_j^*B_i|e_0> A_i\rho A_j^*$$

Define

$$c_{ji} = <e_0|B_j^*B_i|e_0>$$

Then, $((c_{ij}))$ is a positive semidefinite Hermitian matrix and hence has the eigendecomposition

$$c_{ji} = \sum_k \lambda_k u_{jk}\bar{u}_{ik}$$

where $((u_{ij}))$ is a unitary matrix and $\lambda_k \geq 0$. Thus

$$tr_2(U(\rho \otimes |e_0><e_0|)U^*) = \sum_{i,j,k} \lambda_k u_{jk}\bar{u}_{ik} A_i\rho A_j^*$$

$$= \sum_k E_k\rho E_k^*$$

where

$$E_k = \sum_i \sqrt{\lambda_k} u_{ik} A_i$$

Reference:Nielsen and Chuang, Quantum computation and Quantum Information.

[56] Assume that $\psi(t,\mathbf{r})$ satisfies the time dependent Schrodinger wave equation. Substitute $\psi(t,\mathbf{r}) = exp(2\pi i S(t,\mathbf{r})/h)$ and derive the equation satisfied by S. Expand S in powers of $h/2\pi$ and

obtain an approximate solution for S upto $O(h/2\pi)$. Assume that the potential $V(\mathbf{r})$ is time independent.

hint:

$$\nabla\psi = (2\pi i/h)(\nabla S)exp(2\pi i S/h),$$

$$\nabla^2\psi = (2\pi i/h)\nabla^2 S - (4\pi^2/h^2)|\nabla S|^2$$

$$\psi_{,t} = (2\pi i/h)S_{,t}.exp(2\pi i S/h)$$

The Schrodinger equation then gives

$$(-h^2/8\pi^2 m)((2\pi i/h)\nabla^2 S - (4\pi^2/h^2)|\nabla S|^2) + V + S_{,t} = 0$$

or

$$|\nabla S|^2/2m - (ih/4\pi m)\nabla^2 S + V + S_{,t} = 0$$

We then expand the function S in powers of h so that

$$S = S_0 + hS_1 + h^2 S_2 + ...$$

and equate equal powers of h to zero giving for the zeroth order term

$$|\nabla S_0|^2/2m + V + S_{0,t} = 0$$

This is simply the Hamilton-Jacobi equation in classical mechanics:

$$H(\mathbf{r}, \nabla S) + S_{,t} = 0, H(\mathbf{r}, \mathbf{p}) = p^2/2m + V(\mathbf{r})$$

Now by equating the coefficient of h to zero gives us a differential equation for the first order term S_1.

[57] Give a numerical algorithm for determining the eigenvalues and eigenfunctions of the Dirac relativistic wave equation for a particle moving in a potential $V(\mathbf{r})$.

[58] Derive a numerical algorithm for solving the time dependent Dirac equation for an electron in an electromagnetic field.

hint: Let A_μ be the four potential corresponding to the Electromagnetic field. Then the Dirac equation reads

$$(\gamma^\mu(i\partial_\mu + e.A_\mu) - m)\psi = 0$$

where e is the electronic charge and m is the mass of the electron. γ^μ are the 4×4 Dirac matrices. Now discretize this equation with respect to the time and spatial arguments.

[59] Define a projective unitary representation of a Lie group. What do the commutation relations for the generators of the corresponding Lie algebra look like.

hint.

$$U(T_1 T_2) = exp(i\phi(T_1, T_2))U(T_1)U(T_2), T_2, T_2 \in G$$

This equation defines a projective unitary representation of G. $U(T), T \in G$ are unitary operators in the Hilbert space of states. $\phi : G \times G \to \mathbb{R}$ is a smooth map. We have

$$U(T_1 T_2 T_3) = exp(i\phi(T_1 T_2, T_3))U(T_1 T_2)U(T_3) = exp(i\phi(T_1 T_2, T_3))exp(i\phi(T_1, T_2))U(T_1)U(T_2)U(T_3)$$

on the one hand and on the other,

$$U(T_1 T_2 T_3) = exp(i\phi(T_1, T_2 T_3))U(T_1)U(T_2 T_3) = exp(i\phi(T_1, T_2 T_3))exp(i\phi(T_2, T_3))U(T_1)U(T_2)U(T_3)$$

It follows that

$$\phi(T_1 T_2, T_3) + \phi(T_1, T_2) = \phi(T_1, T_2 T_3) + \phi(T_2, T_3)$$

modulo an integer multiple of 2π.

[60] Construct four 4×4 matrices $\gamma^\mu, \mu = 0, 1, 2, 3$ such that if $p^\mu, \mu = 0, 1, 2, 3$ and m are scalar variables, then

$$(\gamma^0 p^0 + \sum_{i=1}^{3} \gamma^i p^i + m)(\gamma^0 p^0 - \sum_{i=1}^{3} \gamma^i p^i - m)$$

$$= (p^0)^2 - \sum_{i=1}^{3}(p^i)^2 - m^2$$

Using this derive Dirac's relativistic wave equation for the electron.

[61] Why is time ordering of operators important in quantum mechanics. Illustrate this by giving an example of computing the evolution operator for a time dependent Hamiltonian.

hint.

Let $H(t), t \geq 0$ be the time dependent Hamiltonian operator of a quantum mechanical system. Schrodinger's wave equation for the wave function $\psi(t)$ is given by

$$\psi'(t) = -iH(t)\psi(t)$$

Since the $H(t)'s$ for different $t's$ do not commute, we cannot write the solution as

$$\psi(t) = exp(-i \int_0^t H(s)ds)\psi(0)$$

The correct solution is given by

$$\psi(t) = (I + \sum_{n=1}^{\infty}(-i)^n \int_{0<t_1<...<t_n<t} H(t_n)H(t_{n-1})...H(t_2)H(t_1)dt_1 dt_2...dt_n)\psi(0)$$

If T denotes the time ordering operator, so that for $t_1 < ... < t_n$ and any permutation σ of $\{1, 2, ..., .n\}$, we have

$$T\{H(t_{\sigma 1})H(t_{\sigma 2})...H(t_{\sigma n})\} = H(t_1)H(t_2)...H(t_n)$$

then we can write the evolution operator as

$$U(t) = T\{exp(-i \int_0^t H(s)ds)\}$$

[62] Let $\mathcal{H}(x)$ be the Hamiltonian density for a set of quantum fields. Here, $x = (x^0, x^1, x^2, x^3)$ are the space-time coordinates of an event. x^0 is the time coordinate and x^1, x^2, x^3 are the spatial coordinates. The Hamiltonian is given by

$$H(x^0) = \int \mathcal{H}(x)dx^1 2dx^2 dx^3$$

and the evolution operator from time $x^0 = -\infty$ to time $x^0 = \infty$ is given by

$$U(\infty, -\infty) = T\{exp(-i \int \mathcal{H}(x)d^4x)\}$$

Let $|\Phi_0 >$ denote the vacuum state and $|p_1, ..., p_n >$ the state in which there are n particles of four momenta $p_1, p_2, ..., p_n$. Let $a(p)$ denote the annihilation operator of a particle of four momentum p and $a(p)^*$ the creation operator of a particle of four momentum p. Then,

$$|p_1, ..., p_n >= a(p_1)^* a(p_2)^* ... a(p_n)^* |\Phi_0 >$$

and

$$< p_1', p_2', ..., p_n'| =< \Phi_0|a(p_1')a(p_2')...a(p_n')|$$

so that the probability of scattering of n particles with four momenta $p_i, i = 1, 2, ..., n$ into particles with four momenta $p_i', i = 1, 2, ..., n$ is given by

$$< \Phi_0|a(p_1')...a(p_n')T\{exp(-i \int \mathcal{H}(x)d^4x)\}a(p_1)^*...a(p_n)^*|\Phi_0 >$$

Explain using examples how this equation in the interaction picture can be used to construct Feynman diagrams for computing the scattering amplitudes.

[63] Interaction picture in quantum mechanics. The Hamiltonian can be split as

$$H(t) = H_0 + V(t)$$

where H_0 is the free particle Hamiltonian and $V(t)$ is its interaction energy with some other particle. In the interaction picture, the states evolve according to the Hamiltonian $V_1(t) = exp(itH_0)V(t)exp(-itH_0)$ while the observables evolve according to the free particle Hamiltonian H_0. Suppose $\psi(t)$ is the state at time t as per the Schrodinger picture. Then, $\psi(t)$ evolves according to $H(t)$ and hence

$$i\psi'(t) = H(t)\psi(t)$$

Define

$$\phi(t) = exp(itH_0)\psi(t)$$

Then,

$$i\phi'(t) = -exp(itH_0)H_0\psi(t) + exp(itH_0)i\psi'(t) = exp(itH_0)(H(t) - H_0)\psi(t)$$

$$= exp(itH_0)V(t)\psi(t) = V_1(t)\phi(t)$$

The state $\phi(t)$ thus evolves according to the Hamiltonian $V_1(t)$. Let X be an observable. In the Schrodinger picture, X does not vary with time while in the Heisenberg picture, X evolves in accord with the Heisenberg equations of motion.

$$X'(t) = i[H(t), X(t)]$$

Define

$$X_1(t) = exp(itH_0)X exp(-itH_0)$$

Then $X_1(t)$ describes the evolution of X in the interaction picture. We compute the rate of change of the quantity $< \phi(t)|X_1(t)|\phi(t) >$.

$$i\frac{d}{dt} < \phi(t)|X_1(t)|\phi(t) >=< -i\phi'(t)|X_1(t)|\phi(t) > + < \phi(t)|iX_1'(t)|\phi(t) > + < \phi(t)|X_1(t)|i\phi'(t) >$$

$$= - < V_1(t)\phi(t)|X_1(t)|\phi(t) > - < \phi(t)|[H_0, X_1(t)]|\phi(t) > + < \phi(t)|X_1(t)|V_1(t)\phi(t) >$$

$$= - < \phi(t)|[V_1(t), X_1(t)]|\phi(t) > - < \phi(t)|[H_0, X_1(t)]|\phi(t) >= - < \phi(t)|[H_0 + V_1(t), X_1(t)]|\phi(t) >$$

$$= - < \phi(t)|exp(itH_0)[H(t), X]exp(-itH_0)|\phi(t) >= - < \psi(t)|[H(t), X]|\psi(t) >= i\frac{d}{dt} < \psi(t)|X|\psi(t) >$$

From this, it follows that

$$< \phi(t)|X_1(t)|\phi(t) >=< \psi(t)|X|\psi(t) >$$

This proves the consistency of the interaction picture with the Schrodinger/Heisenberg picture.

[64] Consider a real scalar field $\phi(x), x \in \mathbb{R}^4$. In the absence of the perturbation $V(\phi)$, ϕ satisfies the Klein-Gordon equation

$$\partial_t^2\phi - \nabla^2\phi + m^2\phi = 0$$

This corresponds to the energy momentum relation

$$E^2 = |\mathbf{p}|^2 + m^2$$

For taking $E = i\partial_t$ and $p = -i\nabla$, and operating both sides on the wave function ϕ, we get

$$(\nabla^2 - \partial_t^2 - m^2)\phi = 0$$

The solution can be written as

$$\phi(x) = \int (u(p)a(p)exp(ip.x) + u^*(p)a(p)^*exp(-ip.x))d^3p$$

where

$$p.x = Et - (\mathbf{p}, \mathbf{x}) = \mathbf{p}_\mu\mathbf{x}^\mu, \mathbf{E} = \mathbf{p}^0 = \sqrt{|\mathbf{p}|^2 + \mathbf{m}^2}$$

$a(p)$ annihilates a particle having four momentum p while $a^*(p)$ creates a particle having four momentum p. $u(p)$ is a scalar function of the four momentum p. $u^*(p)$ is the complex conjugate of $u(p)$. The commutation relations satisfied by the operators $a(p), a^*(p)$ are

$$[a(p), a(p')] = [a^*(p), a^*(p')] = 0, [a(p), a^*(p')] = \delta^3(\mathbf{p} - \mathbf{p'})$$

Let $|0>$ denote the vacuum state. Then $a^*(p)|0> = |p>$ is a state in which there is one particle having four momentum p. We want to compute the propagator

$$\Delta(x, y) = < 0|T\{\phi(x).\phi(y)\}|0 >$$

Where $T\{\phi(x)\phi(y)\} = \phi(x)\phi(y)$ if $x^0 > y^0$ and $T\{\phi(x)\phi(y)\} = \phi(y)\phi(x)$ if $y^0 > x^0$. We have for $x^0 > y^0$,

$$\Delta(x, y) = \int |u(p)|^2 exp(ip.(x - y))d^3p$$

Thus for all x, y, we have

$$\Delta(x, y) = \theta(x^0 - y^0) \int |u(p)|^2 exp(ip.(x - y))d^3p + \theta(y^0 - x^0) \int |u(p)|^2 exp(ip.(y - x))d^3p$$

Suppose on the other hand, the scalar field ϕ is complex with antiparticles, then we can write

$$\phi(x) = \int (a(p)u(p)exp(ip.x) + b(p)^*v(p)exp(-ip.x))d^3p$$

where $u(p), v(p)$ are numerical functions of the four momentum, $a(p)$ is the annihilation operator of a particle having four momentum p and $b(p)$ is the annihilation operator of an antiparticle having four momentum p. $b(p)^*$ is thus the creation operator of an antiparticle having four momentum p. The commutation relations are

$$[a(p), a^*(p')] = \delta^3(\mathbf{p} - \mathbf{p'}), [b(p), b^*(p')] = \delta^2(\mathbf{p} - \mathbf{p'})$$

All the other commutators are zero. Then, the propagator of the complex scalar field is given by

$$\Delta_1(x, y) = < 0|T\{\phi(x)\phi^*(y)|0 >$$

where

$$\phi^*(x) = \int (a^*(p)u^*(p)exp(-ip.x) + b(p)v^*(p)exp(ip.x))d^3p$$

We have for $x^0 > y^0$,

$$\Delta_1(x, y) = \int |u(p)|^2 exp(ip.(x - y))d^3p$$

and for $y^0 > x^0$,

$$\Delta_1(x, y) = \int |v(p)|^2 exp(ip.(y - x))d^3p$$

so that the propagator for all x, y is given by

$$\Delta_1(x, y) = \theta(x^0 - y^0) \int |u(p)|^2 exp(ip.(x - y))d^3p + \theta(y^0 - x^0) \int |v(p)|^2 exp(ip.(y - x))d^3p$$

Vector fields and spin may also be taken into account in these computations by considering the field components to have the form

$$\phi_l(x) = \sum_{-s \leq \sigma \leq s} \int (u_l(p,\sigma)a(p,\sigma)exp(ip.x) + v_l(p,\sigma)b^*(p,\sigma)exp(-ip.x))d^3p, l = 1,2,...,N$$

where s is the spin of the particle.

Reference:The quantum theory of fields, vol.1 by Steven Weinberg.

[65] Suppose $\psi_1(x)$ and $\psi_2(y)$ are two quantum fields satisfying the commutation relation

$$[\psi_1(x), \psi_2(y)] = \eta(x,y)$$

Then,

$$[\psi_{1,i}(x), \psi_{2,j}(y)] = \frac{\partial^2}{\partial x^i \partial y^j}\eta(x,y)$$

[66] Let H be the Hamiltonian operator of a quantum mechanical system. The corresponding unitary evolution operator is given by $U(t,s) = exp(-i(t-s)H)$. We set

$$G(t) = U(t)\theta(t)$$

where $\theta(t) = 1$ for $t > 0$ and $\theta(t) = 0$ for $t \leq 0$. Then,

$$iG'(t) = iU'(t)\theta(t) + iU(t)\delta(t) = HU(t)\theta(t) + i\delta(t)$$

since $U(0) = I$. It follows that

$$iG'(t) - HG(t) = i\delta(t)$$

or

$$G'(t) + iHG(t) = \delta(t)$$

$G(t)$ is the Green's function of the quantum system. Define

$$\hat{G}(\omega) = \int_{-\infty}^{\infty} G(t).exp(-i\omega t)dt$$

Then,

$$\hat{G}(\omega) = (\omega + H)^{-1}$$

This is defined provided that ω is not in the spectrum of H.

[67] Copmute the propagator of a free particle of mass m.

hint:

$$K(\mathbf{r},t,\mathbf{r}',t') = C \int exp(-i|\mathbf{p}|^2(t-t')/2m)exp(i\mathbf{p}.(\mathbf{r} - \mathbf{r}'))d^3p$$

[68] The Lagrangian density of the Dirac field is given by

$$L = K.\bar{\psi}(i\gamma^\mu \partial_\mu - m)\psi$$

where K is a real constant, ψ is a four component wave function and γ^μ are the Dirac gamma matrices. Calculate the corresponding Hamiltonian density. Here, $\bar{\psi} = \psi^* \gamma^0$. We can take

$$\gamma^r = \begin{pmatrix} 0 & \sigma_r \\ -\sigma_r & 0 \end{pmatrix}, r = 1, 2, 3,$$

$$\gamma^0 = \begin{pmatrix} I & 0 \\ 0 & -I \end{pmatrix}$$

Then,

$$\gamma^{02} = I, \gamma^{r2} = -I, r = 1, 2, 3, \gamma^\mu \gamma^\nu + \gamma^\nu \gamma^\mu = 0, \mu \neq \nu$$

These equations can be compressed into

$$\gamma^\mu \gamma^\nu + \gamma^\nu \gamma^\mu = 2g^{\mu\nu}$$

The Dirac equation

$$(i\gamma^\mu \partial_\mu - m)\psi = 0$$

implies

$$(i\gamma^\mu \partial_\mu + m)(i\gamma^\nu \partial_\nu - m)\psi = 0$$

which gives

$$(\nabla^2 - \partial_0^2 - m^2)\psi = 0$$

which is the Klein-Gordon equation. Now

$$\bar{\psi}(i\gamma^\mu \partial_\mu)\psi = i\psi^* \partial_0 \psi + \psi^*(i\gamma^0 \gamma . \nabla)\psi$$

Use these results and the Legendre transformation to compute the Dirac Hamiltonian. Note that the canonical position variables are the components of ψ and ψ^* while the canonical momenta conjugate to ψ are

$$\pi = \frac{\partial L}{\partial \partial_0 \psi} = iK\psi^*$$

Thus, the Hamiltonian density is given by

$$H = \pi.\partial_0 \psi - L = -K\psi^*(i\gamma^0 \gamma . \nabla)\psi + K\psi^*(\gamma^0 m)\psi$$

Taking $K = 1$ and $\alpha^r = \gamma^0 \gamma^r, r = 1, 2, 3$ gives

$$H = \psi^*(m\gamma^0 + (\alpha, p))m)\psi, p = -i\nabla$$

Let $a(p, \sigma)$ be the operator of annihilation of an electron of four momentum p and helicity σ and let $a(p, \sigma)^*$ be the corresponding creation operator. Let $b(p, \sigma), b(p, \sigma)^*$ be respectively the operators of annihilation and creation of a positron of four momentum p and helicity σ. Let $u(p, \sigma)$ be the position

space wave function for the electron and $v(p,\sigma)$ that of the positron. They satisfy the Dirac equation.
We have

$$\psi(x) = \int (2E(p))^{-1/2}(u(p,\sigma)a(p,\sigma).exp(-ip.x) + \bar{v}(p,\sigma)^T b(p,\sigma)^* exp(ip.x))d^3p$$

where

$$E(p) = \sqrt{|\mathbf{p}|^2 + m^2}$$

The equation $(i\gamma^\mu \partial_\mu - m)\psi = 0$ implies,

$$(\gamma^\mu p_\mu - m)u(p,\sigma) = 0$$

and

$$(\gamma^\mu p_\mu + m)\bar{v}(p,\sigma)^T = 0$$

Note that

$$\bar{v}(p,\sigma) = v(p,\sigma)^* \gamma^0$$

so that

$$(\gamma^\mu p_\mu + m)\gamma^{0T} conj(v(p,\sigma)) = 0$$

Thus,

$$(\gamma^0 \gamma^\mu \gamma^0 p_\mu + m)conj(v(p,\sigma)) = 0$$

Now

$$\gamma^0 \gamma^0 \gamma^0 = \gamma^0, \gamma^0 \gamma^r \gamma^0 = -\gamma^r, r = 1,2,3$$

so that

$$(\gamma^0 p^0 + \gamma^r p^r + m)conj(v(p,\sigma)) = 0$$

Note that

$$p^0 = \sqrt{\sum_r p^{r2} + m^2}$$

so that changing p^r to $-p^r$ gives

$$(\gamma^\mu p_\mu + m)conj(v(q,\sigma)) = 0$$

where

$$q = (q^0, q^1, q^2, q^3) = (p^0, -p^1, -p^2, -p^3)$$

Note also that using $\gamma^{r*} = -\gamma^r$ and $\gamma^{0*} = \gamma^0$, we get

$$v(p,\sigma)^T(\gamma^\mu p_\mu + m) = 0$$

[69] Define the Pauli spin matrices

$$X = \begin{pmatrix} 0 & 1 \\ 1 & 0 \end{pmatrix}, Y = \begin{pmatrix} 0 & -i \\ i & 0 \end{pmatrix}$$

$$Z = \begin{pmatrix} 1 & 0 \\ 0 & -1 \end{pmatrix}$$

Show that

$$X^2 = Y^1 = Z^2 = I, XY = -YX = iZ, YZ = -ZY = iX, ZX = -XZ = iY$$

Let $\mathbf{n} = (n_1, n_2, n_3)$ be a unit vector in $\mathbb{R}^3$ and let $\sigma = (X, Y, Z)$. For real θ define

$$R_n(\theta) = exp(i\theta(\mathbf{n}, \sigma)/2)$$

where

$$(\mathbf{n}, \sigma) = n_1 X + n_2 Y + n_3 Z$$

Show that

$$(\mathbf{n}, \sigma)^2 = I$$

and hence

$$R_n(\theta) = cos(\theta/2).I + i.sin(\theta/2).(\mathbf{n}, \sigma)$$

Show that if $\mathbf{m} = (m_1, m_2, m_3)$ is any vector in $\mathbb{R}^3$, then

$$(\mathbf{n}, \sigma).(\mathbf{m}, \sigma) = (\mathbf{n}, \mathbf{m}) + i(\mathbf{n} \times \mathbf{m}, \sigma)$$

Deduce that

$$(I + i\delta\theta.(\mathbf{n}, \sigma)/2)(\mathbf{m}, \sigma)(I - i\delta\theta.(\mathbf{m}, \sigma))$$

$$= (\mathbf{m}, \sigma) - \delta\theta.(\mathbf{n} \times \mathbf{m}, \sigma) + O(\delta\theta^2)$$

Note that if $\mathbf{m} \in \mathbb{R}^3$, then $\mathbf{m} + \delta\theta.\mathbf{n} \times \mathbf{m}$ is with neglect of $O(\delta\theta^2)$, the vector obtained by applying an infinitesimal rotation about the axis $\mathbf{n}$ by an angle $\delta\theta$ to the vector $\mathbf{m}$. Use the above formula to interpret $R_n(\theta)(\mathbf{m}, \sigma)R_n(-\theta)$ where θ is a finite real number.

[70] Assume that the Hamiltonian of a quantum mechanical system has the form

$$H = H_0 + \epsilon.V$$

where ϵ is a small real number. Determine upto $O(\epsilon^2)$ the stationary state wave functions and corresponding energy levels assuming that the unperturbed Hamiltonian H_0 does not have any degenerate energy levels.

hint: Let $u_n^{(0)}, n = 1, 2, ...$ be the unperturbed normalized stationary state wave functions with corresponding energy levels $E_n^{(0)}, n = 1, 2, ...$ Let the perturbed wave functions be given by

$$u_n = u_n^{(0)} + \epsilon.u_n^{(1)} + \epsilon^2 u_n^{(2)} + O(\epsilon^3),$$

and the corresponding perturbed energy levels by

$$E_n = E_n^{(0)} + \epsilon.E_n^{(1)} + \epsilon^2.E_n^{(2)} + O(\epsilon^3)$$

Plugging these expressions into the eigenvalue equation $Hu_n = E_n u_n$ gives us

$$(H_0 + \epsilon.V)(u_n^{(0)} + \epsilon.u_n^{(1)} + \epsilon^2 u_n^{(2)} + ...)$$

$$= (E_n^{(0)} + \epsilon.E_n^{(1)} + \epsilon^2 E_n^{(2)} + ...)(u_n^{(0)} + \epsilon.u_n^{(1)} + \epsilon^2 u_n^{(2)} + ...)$$

Equating coefficients of $\epsilon^m, m = 0, 1, 2$ on both sides gives

$$H_0 u_n^{(0)} = E_n^{(0)} u_n^{(0)}, H_0 u_n^{(1)} + V u_n^{(0)} = E_n^{(0)} u_n^{(1)} + E_n^{(1)} u_n^{(0)},$$

$$H_0 u_n^{(2)} + V u_n^{(1)} = E_n^{(0)} u_n^{(2)} + E_n^{(1)} u_n^{(1)} + E_n^{(2)} u_n^{(0)}$$

The second equation gives on taking inner products with $u_m^{(0)}$,

$$E_n^{(1)} = < u_n^{(0)} | V | u_n^{(0)} >,$$

$$(E_m^{(0)} - E_n^{(0)}) < u_m^{(0)} | u_n^{(1)} > + < u_m^{(0)} | V | u_n^{(0)} > = 0, m \neq n$$

so that

$$< u_m^{(0)} | u_n^{(1)} > = \frac{V_{mn}}{E_n^{(0)} - E_m^{(0)}}, m \neq n$$

where

$$V_{mn} = < u_m^{(0)} | V | u_n^{(0)} >$$

The third equation gives

$$<< u_n^{(0)} | V | u_n^{(1)} > = E_n^{(1)} < u_n^{(0)} | u_n^{(1)} > + E_n^{(2)}$$

$$(E_m^{(0)} - E_n^{(0)}) < u_m^{(0)} | u_n^{(2)} > = E_n^{(1)} < u_m^{(0)} | u_n^{(1)} >, m \neq n$$

From these equations, we can arrive at the expansions for $u_n^{(1)}, E_n^{(2)}, u_n^{(2)}$.

[71] Let the joint wave function of the system and environment be $\psi(q, x)$ where q are the variables describing the environment and x are the variables describing the system. Let X be a system observable. With respect to the system coordinates x, y X can be viewed as an integral operator with kernel $X(x, y)$. Then the average value of X in the state ψ is given by

$$< X > = < \psi | X | \psi > = \int \bar{\psi}(q, x) X(x, y) \psi(q, y) dq dx dy = \int \rho(y, x) X(x, y) dx dy$$

where

$$\rho(y, x) = \int \psi(q, y) \bar{\psi}(q, x) dq$$

is the kernel of the density operator of the system.

[72] Let $U(r)$ be a radially symmetric potential. A quantum mechanical particle moving in this potential has a stationary state wave function $\psi(r, \theta, \phi)$ satisfying the eigenvalue equation

$$-\frac{h^2}{8\pi^2 m} \nabla^2 \psi + U\psi = E\psi$$

or in terms of the squared angular momentum operator L^2,

$$-\frac{h^2}{8\pi^2 mr^2}(r^2\psi_{,r})_{,r} + (L^2\psi/2mr^2) + U\psi = E\psi$$

In a state of definite squared angular momentum and zero z component of angular momentum (corresponding to azimuthal symmetry of the wave function)

$$\psi(r,\theta) = R(r)P_l(cos(\theta))$$

where $l(l+1)h^2/4\pi^2$ is the eigenvalue of L^2 In other words,

$$L^2 P_l = (l(l+1)h^2/4\pi^2)Y_{lm}, L_z P_l = 0$$

l assumes values $0, 1, 2, ...$ The radial component R of the wavefunction in this state therefore satisfies

$$-\frac{h^2}{8\pi^2 mr^2}(r^2 R'(r))' + l(l+1)h^2 R(r)/8mr^2 + U(r)R(r) = ER(r)$$

E is the energy of the particle and we write $E = h^2 k^2/8\pi^2 m$ and $R(r) = R_{kl}(r)$. The problem in scattering theory is to determine the asymptotic form of the radial wave function $R_{kl}(r)$. The complete wave function then has the form

$$\psi(r,\theta) = \sum_l A_l R_{kl}(r)P_l(cos(\theta))$$

P_l is the Legendre polynomial of order l. In case azimuthal symmetry is absent, the wave function has the form

$$\psi(r,\theta,\phi) = \sum_{l,p} A_{lp} R_{kl}(r)Y_{lp}(\theta,\phi)$$

where $\{Y_{lp} : l = 0, 1, 2, ...p = -l, -l+1, ..., l-1, l\}$ are the spherical harmonics. They satisfy

$$L^2 Y_{lp} = (l(l+1)h^2/4\pi^2)Y_{lp}, L_z Y_{lp} = (ph/2\pi)Y_{lp}$$

In the state $R(r)Y_{lp}(\theta,\phi)$, L^2 has the value $l(l+1)h^2/4\pi^2$ and L_z has the value $ph/2\pi$. For $U = 0$, R satisfies

$$r^2 R'' + 2rR' + (k^2 r^2 - l(l+1))R = 0$$

[73] Computation of the wave function and matrix elements of operators in quantum mechanics using the quasi-classical approximation. The wave function $\psi(\mathbf{r}, t)$ in quantum mechanics satisfies the Schrodinger equation

$$-\frac{h^2}{8\pi^2 m}\nabla^2\psi(\mathbf{r},t) + U(\mathbf{r},t))\psi(\mathbf{r},t) = \frac{ih}{2\pi}\psi_{,t}(\mathbf{r},t)$$

The magnitude square of the wave function has the interpretation of being the probability density of the position of the quantum particle at each time. For most potentials, solving the time dependent Schrodinger equation in closed form is impossible. Hence, we develop computer programmes for obtaining approximate solutions based on the quasi-classical approximation. The idea is to try a wave function of the form

$$\psi(\mathbf{r}, t) = A(\mathbf{r}, t).exp(2\pi i S(\mathbf{r}, t)/h)$$

where the amplitude A and phase S are assumed to be real functions. We have

$$\nabla\psi = (\nabla A + \frac{2\pi i}{h} A\nabla S)exp(2\pi i S/h)$$

$$\nabla^2\psi = (\nabla^2 A + \frac{4\pi i}{h}(\nabla A, \nabla S) + \frac{2\pi i}{h}A\nabla^2 S - \frac{4\pi^2}{h^2}A|\nabla S|^2)exp(2\pi i S/h)$$

$$\psi_{,t} = (A_{,t} + (2\pi i/h)AS_{,t})exp(2\pi i S/h)$$

and substituting this into the wave equation and equating the real and imaginary parts gives us two partial differential equations for the two functions A, S. These are

$$-\frac{h^2}{8\pi^2 m}(\nabla^2 A - (4\pi^2/h^2)A|\nabla S|^2) + UA = -AS_{,t}$$

$$-\frac{h^2}{8\pi^2 m}((4\pi/h)(\nabla A, \nabla S) + (2\pi/h)A\nabla^2 S) = (h/2\pi)A_{,t}$$

In the limit $h \to 0$, the first equation reduces to

$$|\nabla S|^2/2m + U + S_{,t} = 0$$

which is precisely the Hamilton-Jacobi equation for the classical action S. The second equation is the same as

$$A_{,t} + (\nabla A, \nabla S)/m + A\nabla^2 S/2m = 0$$

or

$$(A^2)_{,t} + \nabla.(A^2\nabla S/m) = 0$$

which is the mass conservation corresponding to a fluid with density field $A^2 = |\psi|^2$ and velocity field $\mathbf{v} = \nabla S/m$. Thus to a good degree of approximation, the quantum mechanical wave function may be determined using classical methods. Suppose we are interested in obtaining the stationary state wave function corresponding to the energy level E when the potential $U(\mathbf{r})$ is independent of time. Then $S(\mathbf{r}, t)$ is replaced by a time independent function $S(\mathbf{r})$ and $A(\mathbf{r}, t)$ by a time independent function $A(\mathbf{r})$. Plugging this into the Stationary state Schrodinger equation gives us

$$-\frac{h^2}{8\pi^2 m}\nabla^2 A + A|\nabla S|^2/2m + UA = EA$$

$$\nabla.(A^2\nabla S) = 0$$

If the quantum mechanical problem is one dimensional, then we get approximately

$$S(x) = \int_{x_0}^{x} \sqrt{2m(E - U(\xi))}d\xi = \int_{x_0}^{x} p(\xi)d\xi,$$

$$A(x) = C/\sqrt{S'} = C/\sqrt{p(x)}$$

In these equations, $p(x) = \sqrt{2m(E - U(x))}$ is the classical momentum. However, if $U(x) > E$, the classical momentum $p(x)$ becomes imaginary and classically, a particle of energy E cannot be present there. However, quantum mechanically, there is a nonzero finite probability of the particle existing in such a region and the resulting wave function in such a region is an exponentially decaying one. To get more accurate information about such a wave function, we must use contour integration as is given in the book "Quantum Mechanics" by Landau and Lifshitz. The resulting quasi classical wave functions corresponding to different energy levels E can be used to compute the matrix elements of observables which are important in determining the transition probabilities of the system from one state to another under the influence of a perturbing potential.

[74]. Simulation of quantum gates and channels using physical systems and scattering experiments. A quantum gate is a finite dimensional unitary matrix. The notion of a quantum gate can be extended to infinite dimensions. One wave to realize and infinite dimensional quantum gate is by scattering experiments in which a projectile interacts with a central potential and gets scattered into a solid angle. Suppose for example that the central potential is $U(\mathbf{r})$. The incident beam of particles are described by the wave function $\psi_0(\mathbf{r}) = exp(i\mathbf{k}.\mathbf{r})$ where $h\mathbf{k}/2\pi$ is the momentum of the incident particle. Note that since $|\psi_0|^2 = 1$, the incident beam has a uniform density. After interaction with the potential the wave function has the form

$$\psi(\mathbf{r}) = \psi_0(\mathbf{r}) + \psi_1(\mathbf{r})$$

where ψ_1 is small in magnitude as compared with ψ_0. The stationary Schrodinger equation corresponding to the energy $E = h^2 k^2/8\pi^2 m$ reads

$$-\frac{h^2}{8\pi^2 m}\nabla^2\psi + U\psi = h^2 k^2\psi/8\pi^2 m$$

or

$$(\nabla^2 + k^2)\psi = \frac{8\pi^2 m}{h^2}U\psi$$

We assume that the potential is weak. Then the scattered wave function ψ_1 is small in magnitude as compared with ψ_0 and hence ψ_1 approximately satisfies

$$(\nabla^2 + k^2)\psi_1 = \frac{8\pi^2 m}{h^2}U\psi_0$$

which has the solution

$$\psi_1(\mathbf{r}) = \frac{8\pi^2 m}{h^2}\int U(\mathbf{r}')\psi_0(\mathbf{r}')\frac{exp(ik|\mathbf{r} - \mathbf{r}'|)}{|\mathbf{r} - \mathbf{r}'|}d^3 r'$$

The Born approximation involves considering $r >> r'$ to get

$$\psi_1(\mathbf{r}) \approx \frac{8\pi^2 m.exp(ikr)}{h^2 r}\int U(\mathbf{r}')exp(i(\mathbf{k}.\mathbf{r}' - k\hat{r}.\mathbf{r}'))d^3 r'$$

This approximation determines the angular distribution of scattered particles in terms of the momentum of the incident particle. More generally, the state of the incident particle can be taken as a superposition

$\psi_0(\mathbf{r}) = \int F(\mathbf{k})exp(i\mathbf{k.r})d^3k$ where $|F(\mathbf{k})|^2 d^3k$ is proportional to the number of incident particles in the momentum range d^3k about $\mathbf{k}$. Then the scattered wave function will have the form

$$\psi_1(\mathbf{r}) = \frac{8\pi^2 m}{h^2} \int U(\mathbf{r}')F(\mathbf{k})exp(i\mathbf{k.r}')\frac{exp(ik|\mathbf{r}-\mathbf{r}'|)}{|\mathbf{r}-\mathbf{r}'|}d^3r'd^3k$$

and in the momentum domain,

$$G(\mathbf{k_1}) = \int \psi_1(\mathbf{r})exp(-i\mathbf{k_1.r})d^3r$$

$$= \frac{8\pi^2 m}{h^2} \int U(\mathbf{r}')\frac{exp(ik|\mathbf{r}-\mathbf{r}'|)}{|\mathbf{r}-\mathbf{r}'|}exp(i(\mathbf{k.r}'-\mathbf{k_1.r}))d^3r'd^3rd^3k$$

$$= \int S(\mathbf{k_1},\mathbf{k})F(\mathbf{k})d^3k$$

where S is the scattering operator kernel in the momentum domain. By controlling the potential U we can control the scattering operator kernel S and hence design a variety of quantum gates. Note that $F(\mathbf{k})$ is the momentum space wave function before the interaction with the potential and $F(\mathbf{k})+G(\mathbf{k}) = (I+S)F(\mathbf{k})$ is the wave approximate wave function of the particles after the scattering.

[75]. **Random perturbations of quantum systems.** Suppose $V(\mathbf{r})$ is the potential function and its amplitude is modulated by a random process $\xi(t), t \geq 0$. Then the quantum system evolves in a noisy environment defined by the time varying random potential $V(\mathbf{r},t) = \xi(t)V(\mathbf{r})$. Using time dependent perturbation theory, the transition probabilities from one stationary state to another stationary state of the unperturbed system can be obtained in the usual way and then averaged out with respect to the probability law of the random process $\xi(.)$. Assume that the system starts in the state $|n >$ which is a stationary state of the unperturbed Hamiltonian H_0 corresponding to the eigenvalue E_n, ie, $H_0|n >= E_n|n >$. After the perturbation is switched on, the state evolves according to the law

$$\psi'(t) = (-2\pi i/h)(H_0 + \xi(t)V(\mathbf{r}))\psi(t)$$

Define

$$\phi(t) = exp(2\pi it H_0/h)\psi(t)$$

Then

$$\phi'(t) = (2\pi i/h)exp(2\pi it H_0/h)H_0\psi(t) + exp(2\pi it H_)/h)\psi'(t) = (-2\pi i/h)\xi(t)exp(2\pi it H_0)/h)V\psi(t)$$

or

$$\phi'(t) = -(2\pi i/h)\xi(t)\tilde{V}(t)\phi(t)$$

where

$$\tilde{V}(t) = exp(2\pi it H_0/h).V.exp(-2\pi it H_0/h)$$

This description of the state evolution is called the interaction picture. The solution is given by

$$\phi(t) = U(t)\phi(0)$$

where

$$U(t) = I + \sum_{n=1}^{\infty}(-2\pi i/h)^n \int_{0<t_n<t_{n-1}<...<t_1<t} \xi(t_1)\xi(t_2)...\xi(t_n)\tilde{V}(t_1)\tilde{V}(t_2)..\tilde{V}(t_n)dt_n dt_{n-1}...dt_1$$

is the random evolution operator. To make the dependence of $U(t)$ on the random process $\xi(t,\omega)$ explicit, we write $U(t,\xi(.,\omega))$. Let P be the probability measure of the probability space on which the random process ξ is defined. Then the ensemble averaged evolution operator in the interaction picture is given by

$$<U>(t) = \int U(t,\xi(.,\omega))dP(\omega)$$

If the starting state is $|n>$, then the probability that in time t, the system will make a transition to the state $|m>$ averages out over all the ensembles of ξ to give

$$P_t(n \to m) = \int |<m|U(t,\xi(.,\omega))|n>|^2 dP(\omega)$$

The problem is to compute such ensemble averaged transition probabilities for different random processes xi like Brownian motion, Poisson process and diffusion processes with specified drift and diffusion coefficients. More generally, if X is an observable on the system, its average value at time t given that the system started in the stationary state $|n>$ is given by

$$<X>(t) = \int <\psi(t)|X|\psi(t)> dP$$

where

$$\psi(t) >= exp(-2\pi it H_0/h)\phi(t), \phi(t) = U(t)|n>$$

This gives

$$<X>(t) = \int <n|U(t,\xi(.,\omega))^* exp(2\pi it H_0/h).X.exp(-2\pi it H_0/h)U(t,\xi(.,\omega))|n> dP(\omega)$$

[76] An atom in a uniform magnetic field. Let the magnetic field be given $\mathbf{B}$, a constant vector in space. Then we can take as vector potential $\mathbf{A} = \frac{1}{2}\mathbf{B} \times \mathbf{r}$. Indeed for this $\mathbf{A}$, we get

$$A_x = \frac{1}{2}(B_y z - B_z y), A_y = \frac{1}{2}(B_z x - B_x z), A_z = \frac{1}{2}(B_x y - B_y x)$$

so that

$$A_{z,y} - A_{y,z} = B_x, A_{x,z} - A_{z,x} = B_y, A_{y,x} - A_{x,y} = B_z$$

The Hamiltonian of the electron is

$$H = \frac{1}{2m}(\mathbf{p} + e\mathbf{A})^2 = \frac{p^2}{2m} + \frac{e^2 A^2}{2m} + \frac{e}{2m}(\mathbf{p}.\mathbf{A} + \mathbf{A}.\mathbf{p})$$

$$= -\frac{h^2}{8\pi^2 m}\nabla^2 + \frac{e^2}{8m}(\mathbf{B}\times\mathbf{r})^2 - \frac{ieh}{4\pi m}(\nabla.\mathbf{A}) - \frac{ieh}{2\pi m}(\mathbf{A}.\nabla)$$

Now,

$$\nabla.\mathbf{A} = A_{x,x} + A_{y,y} + A_{z,z} = 0$$

and

$$-\frac{ieh}{2\pi m}(\mathbf{A}.\nabla) = \frac{e}{m}\mathbf{A}.\mathbf{p} = \frac{e}{2m}\mathbf{B}\times\mathbf{r}.\mathbf{p} = \frac{e}{2m}\mathbf{B}.\mathbf{r}\times\mathbf{p}$$

$$= \frac{e}{2m}\mathbf{B}.\mathbf{L}$$

where $\mathbf{L} = \mathbf{r}\times\mathbf{p}$ is the angular momentum of the electron. Thus,

$$H = \frac{p^2}{2m} + \frac{e^2}{8m}(\mathbf{B}\times\mathbf{r})^2 + \frac{e}{2m}\mathbf{B}.\mathbf{L}$$

[77] The Hamiltonian of a quantum system has the form

$$H = H_1 \otimes I + I \otimes H_2 + V$$

where the underlying Hilbert space is $\mathcal{H}_1 \otimes \mathcal{H}_2$ with $\mathcal{H}_i, i = 1, 2$ being two Hilbert spaces. H_1 acts on $\mathcal{H}_1$, H_2 acts on $\mathcal{H}_2$ and V acts on $\mathcal{H}_1 \otimes \mathcal{H}_2$. Assume a trial wave function

$$\psi = \psi_1 \otimes \psi_2 - \psi_2 \otimes \psi_1$$

where ψ_i is a vector in $\mathcal{H}_i$. The condition for $\parallel \psi \parallel = 1$ is that

$$2 \parallel \psi_1 \parallel^2 . \parallel \psi_2 \parallel^2 - 2| < \psi_1|\psi_2 > |^2 = 1$$

We compute

$$< \psi|H|\psi > = < \psi_1 \otimes \psi_2 - \psi_2 \otimes \psi_1|H_1 \otimes I|\psi_1 \otimes \psi_2 - \psi_2 \otimes \psi_1 >$$
$$+ < \psi_1 \otimes \psi_2 - \psi_2 \otimes \psi_1|I \otimes H_2|\psi_1 \otimes \psi_2 - \psi_2 \otimes \psi_1 >$$
$$+ < \psi_1 \otimes \psi_2 - \psi_2 \otimes \psi_1|V|\psi_1 \otimes \psi_2 - \psi_2 \otimes \psi_1 >$$
$$= < \psi_1|H_1|\psi_1 >< \psi_2|\psi_2 > + < \psi_2|H_1|\psi_2 >< \psi_1|\psi_1 >$$
$$- < \psi_1|H_1|\psi_2 >< \psi_2|\psi_1 > - < \psi_2|H_1|\psi_1 >< \psi_1|\psi_2 >$$
$$+ < \psi_2|H_2|\psi_2 >< \psi_1|\psi_1 > + < \psi_1|H_2|\psi_1 >< \psi_2|\psi_2 >$$
$$- < \psi_2|H_2|\psi_1 >< \psi_1|\psi_2 > - < \psi_1|H_2|\psi_2 >< \psi_2|\psi_1 >$$
$$+ < \psi_1 \otimes \psi_2|V|\psi_1 \otimes \psi_2 > + < \psi_2 \otimes \psi_1|V|\psi_2 \otimes \psi_1 >$$
$$- < \psi_1 \otimes \psi_2|V|\psi_2 \otimes \psi_1 > - < \psi_2 \otimes \psi_1|V|\psi_1 \otimes \psi_2 >$$

Using these formulas, determine the equations satisfied by ψ_1 and ψ_2 so that $< \psi|H|\psi >$ is minimized subject to the constraint $\parallel \psi \parallel = 1$.

Reference:Landau and Lifshitz, Quantum Mechanics.

[78] Write short notes on the following.

[1] Wave function in quantum mechanics

hint: The wave function is a unit vector in a Hilbert space. More precisely it is a representation of a quantum state relative to a complete set of commuting observables. If the Hilbert space is $L^2(\mathbb{R}^{3N})$, then the wave function can be interpreted as the state of a system of N particles in the position representation. If the wave function $\psi \in L^2(\mathbb{R}^{3N})$, then the normalization condition gives $\int |\psi(x_1, ..., x_{3N})|^2 dx_1 dx_2...dx_{3N} = 1$. $|\psi(x_1, ..., x_{3N})|^2$ gives the joint probability density of the positions of the $3N$ particles. More generally, if $O_1, ..., O_n$ are commuting operators in a Hilbert space and ψ is a unit vector in this Hilbert space, $\psi(\xi_1, ..., \xi_n) = < \xi_1, ...\xi_n|\psi >$ is a wave function with $|\psi(\xi_1, ..., \xi_n)|^2$ being the joint probability density of $O_1, ..., O_n$ in the state ψ. Here, we are assuming that $|\xi_1, ..., \xi_n >$ is an eigenvector of O_i with eigenvalue ξ_i in the continuous spectrum. (To be mathematically precise, we must interpret $\int_E |\xi_1, ..., \xi_n > d\xi_1...dx1_n < \xi_1, ..., \xi_n|$ as the orthogonal projection operator on the subspace of spanned by the states in which $(O_1, ..., O_n)$ assumes values in the set $E \subset \mathbb{R}^n$. Suppose $O_1, ..., O_n$ has a set of joint eigenvectors $|\xi_{d1}, ..., \xi_{dn} >$ in the discrete spectrum and joint eigenvectors $|\xi_1, ..., \xi_n >$ in the continuous spectrum. Then $\psi(\xi_{d1}, ..., m\xi_{dn}) = < \xi_{d1}, ..., \xi_{dn}|\psi >$ is the discrete part of the wave function corresponding to the quantum state ψ and $< \xi_1, ..., \xi_r|\psi >$ is the continuous part. $|\psi(\xi_{d1}, ..., \xi_{dn})|^2$ is the probability that O_i will assume the value ξ_{di} for each $i = 1, 2, ..., r$ when the quantum system is in the state ψ. $|\psi(\xi_1, ..., \xi_n)|^2$ is the joint probability density of $O_1, ..., O_n$ at $(\xi_1, ..., \xi_n)$.

[2] pure state and mixed state of a quantum system

hint: A pure state is an orthogonal projection operator in the Hilbert space having rank one while a mixed state has the spectral representation $\rho = \sum_{i=1}^{\infty} p_i E_i$ where $p_i \geq 0, \sum_i p_i = 1$ and $\{E_i\}$ is a resolution of the identity, ie a family of orthogonal projection operators onto mutually orthogonal subspaces. A mixed state is thus a convex linear combination of pure states.

[3] Observable in quantum mechanics:

hint: An observable is a Hermitian operator X in the Hilbert space of quantum states and hence its eigenvalues are all real. Its eigenvalues give the possible outcomes of the experiment in which X is measured. If X has the spectral representation

$$X = \int_{\mathbb{R}} x.dE(x)$$

and the state of the system is ρ ($\rho \geq 0, Tr(\rho) = 1$), then for any Borel set B, the probability of the outcome of the experiment involved in measuring X is given by $Tr(\rho.E(B))$. If $\rho = |\psi >< \psi|$ is a pure state, then this probability reduces to $< \psi|E(B)|\psi >$. The average value of X in the state ρ is then $\int_{\mathbb{R}} x.Tr(\rho.E(dx)) = Tr(\rho X)$.

[4] Quasi-classical wave function

[5] Tunneling through a potential barrier and the reflection and transmission coefficients

[6] Symmetry and conservation laws in quantum mechanics

hint: Suppose H is the Hamiltonian of the quantum system and Q is an observable that commutes with H, ie, $[H, Q] = 0$. Then if follows that H is invariant under the adjoint action of the unitary group $\{exp(itQ), t \in \mathbb{R}\}$, ie,

$$Ad(exp(itQ))(H) = exp(itQ).H.exp(-itQ) = exp(it.ad(Q))(H) = H$$

Conversely, if Q is a Hermitian operator and H is invariant under the adjoint action of $\{exp(itQ), t \in \mathbb{R}\}$, then $[H, Q] = 0$ so that Q is conserved, ie, $Q(t) = exp(itH)Q.exp(-itH) = Q$ for all $t \in \mathbb{R}$. This fact shows how the invariance of the Hamiltonian under a group leads to the existence of a conserved quantity.

[7] Antisymmetric and symmetric wave functions and Bose and Fermi particles

[8] position-momentum and energy-time uncertainty relations

[9] Transformation of the position space wave function to the momentum space wave function

[10] Radial component of Schrodinger's equation in a radially symmetric potential

[11] Tensor product of two states

[12] Application of elliptic coordinates for the solution of Schrodinger's equation for the H_2^+ ion.

[13] Simultaneous diagonability of commuting observables

[14] Simultaneous measurement for two noncommuting observables is impossible

hint: Let X, Y be two commuting observables in a finite dimensional Hilbert space. Then X, Y can be simultaneously diagonalized, ie, there exists a complete family of orthonormal states $|i, j >, i = 1, 2, ..., n, j = 1, 2, ..., m$ such that $X|i, j >= x_i|i, j >, Y|i, j >= y_j|i, j >$ with x_i, y_j real. Thus in the state $|i, j > X$ assumes the value x_i and Y the value y_j. If however the observables X, Y do not commute, then they cannot be simultaneously diagonalized and hence we cannot speak of X, Y assuming definite values in a state.

[15] Number operator for systems of identical particles (For both cases, Bosons and Fermions)

[16] Evolution of the mixed state of a quantum system

[17] Born approximation of the scattering amplitude

[18] Decomposition of the tensor product of two irreducible representations of a group into a direct sum of irreducibles with applications to addition of angular momenta.

[19] Second quantization and the Fock spaces (Free Fock space, Boson Fock space, Fermion Fock space).

[20] Time independent perturbation theory upto second order.

[21] In states and out states in scattering theory, Moller and wave operators.

[22] Application of group representation theory to selection rules.

[23] Splitting of the energy levels of the Hydrogen atom in a magnetic field.

[24] Spherical harmonics as eigenfunctions of L^2 and L_z where $\mathbf{L}$ is the orbital angular momentum operator.

[25] Fock space stochastic calculus. Let $\mathcal{H}$ be a Hilbert space and let $\otimes$ be a tensor product. Consider the Hilbert spaces

$$\mathcal{H}^{\otimes n}, n = 1, 2, \ldots$$

and consider the Hilbert space

$$\Gamma(\mathcal{H}) = \mathbb{C} \oplus \oplus_{n=1}^{\infty} \mathcal{H}^{\otimes n}$$

$\Gamma(\mathcal{H})$ is called the Free Fock space over $\mathcal{H}$. Let $\otimes_s$ be the symmetric tensor product. Thus, if $x_1, \ldots, x_n \in \mathcal{H}$, then

$$x_1 \otimes_s x_2 \otimes \ldots \otimes_s x_n = \frac{1}{n!} \sum_{\sigma \in S^n} x_{\sigma 1} \otimes \ldots \otimes x_{\sigma n}$$

where S_n is the permutation group of order n. Let

$$\Gamma_s(\mathcal{H}) = \mathbb{C} \oplus \oplus_{n=1}^{\infty} \mathcal{H}^{\otimes_s n}$$

Then $\Gamma_s(\mathcal{H})$ is called the symmetric Fock space or Boson Fock space over $\mathcal{H}$. Let $\otimes_a$ be the antisymmetric tensor product. Thus,

$$x_1 \otimes_a x_2 \otimes_a \ldots \otimes_a x_n = \frac{1}{n!} \sum_{\sigma \in S_n} sgn(\sigma) x_{\sigma 1} \otimes \ldots \otimes x_{\sigma n}$$

Let

$$\Gamma_a(\mathcal{H}) = \mathbb{C} \oplus \oplus_{n=1}^{\infty} \mathcal{H}^{\otimes_a n}$$

Then, $\Gamma_a(\mathcal{H})$ is called the antisymmetric Fock space or Fermion Fock space over $\mathcal{H}$.

[26] Quantum characteristic function of a finite set of observables.

[79] Let $H = H_0 + V$ be the Hamiltonian of a quantum mechanical system. H_0 is of zeroth order of smallness and V is of first order of smallness. Let $\psi_n^{(0)}, n = 1, 2, \ldots$ be the normalized eigenfunctions of H_0 with corresponding eigenvalues $E_n^{(0)}, n = 1, 2, \ldots$. Let ψ be an eigenfunction of H with eigenvaluie E. Then,

$$(H - E)\psi = 0$$

or

$$(H_0 - E)\psi = -V\psi$$

Expand ψ in terms of $\psi_n^{(0)}, n = 1, 2, \ldots$ so that

$$\psi = \sum_n c_n \psi_n^{(0)}$$

Substituting this into the above eigenvalue equation gives

$$\sum_n c_n(E_n^{(0)} - E)\psi_n^{(0)} = -\sum_n c_n V \psi_n^{(0)}$$

The wave functions $\{\psi_n^{(0)}\}$ form an orthonormal basis for the underlying Hilbert space on which H_0, V, H are linear operators. Then from the above we get

$$c_n(E_n^{(0)} - E) = -\sum_k c_k V_{kn}$$

where

$$V_{kn} = <\psi_k^{(0)}|V|\psi_n^{(0)}>$$

We set

$$E = E_n^{(0)} + E_n^{(1)} + E_n^{(2)} + ...$$

$$c_k = \delta_{kn} + c_k^{(1)} + c_k^{(2)} + ...$$

where $c_k^{(1)}$ is of the first order of smallness, $c_k^{(2)}$ is of the second order of smallness and so on and likewise $E_n^{(1)}$ is of the first order of smallness and $E_n^{(2)}$ is of the second order of smallness etc. Then

$$(\delta_{kn} + c_k^{(1)} + c_k^{(2)} + ...)(E_k^{(0)} - E_n^{(0)} - E_n^{(1)} - E_n^{(2)} - ...) = -\sum_m (\delta_{mn} + c_m^{(1)} + c_m^{(2)} + ...)V_{mk}$$

Put $k = n$ and equate terms of first order of smallness on both sides to get

$$E_n^{(1)} = V_{nn}$$

Now let $k \neq n$ and equate coefficients of first order of smallness on both sides to get

$$c_k^{(1)}(E_k^{(0)} - E_n^{(0)}) = -V_{nk}$$

or

$$c_k^{(1)} = \frac{V_{nk}}{E_n^{(0)} - E_k^{(0)}}, k \neq n$$

Note that $c_n^{(1)}$ cannot be determined. The wave function upto first order of smallness is then

$$\psi_n^{(0)} + \sum_{k \neq n} \frac{V_{nk}}{E_n^{(0)} - E_k^{(0)}} + c_n^{(1)}\psi_n^{(0)}$$

The norm square of this wave function is

$$|1 + c_n^{(1)}|^2 + \sum_{k \neq n} \frac{|V_{nk}|^2}{|E_n^{(0)} - E_k^{(0)}|^2}$$

This differs from unity by terms of second order of smallness iff $c_n^{(1)} = 0$. If $c_n^{(1)} \neq 0$, then the norm square of this wave function will be unity plus terms of first order of smallness. So we must take $c_n^{(1)} = 0$

to get the correct normalization. Setting $k = n$ and equating terms of second order of smallness on both sides gives us

$$E_n^{(2)} + c_n^{(1)} E_n^{(1)} = \sum_m c_m^{(1)} V_{mn}$$

Thus,

$$E_n^{(2)} = \sum_m c_m^{(1)} V_{mn} = \sum_{m \neq n} \frac{|V_{nm}|^2}{E_n^{(0)} - E_m^{(0)}}$$

Thus the energy upto second order is given by

$$E_n \approx E_n^{(0)} + V_{nn} + \sum_{m \neq n} \frac{|V_{nm}|^2}{E_n^{(0)} - E_m^{(0)}}$$

E_n is the energy level of the system obtained by perturbing the energy level $E_n^{(0)}$ of the unperturbed system having Hamiltonian H_0 with the perturbation Hamiltonian being V.

[80] Consider the Hilbert space $\mathbb{C}^2$ and the Pauli spin observables $\sigma_x, \sigma_y, \sigma_z$ defined on this space. Let ρ be a 2×2 density matrix, ie, $Tr(\rho) = 1, \rho \geq 0$. For t_1, t_2, t_3 real, compute the joint quantum characteristic function of the three Pauli spin observables, ie,

$$\psi(t_1, t_2, t_3) = Tr(\rho.exp(i(t_1 \sigma_x + t_2 \sigma_y + t_3 \sigma_z)))$$

Show that there exist density matrices ρ such that ψ cannot be a characteristic function of classical probability.

[81] The quantum mechanical Hamiltonian of a charged particle in a constant magnetic field is given by

$$H = \frac{1}{2m}((ih/2\pi)\nabla + (e/2)(\mathbf{B} \times \mathbf{r})^2$$

$$= -\frac{h^2}{8\pi^2 m}\nabla^2 + (e^2 B^2/8m)(x^2 + y^2) - (eBL_z/2m)$$

where the direction of the constant magnetic field $\mathbf{B}$ is the z axis. $L_z = -ih/2\pi)\mathbf{r} \times \nabla$. In cylindrical coordinates (ρ, ϕ, z), this Hamiltonian is given by

$$H\psi = -\frac{h^2}{8\pi^2 m}(\psi_{,\rho\rho} + \rho^{-1}\psi_{,\rho} + \rho^{-2}\psi_{,\phi\phi} + \psi_{,zz}) + (e^2 B^2 \rho^2/8m)\psi$$

$$+(ieBh/4\pi m)\psi_{,\phi}$$

Solve the eigenvalue equation

$$H\psi = E\psi$$

by using the separation of variables in cylindrical coordinates.

[82] The Hartree-Fock equations for a two electron atom. Let $\psi_1(\mathbf{r}_1)$ be the wave function of the first electron and $\psi_2(\mathbf{r}_2)$ the wave function of the second electron. The Hamiltonian of the system is $(with\, h/2\pi = 1)$

$$H = -\Delta_1 \psi_1(\mathbf{r}_1)/2m - \Delta_2 \psi_2(\mathbf{r}_2)/2m - 2e^2/r_1 - 2e^2/r_2 + e^2/r_{12}$$

where

$$H = \Delta_i = \nabla^2_{\mathbf{r}_i}, r_i = |\mathbf{r}_i|, i = 1, 2, r_{12} = |\mathbf{r}_1 - \mathbf{r}_2|$$

We choose the wave functions $\psi_i, i = 1, 2$ so that $< \psi_1 \otimes \psi_2 | H | \psi_1 \otimes \psi_2 >$ is minimized subject to the constraints $\| \psi_i \|^2 = 1$. Then we get the Hartree-Fock equations for $\psi_i, i = 1, 2$.

$$(-\Delta_1/2m - 2e^2/r_1 + e^2 \int |\psi_2(\mathbf{r}_2)|^2 d^3 r_2/r_{12})\psi_1(\mathbf{r}_1 = E_1\psi_1(\mathbf{r}_1),$$

$$(-\Delta_2/2m - 2e^2/r_2 + e^2 \int |\psi_1(\mathbf{r}_1)|^2 d^3 r_1/r_{12})\psi_2(\mathbf{r}_2 = E_2\psi_2(\mathbf{r}_2),$$

[83] Derive Schrodinger's wave equation from an action principle. Let $\psi(\mathbf{r}, t)$ be the complex wave function and $V(\mathbf{r}, t)$ the potential. Then, Schrodinger's wave equation is

$$i\psi_{,t} = -\frac{1}{2m}\nabla^2\psi + V\psi$$

Let $\bar{\psi}$ denote the complex conjugate of ψ. Then, taking the conjugate of Schrodinger's equation and using the fact that the potential V is real, we get

$$-i\bar{\psi}_{,t} = -\frac{1}{2m}\nabla^2\bar{\psi} + V\bar{\psi}$$

Consider the action

$$S[\psi, \bar{\psi}] = -i \int \bar{\psi}.\psi_{,t} d^3 r dt + \frac{1}{2m} \int (\nabla\psi, \nabla\bar{\psi}) d^3 r dt + \int V\psi\bar{\psi} d^3 r dt$$

The corresponding Lagrangian density is then

$$\mathcal{L}(\psi, \bar{\psi}, \nabla\psi, \nabla\bar{\psi}, \psi_{,t}) =$$

Then,

$$\frac{\partial \mathcal{L}}{\partial \psi_{,t}} = i\bar{\psi},$$

$$\frac{\partial \mathcal{L}}{\partial \nabla\psi} = \nabla\bar{\psi}/2m, \frac{\partial \mathcal{L}}{\partial \psi} = V\bar{\psi}$$

so that setting $\delta S[\psi, \bar{\psi}] = 0$ gives

$$-i\bar{\psi}_{,t} = -\frac{1}{2m}\nabla^2\bar{\psi} + V\bar{\psi}$$

Also

$$\frac{\partial \mathcal{L}}{\partial \bar{\psi}_{,t}} = 0,$$

$$\frac{\partial \mathcal{L}}{\partial \nabla \bar{\psi}} = \frac{1}{2m} \nabla \psi, \frac{\partial \mathcal{L}}{\partial \bar{\psi}} = -i\psi_{,t} + V\psi$$

so that $\delta S[\psi, \bar{\psi}] = 0$ gives

$$-i\psi_{,t} + V\psi - \frac{1}{2m} \nabla^2 \psi$$

or

$$i\psi_{,t} = -\frac{1}{2m} \nabla^2 \psi + V\psi$$

Note that the Euler-Lagrange equations obtained from the action principle $\delta S[\psi, \bar{\psi}] = 0$ are

$$\frac{\partial}{\partial t} \frac{\partial \mathcal{L}}{\partial \psi_{,t}} = \frac{\partial \mathcal{L}}{\partial \psi} - \nabla \cdot \frac{\partial \mathcal{L}}{\partial \nabla \psi}$$

and a similar equation with ψ replaced by $\bar{\psi}$. The above action S or equivalently, the Lagrangian density $\mathcal{L}$ is not real. However, the Lagrangian density can be made real by adding to it $(i/2)(\bar{\psi}\psi)_{,t}$ without affecting the field equations. Using the fact that

$$-i\bar{\psi}\psi_{,t} + (i/2)(\bar{\psi}\psi)_{,t} = (i/2)(\psi\bar{\psi}_{,t} - \bar{\psi}\psi_{,t}) = Im(\bar{\psi}\psi_{,t})$$

the new Lagrangian density is

$$\mathcal{L}_1 = \mathcal{L} + (i/2)(\bar{\psi}\psi)_{,t}$$

$$= Im(\bar{\psi}\psi_{,t}) + \frac{1}{2m}(\nabla\psi, \nabla\bar{\psi}) + V\psi\bar{\psi}$$

$\mathcal{L}_1$ is real and if we define $S_1 = \int \mathcal{L}_1 d^3r dt$, then it can be verified by writing down the Euler-Lagrange equations for $\mathcal{L}_1$, that the Schrodinger equation is obtained.

[84] Nonlinear Schrodinger equation from an action principle. Let

$$\mathcal{L} = Im(\bar{\psi}\psi_{,t}) + \frac{1}{2m}(\nabla\psi, \nabla\bar{\psi}) + V(\psi, \bar{\psi}, \mathbf{r}, t)|\psi|^2$$

where V is real. Write down the Euler-Lagrange equations for the corresponding action $S[\psi, \bar{\psi}] = \int \mathcal{L}d^3r dt$. We have

$$\delta \int \mathcal{L}d^4x =$$

$$(-i/2) \int \delta(\bar{\psi}\psi_{,t} - \psi\bar{\psi}_{,t})d^4x + (1/2m) \int ((\nabla\psi, \delta\nabla\bar{\psi}) + (\delta\nabla\psi, \nabla\bar{\psi}))d^4x$$

$$+ \int |\psi|^2((\partial V/\partial\psi)\delta\psi + (\partial V/\partial\bar{\psi})\delta\bar{\psi})d^4x + \int V(\psi\delta\bar{\psi} + \bar{\psi}\delta\psi)d^4x$$

Noting that

$$\delta(\bar{\psi}\psi_{,t} - \psi\bar{\psi}_{,t})$$

$$= \bar{\psi}\delta\psi_{,t} + \psi_{,t}\delta\bar{\psi} - \psi\delta\bar{\psi}_{,t} - \bar{\psi}_{,t}\delta\psi$$

and integrating this by parts we get

$$\delta \int \mathcal{L}d^4x =$$

$$-i\int(-\bar{\psi}_{,t}.\delta\psi + \psi_{,t}\delta\bar{\psi})d^4x$$

$$+(1/2m)\int((-\nabla^2\psi)\delta\bar{\psi} + (-\nabla^2\bar{\psi})\delta\psi)d^4x$$

$$+\int|\psi|^2((\partial V/\partial\psi)\delta\psi + (\partial V/\partial\bar{\psi})\delta\bar{\psi})d^4x + \int V.(\psi\delta\bar{\psi} + \bar{\psi}\delta\psi)d^4x$$

Equating the coefficients of $\delta\psi$ and $\delta\bar{\psi}$ to zero then gives us

$$-i\psi_{,t} - (1/2m)\nabla^2\psi + (\partial V/\partial\bar{\psi}) + V.\psi = 0$$

or

$$i\psi_{,t} = -(1/2m)\nabla^2\psi + V\psi + (\partial V/\partial\bar{\psi})|\psi|^2$$

which is the nonlinear Schrodinger equation. Suppose there is just one spatial variable x. Then, the above action becomes

$$S = \int(Im(\bar{\psi}\psi_{,t}) + (1/2m)|\psi_{,x}|^2 + V|\psi|^2)dxdt$$

The one dimensional nonlinear Schrodinger equation

$$i\psi_{,t} = -(1/2m)\psi_{,xx} + V\psi + (\partial V/\partial\bar{\psi})|\psi|^2$$

is then obtained from the action principle $\delta S = 0$.

[85] Solution to the nonlinear Schrodinger equation using the method of moments. The nonlinear Schrodinger equation is given by

$$i\psi_{,t} = -(1/2m)\psi_{,xx} + (a|\psi|^2 + b(|\psi|^2)_{,x})\psi$$

The initial condition $\psi(0,x) = \psi_0(x)$ is given. We expand the wave function as a linear combination of test functions $\phi_n(x), n = 1, 2, ..., N$ with the coefficients assumed to be complex functions of time. Thus,

$$\psi(t,x) = \sum_{n=1}^{N} a_n(t)\phi_n(x)$$

Substituting this into the nonlinear Schrodinger equation gives us

$$i\sum_n a'_n(t)\phi_n(x) = -(1/2m)\sum_n a_n(t)\phi''_n(x) + (a|\sum_n a_n(t)\phi_n(x)|^2 + b(|\sum_n a_n(t)\phi_n(x)|^2)_{,x})\sum_n a_n(t)\phi_n(x)$$

Take the inner product on both sides with $\phi_k(x)$ so that

$$i\sum_n(\int \bar{\phi}_k(x)\phi_n(x)dx)a'_n(t) = -(1/2m)\sum_n(\int \bar{\phi}_k(x)\phi''_n(x)dx)a_n(t)$$

$$+a \sum_{n,r,m} a_n(t)\bar{a}_r(t)a_m(t) \int \phi_n(x)\bar{\phi}_r(x)\bar{\phi}_k(x)\phi_m(x)dx$$

$$+b \sum_{n,r,m} a_n(t)\bar{a}_r(t)a_m(t) \int (\phi_n(x)\bar{\phi}_r(x))_{,x}\phi_m(x)\bar{\phi}_k(x)dx$$

Define

$$c_1(k,n) = \int \bar{\phi}_k(x)\phi_n(x)dx, \; c_2(k,n) = \int \bar{\phi}_k(x)\phi_n''(x)dx,$$

$$c_3(k,n,r,m) = \int \bar{\phi}_k(x)\phi_n(x)\bar{\phi}_r(x)\phi_m(x)dx,$$

$$c_4(k,n,r,m) = \int \bar{\phi}_k(x)(\phi_n(x)\bar{\phi}_r(x))_{,x}\phi_m(x)dx$$

Then the above equation can be expressed as

$$i \sum_n c_1(k,n)a_n'(t) = -(1/2m) \sum_n c_2(k,n)a_n(t) + a \sum_{n,r,m} c_3(k,n,r,m)a_n(t)\bar{a}_r(t)a_m(t)$$

$$+b \sum_{n,r,m} c_4(k,n,r,m)a_n(t)\bar{a}_r(t)a_m(t)$$

Assuming that the matrix $((c_1(k,n)))$ is non-singular, this equation can be expressed in the form

$$a_k'(t) = \sum_n B_1(k,n)a_n(t) + \sum_{n,r,m} B_2(k,n,r,m)a_n(t)\bar{a}_r(t)a_m(t) + \sum_{n,r,m} B_3(k,n,r,m)a_n(t)\bar{a}_r(t)a_m(t)$$

where $B_1(k,n), B_2(k,n,r,m)$ and $B_3(k,n,r,m)$ are known complex constants. This equation can be solved for $a_n(t)$ approximately by the finite difference method.

[86] Let $\mathcal{H}$ be a Hilbert space and $\otimes$ a tensor product on this space. Consider the Hilbert space $\mathcal{H}_n = \mathcal{H}^{\otimes n}$, ie, $\mathcal{H}_n$ is the n-fold tensor product of $\mathcal{H}$ with itself. Let $\otimes_s$ and $\otimes_a$ denote respectively the symmetric and antisymmetric tensor products on this space, ie, if $u_1, ..., u_n \in \mathcal{H}$, then

$$\otimes_{ak=1}^n u_k = \frac{1}{n!} \sum_{\sigma \in S_n} sgn(\sigma) \otimes_{k=1}^n u_{\sigma k}$$

$$\otimes_{sj=1}^n u_k = \frac{1}{n!} \sum_{\sigma \in S_n} \otimes_{k=1}^n u_{\sigma k}$$

Then note that

$$< \otimes_{k=1}^n u_k, \otimes_{p=1}^n v_p > = \Pi_{k=1}^n < u_k, v_k >$$

$$< \otimes_{ak=1}^n u_k, \otimes_{ap=1}^n v_p > = \frac{1}{(n!)^2} \sum_{\sigma,\tau \in S_n} sgn(\sigma\tau) < \otimes_{k=1}^n u_{\sigma k}, \otimes_{p=1}^n v_{\tau p} >$$

$$= \frac{1}{(n!)^2} \sum_{\sigma,\tau} sgn(\sigma\tau)\Pi_{k=1}^n < u_{\sigma k}, v_{\tau k} >$$

$$= \frac{1}{(n!)^2} \sum_{\rho,\tau \in S_n} sgn(\rho) \Pi_{k=1}^n < u_{\rho \tau k}, v_{\tau k} >$$

$$= \frac{1}{n!} \sum_{\rho \in S_n} sgn(\rho) \Pi_{k=1}^n < u_{\rho k}, v_k > = \frac{1}{n!} det((< u_i, v_j >))$$

[87] Let $\mathbf{E}$ be the electric field in space and $\mathbf{B}$ the magnetic field. Show that the total energy of the field is given by

$$U = \int_{\mathbb{R}^3} (\frac{\epsilon_0}{2} |\mathbf{E}|^2 + \frac{1}{2\mu_0} |\mathbf{B}|^2) d^3 x$$

Use this equation to quantize the electromagnetic field by expressing the total energy as a sum of the energies of harmonic oscillators.

[88] Let ψ be the Dirac wave function satisfying the Dirac equation

$$(i\gamma^\mu \partial_\mu - m)\psi = 0$$

where $\gamma^\mu, \mu = 0,1,2,3$ are the 4×4 Dirac gamma matrices and m is the mass of the particle. Derive in terms of ψ and γ^μ an expression for the four current density J^μ of the Dirac field.

hint: Define $\bar{\psi} = \psi^* \gamma^0$ and consider the four divergence of $J^\mu = \bar{\psi} \gamma^\mu \psi$.

[89] Show that Dirac's equation for the electron can be expressed as

$$(E - (\alpha, \mathbf{p}) - \beta m)\psi = 0$$

where $E = i\partial_t$ is the energy operator, $\mathbf{p} = -i\nabla$ is the three momentum vector operator and α, β are Hermitian 4×4 matrices.

[90] Calculate the Feynman path integral for the forced harmonic oscillator, ie,

$$\int (\Pi_{t \in [0,T]} dx(t)).exp(i \int_0^T (mx'^2(t) - m\omega^2 x^2(t))dt - i \int_0^T J(t)x(t)dt)$$

where $x(0) = a, x(T) = b$ and $J(t)$ is an external forcing function.

hint: Discretize the integral so that it can be expressed as the limit of

$$\int_{\mathbb{R}^{N-1}} (\Pi_{n=1}^{N-1} dx[n]).exp(i \sum_{n=0}^{N-1} (m(x[n+1] - x[n])^2/2\Delta - m\omega^2 x^2[n]\Delta) - i\Delta \sum_{n=0}^{N-1} J[n]x[n])$$

where $J[n] = J(n\Delta)$. Then, use the well known results about Gaussian integrals.

[91] Consider the classical Klein-Gordon field with a nonlinear potential $V(\phi)$. The field equatio can be expressed as

$$\partial_t^2 \phi - \nabla^2 \phi + m^2 \phi + V'(\phi) = 0$$

Derive this equation from an action principle. From the Lagrangian density of the field, derive the Hamiltonian of the field and express the field equations in Hamiltonian form.

[92] Let $H(u(t))$ be the Hamiltonian of a quantum mechanical system dependent on an input signal $u(t)$. Let $\psi(t)$ be the state after time t. It satisfies the Schrodinger equation

$$i\psi'(t) = H(u(t))\psi(t), t \geq 0$$

Let $\psi(0)$ be given and let X be a given observable. How would you determine the input $u(t), t \in [0, T]$ so that $\int_0^T < \psi(t)|X|\psi(t) > dt$ is a minimum. Note that if T denotes the time ordering operator, then

$$\psi(t) = T\{exp(-i \int_0^t H(u(s))ds)\}\psi(0)$$

Chapter 5

Electromagnetic Field Theory

Introduction: This chapter discusses coordinate transformation of vector fields, electrostatics and magnetostatics involving computation of the electric and magnetic fields given respectively the charge and current distributions, nonlinear problems in electromagnetic field theory, in particular how to arrive at the approximate solution to the Maxwell equations when the permittivity and permeability of the medium are field dependent, approximate determination of the modes of wave propagation in a waveguide, basic antenna theory and the solution to the Maxwell equations in a curved space-time given the metric tensor.

[1] .Determine an algorithm for estimating the current through a finite straight wire based on measurements taken on the magnetic field at a set of points in space.

hint:Use the Biot-Savart law combined with discretization of integrals and the least square method:

$$B(\mathbf{r}) = \frac{\mu}{4\pi} \int_a^b I(z')dz'\hat{z} \times (\mathbf{r} - z'\hat{z})/|\mathbf{r} - z'\hat{z}|^3$$

This is the Biot-Savart law. Discretize this integral to get

$$B(\mathbf{r}) \approx \frac{\mu}{4\pi} \sum_{k=p}^{q} I(k\Delta)\hat{z} \times (\mathbf{r} - k\Delta\hat{z})\Delta/|\mathbf{r} - k\Delta\hat{z}|^3$$

[2].Explain how you would formulate an approximate second order Volterra model for the electrostatic potential produced by a charge distribution when the permittivity is potential dependent. For simplicity, consider a one dimensional model.

hint:

$$(\epsilon(V(x))V'(x))' = -\rho(x)$$

Write

$$\epsilon(V) = 1 + \delta.V$$

where δ is a small parameter. Then, the above equation becomes

$$V' + \delta(VV')' = -\rho$$

Write

$$V = V_0 + \delta.V_1$$

so that upto $O(\delta)$ terms, we have

$$V_0' = -\rho, V_1' + (V_0V_0')' = 0$$

This gives

$$V_0 = -\int \rho dx, V_1 = -\int (V_0V_0')'dx$$

It is clear that V_1 can be expressed as a quadratic functional of ρ.

[3]. Summary of the research carried out by one of the author's colleagues Arathi. This work deals with numerical techniques for solving electromagnetic wave propagation problems in a waveguide. The kinds of problems in this work include (1) Determination of the modes of propagation in a waveguide having inhomogeneous permittivity using the method of moments, (2) determination of the modes of propagation in a waveguide using the finite element method, (3) determination of the modes of propagation in a waveguide placed close to a massive gravitating body inducing general relativistic corrections using the finite element method. The equation that describes the last problem is the Laplace Beltrami equation for the propagation of a scalar field in a curved metric. The Laplace Beltrami equation can be derived from an action principle and this immediately leads one to formulating the finite element method for mode determination. The shift in the modes produced by gravitational effects has been depicted in this work on a graph. The Scalar wave propagation equation is only an approximation to the exact Maxwell equations in a curved space-time metric. To determine the modes more accurately, one must ideally apply the finite element method to the action for the electromagnetic field in a curved space time metric background. Then the modes are obtained from homegeneous linear equations for the electromagnetic four potential and this leads one to the next phase of Arathi's work. The metric chosen is Newtonian for a given gravitational field and it involves the gravitational potential. The resulting action integral is approximated by integrations over triangular elements with unknown modal potentials and on carrying out the optimization of the resulting quadratic form, we obtain generalized eigenvalue equations for the modes.

[4]. Waveguide filled with an inhomogeneous conducting medium. We wish to compute the modes of propagation. Here the guide axis is the z axis and the cross sections of the waveguide are parallel to the xy plane. We must first of all express the transverse components of the electric and magnetic fields in terms of the z components. The relevant equations are

$$E_{z,y} + \gamma E_y = -j\omega\mu(x,y)H_x, -\gamma E_x - E_{z,x} = -j\omega\mu(x,y)H_y, E_{y,x} - E_{x,y} = -j\omega\mu(x,y)H_z$$

$$H_{z,y} + \gamma H_y = (\sigma(x,y) + j\omega\epsilon(x,y))E_x,$$

$$H_{z,x} + \gamma H_x = -(\sigma(x,y) + j\omega\epsilon(x,y))E_y,$$

$$H_{y,x} - H_{x,y} = (\sigma(x,y) + j\omega\epsilon(x,y))E_z$$

In deriving these equations, we are assuming that the z dependence of the fields is $exp(-\gamma z)$. The equations satisfied by E_y and H_x are

$$-(\sigma + j\omega\epsilon)E_y - \gamma H_x = H_{z,x},$$

$$\gamma E_y + j\omega\mu H_x = -E_{z,y}$$

The second implies

$$E_y = (-E_{z,y} - j\omega\mu H_x)/\gamma$$

and substituting this into the first, we get

$$-(\sigma + j\omega\epsilon)(-E_{z,y} - j\omega\mu H_x)/\gamma - \gamma H_x = H_{z,x}$$

or

$$[j\omega\mu(\sigma + j\omega\epsilon)/\gamma - \gamma]H_x = H_{z,x} - (\sigma + j\omega\epsilon)E_{z,y}/\gamma$$

or

$$H_x = (-\gamma^2 + j\omega\mu(\sigma + j\omega\epsilon))H_x = -(\sigma + j\omega\epsilon)E_{z,y} + \gamma H_{z,x}$$

from which we get the desired expression for H_x:

$$H_x = \frac{-\gamma H_{z,x} + (\sigma + j\omega\epsilon)E_{z,y}}{\gamma^2 + \omega^2\epsilon\mu - j\omega\mu\sigma}$$

We can now obtain an expression for E_y:

$$E_y = -\gamma^{-1}E_{z,y} - (j\omega\mu/\gamma)\frac{-\gamma H_{z,x} + (\sigma + j\omega\epsilon)E_{z,y}}{\gamma^2 + \omega^2\epsilon\mu - j\omega\mu\sigma}$$

$$= E_{z,y}(-\gamma^{-1} + \frac{(\omega^2\mu\epsilon - j\omega\mu\sigma)/\gamma}{\gamma^2 + \omega^2\mu\epsilon - j\omega\mu\sigma})$$

$$+ \frac{j\omega\mu}{\gamma^2 + \omega^2\mu\epsilon - j\omega\mu\sigma}H_{z,x}$$

$$= \frac{-\gamma E_{z,y} + j\omega\mu H_{z,x}}{\gamma^2 + \omega^2\mu\epsilon - j\omega\mu\sigma}$$

We need to also solve for H_y and E_x. The equations satisfied by these are

$$-\gamma E_x - E_{z,x} = -j\omega\mu(x,y)H_y,$$

$$H_{z,y} + \gamma H_y = (\sigma(x,y) + j\omega\epsilon(x,y))E_x,$$

From the first of these, we get

$$H_y = \frac{\gamma E_x + E_{z,x}}{j\omega\mu}$$

Plugging this into the second results into

$$H_{z,y} + (\gamma/j\omega\mu)(\gamma E_x + E_{z,x}) = (\sigma + j\omega\epsilon)E_x$$

which can be rearranged as

$$(\sigma + j\omega\epsilon - \gamma^2/j\omega\mu)E_x = (\gamma/j\omega\mu)E_{z,x} + H_{z,y}$$

so that

$$E_x = \frac{(\gamma/j\omega\mu)E_{z,x} + H_{z,y}}{\sigma + j\omega\epsilon - \gamma^2/j\omega\mu}$$

$$= \frac{\gamma E_{z,x} + j\omega\mu H_{z,y}}{j\omega\mu\sigma - \omega^2\epsilon\mu - \gamma^2}$$

and hence

$$H_y = (1/j\omega\mu)E_{z,x} + \frac{\gamma(\gamma E_{z,x} + j\omega\mu H_{z,y})}{j\omega\mu(j\omega\mu\sigma - \omega^2\mu\epsilon - \gamma^2)}$$

$$= \frac{(\sigma + j\omega\epsilon)E_{z,x} + j\omega\mu H_{z,y}}{j\omega\mu\sigma - \omega^2\mu\epsilon - \gamma^2}$$

We now consider the equation

$$E_{y,x} - E_{x,y} = -j\omega\mu H_z$$

and substitute for E_x and E_y from the above to get

$$(\frac{-\gamma E_{z,y} + j\omega\mu H_{z,x}}{\gamma^2 + \omega^2\mu\epsilon - j\omega\mu\sigma})_{,x} -$$

$$(\frac{\gamma E_{z,x} + j\omega\mu H_{z,y}}{j\omega\mu\sigma - \omega^2\epsilon\mu - \gamma^2})_{,y}$$

$$= -j\omega\mu H_z$$

Note that this equation is valid for all functions μ, ϵ, σ of x, y. In the special case when these are all constant, it reduces to

$$H_{z,xx} + H_{x,yy} + (\gamma^2 + \omega^2\mu\epsilon - j\omega\mu\sigma)H_z = 0$$

which is the standard two dimensional Helmholtz equation. Likewise, we consider the equation

$$H_{y,x} - H_{x,y} = (\sigma + j\omega\epsilon)E_z$$

and substitute for H_x, H_y to get

[5] Syllabus for Electromagnetic Field Theory.

1.Vector algebra and calculus fundamentals

The representation of a vector as a real linear combination of the cartesian unit vectors is to be discussed here. The expression for the cylindrical and polar coordinate variables in terms of the cartesian variables and the corresponding inversion formula is also to be discussed here. Unit vectors in the cylindrical and spherical polar systems are then derived by differentiating this transformation formula. The expressions for the unit vectors in the cylindrical and spherical polar systems in terms of those in the cartesian system are to be derived. Then given a vector field in any one of these three systems, the expressions for the same field in the other systems is derived.

1.Cartesian, cylindrical and spherical polar coordinate systems, gradient, divergence and curl operations in the three systems, computation of line, surface and volume integrals, Gauss', Stokes and Green's theorems. By using the chain rule for partial derivatives, we derive the expressions for the gradient of a scalar field and curl and divergence of vector fields in the different systems.

1.Definition of charge and linear, surface and volume charge densities.

Charge may be defined via the electric force that it exerts on another charge due to Coulomb's law. A charge Q is said to have unit value in the CGS system if the repulsive force between two such identical charges separated by a distance of one centimetre equals $1/4\pi$.

2.Definition of electric field.

3.Electric field produced by a charge distribution in space, both discrete and continuous

4.Coulomb's and Gauss' laws for electrostatics, Gauss' law in differential and integral form.

5.Laplace's and Poisson's equations for electrostatics

6.Uniqueness theorem for the Poisson equation

7.Dielectrics, polarization and electric displacement vector fields.

8. Definition of current and current density

9.Ampere's force law on a current

10.Ampere's circuital law in integral form for magnetostatics

11.Ampere's circuital law in differential form for magnetostatics.

12.The continuity equation for charge conservation.

13.Faraday's law of induction and its applications.

14.Displacement current and Maxwell's equations.

15.Derivation of the wave equation from Maxwell's equations.

16.Lorentz force equation and motion of charged particles in an electromagnetic field.

17.Reflection and refraction of electromagnetic waves at dielectic and conductor interfaces.

18.Cylindrical and spherical waves

19.Solution to the Maxwell equations using retarded potentials.

20.Basic radiation theory

21.Poynting's theorem and the flow of power, energy density in the electromagnetic field.

[6]. Syllabus for Antennas and wave propagation

1.Maxwell's equations incoprorating magnetic current density and magnetic charge density.

2.Basics of radiation theory from sources, expression for the far field electric and magnetic fields in terms of electric and magnetic current densities. Notion of magnetic scalar potential and electric vector potential.

3.Equivalent sources for apertures. Given an aperture on which there is an electric and magnetic field, one can determine expressions for the equivalent surface electric and magnetic current densities on this surface which can be used to obtain integral expressions for the far field radiation pattern caused by the aperture fields. These will be used later in the discussion of aperture antennas.

Given the electric and magnetic fields on a surface, how to compute the equivalent surface electric and magnetic current densities.

4.Reciprocity theorem.

5.Notion of loss resistance, radiation resistance, load resistance and reactance of an antenna.

6.scattering area, loss area, power density incident on an antenna, maximum effective aperture area.

7.Radar equation for transmitting and receiving antennas

8.Directivity, polarization loss

9.Analysis of the radiation pattern for infinitesimal dipole, finite straight wire, circular current loop, rectangular loop.

10.Analysis of the radiation pattern for surfaces carrying current.

11.Method of images in antenna analysis:dipole above the ground plane when (a) the ground plane is an infinite conducting sheet, (b) the ground plane is a dielectric.

12. Antenna pattern for a straight wire with current expanded in Fourier series.

13.Woodward-Lawson method of design of antennas from location of nulls.

14.Pocklington's integral equation for computing current distribution on an antenna when the incident field is known.

15.Hallen's integral equation for determination of the current distribution on an antenna

16.Generalization of Pocklington's integral equation to surface antennas.

17. Antenna arrays:The pattern multiplication theorem

18. Design of antenna arrays:Chebyshev array, binomial array, circular array.

19. Analysis of the Yagi-Uda array using Pocklington's integral equation.

20. Wave propagation equation in different coordinate systems and solution using special functions.

21. Aperture antennas, analysis using Fourier transform methods and equivalent sources.

22.Method of stationary phase for aperture antenna analysis

23. Parabolic reflector:A paraboloid is obtained by rotating a parabola around its axis of symmetry. The equation of a parabola is first derived using its definition as the locus of a point the ratio of whose distance from a line and from a point remains fixed. One then proves using Snell's law of reflection that if a ray is incident on the parabola and this ray is parallel to the axis of symmetry then the reflected ray will pass through the focus of the parabola.

24. Helical antenna analysis and design

25. Frequency independent antennas.

26. Biconical antenna

27. Horn antenna

28. Analysis of microstrip antennas for rectangular and circular shapes.

[7]. Syllabus for Transmission lines and waveguides

1. Maxwell's equations and wave propagation

2. Maxwell's equations for sinusoidal fields

3. Maxwell's equations in a rectangular waveguide with exponential dependence on z coordinate.

4. Expressing the transverse components of the electromagnetic fields in terms of the longitudinal components.

5. TE and TM modes in a rectangular and circular waveguide.

6. Notion of cutoff frequency

7. Power absorbed by conducting walls of a waveguide.

8. Wave guide with nonuniform permittivity, permeability and conductivity

9. Wave guide with anisotropic permittivity, permeability and conductivity

10.

Course content of transmission lines and waveguides:

1. Summary of the Maxwell equations (Five lectures)

(a) Gauss' law

$$div\mathbf{E} = \rho/\epsilon_0$$

(b) Ampere's law with Maxwell's displacement current correction term

$$\nabla \times \mathbf{H} = \mathbf{J} + \frac{\partial \mathbf{D}}{\partial t}$$

(c) Faraday's law

$$\nabla \times \mathbf{E} = -\mu\frac{\partial \mathbf{H}}{\partial t}$$

(d) No magnetic monopole condition

$$\nabla.\mathbf{B} = 0$$

2.Specialization to electrostatics and magneto statics.

Electrostatics:

$$\nabla.\mathbf{E} = 0, \nabla \times \mathbf{E} = \mathbf{0}$$

Magnetostatics:

$$\nabla \times \mathbf{B} = \mu_0 \mathbf{J}, \nabla.\mathbf{B} = 0$$

3.Solution to specific problems in electrostatics and magnetostatics. This involves determining the electric field in space produced by static charge distributions under various symmetry conditions and the magnetic field in space produced by current distributions under various symmetry conditions. This also involves computation of the capacitance between two conducting surfaces using general methods. Given two conducting surfaces, we assume that the first surface is at a potential $V_0/2$ and the second surface at a potential $-V_0/2$. The potential in the region between the two surfaces satisfies Laplace's equation and we solve this equation giving the resulting potential as $V_0\phi(x, y, z)$ where ϕ is a function that depends only on the geometry of the conductors, ie, on the relative shape and location of the conductors. The normal component of the electric field on one of the surfaces is then computed by projecting the gradient of the potential along the unit normal to the surface and then multiplying by ϵ_0 and then integrating the resulting over the surface. This yields the charge Q on the surface and the capacitance can then be determined as Q/V_0.

Examples include (a) determination of the electric field produced by a sphere having radially symmetric charge distribution, (b) an infinitely long cylinder having uniform charge distribution. The method of separation of variables used to solve Poisson's equation with specified boundary conditions. Determination of the magnetic field produced by a current carrying conductor of arbitrary shape using the Biot-Savart law. Determining the inductance of a coil and of a coaxial cable using Ampere's circuital law. Separation of variables in the cylindrical and spherical polar coordinate systems. General method of determining the capacitance between two perfectly conducting surfaces.

4.Derivation of the wave equation satisfied the electric and magnetic fields in free space.

5.Derivation of the retarded potential solution for the electrostatic and magnetic vector potentials.

6.The wave equation inside a waveguide when the field amplitudes depend exponentially on the z direction with the z-direction along the axis of the guide.

7.Notion of transverse electric and transverse magnetic field modes. In the transverse electric mode TE_{mn}, the z component of the magnetic field is set equal to zero while in the transverse magnetic mode TM_{mn}, the z component of the magnetic field is set equal to zero. The Maxwell equations are written down in component form within the waveguide and what results is a set of four linear algebraic equations for the four transverse components of the electric and magnetic field in terms of partial derivatives of the longitudinal components of the electric and magnetic fields. These are solved and boundary conditions are applied to arrive at the modes.

8.Plotting of the field lines inside a waveguide. Having determined the modes as above, and the fields, an important problem is to plot the electric and magnetic field lines at a given cross section of the guide. For example, consider the TE_{mn} mode in which $E_z = 0$ and E_x, E_y, H_x, H_y are known functions of (x, y). The electric field lines are plotted by numerically solving the differential equations

$$\frac{dx}{E_x(x,y)} = \frac{dy}{E_y(x,y)}$$

Numerical solutions to this equation can be obtained via the iterative scheme

$$x[n+1] = x[n] + \Delta.E_x(x[n], y[n]), y[n+1] = y[n] + \Delta.E_y(x[n], y[n])$$

9.Problems dealing with the cutoff frequency. For each mode, we determine the expression for the cutoff frequency and hence an expression for the phase and group velocities of waves propagating in the guide. For example, for the TE_{nm} mode, when the operating frequency is ω, the phase constant is given by

$$\beta = (\omega^2 \mu_0 \epsilon_0 - (n\pi/a)^2 - (m\pi/b)^2)^{1/2}$$

and the phase velocity is simple $v_p = \omega/\beta$ while the group velocity is $v_g = (\frac{d\beta}{d\omega})^{-1/2}$.

10.Circular waveguides: To determine the modes here, we write down the two Maxwell curl equations in the cylindrical coordinate system and solve for the transverse components of the electromagnetic field in terms of partial derivatives of the longitudinal component. The longitudinal component satisfies the two dimensional Helmholtz equation. On writing this equation down in the cylindrical coordinate system and separating out the ρ and ϕ variables, we get a Bessel equation which on applying the appropriate boundary conditions, gives the modes that can propagate inside the guide. These modes are expressed in terms the zeroes of the Bessel function or its derivative.

11.Waveguides of arbitrary cross section. In order to obtain the basic equations, we split the gradient operator into two parts, a transverse part and a longitudinal part and then write down the Maxwell curl equations. Assuming exponential dependence on the longitudinal variable (the z variable), we can solve for the transverse components of the fields in terms of the partial derivatives of the longitudinal components. Then boundary conditions are applied to the longitudinal components (vanishing of the z component of the electric field and of the component of the magnetic field that is normal to the boundary. In principle, this method yields the possible modes that can propagate. For the solution of the Helmholtz equation with the boundary conditions, numerical techniques must be developed.

12.Computation of the intrinsic impedance of a given mode in a rectangular waveguide:For example, suppose we are dealing with TM modes. Then the field components can be derived from $E_z(x, y)$ where

$$E_z(x, y) = A.sin(m\pi x/a)Sin(n\pi y/b)$$

Note that the dependence of the fields on z is given by $exp(-\gamma z)$, where

$$\gamma = ((m\pi/a)^2 + (n\pi/b)^2 - \omega^2 \mu\epsilon)^{1/2} = \gamma_{m,n}$$

Using this, we can in principle compute E_x, H_y, E_y, H_x. The intrinsic impedance of the mode is defined as the ratio $E_x/H_y = -E_y/H_x$.

13. Cavity resonators:Here we have a cuboid inside which an electromagnetic field is oscillating. The boundary conditions are that the field vanishes at all the six walls. Thus, the fields have the form

$$sin(n\pi x/a)sin(m\pi y/b)sin(p\pi z/c)$$

where n, m, p are positive integers. This equation can be derived directly from the Helmholtz equation

$$\psi_{,xx} + \psi_{,yy} + \psi_{,zz} + h^2\psi = 0$$

which is an eigenvalue problem. The possible values of h are then

$$h[m, n, p] = \sqrt{(m\pi/a)^2 + (n\pi/b)^2 + (p\pi/c)^2}$$

14.Variational formulation of the Maxwell equations showing how it leads to the finite element method for computation of the modes of propagation in a waveguide. For example, suppose we wish to solve the Helmholtz equation inside the guide with the Dirichlet boundary condition. Then the action principle

$$\delta S(\phi) = 0, S(\phi) = \frac{1}{2}\int |\nabla\phi|^2 dxdy - \frac{h^2}{2}\int \phi^2 dxdy$$

works. By discretizing the integral and carrying out a numerical optimization, the finite element algorithm can be derived. Alternately, we can expand ϕ as a linear combination of test functions and carry out the optimization with resepect to the coefficients in the linear combination. Suppose $\psi_i, i = 1, 2, ..., N$ are given test functions. We expand ϕ as a linear combination of these test functions:

$$\phi(x, y) = \sum c_i\psi_i(x, y)$$

Then,

$$\int |\nabla\phi|^2 dxdy = \sum_{i,j} c_ic_j \int (\nabla\psi_i, \nabla\psi_j)dxdy$$

and

$$\int \phi^2 dxdy = \sum_{i,j} c_ic_j \int \psi_i\psi_j dxdy$$

Thus the problem of mode determination approximates to extermizing a quadratic function of the coefficients (c_i) given by

$$Q(\mathbf{c}) = \frac{1}{2}\mathbf{c}^T\mathbf{Q}\mathbf{c} - \frac{h^2}{2}\mathbf{c}^T\mathbf{R}\mathbf{c}$$

where both $\mathbf{Q}, \mathbf{R}$ are positive definite matrices.

15.A comparison between the finite element and finite difference method for solving the Poisson equation for the potential in a two dimensional region. The finite difference method involves replacing the Poisson equation in two variables by

$$\phi[i + 1, j] + \phi[i - 1, j] + \phi[i, j + 1] + \phi[i, j - 1] - 4\phi[i, j] = -\Delta^2\rho[i, j]$$

This leads to the iterative scheme

$$\phi^{(n+1)}[i, j] = \frac{1}{4}(\phi^{(n)}[i + 1, j] + \phi^{(n)}[i - 1, j] + \phi^{(n)}[i, j + 1] + \phi^{(n)}[i, j - 1]) + \frac{\Delta^2}{4}\rho[i, j]$$

16.Derivation of the basic equations for wave propagation inside a waveguide having inhomogeneous permittivity, permeability and conductivity: $\epsilon(x, y), \mu(x, y), \sigma(x, y)$. Demonstration of the impossibility of separation of the modes into transverse electric and transverse magnetic modes.

17.Derivation of the basic equations of wave propagation in a medium having homogeneous but anisotropic permittivity, permeability and conductivity. In this case, the equations connecting the electric displacement and electric field are

$$D_x = \epsilon_{xx} E_x + \epsilon_{xy} E_y + \epsilon_{xz} E_z,$$

$$D_y = \epsilon_{yx} E_x + \epsilon_{yy} E_y + \epsilon_{yz} E_z,$$

$$D_z = \epsilon_{zx} E_x + \epsilon_{zy} E_y + \epsilon_{zz} E_z$$

or equivalently, in matrix notation,

$$\begin{pmatrix} D_x \\ D_y \\ D_z \end{pmatrix} =$$

$$\begin{pmatrix} \epsilon_{xx} & \epsilon_{xy} & \epsilon_{xz} \\ \epsilon_{yx} & \epsilon_{yy} & \epsilon_{yz} \\ \epsilon_{zx} & \epsilon_{zy} & \epsilon_{zz} \end{pmatrix} \begin{pmatrix} E_x \\ E_y \\ E_z \end{pmatrix}$$

The equations connecting the magnetic flux density **B** and the magnetic field intensity **H** are

$$B_x = \mu_{xx} H_x + \mu_{xy} H_y + \mu_{xz} H_z,$$

$$B_y = \mu_{yx} H_x + \mu_{yy} H_y + \mu_{yz} H_z,$$

$$B_z = \mu_{zx} H_x + \mu_{zy} H_y + \mu_{zz} H_z$$

or equivalently in matrix notation,

$$\begin{pmatrix} B_x \\ B_y \\ B_z \end{pmatrix}$$

$$= \begin{pmatrix} \mu_{xx} & \mu_{xy} & \mu_{xz} \\ \mu_{yx} & \mu_{yy} & \mu_{yz} \\ \mu_{zx} & \mu_{zy} & \mu_{zz} \end{pmatrix}$$

$$\begin{pmatrix} H_x \\ H_y \\ H_z \end{pmatrix}$$

Finally, the equations connecting the current density **J** and the electric field **E** are

$$J_x = \sigma_{xx} E_x + \sigma_{xy} E_y + \sigma_{xz} E_z$$

$$J_y = \sigma_{yx} E_x + \sigma_{yy} E_y + \sigma_{yz} E_z$$

$$J_z = \sigma_{zx} E_x + \sigma_{zy} E_y + \sigma_{zz} E_z$$

or equivalently, in matrix notation,

$$\begin{pmatrix} J_x \\ J_y \\ J_z \end{pmatrix} =$$

$$\begin{pmatrix} \sigma_{xx} & \sigma_{xy} & \sigma_{xz} \\ \sigma_{yx} & \sigma_{yy} & \sigma_{yz} \\ \sigma_{zx} & \sigma_{zy} & \sigma_{zz} \end{pmatrix}$$

$$\begin{pmatrix} E_x \\ E_y \\ E_z \end{pmatrix}$$

The wave equation for the electric and magnetic fields must be derived from the Maxwell equations

$$\nabla \times \mathbf{E} = -j\omega\mu.\mathbf{H}, \nabla \times \mathbf{H} = (\sigma + j\omega\epsilon).\mathbf{E},$$

$$\nabla.\mathbf{E} = 0, \nabla.\mathbf{H} = 0$$

The first equation gives

$$H = (j/\omega)\mu^{-1}.\nabla \times E$$

and hence taking the curl and using the second equation, we get

$$(j/\omega)\nabla \times (\mu^{-1}\nabla \times E) = \nabla \times H = (\sigma + j\omega\epsilon).E$$

This equation can be expressed in component form.

18. General solution of the Helmholtz equation in cylindrical coordinates using the separation of variables method:

$$\frac{1}{\rho}\frac{\partial}{\partial\rho}(\rho\frac{\partial\psi}{\partial\rho}) + \frac{1}{\rho^2}\frac{\partial^2\psi}{\partial\phi^2}$$
$$+h^2\psi = 0, h^2 = \gamma^2 + \omega^2\epsilon\mu$$

19. Derivation of the expression for the gradient, divergence and curl of a vector field in cylindrical coordinates. The gradient is computed first using the chain rule for partial derivatives. Specifically, let ρ, ϕ, z denote the cylindrical coordinates. Then

$$\rho = \sqrt{x^2 + y^2}, \phi = tan^{-1}(y/x),$$

with inverse formulas

$$x = \rho.cos(\phi), y = \rho.sin(\phi)$$

Given a scalar field $f(\rho, \phi, z)$, its gradient can be calculated as

$$\nabla f = \nabla\rho.f_{,\rho} + \nabla\phi.f_{,\phi} + \nabla z f_{,z}$$

and

$$\nabla\rho = \hat{\rho}, \nabla\phi = \hat{\phi}/\rho, \nabla z = \hat{z}$$

20.Formulation of the transmission line equations in which we assume that the line has a definite resistance, capacitance, inductance and conductance per unit length. We obtain linear partial differential equations for the current and voltage. The voltage $v(t, z)$ along the line obeys the pde

$$-v_{,z}(t, z) = R(z)i(t, z) + L(z)i_{,t}(t, z), -i_{,z}(t, z) = Cv_{,t}(t, z) + Gv(t, z)$$

Note that in the transmission line model, $R(z), L(z)$ are series resistances and inductances per unit length while $C(z), G(z)$ are parallel capacitance and conductance per unit length. In the special case of sinusoidal excitation, these equations become equations for the complex phasor $V(z)$ and $I(z)$ of voltage and current respectively:

$$-V'(z) = (R(z) + j\omega L(z))I(z), \quad -I'(z) = (G(z) + j\omega C(z))V(z)$$

We can from these two equations derive second order differential equations for $V(z)$ and $I(z)$ respectively. These are respectively

$$(V'/(R + j\omega L))' = (G + j\omega C)V$$

and

$$(I'/(G + j\omega C))' = (R + j\omega L)I$$

21.Special case of transmission lines having constant resistance, conductance, capacitance and inductance per unit length. In this case, the voltage and current phasors can be solved for exactly leading to forward and backward propagating waves along the line.

$$V(z) = V_+ exp(-\gamma z) + V_- exp(\gamma z),$$

$$I(z) = Z_0^{-1}(V_+ exp(-\gamma z) - V_- exp(\gamma z))$$

where $Z_0 = \sqrt{\frac{R+j\omega L}{G+j\omega C}}$.

22.Determination of the voltage and current waveforms in a transmission line when excited at one end by an arbitrary specified voltage as a function of time. Suppose that a transmission line is connected to a voltage source $v_s(t)$ having a source impedance $Z_s(\omega)$. Assume that the length of the line is L and that at the other end, there is a load of impedance $Z_L(\omega)$. The problem is then to determine the voltage $v(t,z)$ and current $i(t,z)$ along the line. Let $V_s(\omega)$ be the Fourier transform of $v_s(t)$. and let $V(\omega, z)$ denote the Fourier transform of $v(t, z)$. Then we have

$$V(\omega, z) = V_+(\omega)exp(-j\omega z/u) + V_-(\omega)exp(j\omega z/u)$$

$$I(\omega, z) = R_0^{-1}(V_-(\omega)exp(-j\omega z/u) - V_-(\omega)exp(j\omega z/u))$$

where we are assuming the line to be lossless. The functions $V_+(\omega)$ and $V_-(\omega)$ are determined from the conditions

$$V_s(\omega) - Z_s(\omega)I(\omega, 0) = V(\omega, 0),$$

$$V(\omega, L) = I(\omega, L)Z_L(\omega)$$

This determines the functions $V(\omega, z)$ and $I(\omega, z)$ and by taking the inverse Fourier transform of these two functions, we obtain $v(t, z), i(t, z)$.

23. Quarter wave transformer: By connecting a transmission line to a load, we can transform the load impedance. A particular case arises when the length of the line is $\lambda/4$ where $\lambda = 2\pi/\beta$ with $\beta = \omega\sqrt{LC} = \omega/u$ as the propagation constant. The line is assumed to be lossless so that $R = 0, G = 0$. In this case, the transformed impedance is given by Z_0^2/Z_L where Z_0 is the characteristic impedance of the line.

24.Use of the smith chart in transmission line computations. For a lossless line,

$$V(z) = V_+ exp(-j\beta z) + V_- exp(j\beta z), = V_+(z) + V_-(z)$$

$$I(z) = R_0^{-1}(V_+ exp(-j\beta z) - V_- exp(j\beta z)) = R_0^{-1}(V_+(z) - V_-(z))$$

The load condition gives

$$V(L) = I(L)Z_L$$

so that

$$V_+(L) + V_-(L) = (Z_L/R_0)(V_+(L) - V_-(L))$$

This gives

$$\frac{V_-(L)}{V_+(L)} = \frac{Z_L - R_0}{Z_L + R_0} = \Gamma(L)$$

This is the reflection coefficient at the load end. The reflection coefficient at a distance z from the load is given by

$$\Gamma(L - z) = \Gamma(L)exp(-j2\beta z)$$

The input impedance at this point $Z_{in}(z)$ is then given by

$$Z_{in}(z) = \frac{\Gamma(L-z) - 1}{\Gamma(L-z) + 1} = \frac{\Gamma(L)exp(-j2\beta z) - 1}{\Gamma(L)exp(j2\beta z) + 1}$$

Problem:Express $Z_{in}(z)$ in terms of Z_L.

Hint:In the above expression for $Z_{in}(z)$, express $\Gamma(L)$ in terms of Z_L.

25.Write down Maxwell's equations in the cylindrical coordinate system and assuming exponential dependence of the fields on the z coordinate and sinusoidal time dependence, express the transverse components $E_\rho, E_\phi, H_\rho, H_\phi$ in terms of the longitudinal components E_z, H_z.

Hint:The Lame's coefficients for the cylindrical coordinate system are

$$H_\rho = 1, H_\phi = \rho, H_z = 1$$

and hence the curl of a vector field $\mathbf{F} = F_\rho \hat{\rho} + F_\phi \hat{\phi} + F_z \hat{z}$ in the cylindrical system is given by

$$\nabla \times \mathbf{F} = \begin{pmatrix} \hat{\rho}/\rho & \hat{\phi} & \hat{z}/\rho \\ \frac{\partial}{\partial \rho} & \frac{\partial}{\partial \phi} \frac{\partial}{\partial z} \\ F_\rho & \rho F_\phi & F_z \end{pmatrix}$$

Use this to formulate the two Maxwell curl equations in the cylindrical coordinate system.

26. Write down the Maxwell equations in the spherical polar coordinate system and using the separation of variables applied to the wave equation, derive the modes inside a circular cavity resonator.

hint:Make use of the chain rule for partial derivatives applied to the transformation formulae

$$x = r.cos(\phi)sin(\theta), y = r.sin(\phi)sin(\theta), z = r.cos(\theta)$$

or equivalently, the inverse formulas

$$r = \sqrt{x^2 + y^2 + z^2}, \theta = cos^{-1}(z/r), \phi = tan^{-1}(y/x)$$

[8].Questions on transmission lines and waveguides for examination

[1] Derive the Maxwell equations for wave propagation inside a rectangular waveguide having dimensions $a \times b$ with the axis of the guide parallel to the z axis. Assume that the dependence of the fields on the z coordinate is $exp(-\gamma z)$ and the dependence of the fields on time is $exp(j\omega t)$.

[2] State the boundary conditions on the electric and magnetic fields inside a rectangular waveguide for both the TE and TM modes.

[3] Derive the expression for the power flowing across a cross section of a rectangular waveguide in the TE_{mn} mode.

[4] A transmission line having length L, complex characteristic impedance Z_0 and complex propagation constant γ is connected to a load having impedance Z_L. Derive the formulae from first principles for the input impedance of the line. Your starting point must be the differential equations satisfied by the voltage and current phasors as functions of the distance from the source.

[5] Considering a transmission line of length L, characteristic impedance Z_0 and propagation constant γ, write down the relations between input voltage and current and output voltage and current in matrix form and deduce the ABCD parameters of the line.

[6] What do you understand by reflection coefficient at a point along a transmission line ? How is the reflection coefficient related to the input impedance at that point? How does the reflection coefficient vary along the line ?

[7] Prove that for a lossless transmission line, the points along the line where voltage maxima occur are also the points where the current minima occur.

[8] Solve the wave equation in cylindrical coordinates using the separation of variables method. Show as a result of this computation that the Bessel function occurs in the radial part and explain how application of the boundary conditions for the TM modes inside a cylindrical waveguide results in the expression for the cutoff frequencies.

[9] Suppose that we have obtained the solution $G(\mathbf{r}|\mathbf{r}')$ to the equation

$$(\nabla^2 + k^2)G(\mathbf{r}|\mathbf{r}') = \delta(\mathbf{r} - \mathbf{r}')$$

in free space. Explain how using this function G in Green's theorem, one can obtain a solution to the Helmholtz equation with source

$$(\nabla^2 + k^2)\psi(\mathbf{r}) = \rho(\mathbf{r})$$

10.Write down the retarded potential solution to the Maxwell equations in terms of the current density alone. By differentiating these potentials with respect to the spatial and temporal coordinates, write down the expressions for the electromagnetic fields.

11.Determine the electromagnetic fields produced by a point charge Q moving along an arbitrary trajectory $\mathbf{R}(t)$. Your answer may be expressed in terms of the retarded time $t' = t'(t, \mathbf{r})$ defined by the equation

$$t' = t - |\mathbf{r} - \mathbf{R}(t')|/c$$

12.Determine the differential equations for the electric and magnetic field lines for the TE_{mn} modes over a cross section of a rectangular waveguide.

13. A transmission line of length L_1 and characteristic impedance Z_{01} is connected in tandem to another transmission line of length L_2 and characteristic impedance Z_{02}. A load of impedance Z_L is connected to the second line. Determine the overall input impedance of the line.

14. N transmission lines of lengths $L_1, L_2, ..., L_N$ having respective characteristic impedances $Z_{0i}, i = 1, 2, ..., N$ are connected in tandem. The final line is terminated with a load of impedance Z_L. Determine the input impedance looking into the first line.

15.Let $\Gamma = \Gamma_r + j\Gamma_i$ be the reflection coefficient at a point on a transmission line and $Z/Z_0 = r + jx$ the normalized input impedance at the same point. Determine the locus of the point (Γ_r, Γ_i) for r constant and for x constant. Explain how these loci are used in the construction of the Smith chart.

16.Decompose the ∇^2 operator in the spherical polar coordinate system and determine the radial and angular parts of this operator respectively.

17. Prove using Green's theorem the identity

$$(\nabla^2 + k^2)(exp(-jkr)/4\pi r) = -\delta(\mathbf{r})$$

18.Explain how you would use the perturbation method to solve the partial differential equation

$$(\nabla^2 + k_0^2 + \epsilon.k(x)^2)\psi(x) = 0$$

Here, k_0 is a constant, ϵ is a small parameter and $k(x)$ is a function of the spatial location x.

[9]. Determine the electromagnetic fields inside a rectangular waveguide when the guide is excited at one end with a sinusoidal current source. This problem must be solved by obtaining the Green's function for the waveguide, ie a solution to the equation $(\nabla^2 + k^2)G(\mathbf{r}|\mathbf{r}') = \delta(\mathbf{r} - \mathbf{r}')$ for $\mathbf{r}, \mathbf{r}'$ inside the guide with G vanishing on the bounding walls of the guide.

[10]. Determine the magnetic field produced by a surface current distribution on a torus. Assume that the torus is obtained by rotating a circle of radius a around the z axis. Let R denote the distance of the centre of the circle being rotated from the origin. Then a point on the torus can be represented as

$$\mathbf{r}' = (R + a.cos(\theta))cos(\phi)\hat{x} + (R + a.cos(\theta))sin(\phi)\hat{y} + a.sin(\theta)\hat{z}$$

Leta $\mathbf{J}_s$ denote the surface current density. Then,

$$\mathbf{J}_s dS' = J_0(R + a.cos(\theta))d\phi.a.d\theta.(-sin(\phi)\hat{x} + cos(\phi)\hat{y})$$

and hence the magnetic vector potential at a point (x, y, z) due to this surface current density is given by

$$\mathbf{A}(x, y, z) = \frac{\mu}{4\pi} \int \frac{\mathbf{J}_s dS'}{|\mathbf{r} - \mathbf{r}'|}$$

$$= \frac{\mu J_0}{4\pi} \int_0^{2\pi} \int_0^{2\pi} (R + a.\cos(\theta))a.d\theta.d\phi(-\hat{x}.\sin(\phi) + \hat{y}.\cos(\phi))/|\mathbf{r} - \mathbf{r}'|$$

where

$$|\mathbf{r} - \mathbf{r}'|^2 = (x - (R + a.\cos(\theta))\cos(\phi))^2 + (y - (R + a.\cos(\theta))\sin(\phi))^2 + (z - a.\sin(\theta))^2$$

[11]. Radiation fields emitted by charged particles moving along arbitrary trajectories. Assume that a point charge Q is moving along a trajectory $\mathbf{R}(t)$. The charge and current densities associated with this moving point charge are respectively given by

$$\rho(\mathbf{r}, t) = Q.\delta(\mathbf{r} - \mathbf{R}(t)), \mathbf{J}(\mathbf{r}, t) = Q\mathbf{V}(t)\delta(\mathbf{r} - \mathbf{R}(t))$$

where $\mathbf{V}(t) = d\mathbf{R}(t)/dt$. The magnetic vector potential is then given by

$$\mathbf{A}(\mathbf{r}, t) = \frac{\mu Q}{4\pi} \int \mathbf{V}(t')\delta(t' - t + |\mathbf{r} - \mathbf{r}'|/c)\delta(\mathbf{r} - \mathbf{R}(t'))dt'd^3r'/|\mathbf{r} - \mathbf{R}(t')|$$

Now we use a general theorem on the Dirac δ functions. Let $\psi(t)$ be a function on $\mathbb{R}$ having zeroes precisely at the points $t_k, k = 1, 2, ...$ Then,

$$\int_{-\infty}^{\infty} \phi(t)\delta(\psi(t))dt = \sum_k \frac{\phi(t_k)}{|\psi'(t_k)|}$$

This is equivalent to saying that

$$\delta(\psi(t)) = \sum_k \frac{\delta(t - t_k)}{|\psi'(t_k)|}$$

Using this formula, it follows that

$$\mathbf{A}(\mathbf{r}, t) = \frac{\mu Q}{4\pi|\mathbf{r} - \mathbf{R}(t')|(1 - (\mathbf{V}(t'), \mathbf{n}(t'))/c)}$$

where t' is determined by the equation

$$t' = t - |\mathbf{r} - \mathbf{R}(t')|/c$$

We are assuming that t' is the only root of this equation and

$$\mathbf{n}(t') = \frac{\mathbf{r} - \mathbf{R}(t')}{|\mathbf{r} - \mathbf{R}(t')|}$$

t' is called the retarded time. If an electromagnetic signal is emitted by the charged particle at time t', then it will arrive at the point $\mathbf{r}$ at time t provided that

$$t = t' + |\mathbf{r} - \mathbf{R}(t')|/c$$

[12]. Propagation of electromagnetic waves in a waveguide of arbitrary cross section. Let $\nabla_\perp$ denote the transverse component of the gradient operator. Then the total gradient operator can be expressed as

$$\nabla = \nabla_\perp + \frac{\hat{\partial}}{\partial z}$$

The fields have the dependence $exp(-\gamma z)$ on the z coordinate. Then the gradient operator acting on the fields has has the form

$$\nabla = \nabla_\perp - \gamma \hat{z}$$

so that the Maxwell curl equations are given by

$$(\nabla_\perp - \gamma \hat{z}) \times (\mathbf{E}_\perp + E_z \hat{z}) = -j\omega\mu(\mathbf{H}_\perp + H_z \hat{z})$$

$$(\nabla_\perp - \gamma \hat{z}) \times (\mathbf{H}_\perp + H_z \hat{z}) = j\omega\epsilon(\mathbf{E}_\perp + E_z \hat{z})$$

Equating the transverse and longitudinal components in these equations, we get

$$\nabla_\perp \times \mathbf{E}_\perp = -j\omega\mu H_z \hat{z}$$

$$\nabla_\perp E_z \times \hat{z} - \gamma \hat{z} \times \mathbf{E}_\perp = -j\omega\mu \mathbf{H}_\perp$$

$$\nabla_\perp \times \mathbf{H}_\perp = j\omega\epsilon E_z \hat{z}$$

$$\nabla_\perp H_z \times \hat{z} - \gamma \hat{z} \times \mathbf{H}_\perp = j\omega\epsilon \mathbf{E}_\perp$$

These equations can be solved for $\mathbf{E}_\perp$ and $\mathbf{H}_\perp$ in terms of the transverse partial derivatives of E_z and H_z. The solutions are given by

[13] Determine the partial differential equation and boundary conditions that define waves inside a cuboidal cavity resonator. What about cylindrical and spherical cavity resonators. Obtain an expression for the characteristic frequencies of oscillations.

hint: The differential equation is the Helmholtz equation with the boundary conditiona that the field vanishes on the boundary of the resonator. The modes are the eigen modes of the Laplacian operator:

$$(\nabla^2 + k^2)\psi(\mathbf{r}) = 0, \mathbf{r} \in D,$$

$$\psi(\mathbf{r}) = 0, \mathbf{r} \in \partial D$$

In a cuboid with side lengths a, b, c, we thus have

$$\left(\frac{\partial^2}{\partial x^2} + \frac{\partial^2}{\partial y^2} + \frac{\partial^2}{\partial z^2} + k^2\right)\psi(x, y, z) = 0,$$

$$\psi(x, y, 0) = \psi(x, y, c) = \psi(0, y, z) = \psi(a, y, z) = \psi(x, 0, z) = \psi(x, b, z) = 0$$

[14] Explain how you can obtain the Helmholtz equation using a variational principle, ie by extremizing a functional of the field. Explain how this variational principle can be used to derive a finite element method for approximate calculation of the modes.

hint: Define

$$S[\phi] = \frac{1}{2} \int_D |\nabla\phi(x,y)|^2 dxdy - \frac{k^2}{2} \int_D \phi^2 dxdy$$

where

$$\nabla^2 = \frac{\partial^2}{\partial x^2} + \frac{\partial^2}{\partial y^2}$$

The action principle $\delta S[\phi] = 0$ with the boundary condition that $\phi = 0$ on ∂D gives the Helmholtz equation

$$(\nabla^2 + k^2)\phi = 0$$

To implement this as a finite element method, we expand the field $\phi(x,y)$ as a linear combination of test functions:

$$\phi(x,y) = \sum_{i=1}^{N} c_i \phi_i(x,y)$$

and plug this into the action functional. Then the variational problem gets translated into extermizing a quadratic form

$$F(\mathbf{c}) = \frac{1}{2}\mathbf{c}^T\mathbf{Q}\mathbf{c} - \frac{\lambda}{2}\mathbf{c}^T\mathbf{R}\mathbf{c}$$

where $\mathbf{Q}$ and $\mathbf{R}$ are finite matrices. The equation $\nabla_{\mathbf{c}}F(\mathbf{c}) = \mathbf{0}$ then leads to the generalized eigenvalue problem

$$(\mathbf{Q} - \lambda\mathbf{R})\mathbf{c} = \mathbf{0}$$

which means that the modes λ are obtained by solving the polynomial equation

$$det(\mathbf{Q} - \lambda\mathbf{R}) = 0$$

This method can be generalized to the situation of inhomogeneous permittivity and/or permeability. For example suppose $f(x,y), g(x,y)$ are two functions and consider the action functional

$$S[\phi] = \frac{1}{2} \int_D f(\phi_{,x}^2 + \phi_{,y}^2)dxdy - \frac{\lambda}{2} \int_D g\phi^2 dxdy$$

The variational principle $\delta S[\phi] = 0$ then gives the generalized Helmholtz equation

$$\frac{\partial}{\partial x}(f\frac{\partial\phi}{\partial x}) + \frac{\partial}{\partial y}(f\frac{\partial\phi}{\partial y}) + \lambda g\phi = 0$$

[15] Suppose that R, G, L, C for a transmission line are functions of the point z on the line. Then formulate the differential equations that govern the propagation of voltage and current along the line and using perturbation theory, obtain series expansions for the solution. Specifically assume that

$$R(z) = R_0 + \delta.R_1(z), G(z) = G_0 + \delta.G_1(z), L(z) = L_0 + \delta.L_1(z), C(z) = C_0 + \delta.C_1(z)$$

where δ is a small parameter. Expand the voltage and current along the line in powers of δ, equate coefficients of equal powers of δ and hence derive a series of differential equations for each approximant.

[16] Derive an expression for the skin depth for penetration of electromagnetic waves inside a conductor having conductivity σ. Specifically start with Maxwell's equations for sinusoidal sources and derive the modified Helmholtz equation for the electric and magnetic fields. The propagation constant will come out to be complex and the reciprocal of its real part is the skin depth. Derive expressions for R, L, G, C along a two wire transmission line in which each wire is a cylinder of radius a and the distance between the two wires is d. The formulae for R, G will be expressed in terms of the skin depth.

[17] Suppose that the permittivity of the medium that fills the waveguide is a function of the field. Formulate the Maxwell equations and derive the nonlinear wave propagation equation for such a medium. Explain how this equation can be solved using perturbation theory for differential equations. Show how the solution leads to nonlinear frequency coupling between the modes.

[18] Derive expressions for phase velocity, group velocity and intrinsic impedance of the TE_{mn} and TM_{mn} modes for electromagnetic wave propagation inside a rectangular waveguide. Note that if the imaginary part of the propagation constant γ is β, then the phase velocity is ω/β while the group velocity is $d\omega/d\beta$. The intrinsic impedance of a mode is $|E_x/H_y|$ which will be the same as $|E_y/H_x|$.

[19] Analysis of the waveguide equations in the presence of a gravitational field. The scalar wave propagation equation in a curved space-time metric is formulated and approximately solved using the method of moments. Specifically, let $((g_{ij}(x)))$ be the metric tensor and $\phi(x)$ a scalar field. $\phi_{,i}$ is a covariant four vector and $g^{ij}\phi_{,j}$ is a contravariant four vector. Let $g = det((g_{ij}))$. Then $g < 0$ and the covariant divergence of the four vector $g^{ij}\phi_{,j}$ is given by $(g^{ij}\phi_{,j}\sqrt{-g})_{,i}$. This is set equal to zero resulting in the scalar wave equation in curved space-time. The finite element method for analyzing the modes of such a field is based on the fact that the variational principle

$$\delta S[\phi] = 0, S[\phi] = \int g^{ij}\sqrt{-g}\phi_{,i}\phi_{,j}d^4x$$

leads to the wave equation

$$(g^{ij}\sqrt{-g}\phi_{,j})_{,i} = 0$$

[20] Derivation of the expressions for R, G, L, C for two wire and coaxial cable transmission lines. For the two wire line, we assume that each wire is a cylinder of radius a and that the two conductors are parallel to each other with a spacing of d between their axes. Assume that a current I flows in the first wire and $-I$ in the second wire. Application of Ampere's law gives the magnetic field in the region between the two wires as well as the field inside the wires. The flux of the magnetic field through a rectangle bounded by the two wires is computed and from this the expression for the inductance L per unit length of the line is derived. At a distance ρ from the axis of the first line, the magnetic field in the plane bounded by the two wires is given by

$$H = \frac{\mu I}{2\pi\rho} + \frac{\mu I}{2\pi(d-\rho)}$$

provided that $a < \rho < d - a$. For $0 < \rho < a$, the field is

$$H = \frac{\mu I \rho}{2\pi a^2} + \frac{\mu I}{2\pi(d-\rho)}$$

while for $d - a < \rho < d$, the field is

$$H = \frac{\mu I(d - \rho)}{2\pi a^2}$$

The flux of the magnetic field through a rectangle of length l and width d is thus given by

$$\Phi = \int_0^d H.l.d\rho = \int_0^a H.l.d\rho + \int_a^{d-a} H.l.d\rho + \int_{d-a}^d H.l.d\rho$$

$$= \frac{2\mu Il}{2\pi a^2}(a^2/2 + log(d/(d - a))) + \frac{\mu I}{2\pi}$$

[21] Derivation of the dispersion relation for electromagnetic waves propagating in a homogeneous anisotropic medium. Here ϵ is a 3×3 matrix and the relevant Maxwell equations are

$$\nabla.(\epsilon.E) = 0, \nabla \times E = -j\omega\mu H, \nabla \times H = j\omega\epsilon.E$$

so that

$$\nabla(\nabla.E) - \nabla^2 E = \omega^2 \mu\epsilon.E$$

Assuming that

$$E = E_0 exp(jk.r)$$

we thus have

$$\mathbf{k}.(\epsilon\mathbf{E_0}) = 0, \mathbf{k}(\mathbf{k}.\mathbf{E_0}) - k^2\mathbf{E_0} = \omega^2\mu\epsilon.\mathbf{k}$$

from which the dispersion relation can be derived. Note that the three components of the second equation above are linearly dependent.

[22].Solve numerically the equations of motion of a charged particle moving in the electromagnetic field of a plane electromagnetic wave.

hint: The electric field can be taken as

$$E = A.cos(\omega t - kz)\hat{x}$$

and the magnetic field as

$$H = \eta A.cos(\omega t - kz)\hat{y}$$

Now write down the Lorentz equations of motion

$$\mathbf{r'} = \mathbf{v}, \mathbf{v'} = \frac{e}{m}(\mathbf{E} + \mathbf{v} \times \mathbf{B}), \mathbf{B} = \mu\mathbf{H}$$

[23].Formulate the Maxwell equations in a wave guide when the space-time metric corresponds to that of a black hole, ie, the Schwarzchild metric.

hint: The metric of space time is

$$d\tau^2 = (1 - 2GM/rc^2)dt^2 - \frac{1}{c^2}((1 - 2GM/rc^2)^{-1}dr^2 + r^2 d\theta^2 + r^2 sin^2(\theta)d\phi^2)$$

Take the coordinates as

$$x^0 = t, x^1 = r, x^2 = \theta, x^3 = \phi$$

so that the metric coefficients in this coordinate system are given by

$$g_{00} = 1 - 2GM/rc^2, g_{11} = -c^{-2}(1 - 2GM/rc^2)^{-1}, g_{22} = -c^{-2}r^2, g_{33} = -c^{-2}sin^2(\theta)$$

The components of the contravariant electromagnetic four potential are denoted by

$$A^0 = V, A^1 = A_r, A^2 = A_\theta, A^3 = A_\phi$$

Calculate the covariant components

$$A_0 = g_{00}A^0, A_1 = g_{11}A^1, A_2 = g_{22}A^2, A_3 = g_{33}A^3$$

and then compute the electromagnetic antisymmetric field tensor

$$F_{\mu\nu} = A_{\nu,\mu} - A_{\mu,\nu}$$

Formulate the Maxwell equations

$$(F^{\mu\nu}\sqrt{-g})_{,\nu} = 0$$

and simplify it using the Lorentz gauge condition

$$(A^{\mu}\sqrt{-g})_{,\mu} = 0$$

Show that this equation in the limit $M \to 0$ reduces to the classical wave equation for the electromagnetic four potential.

[24] Problems

[1] Assume that the propagation constant along the z direction in a waveguide is γ. Derive expressions for E_x, E_y, H_x, H_y in terms of E_z for the TM mode.

[2] A rectangular waveguide of dimensions $a \times b$ carries a TM_{mn} mode. Derive expressions for the power transmitted through a cross section of the guide from first principles.

[3] A transmission line of length L has characteristic impedance Z_0. Derive from first principles (KVL and KCL) the input impedance at a distance z from the load. Assume the load impedance is Z_L and the transmission line is lossless with propagation constant $\gamma = j\beta$.

[4] A waveguide has anisotropic permittivity tensor

$$\epsilon = \begin{pmatrix} \epsilon_{xx} & \epsilon_{xy} & 0 \\ epsilon_{yx} & \epsilon_{yy} & 0 \\ 0 & 0 & \epsilon_{zz} \end{pmatrix}$$

Derive the wave equations satisfied by (E_z, H_z) assuming dependence on z to be proportional to $exp(-\gamma z)$ and dependence on t to be $exp(j\omega t)$.

[5] A lossless transmission line of length L is terminated with a load impedance $Z_L = R_L + jX_L$. Derive an expression for $\frac{V_{max}}{V_{min}}$ at a distance z from the load assuming that the characteristic impedance of the line is R_0 and the propagation constant is $j\beta$.

[25].The transmission parameters of a transmission line. Let d be the length of the line. Then, we have

$$V_1 = V_+ + V_-, I_1 = Z_0^{-1}(V_+ - V_-), V_2 = V_+ exp(-\gamma d) + V_- exp(\gamma d), I_2 = Z_0^{-1}(V_+ exp(-\gamma d) - V_- exp(\gamma d))$$

The first two equations give

$$V_+ = (V_1 + Z_0 I_1)/2, V_- = (V_1 - Z_0 I_1)/2$$

Substituting these into the second and third equations, we get

$$V_2 = (V_1 + Z_0 I_1)exp(-\gamma d)/2 + (V_1 - Z_0 I_1)exp(\gamma d)/2, = V_1 cosh(\gamma d) - I_1 Z_0 sinh(\gamma d)$$

$$I_2 = Z_0^{-1}((V_1 + Z_0 I_1)exp(-\gamma d)/2 - (V_1 - Z_0 I_1)exp(\gamma d)/2 = -V_1 Z_0 sinh(\gamma d) + I_1 Z_0 cosh(\gamma d)$$

These equations can be expressed in matrix notation as

$$\begin{pmatrix} V_2 \\ I_2 \end{pmatrix} = \begin{pmatrix} cosh(\gamma d) & -Z_0 sinh(\gamma d) \\ -Z_0 sinh(\gamma d) & Z_0 cosh(\gamma d) \end{pmatrix} \times \begin{pmatrix} V_1 \\ I_1 \end{pmatrix}$$

[26]. Implement the equations of energy and momentum conservation in a collision between two particles moving with relativistic velocities.

hint:Let $\mathbf{p}_1, \mathbf{p}_2$ denote respectively the three momenta of the two particles having masses m_1, m_2 respectively before the collision and $\mathbf{p}_1', \mathbf{p}_2'$ the three momenta of the two partiles after the collision. Then, momentum conservation gives

$$\mathbf{p}_1 + \mathbf{p}_2 = \mathbf{p}_1' + \mathbf{p}_2'$$

and energy conservation gives

$$\sqrt{m_1^2 + |\mathbf{p}_1|^2} + \sqrt{m_2^2 + |\mathbf{p}_2|^2} = \sqrt{m_1^2 + |\mathbf{p}_1'|^2} + \sqrt{m_2^2 + |\mathbf{p}_2'|^2}$$

These are four equations in all and the number of unknowns is six, ie, $\mathbf{p}_1', \mathbf{p}_2'$. So to solve for the final momenta, we need two more equations. These can be obtained by specifying the direction of either $\mathbf{p}_1'$ or $\mathbf{p}_2'$.

[27]. Computation of the interaction energy between charges and the electromagnetic field. Let A^μ denote the contravariant electromagnetic four potential and A_μ the covariant electromagnetic four potential. The antisymmetric electromagnetic field tensor is then given by

$$F_{\mu\nu} = A_{\nu,\mu} - A_{\mu,\nu} = \partial_\mu A_\nu - \partial_\nu A_\mu$$

The Lagrangian for the field is given by

$$\mathcal{L}_{field} = C_1 . F_{\mu\nu} F^{\mu\nu}$$

where C_1 is a constant. The Lagrangian for the interaction between charges and the field is

$$\mathcal{L}_{int} = C_2 . J_\mu A^\mu$$

where J_μ is the electromagnetic four current density. We assume that the current density J^μ is known and then the total Lagrangian density for the field and its interaction with charged matter is

$$\mathcal{L} = \mathcal{L}_{field} + \mathcal{L}_{int}$$

Let us compute the corresponding Hamiltonian density. Let A^μ denote the canonical positions and π_μ the canonical momenta. We have

$$\pi_\mu = \frac{\partial L}{\partial A^\mu_{,0}} = \frac{\partial \mathcal{L}_{field}}{\partial A^\mu_{,0}}$$

Thus, the canonical Hamiltonian density is given by

$$\mathcal{H} = \pi_\mu A^\mu_{,0} - \mathcal{L} = A^\mu_{,0} \frac{\partial \mathcal{L}_{field}}{\partial A^\mu_{,0}} - \mathcal{L}_{field} - \mathcal{L}_{int}$$

$$= \mathcal{H}_{field} + \mathcal{H}_{int}$$

where $\mathcal{H}_{field}$ is the Hamiltonian density of the field and is given by

$$\mathcal{H}_{field} = A^\mu_{,0} \frac{\partial \mathcal{L}_{field}}{\partial A^\mu_{,0}} - \mathcal{L}_{field}$$

and $\mathcal{H}_{int}$ is the Hamitlonian density of the interaction between charged matter and the field and is given by

$$\mathcal{H}_{int} = -C_2 J_\mu A^\mu$$

This interaction Hamiltonian density is used in quantum electrodynamics to calculate transition amplitudes for scattering of electrons and positrons with photons. Note the constraints: Since $A^0_{,0}$ does not appear in the Lagrangian density, it follows that $\pi_0 = 0$. Also,

$$\pi_i = \frac{\partial \mathcal{L}}{\partial A^i_{,0}} = -\frac{\partial \mathcal{L}}{\partial A_{i,0}}$$

$$= \frac{\partial \mathcal{L}}{\partial A_{0,i}} = \frac{\partial \mathcal{L}}{\partial A^0_{,i}}, i = 1, 2, 3$$

and hence from the equations of motion,

$$\sum_{i=1}^{3} \pi_{i,i} = \frac{\partial \mathcal{L}}{\partial A^0} = C_2 J_0$$

Note that the equation of motion for A^0 is

$$\frac{\partial \mathcal{L}}{\partial A^0} = \sum_{i=1}^{3} \left(\frac{\partial}{\partial x^i} \frac{\partial \mathcal{L}}{\partial A^0_{,i}} \right)$$

$$+ \frac{\partial}{\partial x^0} \frac{\partial \mathcal{L}}{\partial A^0_{,0}}$$

and the second term on the right drops out since $A^0_{,0}$ does not appear in $\mathcal{L}$.

[28]. Using MATLAB implement the computation of the retarded potential solution for the electromagnetic fields generated by moving charges.

hint:Given a current density $J(t, r)$, the magnetic vector potential generated by this is given by

$$A(t, r) = \frac{\mu}{4\pi} \int \frac{J(t - |r - r'|/c, r')}{|r - r'|} d^3 r'$$

From the current density J, we can obtain the charge density by using the charge conservation equation

$$\frac{\partial \rho}{\partial t} + div\mathbf{J} = 0$$

Thus,

$$\rho(t, r) = C(r) - \int_0^t div J(t', r) dt'$$

where $C(r)$ is any function of only the spatial coordinates. If $C(r)$ is known, the electrostatic scalar potential $V(t, r)$ can be found and it is given by

$$V(t, r) = \frac{1}{4\pi\epsilon} \int \frac{\rho(t - |r - r'|/c, r')}{|r - r'|} d^3 r'$$

Suppose, for example, that the charges and currents are confined to a rectangular box. Then these integrals can be evaluated using the discretization process. Let τ be the time discretization step sized and Δ the spatial discretization step size. Then the discretized current density is $J[n, m_x, m_y, m_z] = J(n\tau, m_x\Delta, m_y\Delta, m_z\Delta)$ where n, m_x, m_y, m_z are integers. The discretized magnetic vector potential is given by $A[n, m_x, m_y, m_z] = A(n\tau, m_x\Delta, m_y\Delta, m_z\Delta)$ and is obtained approximately by discretizing the above integrals.

$$A[n, m_x, m_y, m_z] = \frac{\mu}{4\pi} \sum_{p_x, p_y, p_z} \frac{J[n - \frac{\Delta}{c\tau}((m_x - p_x)^2 + (m_y - p_y)^2 + (m_z - p_z)^2)^{1/2}, p_x, p_y, p_z]}{((m_x - p_x)^2 + (m_y - p_y)^2 + (m_z - p_z)^2)^{1/2}} \Delta^2$$

For a charge Q moving along the trajectory $\mathbf{R}(t)$, we set

$$\rho(t, \mathbf{r}) = Q.\delta(\mathbf{r} - \mathbf{R}(t)), \mathbf{J}(t, \mathbf{r}) = Q.\mathbf{R}'(t)\delta(\mathbf{r} - \mathbf{R}(t))$$

[29]. Write down the discretized form of the Maxwell equations in a waveguide corresponding to the TE and TM modes.

hint: The dependence of the fields on the z-coordinate is given by $exp(-\gamma z)$. It follows that the partial derivative $\frac{\partial}{\partial z}$ can be replaced by multiplication with $-\gamma$. This results in the equations

$$E_{z,y} + \gamma E_y = -j\omega\mu H_x, \gamma E_x + E_{z,x} = j\omega\mu H_y, E_{y,x} - E_{x,y} = -j\omega\mu H_z,$$

$$H_{z,y} + \gamma H_z = (\sigma + j\omega\epsilon)E_x, \gamma H_x + H_{z,x} = -(\sigma + j\omega\epsilon)E_y, H_{y,x} - H_{x,y} = (\sigma + j\omega\epsilon)E_z$$

We assume now that the waveguide has a rectangular cross section and discretize along the x and y directions. Suppose then that the length of the guide along the x direction is a and along the y direction is b. Let Δ_x and Δ_y denote respectively the discretization step size along the x and y directions so that for some large positive integers N_x, N_y, we have $N_x\Delta_x = a, N_y\Delta_y = b$. Then the discretized fields are given by

$$E_x[n,m] = E_x(n\Delta_x, m\Delta_y), E_y[n,m] = E_y(n\Delta_x, m\Delta_y), E_z[n,m] = E_z(n\Delta_x, m\Delta_y),$$

$$H_x[n,m] = H_x(n\Delta_x, m\Delta_y), H_y[n,m] = H_y(n\Delta_x, m\Delta_y), H_z[n,m] = H_z(n\Delta_x, m\Delta_y)$$

The partial derivatives with respect to x and y are then replaced by partial differences with respect to the n and m variables. For example, $E_{z,y}(x,y)$ is replaced by $(E_z[n,m+1] - E_z[n,m])/\Delta_y$ and $E_{z,x}$ is replaced by $(E_z[n+1,m] - E_z[n,m])/\Delta_x$ and likewise for the other field components. The Helmholtz equations for E_z and H_z are then replaced by discretized Helmholtz equations.

[30].Explain how you would derive the energy-momentum tensor of the electromagnetic field by considering variations of the action with respect to the metric tensor of curved space-time.

[31].Determine the Green's function for the Laplacian operator and for the Helmholtz operator.

hint. The Green's function $G(r|r')$ satisfies

$$\nabla^2 G(r|r') = \delta(r - r')$$

for the Laplacian operator and for the Helmholtz operator, it satisfies

$$(\nabla^2 + k^2)G(r|r') = \delta(r - r')$$

[32]. Explain how you would solve the wave equation in one dimension approximately by discretization of the spatial and time variables.

hint: The wave equation is

$$\psi_{,tt}(t,x) = c^2\psi_{,xx}(t,x)$$

Discretization with respect to the time and spatial variables leads to the partial differential equation

$$\psi[n+2,k] - 2\psi[n+1,k] + \psi[n,k] = \alpha^2(\psi[n,k+2] - 2\psi[n,k+1] + \psi[n,k])$$

where

$$\alpha = c\tau/\Delta$$

Here, τ is the discretization step size for the time variable while Δ is the discretization step size for the spatial variable. Now define a vector ψ_n whose components are $\psi[n, k], k = -N, -N+1, ..., N-1, N$ and show that the above difference equation can be cast as

$$\psi_{n+2} - 2\psi_{n+1} + \psi_n = A\psi_n$$

where A is a $2N+1 \times 2N+1$ tridiagonal matrix. To find the modes, ie characteristic frequencies of the system, let $\psi_n = \lambda^n \phi$ where ϕ is a $2N+1 \times 2N+1$ vector. We then end up with a generalized eigen-equation

$$((\lambda^2 - 2\lambda + 1)I - A)\phi = 0$$

which implies that the possible values of λ which give nonzero solutions ϕ are obtained as the solution to the polynomial equation

$$det((\lambda^2 - 2\lambda + 1)I - A) = 0$$

[33]. Solve the wave equation in two dimensions by discretization using the Discrete Fourier transform.

hint:

$$\psi_{,tt}(t, x, y) = c^2(\psi_{,xx}(t, x, y) + \psi_{,yy}(t, x, y))$$

Discretization gives

$$\tau^{-2}(\psi[n+2, k, m] - 2\psi[n+1, k, m] + \psi[n, k, m]) =$$

$$c^2\Delta^{-2}(\psi[n, k+2, m] - 2\psi[n, k+1, m] + \psi[n, k, m] + \psi[n, k, m+2] - 2\psi[n, k, m+1] + \psi[n, k, m])$$

Now assume that k, m have finite ranges and $\psi[n, k, m]$ is set equal to zero at all points (k, m) that fall outside the range. Then, determine the possible modes of oscillation by forming the two dimensional DFT

$$\hat{\psi}[n, p, q] = \sum_{k=-N}^{N} \sum_{m=-N}^{N} \psi[n, k, m] exp(-2\pi i(kp + mq)/(2N+1))$$

We write this equation as

$$\hat{\psi} = F(\psi)$$

or

$$\hat{\psi}[n, p, q] = F(\psi[n, k, m])$$

Then,

$$F(\psi[n, k+2, m]) = exp(4\pi ip/(2N+1))\hat{\psi}[n, p, q]$$

and similarly for other shifts. More precisely, we can define the operators Z_1, Z_2 by

$$Z_1(\psi)[n, k, m] = \psi[n, k+1, m], Z_2(\psi)[n, k, m] = \psi[n, k, m+1]$$

Then,

$$FZ_1(\psi)[n,p,q] = exp(2\pi ip/(2N+1))F(\psi)[n,p,q]$$

$$FZ_2(\psi)[n,p,q] = exp(2\pi iq/(2N+1))F(\psi)[n,p,q]$$

and more generally,

$$(FZ_1^r Z_2^s \psi)[n,p,q] = exp(2\pi i(rp+sq)/(2N+1))F(\psi)[n,p,q]$$

Thus, we can define the multiplication operators M_1, M_2 so that

$$M_1\psi[n,k,m] = exp(2\pi ik/(2N+1))\psi[n,k,m], \quad M_2\psi[n,k,m] = exp(2\pi im/(2N+1))\psi[n,k,m]$$

and then the above equations can be expressed as operator equations

$$FZ_1^r Z_2^s = M_1^r M_2^s F$$

[34]. Problems on transmission lines and waveguides.

[1] Two lossless parallel wire transmission lines having lengths L_1, L_2, propagation constants β_1, β_2 and characteristic impedances Z_{01}, Z_{02} respectively are connected to each other. The line having length L_2 is terminated by a load Z_L. Determine the input impedance as seen from the free end of the line having length L_1.

[2] In a cylindrical waveguide, the $TM_{m,n}$ mode is propagating. Write down the partial differential equation satisfied by H_z and state the boundary conditions. Also derive expressions for the phasors $E_\rho(\rho,\phi), E_\phi(\rho,\phi), H_\rho(\rho,\phi)$ and $H_\phi(\rho,\phi)$ in terms of $H_z(\rho,\phi)$.

[3] A rectangular waveguide having cross sectional dimensions a, b is oriented with its axis along the z axis so that the bounding walls of the guide are parallel to the xz and yz planes respectively. The permittivity and permeability are $\epsilon(x,y), \mu(x,y)$, ie, functions of x, y. Formulate the Maxwell equations assuming propagation constant γ and solve for E_x, E_y, H_x, H_y in terms of E_z, H_z. Hence deduce the partial differential equations satisfied by $E_z(x,y), H_z(x,y)$.

[4] Derive the expressions for the phase and group velocity of $TE_{m,n}$ waves inside a rectangular waveguide. Assume that the frequency is ω and it is greater than the cutoff frequency.

[5] Define the reflection coefficient in a transmission line as a function of the distance from the load. Derive an expression for the reflection coefficient as a function of the load and the distance from the load. From the expression for the reflection coefficient as a function of the input impedance at the given point, derive the relationships between the real and imaginary parts of the reflection coefficient corresponding to constant r and constant x, where $r + jx$ is the input impedance at the given point on the line. Show that these are circles.

[6] Derive the differential equations satisfied by the voltage and current along a transmission line when R, L, G, C are functions of the position on the line. Specialize to the case when R, L, G, C are constants and by solving the resulting differential equations show that the voltage and current along the line can be expressed as

$$V(z) = V_+ exp(-\gamma z) + V_- exp(\gamma z),\, I(z) = Z_0^{-1}(V_+ exp(-\gamma z) - V_- exp(\gamma z))$$

Obtain expressions for the propagation constant γ and the characteristic impedance Z_0 as functions of R, L, G, C.

[7] Consider a lossless transmission line with propagation constant $\gamma = j\beta$. Obtain the points on the line where the voltage is a maximum and the points where the current is a minimum. Take the characteristic impedance as R_0. Assume that the source is sinusoidal of frequency ω and the load is resistive with resistance R.

[8] Show that the average power flow across a cross section of a waveguide is given by

$$P = \frac{1}{2} \int_S Re(E_x(x,y)H_y^*(x,y) - E_y(x,y)H_x^*(x,y))dxdy$$

where S is the waveguide cross section. Calculate this for a cylindrical waveguide of radius R corresponding to a given mode.

[35] Problems in antenna theory.

[1] Show that the far field magnetic vector potential due to a current density $\mathbf{J}(\mathbf{r})$ can be expressed as

$$\mathbf{A}(\mathbf{r}) = \frac{\mu_0 . exp(-jkr)}{4\pi r} \int \mathbf{J}(\mathbf{r}')exp(jk\hat{r}.\mathbf{r}')d^3r'$$

Determine the electric and magnetic fields corresponding to this vector potential.

hint:Apply the Lorentz gauge condition to determine the electric vector potential from the magnetic vector potential and then differentiate to obtain the electric and magnetic fields.

[2] Assume that a wire in the shape of a curve $\mathbf{R}(s), 0 \le s \le 1$ carries a sinusoidal current of frequency $\omega = kc$. Show that the far field magnetic vector potential phasor produced by this wire can be expressed as

$$\mathbf{A}(\mathbf{r}) = \frac{\mu I . exp(-jkr)}{4\pi r} \int_0^1 \mathbf{R}'(s).exp(jk\hat{r}.\mathbf{R}(s))ds$$

[3] Suppose that the equation of a two dimensional surface in three dimensional space is given in parametric form by $(u, v) \to \mathbf{R}(u, v)$. Show that the differential surface area element on this surface can be expressed as

$$dS(u, v) = |\mathbf{R}'(u) \times \mathbf{R}'(v)|dudv$$

Using this formula, show that if the equation of the surface in cartesian coordinates is given by $z = f(x, y)$, then the surface element is given by

$$dS(x, y) = (1 + (\frac{\partial f}{\partial x})^2 + (\frac{\partial f}{\partial y})^2)^{1/2} dxdy$$

Use this formula to calculate the surface area of a sphere of radius R.

[4] Suppose that a surface having parametric equation $(u, v) \to \mathbf{R}(u, v), u_1 \le u \le u_2, v_1 \le v \le v_2$ carries a surface current phasor of $\mathbf{K}(u, v)$ amperes per unit length. Show that the far field magnetic vector potential phasor produced by this sheet can be expressed as

$$\mathbf{A}(\mathbf{r}) = \frac{\mu.exp(-jkr)}{4\pi r} \int_{u_1}^{u_2} \int_{v_1}^{v_2} \mathbf{K}(u, v) exp(jk\hat{r}.\mathbf{R}(u, v)) |\mathbf{R}'(u) \times \mathbf{R}'(v)| dudv$$

[5] Consider a linear antenna array with sensors placed along the z axis with the nth sensor being located at $d_n \hat{z}, n = 1, 2, ..., N$. If each sensor has the same far field amplitude pattern $f(\theta, \phi)$, then show that the overall pattern of the array is given by

$$F(\theta, \phi) = f(\theta, \phi) \sum_{n=1}^{N} I_n.exp(jkd_n cos(\theta))$$

where I_n is the excitation current of the nth sensor.

[6] Consider isotropic antenna sources placed along a curve having parametric equation $s \to \mathbf{R}(s)$ for $0 \le s \le 1$ at the points $\mathbf{R}|(s_k), k = 1, 2, ..., N$. Show that the far field amplitude pattern is given by

$$F\mathbf{r}) = \frac{exp(-jKr)}{r} \sum_{k=1}^{N} exp(jK(\mathbf{R}(s_k), \hat{r}))$$

where $K = \omega/c$ is the wavenumber with ω as the frequency of the currents in the individual senors and c the velocity of light in the medium under consideration.

[7] Consider a linear current excitation along the z axis with the current phasor given by $I(z), 0 \le z \le L$. Show that the far field amplitude pattern is given by

$$F(\theta, \phi) = \frac{exp(-jKr)}{r} \int_0^L I(\xi).exp(jK\xi.cos(\theta))d\xi$$

[8] Suppose that linear current sources with phasors $I_n(z), 0 \le z \le L_n$ parallel to the z axis are located with one end in the xy plane so that the nth source is at $(x_n, y_n, 0)$. Show that the far field amplitude pattern is given by

$$F(\theta, \phi) = \frac{exp(-jKr)}{r} sum_{n=1}^{N}(\int_0^{L_n} I_n(\xi).exp(jK\xi.cos(\theta)d\xi) exp(jK(x_n cos(\phi) + y_n sin(\phi)) sin(\theta))$$

[8] In the presence of magnetic current sources and magnetic charges, the Maxwell equations assume the form

$$\nabla.E = \rho/\epsilon_0, \nabla \times E = -M - \frac{\partial B}{\partial t}, \nabla \times H = J + \epsilon_0 \frac{\partial E}{\partial t},$$

$$\nabla.H = \rho_m/\mu_0$$

where ρ_m is the magnetic charge density and M is the magnetic current density. Obtain the solutions to these equations by using the notion of electric vector potential and magnetic scalar potential.

[36]. Prove the Green's identity

$$u\nabla^2 v - v\nabla^2 u = div(u\nabla v - v\nabla u)$$

in an n-dimensional space.

hint:

$$u\nabla^2 v - v\nabla^2 u = \sum_i (uv_{,ii} - vu_{,ii})$$

on the one hand, while on the other,

$$div(u\nabla v - v\nabla u) = \sum_i (uv_{,i} - vu_{,i})_{,i} = \sum_i (u_{,i}v_{,i} + uv_{,ii} - v_{,i}u_{,i} - vu_{,ii})$$

$$= \sum_i (uv_{,ii} - vu_{,ii})$$

[37] Let D be a volume in $\mathbb{R}^3$ having boundary $S = \partial D$. Explain how the Green's function for the Helmholtz operator $\nabla^2 + k^2$ with Dirichlet boundary conditions can be used to solve the pde $(\nabla^2 + k^2)\phi = \psi$ in D with the Dirichlet boundary condition on ϕ, namely, that ϕ vanishes on S.

[38] Given a vector field F in $\mathbb{R}^3$, show that there exist two vector fields A, B such that $F = A + B$ and $divA = 0, curlB = 0$.

hint: Let ϕ be a scalar field such that $div(F - \nabla\phi) = 0$. In other words, $\nabla^2\phi = divF$. It is known, for example, from electrostatics, that this equation has a solution ϕ. Define

$$B = \nabla\phi, A = F - B$$

Then A, B satisfy the required properties.

[39] Prove the following identities in vector calculus: (a) $\nabla \times (\nabla \times A) = \nabla(\nabla.A) - \nabla^2 A$, (b) $\nabla.(A \times B) = (\nabla \times A).B - A.\nabla \times B$ (c) $\nabla \times (A \times B) = (\nabla.B)A - A.\nabla B + B.\nabla A - (\nabla.A)B$.

[40] Lecture schedule for a course on antennas and wave propagation.

[1] Antenna parameters, load resistance, loss resistance, radiation resistance, polarization loss, directivity, maximum effective area.

[2] Maxwell equations in the presence of magnetic charges and magnetic current density.

$$\nabla \times \mathbf{E} = -\mathbf{M} - j\omega\mu\mathbf{H}, \nabla \times \mathbf{H} = \mathbf{J} + j\omega\epsilon\mathbf{E},$$

$$\nabla.\mathbf{E} = \rho/\epsilon, \nabla.\mathbf{H} = \rho_m/\mu$$

[3] Solution to the Maxwell equations in terms of the retarded potentials for sinusoidal sources.

$$\mathbf{A}(t,\mathbf{r}) = \frac{\mu}{4\pi} \int \frac{\mathbf{J}(t - |\mathbf{r} - \mathbf{r}'|/c, \mathbf{r}')}{|\mathbf{r} - \mathbf{r}'|} d^3 r'$$

and for sinusoidal sources

$$\mathbf{A}(\mathbf{r}) = \frac{\mu}{4\pi} \int_{\mathbb{R}^3} \frac{exp(-jk|\mathbf{r} - \mathbf{r}'|)}{|\mathbf{r} - \mathbf{r}'|} \mathbf{J}(\mathbf{r}')d^3 r'$$

[4] Far field amplitude pattern.

$$\mathbf{A}(\mathbf{r}) = \frac{\mu.exp(-jkr)}{r} \int \mathbf{J}(\mathbf{r}')exp(jk\hat{r}.\mathbf{r}')d^3 r'$$

[5] Infinitesimal dipole, computation of the power pattern and radiation resistance.

[6] Infinitesimal circular loop.

[7] Equivalence of magnetic dipole with infinitesimal current loop.

[8] Magnetic charges and magnetic current density, magnetic scalar potential and electric vector potential.

[9] General solutions to the Maxwell equations in the presence of electric current density and magnetic current density.

[10] Reciprocity theorem.

If two sources (J_1, M_1) and (J_2, M_2) produce electromagnetic fields (E_1, H_1) and (E_2, H_2) respectively, then a relationship between these pairs can be derived from the Maxwell equations.

[11] Mutual impedance between antennas.

[12] Yagi array, Analysis of the Yagi array using Pocklington's integral equation.

[13] General theory of antenna arrays, the pattern multiplication theorem.

[14] Aperture antennas.

[15] Horn antennas. Prior to introducing horn antennas, the theory of rectangular waveguides must be covered. The aperture of a horn antenna is fed by a waveduide and hence spatial Fourier transforms of the electromagnetic field in a waveguide yields the desired amplitude pattern of the radiated fields.

[41] Solve the three dimensional Helmholtz equation in cylindrical coordinates using separation of variables.

hint: The cylindrical Helmholtz equation is

$$\rho^{-1}\frac{\partial}{\partial\rho}(\rho.\frac{\partial\psi}{\partial\rho}) + \frac{1}{\rho^2}\frac{\partial^2\psi}{\partial\phi^2} + \frac{\partial^2\psi}{\partial z^2} + k^2\psi = 0$$

where (ρ, ϕ, z) are the cylindrical coordinates and $\psi(\rho, \phi, z)$ is the wave amplitude.

[42] Suppose isotropic antenna sensors are located at the points $d_n\hat{z}, n = 1, 2, ..., N$ where the $d_n's$ are identically distributed independent random variables having means $d_n^{(0)}$ and variances σ_n^2. Assume that the nth sensor carries a current I_n. Then determine (a) the mean value of the amplitude pattern along a given direction and (b) the covariance between the amplitudes along two directions θ_1, θ_2.

[43] State and prove the result of Pocklington's integral equation for a cylindrical wire antenna and explain how it leads to a numerical method for determining the surface current density on the antenna given an incident electromagnetic field.

[44] Let $I(t, \xi)$ be a random current inside a straight wire antenna along the z axis extending from $\xi = -L/2$ to $\xi = L/2$. The autocorrelation function of this current is given by

$$\mathbb{E}(I(t_1, \xi_1).I_2(t_2, \xi_2)) = \rho(t_1 - t_2, \xi_1 - \xi_2)$$

Determine the ensemble averaged radiated power in the far field zone.

hint: The far field magnetic vector potential is given by

$$A_z(r, \theta, t) = \frac{\mu_0}{4\pi r}\int_{-L/2}^{L/2} I(t - r/c + \xi.\cos(\theta)/c, \xi)d\xi$$

The far field θ component of the electric field is given by

$$E_\theta = \sin(\theta)\frac{\partial A_z}{\partial t} = \frac{\mu_0.\sin(\theta)}{4\pi r}\int_{-L/2}^{L/2} I_t(t - r/c + \xi.\cos(\theta)/c, \xi)d\xi$$

where $I_t(t, \xi)$ denotes $\frac{\partial I(t,\xi)}{\partial t}$. The ensembled average power radiated out into space is

$$P = 2\pi\int_0^\pi (|E_\theta|^2/2\eta)r^2.\sin(\theta)d\theta$$

$$= \frac{\mu_0^2}{16\pi\eta}\int_0^\pi\int_{-L/2}^{L/2}\int_{-L/2}^{L/2} \sin^3(\theta)\mathbb{E}(I_t(t - r/c + \xi_1\cos(\theta, \xi_1)I_t(t - r/c + \xi_2\cos(\theta), \xi_2))d\xi_1 d\xi_2 d\theta$$

$$= -\frac{\mu_0^2}{16\pi\eta}\int_0^\pi\int_{-L/2}^{L/2}\int_{-L/2}^{L/2} \sin^3(\theta)\rho_{tt}((\xi_1 - \xi_2)\cos(\theta), \xi_1 - \xi_2)d\xi_1 d\xi_2 d\theta$$

Note that we have made use of the identity

$$\mathbb{E}(I_t(t_1,\xi_1)I_t(t_2,\xi_2)) = \frac{\partial^2}{\partial t_1 \partial t_2}\mathbb{E}(I(t_1,\xi_1)I(t_2,\xi_2)) = -\rho_{tt}(t_1-t_2,\xi_1-\xi_2)$$

where $\rho_{tt}(t,\xi) = \frac{\partial \rho(t,\xi)}{\partial t^2}$.

[45] A rectangular aperture is fed with a waveguide carrying the TE_{10} mode. Determine the far field radiation pattern and the total power radiated by this aperture in the far field.

hint: Let Assume that the aperture is defined by $0 \le x \le a, 0 \le y \le b$. The electric field in the aperture is given by $\mathbf{E} = E_0 sin(\pi x/a)\hat{y}$ for $0 \le x \le a$. The equivalent surface magnetic current density is given by (on making use of Babinet's principle) $\mathbf{M}_S = -2\hat{z} \times \mathbf{E} = 2E_0 sin(\pi x/a)\hat{x}$ and the far field electric vector potential is then given by

$$\mathbf{F} = \hat{x}\frac{\mu.E_0 exp(-jkr)}{4\pi r} \int_0^a \int_0^b sin(\pi x'/a).exp(jk(x'cos(\phi)sin(\theta) + y'sin(\phi)sin(\theta)))dx'dy'$$

$$= \hat{x}\frac{\mu E_0 exp(-jkr)}{4\pi r}[\int_0^a sin(\pi x'/a)exp(jkx'cos(\phi)sin(\theta))dx'][\int_0^b exp(jky'sin(\phi)sin(\theta))dy']$$

we can write this equation as

$$\mathbf{F} = F_x\hat{x}, F_x = \psi(\theta,\phi)\frac{exp(-jkr)}{r}$$

Then, the spherical polar components of $\mathbf{F}$ are given by

$$F_r = F_x.cos(\phi)sin(\theta), F_\theta = F_x.cos(\phi)cos(\theta), F_\phi = -F_x sin(\phi)$$

From these equations, we can compute $\mathbf{E} = \epsilon^{-1}\nabla \times \mathbf{F}$ and $\mathbf{H}$. It can be shown that as $r \to \infty$, E_r, H_r are $O(1/r^2)$. We must compute $E_\theta, E_\phi, H_\theta, H_\phi$ upto $O(1/r)$ and determine the power radiated by the aperture in the far field:

$$P = \frac{1}{2}\int_0^\pi \int_0^{2\pi} Re(E_\theta H_\phi^* - E_\phi H_\theta^*)r^2 sin(\theta)d\theta d\phi$$

[46] Determine the two dimsensional Green's function $G(x,y|x',y')$ in free space satisfying the equation

$$(\frac{\partial^2}{\partial x^2} + \frac{\partial^2}{\partial y^2} + k^2)G(x,y|x',y') = \delta(x-x')\delta(y-y')$$

ans

$$G(x,y|x',y') = -\frac{1}{4\pi}\int_{-\infty}^\infty \frac{exp(-jk\sqrt{(x-x')^2 + (y-y')^2 + \xi^2})}{\sqrt{(x-x')^2 + (y-y')^2 + \xi^2}}d\xi$$

[47] A charge moves along a trajectory that can be modeled as a diffusion process. Compute the correlations in the magnetic vector potential at a given point in space at two different times. Let $\mathbf{R}(t)$ denote the position of the charge at time t and $\mathbf{V}(t)$ the velocity. They satisfy the stochastic differential equation

$$d\mathbf{R}(t) = \mathbf{V}(t)dt, \, d\mathbf{V}(t) = \mathbf{f}(\mathbf{R}(t), \mathbf{V}(t))dt + \mathbf{G}(\mathbf{R}(t), \mathbf{V}(t))d\mathbf{B}(t)$$

where $\mathbf{B}(t)$ is a standard three component Brownian motion process. The magnetic vector potential at a given point $\mathbf{r}$ in space is approximately given by

$$\mathbf{A}(t, \mathbf{r}) = \frac{\mu Q \mathbf{V}(t)}{4\pi |\mathbf{r} - \mathbf{R}(t)|}$$

and the problem is to obtain an evolution equation for $< A_\alpha(t, \mathbf{r}) >$ and $< A_\alpha(t, \mathbf{r})A_\beta(t, \mathbf{r}) >$. The equations for evolution of these quantities can be obtained

$$dA_\alpha(t) = \frac{\mu.Q}{4\pi} \frac{dV_\alpha(t)}{|\mathbf{r} - \mathbf{R}(t)|} + \frac{\mu.Q}{4\pi} \frac{V_\alpha(t)dX_\beta(t)(x_\beta - X_\beta(t))}{|\mathbf{r} - \mathbf{R}(t)|^3}$$

We have written $A_\alpha(t)$ in place of $A_\alpha(t, \mathbf{r})$. By the Ito differential rule,

$$d(A_\alpha(t).A_\beta(t)) = A_\alpha dA_\beta + A_\beta dA_\alpha + dA_\alpha.dA_\beta$$

Note that

$$dA_\alpha.dA_\beta = \frac{\mu^2 Q^2}{16\pi^2 |\mathbf{r} - \mathbf{R}(t)|^2} \sum_\rho G_{\alpha\rho} G_{\beta\rho} dt$$

Thus,

$$< dA_\alpha.dA_\beta > = \frac{\mu^2 Q^2}{16\pi^2} < |\mathbf{r} - \mathbf{R}(t)|^{-2} \sum_\rho G_{\alpha\rho}(\mathbf{R}(t), \mathbf{V}(t))G_{\beta\rho}(\mathbf{R}(t), \mathbf{V}(t)) > dt$$

[48] Compute $df(X(t), V(t))$ where

$$dX(t) = V(t)dt, \, dV(t) = -\gamma.V(t)dt - U'(X(t)) + \sigma.dB(t)$$

Assume that the second order partial derivatives of $f(x, v)$ exist. $B(.)$ is a standard Brownian motion process.

[49] A linear antenna of length L is fed by a random current $I_0(t)$ at $z = L/2$. The current distribution on the antenna is then $I(t, \xi) = I_0(t)sin(\pi\xi/L)$. Determine the far field correlations in the electromagnetic field at a given point in space at two time instants t_1, t_2. Your answer must be expressed in terms of the current correlations $\rho_0(t, s) = \mathbb{E}(I_0(t).I_0(s))$.

[50] Explain how you would solve the Poisson equation $\nabla^2 V(x) = s(x)$ in a two dimensional domain with specified boundary conditions approximately using the finite element method.

hint:The Poisson equation can be derived from the variational principle $\delta S[V] = 0$, where

$$S[V] = \frac{1}{2}\int_D |\nabla V|^2 dx + \int_D sV\,dx$$

[51] Explain why the electric field inside a perfect conductor must vanish.

hint:If there is a nonzero electric field inside the perfect conductor, then highly mobile charges will orient themselves to cancel this electric field.

[52] Determine the electric and magnetic fields produced by a biconical antenna. In the spherical polar coordinate system, the electric field phasor has the form $\mathbf{E} = E_\theta(r,\theta)\hat{\theta}$ and the magnetic field has the form $\mathbf{H} = H_\phi(r,\theta)\hat{\phi}$. Formulate the Maxwell curl equations

$$\nabla \times \mathbf{E} = -j\omega\mu\mathbf{H}, \nabla \times \mathbf{H} = j\omega\epsilon\mathbf{E}$$

and obtain the general solutions for E_θ and H_ϕ. Note that the curl of a vector field $\mathbf{F}$ in spherical polar coordinates is given by

$$\nabla \times \mathbf{F} = det\begin{pmatrix} \frac{\hat{r}}{r^2 sin(\theta)} & \frac{\hat{\theta}}{r.sin(\theta)} & \frac{\hat{\phi}}{r} \\ \frac{\partial}{\partial r} & \frac{\partial}{\partial \theta} & \frac{\partial}{\partial \phi} \\ F_r & r.F_\theta & r.sin(\theta).F_\phi \end{pmatrix}$$

[53] Derive the partial differential equations (wave equation with source) satisfied by the electric scalar and magnetic vector potentials for a given charge density and current density by making use of the Lorentz gauge transformation.

[54] Give a numerical algorithm for determining the eigenvalues and eigenfunctions for the problem

$$(\nabla^2 + \lambda.f(\mathbf{x}))u(\mathbf{x}) = 0$$

inside a bounded three dimensional volume with vanishing boundary conditions for u on the surface. Discuss both the finite element method and the method of moments.

[55] Discuss linear, circular and elliptic polarization of the electric field.

hint: In linear polarization, the electric field at the given point has the form

$$\mathbf{E} = A.cos(\omega t)\hat{x} + B.cos(\omega t)\hat{y}$$

where A, B are real constants. Then we have

$$\frac{E_y}{E_x} = \frac{B}{A}$$

which means that the graph of E_x versus E_y is a straight line of slope B/A. In circular polarization,

$$\mathbf{E} = A.cos(\omega t)\hat{x} + A.sin(\omega t)\hat{y}$$

where A is a real constant. Then,

$$E_x^2 + E_y^2 = A^2$$

which means that the graph of E_x versus E_y is a circle. In elliptic polarization, after appropriate rotation of the axis, we can write the electric field as

$$\mathbf{E} = A.cos(\omega t \hat{x} + B.sin(\omega t)\hat{y}$$

and then

$$\frac{E_x^2}{A^2} + \frac{E_y^2}{B^2} = 1$$

which is the equation of an ellipse.

[56] Determine the equations of the electric lines of force inside a rectangular waveguide carrying the TE_{10} mode. Note that if the electric field phasor is expressed as $(E_x(x,y)\hat{x} + E_y(x,y)\hat{y})exp(-\gamma z)$, then the differential equation for the lines of force is given by

$$\frac{dx}{E_x(x,y)} = \frac{dy}{E_y(x,y)}$$

[57] Consider the E plane sectoral horn antenna with dimensions of the rectangular waveguide feeding it being a, b. The z axis is the axis of the waveguide. The TE_{10} mode that propagates has the electric field given by

$$E_x = 0, E_y = A.cos(\pi x/a)exp(-j\beta z), -a/2 \leq x \leq a/2, -b/2 \leq y \leq b/2$$

with the propagation constant β obtained from the Helmholtz equation

$$\nabla^2 E_y + \omega^2 \mu \epsilon . E_y = 0$$

or

$$-\beta^2 - (\pi/a)^2 + \omega^2 \mu \epsilon = 0$$

so that

$$\beta = (\omega^2 \mu \epsilon - (\pi/a)^2)^{1/2}$$

The magnetic field is obtained from the Maxwell equation

$$\nabla \times \mathbf{E} = -j\omega\mu\mathbf{H}$$

This gives

$$-j\omega\mu H_x = E_{z,y} - E_{y,z} = j\beta E_y, -j\omega\mu H_y = E_{x,z} - E_{z,x} = 0$$

so that

$$H_x = -(\beta/\omega\mu)E_y = -E_y/Z$$

where

$$Z = \omega\mu(\omega^2\mu\epsilon - (\pi/a)^2)^{-1/2} = (\epsilon/\mu - (\pi^2/(a\omega\mu)^2))^{-1/2}$$

$$= (\mu/\epsilon)^{1/2}(1 - \frac{\mu\pi^2}{\epsilon(a\omega\mu)^2})^{-1/2}$$

Let R denote the radius of curvature of the horn. This curvature is along the y axis. The path difference due to the curvature at a point on the radiating plane having y as the y-coordinate is given by

$$\delta(y) = \sqrt{R^2 + y^2} - R \approx y^2/2R$$

Thus, taking into account this path difference, the electric field phasor at the aperture of the horn is given by

$$E_y = A.cos(\pi x/a).exp(-jk\delta(y)), \, -a/2 \le y \le a/2, -b_1/2 \le y \le b_1/2$$

where b_1 is the width of the aperture along the y axis. Note that $b_1 > b$. We are assuming that the end of the waveguide that feeds the horn is the plane $z = 0$. Here $k = \omega\sqrt{\mu\epsilon}$. The equivalent magnetic current density on the surface of the horn is has only the x component non-vanishing. It is given by

$$\mathbf{M}_S = -\hat{z} \times \mathbf{E} = E_y\hat{x}$$

The electric vector potential in the far field apprxoimation is given by

$$\mathbf{F} = \frac{\mu.exp(-jkr)}{4\pi.r} \int_{-a/2}^{a/2} \int_{-b_1/2}^{b_1/2} \mathbf{M}_S(x', y')exp(jk(x'.cos(\phi)sin(\theta) + y'.sin(\phi)sin(\theta)))dx'dy'$$

We have $\mathbf{F} = F_x\hat{x}$ where

$$F_x(r, \theta, \phi) = \frac{\mu.exp(-jkr)}{4\pi.r} \int_{-a/2}^{a/2} \int_{-b_1/2}^{b_1/2} A.cos(\pi x'/a).exp(-jky'^2/2R)$$

$$exp(jk(x'.cos(\phi)sin(\theta) + y'.sin(\phi)sin(\theta)))dx'dy'$$

[58] In the previous problem, derive the expression for the far field magnetic vector potential by considering the equivalent surface current density due to the magnetic field at the aperture.

[59] The parametric equation of a spiral is given by $x = R.cos(\phi), y = R.sin(\phi), z = a\phi$ where ϕ is the variable parameter and a is a constant. This spiral carries a current phasor I_0 at frequency ω. Calculate the power radiated by this antenna in the far field region.

[60] Describe the Navier-Stokes equation for magnetohydrodynamics, ie, when the fluid carries charge which generates the electromagnetic field.

hint:

$$\rho(\mathbf{v}_{,t} + \mathbf{v}.\nabla\mathbf{v}) = -\nabla p + \rho_e(\mathbf{E} + \mathbf{v}\times\mathbf{B}) + \eta\nabla^2\mathbf{v}$$

where ρ_e is the charge density and $\mathbf{J} = \sigma(\mathbf{E} + \mathbf{v}\times\mathbf{B})$ is the conduction current density in the moving frame of the fluid. The Maxwell equations supplement this in the form

$$\nabla.\mathbf{E} = \rho_e/\epsilon, \nabla\times\mathbf{H} = \mathbf{J} + \epsilon.\mathbf{E}_{,t}, \nabla.\mathbf{H} = 0, \nabla\times\mathbf{E} = -\mu\mathbf{H}_{,t}$$

with $\mathbf{B} = \mu\mathbf{H}$. There is also a convection current density $\rho_e\mathbf{v}$. It is not clear whether this contribution is to be added to the conduction current density in the Maxwell equations. Clarify this concept. Suppose that the displacement current is ignored. Then the second Maxwell equation above gives

$$\nabla\times\mathbf{H} = \sigma(\mathbf{E} + \mathbf{v}\times\mathbf{B})$$

and taking the curl of this equation and using the third Maxwell equation, this gives

$$\nabla^2\mathbf{H} - \mu\sigma\mathbf{H}_{,t} + \mu\sigma\nabla\times(\mathbf{v}\times\mathbf{H}) = 0$$

This is essentially the diffusion equation with a convection term.

[61] Derive the radiation pattern of a microstrip antenna made out of a cylindrical cavity in a substrate such that the flat circular ends of the antenna are perfect electric conductors while the curved side walls are perfect magnetic conductors. Assume that the transverse magnetic (TM) mode is present inside the antenna cavity.

hint:The magnetic vector potential for the transverse magnetic mode has the form $\mathbf{A} = A_z(\rho,\phi,z)\hat{z}$. A_z satisfies the Helmholtz equation and the boundary conditions are that the tangential components of the electric field vanish on the top and bottom circular flat surfaces (since a perfect electric conductor cannot support any surface magnetic current density) and the tangential components of the magnetic field vanish on the side walls (since a perfect magnetic conductor cannot support any surface electric current density).

[62] Consider Maxwell's equations in a medium in which the permittivity ϵ is a function of the electric field $\mathbf{E}$. We write this relation as $\epsilon = \epsilon_0(1 + \chi(\mathbf{E}))$ where χ is small. We make a Taylor expansion of $\chi(\mathbf{E})$ in powers of the electric field as

$$\chi(\mathbf{E}) = \sum_{r_1^{\,\varsigma},r_2,r_3=0}^{\infty} \chi[r_1,r_2,r_3]E_1^{r_1}E_2^{r_2}E_3^{r_3}/r_1!r_2!r_3!$$

We retain only the linear terms so that

$$\chi(\mathbf{E}) \approx \sum_{\alpha=1}^{3}\chi_\alpha E_\alpha = \chi_1 E_1 + \chi_2 E_2 + \chi_3 E_3$$

Maxwell's equations in such a medium are

$$\nabla.(\epsilon\mathbf{E}) = 0, \nabla\times\mathbf{E} = -\mu\mathbf{H}_{,t}, \nabla\times\mathbf{H} = (\epsilon\mathbf{E})_{,t}, \nabla.\mathbf{H} = 0$$

From these equations, we have

$$\nabla.\mathbf{E} = -(\nabla\epsilon/\epsilon, \mathbf{E})$$

and

$$\nabla^2\mathbf{E} - \mu(\epsilon\mathbf{E})_{,tt} + \nabla(\nabla\epsilon/\epsilon, \mathbf{E}) = 0$$

Note that $(\mathbf{A}, \mathbf{B})$ is the same as $\mathbf{A}.\mathbf{B}$, the scalar or dot product of the vector $\mathbf{A}$ with the vector $\mathbf{B}$. Substituting

$$\epsilon = \epsilon_0(1 + \chi_1 E_1 + \chi_2 E_2 + \chi_3 E_3)$$

in this equation, and retaining upto only quadratic terms in the electric field, we get

$$\nabla^2 E_\alpha - \mu\epsilon_0 E_{\alpha,tt} - \mu\epsilon_0 \sum_\beta \chi_\beta (E_\alpha E_\beta)_{,tt}$$

$$+ \sum_{\beta,\rho} \chi_\beta (E_{\beta,\rho}, E_\rho)_{,\alpha} = 0$$

We assume in this equation that the quadratic terms constitute a small perturbation to the linear terms. Thus we introduce a small parameter δ as a multiplier in the quadratic terms and obtain an approximate solution to this equation using first order perturbation theory. The equation is written as

$$\nabla^2 E_\alpha - \mu\epsilon_0 E_{\alpha,tt} - \delta\mu\epsilon_0 \sum_\beta \chi_\beta (E_\alpha E_\beta)_{,tt}$$

$$+\delta \sum_{\beta,\rho} \chi_\beta (E_{\beta,\rho}, E_\rho)_{,\alpha} = 0$$

Set

$$E_\alpha = E_\alpha^{(0)} + \delta.E_\alpha^{(1)}$$

so that $E_\alpha^{(0)}$ is the zeroth order perturbation term and $E_\alpha^{(1)}$ the first order perturbation term. Then equating coefficients of $\delta^0 = 1$ and $\delta^1 = \delta$ in the equation gives us

$$\nabla^2 E_\alpha^{(0)} - \mu\epsilon_0 E_{\alpha,tt}^{(0)} = 0,$$

$$\nabla^2 E_\alpha^{(1)} - \mu\epsilon_0 E_{\alpha,tt}^{(1)} - \mu\epsilon_0 \sum_\beta (E_\alpha^{(0)} E_\beta^{(0)})_{,tt}$$

$$+ \sum_{\beta,\rho} \chi_\beta (E_{\beta,\rho}^{(0)}, E_\rho^{(0)})_{,\alpha} = 0$$

[63] Derive the nonlinear wave equations satisfied by the electromagnetic field in a medium in which the permittivity is a function of the electric field and the permeability is a function of the magnetic field.

hint.

The Maxwell equations are

$$div(\epsilon.\mathbf{E}) = 0, \nabla \times \mathbf{E} = -(\mu\mathbf{H})_{,t},$$

$$\nabla \times \mathbf{H} = (\epsilon \mathbf{E})_{,t}, \; div(\mu \mathbf{H}) = 0$$

where $\epsilon = \epsilon(\mathbf{E})$ and $\mu = \mu(\mathbf{H})$. Take the curl of the second and third equations and make use of the following identities.

$$div(\epsilon \mathbf{E}) = (\nabla \epsilon, \mathbf{E}) + \epsilon.div\mathbf{E},$$

$$div(\mu \mathbf{H}) = (\nabla \mu, \mathbf{H}) + \mu.div\mathbf{H},$$

$$\nabla \times (\epsilon \mathbf{E})_{,t} = \nabla \times (\epsilon_{,t} \mathbf{E} + \epsilon.\mathbf{E}_{,t})$$

$$= \nabla \epsilon_{,t} \times \mathbf{E} + \epsilon_{,t} \nabla \times \mathbf{E} + \nabla \epsilon \times \mathbf{E}_{,t} + \epsilon \nabla \times \mathbf{E}_{,t}$$

$$= \nabla \epsilon_{,t} \times \mathbf{E} - \epsilon_{,t}(\mu \mathbf{H})_{,t} + \nabla \epsilon \times \mathbf{E}_{,t} - \epsilon(\mu \mathbf{H}_{,tt}$$

with a similar equation for $\nabla \times (\mu \mathbf{H})_{,t}$.

[64] Propagation of electromagnetic waves in an anisotropic medium. Assume that the permeability μ is a constant scalar while the permittivity ϵ is a constant 3×3 symmetric matrix which can be assumed to be diagonal by rotating the axes. Thus $\epsilon = diag[\epsilon_x, \epsilon_y, \epsilon_z]$. The Maxwell equations are

$$\nabla \times E = -\mu H_{,t}, \; \nabla \times H = \epsilon.E_{,t}$$

Thus,

$$\nabla(\nabla.E) - \nabla^2 E = -\mu \nabla \times H_{,t} = -\mu\epsilon.E_{,tt}$$

or

$$\nabla^2 E - \mu\epsilon.E_{,tt} - \nabla(\nabla.E) = 0$$

Assuming that

$$E = E_0 exp(i(\omega t - k.r))$$

we get

$$-k^2 E_0 + \omega^2 \mu(\epsilon.E_0) - k(k.E_0) = 0$$

In matrix notation,

$$(-k^2 I + \omega^2 \mu\epsilon - kk^T)E_0 = 0$$

so that the dispersion relation has the form

$$det(-k^2 \mathbf{I} + \omega^2 \mu\epsilon - \mathbf{k}\mathbf{k}^T) = 0$$

[65] The Maxwell equations in a medium having permittivity ϵ and permeability μ and in the absence of external charges and currents are

$$divD = 0, \; divB = 0, \; curlE = -B_{,t}, \; curlH = J + D_{,t}$$

Now form two 4×4 antisymmetric tensors respectively out of (E, B) and (D, H). Denote these tensors by $F^{\mu\nu}$ and $H^{\mu\nu}$ respectively. Show that the second and third Maxwell equations above can be expressed in covariant language as

$$F_{\mu\nu,\sigma} + F_{\nu\sigma,\mu} + F_{\sigma\mu,\nu} = 0$$

Show that the first and fourth Maxwell equations above can be expressed in covariant language as

$$H^{\mu\nu}_{\ ,\nu} = 0$$

What is the covariant version of the constitutive relations $D = \epsilon.E$ and $B = \mu.H$ in terms of the tensors $F^{\mu\nu}, H^{\mu\nu}$ and the four velocity $(u^{\mu}) = (\gamma, \gamma v)$ of the medium where $\gamma = (1 - v^2/c^2)^{1/2}$. In the special case when v/c can be neglected, these covariant constitutive relations should reduce to the noncovariant constitutive relations $D = \epsilon.E, B = \mu.H$.

Reference:Electrodynamics of continuous media, by Landau, Lifshitz and Pitaevskii.

[66] Prove that the energy density of the electromagnetic field is given by

$$U = \frac{\epsilon_0}{2}|\mathbf{E}|^2 + \frac{|\mathbf{B}|^2}{2\mu_0}$$

Also prove that the energy flux, ie, the energy flowing per unit time per unit area in the electromagnetic field is given by $\epsilon_0 c^2 \mathbf{E} \times \mathbf{B}$ where $\mathbf{E}$ and $\mathbf{B}$ are respectively the electric and magnetic fields.

[67] Let $V(t,\mathbf{r})$ be the electric scalar potential and $\mathbf{A}(t,\mathbf{r})$ the magnetic vector potential. Derive the partial differential equations satisfied by these two in the presence of a charge density $\rho(t,\mathbf{r})$ and current density $\mathbf{J}(t,\mathbf{r})$. Assume that the potentials satisfy the Lorentz Gauge condition $div\mathbf{A} + c^{-2}\frac{\partial V}{\partial t} = 0$.

[68] Show that the general solution to the partial differential equation

$$\nabla^2 V(t,\mathbf{r}) - \frac{1}{c^2}V_{,tt}(t,\mathbf{r}) = -\rho(t,\mathbf{r})$$

with vanishing boundary conditions at $r = \infty$ is given by a linear combination of the retarded and advanced potentials. The retarded potential is

$$V_{ret}(t,\mathbf{r}) = \int \frac{\rho(t - |\mathbf{r} - \mathbf{r}'|/c, \mathbf{r}')}{4\pi|\mathbf{r} - \mathbf{r}'|}d^3r'$$

and the advanced potential by

$$V_{adv}(t,\mathbf{r}) = \int \frac{\rho(t + |\mathbf{r} - \mathbf{r}'|/c, \mathbf{r}')}{|\mathbf{r}' - \mathbf{r}'|}d^3r'$$

[69] Consider a conducting fluid which on neglect of the viscosity and displacement current term, is governed by the Navier-Stokes and Maxwell equations

$$\rho(\mathbf{v}.\nabla\mathbf{v} + \mathbf{v}_{,t}) = -\nabla p + \mathbf{J} \times \mathbf{B}$$

$$curl\mathbf{B} = \mu\sigma(\mathbf{E} + \mathbf{v} \times \mathbf{B}) = \mu\mathbf{J}$$

In the limit $\sigma \to \infty$, we get

$$\mathbf{E} = -\mathbf{v} \times \mathbf{B}$$

and

$$\rho(\mathbf{v}.\nabla\mathbf{v} + \mathbf{v}_{,t}) = -\nabla p + (curl\mathbf{B}) \times \mathbf{B}/\mu$$

Combine this with the mass conservation equation

$$\rho_{,t} + div(\rho\mathbf{v}) = 0$$

and show that the Navier-Stokes equation can be expressed as

$$(\rho v_i)_{,t} + \Pi_{ij,j} = 0$$

where Π_{ij} consists of three terms, the momentum flux tensor of the liquid, the pressure tensor and the momentum flux tensor of the magnetic field.

hint: Use the vector identity

$$(curl\mathbf{B}) \times \mathbf{B} = \mathbf{B}.\nabla\mathbf{B} - \frac{1}{2}\nabla(B^2)$$

and show using $div\mathbf{B} = 0$ that the ith component of the right side can be expressed as

$$(B_i B_j - \frac{1}{2}B^2 \delta_{ij})_{,j}$$

Reference:Electrodynamics of Continuous Media, by L.D.Landau, E.M.Lifshitz and Pitaevskii.

[70] Determine the cutoff frequencies for the transverse magnetic modes propagating inside a cylindrical waveguide of radius R when the medium inside has permittivity ϵ and permeability μ.

[71] If the magnetic field $\mathbf{B}$ in space satisfies $div\mathbf{B} \neq 0$, then magnetic monopoles must exist. Justify this statement by taking the example of Gauss' law for the electric field due to a charge distribution.

[72] Derive the finite difference approximation to Poisson's equation in $\mathbb{R}^2$.

hint. Let $\phi(x,y)$ be the potential and $s(x,y)$ the source. Then Poisson's equation for ϕ is given by

$$\phi_{,xx} + \phi_{,yy} = s$$

We discretize with pixel size Δ and set $\phi[n,m] = \phi(n\Delta, m\Delta)$ and $s[n,m] = s(n\Delta, m\Delta)$. Then Poisson's equation approximates to give

$$(\phi[n+1,m] - 2\phi[n,m] + \phi[n-1,m])/\Delta^2 + (\phi[n,m+1] - 2\phi[n,m] + \phi[n,m-1])/\Delta^2 = s[n,m], (n,m) \in D$$

or

$$\phi[n+1,m] + \phi[n,m+1] + \phi[n-1,m] + \phi[n,m-1] - 4\phi[n,m] = \Delta^2 s[n,m]$$

This equation can be implemented recursively using the scheme

$$\phi^{(k+1)}[n,m] = -\frac{\Delta^2}{4}s[n,m] + \frac{1}{4}(\phi^{(k)}[n+1,m] + \phi^{(k)}[n,m+1] + \phi^{(k)}[n-1,m] + \phi^{(k)}[n,m-1])$$

We start with an initial guess $\phi^{(0)}[n,m], (n,m) \in D$ and compute recursively $\phi^{(k)}[n,m], (n,m) \in D$ with $k = 1, 2, ...$ For sufficiently large k, convergence is likely to occur and we stop. Note that this difference equation can be cast in matrix notation as

$$\phi^{(k+1)} = A\phi^{(k)} + s, k = 0, 1, 2, ...$$

where $\phi^{(k)}$ is a vector for each k and A is a square matrix. s is also a vector. If convergence takes place, then clearly

$$\phi^{(\infty)} = (I - A)^{-1}s$$

The condition for convergence is that all the eigenvalues of the matrix A must be smaller than one in magnitude. To show this, we note that if A is nonsingular, then the difference equation yields

$$A^{-k-1}\phi^{(k+1)} - A^{-k}\phi^{(k)} = A^{-k-1}s$$

Summing from $k = 0$ to $k = n - 1$ gives

$$A^{-n}\phi^{(n)} - \phi^{(0)} = \sum_{k=0}^{n-1} A^{-k-1}s = \sum_{k=1}^{n} A^{-k}s$$

or

$$\phi^{(n)} = A^n\phi^{(0)} + \sum_{k=0}^{n-1} A^{n-k-1}s = A^n\phi^{(0)} + (I - A)^{-1}(I - A^n)s$$

This solution is correct even if A is singular. If all the eigenvalues of A are smaller than one in magnitude, then $A^n \to 0$ as $n \to \infty$ and we see that $\phi^{(n)}$ converges to $(I - A)^{-1}s$.

[73] Let $\rho(\mathbf{x})$ be the spatial charge density and $V(\mathbf{x})$ the electrostatic potential. V satisfies the equation

$$\nabla^2 V(\mathbf{x}) = -\rho(\mathbf{x})/\epsilon$$

Solve this equation using the three dimensional spatial Fourier transform and show that it leads to Coulomb's law.

[74] Let $\mathbf{E}(t,\mathbf{r})$ be the electric field in space and $\mathbf{H}(t,\mathbf{r})$ the magnetic field. Derive integral expressions for these fields in terms of the charge density $\rho(t,\mathbf{r})$ and the current density $\mathbf{J}(t,\mathbf{r})$ starting from the Maxwell equations. If $\rho(t,\mathbf{r})$ and $\mathbf{J}(t,\mathbf{r})$ are Gaussian random fields with correlations

$$\mathbb{E}(\rho(t,\mathbf{r})\rho(t',\mathbf{r}')) = R_\rho(t - t', \mathbf{r} - \mathbf{r}'),$$

$$\mathbb{E}(\rho(t,\mathbf{r})J_i(t',\mathbf{r}')) = R_{\rho Ji}(t - t', \mathbf{r} - \mathbf{r}'),$$

$$\mathbb{E}(J_i(t,\mathbf{r})J_k(t',\mathbf{r}')) = R_{Jik}(t - t', \mathbf{r} - \mathbf{r}')$$

then derive integral expressions for the correlations in the electromagnetic field, ie,

$$\mathbb{E}(E_i(t,\mathbf{r}').E_j(t',\mathbf{r}')), \mathbb{E}(E_i(t,\mathbf{r}).H_k(t',\mathbf{r}')), \mathbb{E}(H_i(t,\mathbf{r}).H_k(t',\mathbf{r}'))$$

hint:Express the electrostatic potential V and the magnetic vector potential $\mathbf{A}$ in terms of $\rho, \mathbf{J}$ using the retarded potentials. Then compute from these expressions

$$\mathbf{E} = -\nabla V - \mathbf{A}_{,t}, \mathbf{H} = \mu_0^{-1}\nabla \times \mathbf{A}$$

[75] Suppose $\rho(\mathbf{r})$ is a Gaussian random field with zero mean and correlation

$$\mathbb{E}(\rho(\mathbf{r}_1)\rho(\mathbf{r}_2)) = R_\rho(|\mathbf{r}_1 - \mathbf{r}_2|)$$

calculate the correlation and n^{th} order moment of the electrostatic potential field

$$V(\mathbf{r}) = \frac{1}{4\pi\epsilon}\int \frac{\rho(\mathbf{r}')}{|\mathbf{r} - \mathbf{r}'|}d^3r'$$

where n is an even integer. Note that the nth order moment of the potential field is defined by

$$M_{V,n}(\mathbf{r}_1, ..., \mathbf{r}_n) = \mathbb{E}(V(\mathbf{r}_1)...V(\mathbf{r}_n))$$

[76] Consider a point charge Q moving along a random twice differentiable trajectory $\mathbf{R}(t)$ such that all the moments $\mathbb{E}(X_{i_1}(t_1)...X_{i_r,t}(t_r))$ are known where $\mathbf{R}(t) = (X_1(t), X_2(t), X_3(t))$. Derive an expression for the moments of the electric and magnetic fields by making use of the retarded potentials.

hint: Let $\mathbf{R}'(t) = \mathbf{V}(t)$. Then the magnetic vector potential is given by

$$\mathbf{A}(t,\mathbf{r}) = \frac{\mu Q\mathbf{V}(t')}{4\pi(|\mathbf{r} - \mathbf{R}(t')| - (\mathbf{V}(t'), (\mathbf{r} - \mathbf{R}(t'))/c)}$$

where t' is the retarded time defined as the solution to the equation

$$t' = t - |\mathbf{r} - \mathbf{R}(t')|/c$$

Assume that $|\mathbf{r}| >> |\mathbf{R}(t)|$ and $|\mathbf{V}(t)| << c$ for all t and then make a Taylor expansion.

[77] Assume that a sphere of radius R with centre at the origin has a surface charge density $\sigma(\theta)$. Determine a series expansion for the electrostatic potential produced by this sphere at a point $r > R$.

hint:

$$V(r) = \frac{1}{2\epsilon}\int_0^\pi \frac{\sigma(\theta)sin(\theta)d\theta}{\sqrt{r^2 + R^2 - 2rR.cos(\theta)}}$$

and for $|\xi| < 1$,

$$(1 + \xi^2 - 2\xi.cos(\theta))^{-1/2} = \sum_{l=0}^{\infty} \xi^l.P_l(cos(\theta))$$

where $\{P_l\}$ are the Legendre polynomials.

[78] Assume that the electric field in space is

$$\mathbf{E}(t, \mathbf{r}) = \mathbf{E}_0.cos(\omega t - \mathbf{k}.\mathbf{r})$$

The Maxwell equation $div\mathbf{E} = 0$ implies $\mathbf{k}.\mathbf{E}_0 = 0$. The Maxwell equation

$$\nabla \times \mathbf{E} = -\mu\mathbf{H}_{,t}$$

gives

$$\mathbf{H}_{,t} = -\mu^{-1}\mathbf{k} \times \mathbf{E}_0.sin(\omega t - \mathbf{k}.\mathbf{r})$$

and hence

$$\mathbf{H}(t, \mathbf{r}) = (\omega\mu)^{-1}\mathbf{k} \times \mathbf{E}_0.cos(\omega t - \mathbf{k}.\mathbf{r})$$

The time averaged Poynting vector is

$$< \mathbf{S} >=< \mathbf{E} \times \mathbf{H} >= (\omega\mu)^{-1}\mathbf{E}_0 \times (\mathbf{k} \times \mathbf{E}_0) < cos^2(\omega t - \mathbf{k}.\mathbf{r}) >$$

$$= |\mathbf{E}_0|^2\mathbf{k}/(2\omega\mu)$$

The time averaged power flowing through a patch of the plane $\mathbf{n}.\mathbf{r} = p$ having area A, with $\mathbf{n}$ being the unit normal to the plane is thus given by

$$P =< \mathbf{S} > .\mathbf{n}A = |\mathbf{E}_0|^2\mathbf{n}.\mathbf{k}A/(2\omega\mu)$$

[79] Let $\mathbf{A} : \mathbb{R}^3 \to \mathbb{R}^3$ be such that $\mathbf{k}.\mathbf{A}(\mathbf{k}) = 0$ for all $\mathbf{k} \in \mathbb{R}^3$. Then define

$$\mathbf{E}(t, \mathbf{r}) = \int \mathbf{A}(\mathbf{k})exp(i(|\mathbf{k}|ct - \mathbf{k}.\mathbf{r}))d^3k$$

Show that

$$div\mathbf{E} = 0, c^2\nabla^2\mathbf{E} - \mathbf{E}_{,tt} = 0$$

[80] Let a, b, c be positive real numbers. Solve the generalized Poisson equation

$$a.\phi_{,xx} + b.\phi_{,yy} + c.\phi_{,zz} = -\rho$$

$\phi(x, y, z)$ can be interpreted as the potential produced by the charge density $\rho(x, y, z)$ in an anisotropic medium, ie, in a medium whose permittivity depends on the direction.

hint: Define

$$x' = x/\sqrt{a}, y' = y/\sqrt{b}, z' = z/\sqrt{c}$$

and then show that with respect to these variables, ϕ satisfies the Poisson equation, ie,

$$\phi_{,x'x'} + \phi_{,y'y'} + \phi_{,z'z'} = -\rho$$

[81] Simulate the motion of a conducting fluid in a magnetic field, ie, the magnetohydrodynamic (mhd) equations. Let σ be the conductivity. Then, if $\mathbf{H}$ is the magnetic field, $\mathbf{E}$ the electric field and $\mathbf{v}$ the velocity field of the fluid, $\mathbf{J} = \sigma(\mathbf{E} + \mathbf{v} \times \mathbf{B})$ where $\mathbf{B} = \mu\mathbf{H}$. Then the force per unit volume on this current is given by

$$\mathbf{J} \times \mathbf{B} = \sigma(\mathbf{E} + \mathbf{v} \times \mathbf{B}) \times \mathbf{B}$$

If we neglect the electric field, we get

$$4\mathbf{J} \times \mathbf{B} = \sigma(\mathbf{v} \times \mathbf{B}) \times \mathbf{B}$$

The Navier Stokes equation with this external force reads

$$\rho(\mathbf{v}.\nabla\mathbf{v} + \mathbf{v}_{,t}) = -\nabla p + \nu\nabla^2\mathbf{v} + \sigma(\mathbf{v} \times \mathbf{B}) \times \mathbf{B}$$

We also have the Maxwell equation

$$\nabla \times \mathbf{H} = \sigma(\mathbf{E} + \mathbf{v} \times \mathbf{H}) + \epsilon\mathbf{E}_{,t}$$

Taking the curl of this equation and using the Maxwell equations

$$\nabla \times \mathbf{E} = -\mu\mathbf{H}_{,t}, \nabla.\mathbf{B} = 0$$

gives

$$\nabla^2\mathbf{H} + \sigma(-\mu\mathbf{H}_{,t} + \nabla \times (\mathbf{v} \times \mathbf{H})) - \mu\epsilon\mathbf{H}_{,tt} = \mathbf{0}$$

If the displacement current terms are neglected, then we should use

$$\nabla \times \mathbf{H} = \sigma(\mathbf{E} + \mu\mathbf{v} \times \mathbf{H})$$

which gives on taking curl

$$\nabla^2\mathbf{H} - \mu\sigma\mathbf{H}_{,t} + \mu\nabla \times (\mathbf{v} \times \mathbf{H}) = \mathbf{0}$$

The above Navier-Stokes equation with magnetic force correction term can be expressed as

$$\rho(\Omega \times \mathbf{v} + \nabla v^2/2 + \mathbf{v}_{,t}) = \nabla p + \nu\nabla^2\mathbf{v} + \sigma(\mathbf{v} \times \mathbf{B}) \times \mathbf{B}$$

where

$$\Omega = \nabla \times \mathbf{v}$$

Taking the curl of this equation assuming ρ to be a constant gives us

$$\rho(\nabla \times (\Omega \times \mathbf{v} + \Omega_{,t}) = \nu\nabla^2\Omega + \sigma\nabla \times ((\mathbf{v} \times \mathbf{B}) \times \mathbf{B})$$

Suppose we do not neglect the electric field in the expression for $\mathbf{J}$ but neglect the displacement current. Then

$$\nabla \times \mathbf{H} = \mathbf{J} = \sigma(\mathbf{E} + \mathbf{v} \times \mathbf{B})$$

and the force per unit volume can be expressed as

$$\mathbf{J} \times \mathbf{B} = \mu(\nabla \times \mathbf{H}) \times \mathbf{H} = \mu\mathbf{H}.\nabla\mathbf{H} - \mu\nabla H^2/2$$

The condition for this approximation is that $\epsilon\omega << \sigma$ where ω is the characteristic frequency of the fluid motion. The Navier-Stokes equation with this correction term then becomes

$$\rho(\Omega \times \mathbf{v} + \nabla v^2/2 + \mathbf{v}_{,t}) = -\nabla p + \nu\nabla^2\mathbf{v} + \mu(\mathbf{H}.\nabla\mathbf{H} - \nabla H^2/2)$$

and taking the curl of this equation gives

$$\rho(\nabla \times (\Omega \times \mathbf{v}) + \Omega_{,t}) = \nu\nabla^2\Omega + \mu\nabla \times (\mathbf{H}.\nabla\mathbf{H})$$

[82] Maxwell's equations in a medium having non-uniform magnetic permeability. Assume that the magnetic permeability $\mu(\mathbf{r})$ is nonuniform in space while the electric permittivity ϵ is uniform. The Maxwell equations are

$$\nabla \times \mathbf{H} = j\omega\epsilon\mathbf{E}, \nabla \times \mathbf{E} = -j\omega\mu\mathbf{H}$$

assuming sinusoidal time variation of the fields. We are assuming that the current density $\mathbf{J}$ is zero in the region considered. Note that these equations imply

$$div(\mathbf{E}) = 0, div(\mu\mathbf{H}) = 0$$

From the above equations we get

$$\nabla \times (\nabla \times \mathbf{H}) = j\omega\epsilon\nabla \times \mathbf{E}$$

or

$$\nabla^2\mathbf{H} - \nabla(div\mathbf{H}) + \omega^2\epsilon\mu\mathbf{H} = 0$$

But

$$0 = div(\mu\mathbf{H}) = (\nabla\mu, \mathbf{H}) + \mu.div\mathbf{H}$$

so that

$$div\mathbf{H} = -(\nabla log(\mu), \mathbf{H})$$

and we get

$$\nabla^2\mathbf{H} + \nabla(\nabla log(\mu), \mathbf{H}) + \omega^2\mu\epsilon\mathbf{H} = 0$$

We also have

$$\nabla \times (\nabla \times \mathbf{E}) + j\omega\nabla \times (\mu\mathbf{H}) = 0$$

so that

$$\nabla^2\mathbf{E} - j\omega(\nabla\mu \times \mathbf{H} + \mu\nabla \times \mathbf{H}) = 0$$

or

$$\nabla^2 \mathbf{E} - j\omega \nabla \mu \times \mathbf{H} + \omega^2 \mu \epsilon \mathbf{E} = \mathbf{0}$$

Also

$$\nabla \times (\frac{1}{\mu} \nabla \times \mathbf{E}) = -j\omega \nabla \times \mathbf{H} = \omega^2 \epsilon \mathbf{E}$$

[83] Suppose an electromagnetic wave to be propagating inside an incompressible conducting fluid having velocity field $\mathbf{v}$, net charge density zero, permittivity ϵ, magnetic permeability μ and viscosity η. The relevant Maxwell and fluid dynamical equations are

$$\nabla \times \mathbf{E} = -\mu \mathbf{H}_{,t}, \nabla \times \mathbf{H} = \mathbf{J} + \epsilon \mathbf{E}_{,t},$$

$$\nabla . \mathbf{E} = 0, \nabla . \mathbf{H} = 0,$$

$$\rho(\nabla \times (\Omega \times \mathbf{v}) + \Omega_{,t}) = \eta \nabla^2 \Omega + \mu \nabla \times (\mathbf{J} \times \mathbf{H})$$

$$\mathbf{J} = \sigma(\mathbf{E} + \mu \mathbf{v} \times \mathbf{H}), \Omega = \nabla \times \mathbf{v}$$

$$\nabla . \mathbf{v} = 0$$

Simulate these equations using MATLAB. Note that the Maxwell equations above imply

$$\nabla^2 \mathbf{H} - \mu \sigma \mathbf{H}_{,t} + \mu \nabla \times (\mathbf{v} \times \mathbf{H}) - \mu \epsilon \mathbf{H}_{,tt} = \mathbf{0}$$

This is a damped wave equation for $\mathbf{H}$ with a fluid dynamical correction term.

Suppose the fluid flow is two dimensional with the magnetic field along the z axis. Thus,

$$\mathbf{v} = v_x(t, x, y)\hat{x} + v_y(t, x, y)\hat{y}, \mathbf{H} = H_0(t, x, y)\hat{z}$$

The incompressibility of the fluid implies that

$$v_{x,x} + v_{y,y} = 0$$

and hence there is a scalar function $\psi(t, x, y)$ such that

$$v_x = \psi_{,y}, v_y = -\psi_{,x}$$

We have

$$\Omega = \Omega_0(t, x, y)\hat{z}, \Omega_0 = v_{y,x} - v_{x,y} = -\nabla^2 \psi$$

Also,

$$\mathbf{v} \times \mathbf{H} = (v_x \hat{x} + v_y \hat{y}) \times H_0 \hat{z} = v_y H_0 \hat{x} - v_x H_0 \hat{y}$$

and

$$\nabla \times (\mathbf{v} \times \mathbf{H}) = -\hat{z}((v_x H_0)_{,x} + (v_y H_0)_{,y})$$

$$= -\hat{z}(v_x H_{0,x} + v_y H_{0,y}) = \hat{z}(\psi_{,x} H_{0,y} - \psi_{,y} H_{0,x})$$

The equation satisfied by the magnetic field therefore reduces to a single scalar equation.

$$H_{0,xx} + H_{0,yy} - \mu\sigma H_{0,t} + \mu(\psi_{,x}H_{0,y} - \psi_{,y}H_{0,x}) - \mu\epsilon H_{0,tt} = 0$$

The vorticity equation (which is derived from the Navier-Stokes equation after taking the curl) becomes on neglect of the displacement current term,

$$\rho(\nabla \times (\Omega \times \mathbf{v}) + \Omega_{,t}) = \eta\nabla^2\Omega + \mu\nabla \times ((\nabla \times \mathbf{H}) \times \mathbf{H}))$$

Now,

$$(\nabla \times \mathbf{H}) \times \mathbf{H} = \mathbf{H}.\nabla\mathbf{H} - \frac{1}{2}\nabla(H^2)$$

and hence

$$\nabla \times ((\nabla \times \mathbf{H}) \times \mathbf{H}) = \nabla \times (\mathbf{H}.\nabla\mathbf{H}) = \mathbf{0}$$

since $\mathbf{H}.\nabla\mathbf{H} = \mathbf{0}$. This means that the vorticity equation is independent of the magnetic field but the magnetic field equation is dependent on the fluid velocity field. The vorticity equation can be expressed in terms of ψ. We can thus solve for ψ and use this in the magnetic field equation.

[84] Let $\mathbf{E}$ and $\mathbf{B}$ be respectively the electric and magnetic fields in space. Assuming that $\mathbf{B}$ is constant in space and time while $\mathbf{E}$ depends only on time but not on the spatial coordinates, solve the equations of motion of a charged particle.

$$\mathbf{r}'(t) = \mathbf{v}(t), \mathbf{v}'(t) = (Q/m)(\mathbf{E}(t) + \mathbf{v}(t) \times \mathbf{B})$$

hint:Express the equations of motion in state variable form and compute the state transition matrix.

[85] Derive the Green's function for the Helmholtz operator, ie, by solving

$$(\nabla^2 + k^2)G(\mathbf{r}) = -4\pi\delta(\mathbf{r}), \mathbf{r} \in \mathbb{R}^3$$

[86] Determine the partial differential equations of magnetohydrodynamics in the cylindrical and spherical polar coordinate systems.

hint:Use the expressions for gradient, divergence and curl in these coordinate systems.

[87] Solve the transmission line equations using perturbation theory when the parameters R, L, G, C of the line depend on z, namely they vary along the line.

hint: The transmission line equations are

$$-v_{,z}(t, z) = R(z)i(t, z) + L(z)i_{,t}(t, z), i_{,z}(t, z) = G(z)v(t, z) + C(z)v_{,t}(t, z)$$

Now write

$$R(z) = R_0 + \epsilon.R_1(z), L(z) = L_0 + \epsilon.L_1(z), G(z) = G_0 + \epsilon.G_1(z), C(z) = C_0 + \epsilon.C_1(z)$$

where R_0, L_0, G_0, C_0 are constants. We denote the solution to the above differential equations by $v(t, z, \epsilon)$ and $i(t, z, \epsilon)$ and expand these in powers of ϵ, ie,

$$v(t, z, \epsilon) = \sum_{n=0}^{\infty} v_n(t, z)\epsilon^n, i(t, z, \epsilon) = \sum_{n=0}^{\infty} i_n(t, z)\epsilon^n$$

Substitute these into the above pair of differential equations and equate coefficients of same powers of ϵ on both sides and thus obtain an infinite series solution.

[88] The Lagrangian for a relativistic charged particle of charge Q and rest mass m_0 moving in an electromagnetic field with scalar potential V and magnetic vector potential $\mathbf{A}$ is given by

$$L(\mathbf{r}, \mathbf{v}, t) = -m_0 c^2 \sqrt{1 - v^2/c^2} - Q(V - (\mathbf{v}, \mathbf{A}))$$

Show that the Euler-Lagrange equations

$$\frac{d}{dt}\frac{\partial L}{\partial \mathbf{v}} = \frac{\partial L}{\partial \mathbf{r}}$$

yield the Lorentz equation

$$\frac{d}{dt}\frac{m_0 \mathbf{v}}{\sqrt{1 - v^2/c^2}} = Q(\mathbf{E} + \mathbf{v} \times \mathbf{B})$$

where

$$\mathbf{E} = -\nabla V - \frac{\partial \mathbf{A}}{\partial t}, \mathbf{B} = \nabla \times \mathbf{A}$$

[89] Show that the transmission line equations at a given frequency ω can be written as

$$-dV(z)/dz = (R + i\omega L)I(z), -dI(z)/dz = (G + i\omega C)V(z)$$

where R, L, G, C are respectively the resistance, inductance, conductance and capacitance per unit length of the line. Show that these equations imply

$$V''(z) = \gamma^2 V(z), I''(z) = \gamma^2 I(z)$$

where

$$\gamma^2 = (R + i\omega L)(G + i\omega C)$$

Show that the general solution is thus given by

$$V(z) = V_+ exp(-\gamma z) + V_- exp(\gamma z), I(z) = Z_0^{-1}(V_+ exp(-\gamma z) - V_- exp(\gamma z))$$

where

$$Z_0 = \left(\frac{R + i\omega L}{G + i\omega C}\right)^{1/2}$$

If the line is lossless, ie, $R = 0, G = 0$, then $Z_0 = \sqrt{L/C} = R_0$ say and $\gamma = i\beta$ where $\beta = \omega\sqrt{LC}$. In that case,

$$V(z) = V_+ exp(-i\beta z) + V_- exp(i\beta z), \; I(z) = R_0^{-1}(V_+ exp(-i\beta z) - V_- exp(i\beta z))$$

The forward wave voltage amplitude at the point z is

$$V_+(z) = V_+ exp(-i\beta z)$$

and the backward wave voltage amplitude at z is

$$V_-(z) = V_- exp(i\beta z)$$

The reflection coefficient at z is thus

$$\Gamma(z) = \frac{V_-(z)}{V_+(z)} = \Gamma(0) exp(2i\beta z)$$

Note that $z = 0$ corresponds to the source end and $z = d$ to the load end where d is the length of the line. The reflection coefficient at the load end is

$$\Gamma(d) = \Gamma(0) exp(2i\beta d)$$

and hence

$$\Gamma(z) = \Gamma(d) exp(-2i\beta(d - z)) = \Gamma(d) exp(-2i\beta\xi)$$

where $\xi = d - z$ is the distance of the point z from the load end.

[90] Let (r, θ, ϕ) be spherical polar coordinates with corresponding unit vectors

$$\hat{r} = \hat{x}.cos(\phi)sin(\theta) + \hat{y}.sin(\phi)sin(\theta) + \hat{z}.cos(\theta)$$

$$\hat{\theta} = \hat{x}.cos(\phi)cos(\theta) + \hat{y}.sin(\phi)cos(\theta) - \hat{z}.sin(\theta)$$

$$\hat{\phi} = -\hat{x}.sin(\phi) + \hat{y}.cos(\phi)$$

Consider a vector field

$$\mathbf{A} = A_r(r, \theta, \phi)\hat{r} + A_\theta(r, \theta, \phi)\hat{\theta} + A_\phi(r, \theta, \phi)\hat{\phi}$$

Write down the three components of the Helmholtz equation

$$(\nabla^2 + k^2)\mathbf{A} = 0$$

in spherical polar coordinates.

hint:

$$\nabla^2\mathbf{A} = \frac{1}{r}(r\mathbf{A})_{,rr} + \frac{1}{r^2 sin(\theta)}(sin(\theta)\mathbf{A}_{,\theta})_{,\theta} + \frac{1}{r^2 sin^2(\theta)}\mathbf{A}_{,\phi\phi}$$

In differentiating $\mathbf{A}$, we need the partial derivatives of the unit vector $\hat{r}, \hat{\theta}, \hat{\phi}$ with respect to r, θ, ϕ. For example,

$$(\hat{r})_{,r} = 0, (\hat{r})_{,\theta} = \hat{\theta}, (\hat{r})_{,\phi} = sin(\theta)\hat{\phi},$$

$$(\hat{\theta})_{,r} = 0, (\hat{\theta})_{,\theta} = -\hat{r}, (\hat{\theta})_{,\phi} = cos(\theta)\hat{\phi},$$

$$(\hat{\phi})_{,r} = 0, (\hat{\phi})_{,\theta} = 0, (\hat{\phi})_{,\phi} = -\hat{r}.sin(\theta) - \hat{\theta}.cos(\theta)$$

[91] Dispersion relation in a plasma. The plasma consists of $n(\mathbf{r}, t)$ charges, each of charge e per unit volume and we assume that

$$n(r, t) = N + Re(n(\mathbf{r})exp(i\omega t))$$

where N is a constant and $n(\mathbf{r})$ is of first order of smallness. The velocity field of the plasma is given by

$$\mathbf{v}(\mathbf{r}, t) = Re(\mathbf{v}(\mathbf{r})exp(j\omega t))$$

where $\mathbf{v}(\mathbf{r})$ is of first order of smallness. The electric and magnetic fields are likewise

$$\mathbf{E}(\mathbf{r}, t) = Re(\mathbf{E}(\mathbf{r})exp(j\omega t)), \mathbf{B}(\mathbf{r}, t) = \mathbf{B}_0 + Re(\mathbf{B}(\mathbf{r})exp(j\omega t))$$

The phasors $\mathbf{E}(\mathbf{r})$ and $\mathbf{B}(\mathbf{r})$ are of first order of smallness. The current density is

$$\mathbf{J} = (N + Re(n(\mathbf{r})exp(j\omega t))).e.Re(\mathbf{v}(\mathbf{r})exp(j\omega t))$$

$$= Ne.Re(\mathbf{v}(\mathbf{r})exp(j\omega t))$$

upto first order of smallness. The Lorentz force equation is

$$jm\omega\mathbf{v}(\mathbf{r}) = e(\mathbf{E}(\mathbf{r}) + \mathbf{v}(\mathbf{r}) \times \mathbf{B}_0) - m\nu\mathbf{v}(\mathbf{r})$$

where ν is the velocity damping coefficient and we have retained terms only upto first order of smallness. The Maxwell equations are

$$div\mathbf{E}(\mathbf{r}) = n(\mathbf{r})e/\epsilon, div\mathbf{B}(\mathbf{r}) = 0,$$

$$curl\mathbf{E}(\mathbf{r}) = -j\omega\mathbf{B}(\mathbf{r}), curl\mathbf{B}(\mathbf{r}) = \mu Ne\mathbf{v}(\mathbf{r}) + j\omega\mu\epsilon\mathbf{E}(\mathbf{r})$$

Taking the curl of the third Maxwell equation above and making use of the first and fourth gives us

$$\frac{e}{\epsilon}\nabla(n(\mathbf{r})) - \nabla^2\mathbf{E}(\mathbf{r}) = -j\omega(\mu Ne\mathbf{v}(\mathbf{r}) + j\omega\mu\epsilon\mathbf{E}(\mathbf{r}))$$

If we assume that the electric field, magnetic field, charge density field and velocity field propagate through the plasma as plane waves, then the dependence on $\mathbf{r}$ of these phasors must be of the form $exp(-j\mathbf{k}.\mathbf{r})$, where $\mathbf{k}$ is a complex vector. Then, $\nabla n(\mathbf{r}) = -j\mathbf{k}n(\mathbf{r})$, $\nabla^2\mathbf{E}(\mathbf{r}) = -k^2\mathbf{E}(\mathbf{r})$. From the Lorentz force equation,

$$\mathbf{E}(\mathbf{r}) = jme^{-1}\omega\mathbf{v}(\mathbf{r}) + me^{-1}\nu\mathbf{v}(\mathbf{r}) - \mathbf{v}(\mathbf{r}) \times \mathbf{B}_0$$

and plugging this into the last equation gives

$$\frac{-je}{\epsilon}\mathbf{k}n(\mathbf{r})) + k^2(jme^{-1}\omega\mathbf{v}(\mathbf{r}) + m\nu e^{-1}\mathbf{v}(\mathbf{r}) - \mathbf{v}(\mathbf{r}) \times \mathbf{B}_0)$$

$$= -j\omega\mu Ne\mathbf{v}(\mathbf{r}) + \omega^2\mu\epsilon(jme^{-1}\omega\mathbf{v}(\mathbf{r}) + m\nu\mathbf{v}(\mathbf{r}) - \mathbf{v}(\mathbf{r}) \times \mathbf{B}_0)$$

This is one vector relation between $n(\mathbf{r})$ and $\mathbf{v}(\mathbf{r})$. The other relation is the charge conservation equation which comes by taking the divergence of the fourth Maxwell equation and using the first Maxwell equation.

$$N\mathbf{k}.\mathbf{v}(\mathbf{r}) + \omega n(\mathbf{r}) = 0$$

The dispersion relation between $\mathbf{k}$ and ω can be obtained from the last two equations.

[92] Suppose $\mathbf{f}$ is a vector field in $\mathbb{R}^3$ and let S be a closed surface in $\mathbb{R}^3$ that encloses a volume V. Let $\mathbf{n}$ be the unit normal on the surface and let $\mathbf{a}$ be a constant vector. Then,

$$\mathbf{a}.\int_S \mathbf{n} \times \mathbf{f}dS = \int_S \mathbf{n}.\mathbf{f} \times \mathbf{a}dS$$

$$= \int_V \nabla.(\mathbf{f} \times \mathbf{a})dV$$

by Gauss' theorem. Now,

$$\nabla.(\mathbf{f} \times \mathbf{a}) = \mathbf{a}.\nabla \times \mathbf{f}$$

and hence, since $\mathbf{a}$ is arbitrary, it follows that

$$\int_S \mathbf{n} \times \mathbf{f}dS = \int_V \nabla \times \mathbf{f}dV$$

[93] The electric and magnetic fields in space are given in terms of the electric potential V and the magnetic vector potential $\mathbf{A}$ by the formulas

$$\mathbf{E} = -\nabla V - \mathbf{A}_{,t}, \mathbf{B} = \nabla \times \mathbf{A}$$

Obtain explicit expressions for these in terms of the charge and current densities based on the retarded potentials.

[94] Derive the voltage-current relation for an ideal inductor $v(t) = L.di(t)/dt$ starting from Ampere's law and Faraday's law in integral form.

hint:Ampere's law determines the magnetic field in terms of the current. From the magnetic field, the magnetic flux is computed and from Faraday's law, the electromotive force is determined from the magnetic flux as the negative of its second derivative.

[95] Explain how you can iteratively solve the Poisson equation in two dimensions using a grid.

hint: Let $V(x,y)$ be the potential and $\rho(x,y)$ the charge density in the plane. Then, Poisson's equation reads

$$V_{,xx}(x,y) + V_{,yy}(x,y) = -\rho(x,y)/\epsilon$$

We assume that this equation is valid inside the square region $0 \leq x, y \leq L$ and V is specified on the boundary of the square. Then, divided the square region using a grid with each pixel of the grid being of size $\Delta \times \Delta$ where $\Delta = L/N$ with N a large integer. Poisson's equation on approximation using this grid then reads

$$V[m+1,n] + V[m-1,n] + V[m,n+1] + V[m,n-1] - 4V[m,n] = -\Delta^2 \rho[m,n]/\epsilon$$

where

$$V[m,n] = V(m\Delta, n\Delta)$$

This is the leads us to the recursion

$$V[m,n]^{(k+1)} = \frac{1}{4}(V^{(k)}[m+1,n] + V^{(k)}[m-1,n] + V^{(k)}[m,n+1] + V^{(k)}[m,n-1]) + \Delta^2 \rho[m,n]/\epsilon$$

[96] Let $\mathbf{C} = ((c_{ij}))$ be a real, symmetric, positive definite 3×3 matrix. Solve the generalized Poisson equation

$$\sum_{i,j=1}^{3} c_{ij} V_{,ij}(\mathbf{x}) = -\rho(\mathbf{x}), \mathbf{x} \in \mathbb{R}^3$$

where $\mathbf{x} = (x_1, x_2, x_3)$ and $V_{,i} = V_{,x_i}$.

hint: We first diagonalize $\mathbf{C}$ so that

$$\mathbf{C} = \mathbf{U}.\Lambda.\mathbf{U}^T$$

where Λ is diagonal with positive diagonal entries and $\mathbf{U}$ is a real, orthogonal matrix, ie, $\mathbf{UU}^T = \mathbf{U}^T\mathbf{U} = \mathbf{I}$ Let $\Lambda = diag[\lambda_1, \lambda_2, \lambda_3]$ and $\mathbf{U} = ((u_{ij}))$. Then, the generalized Poisson equation can be written as

$$\sum_{i,j,k} u_{ik} \lambda_k u_{jk} V_{,ij} = -\rho$$

Define

$$y_i = \sum_k u_{ki} x_k$$

or equivalently, $\mathbf{y} = \mathbf{U}^T\mathbf{x}$. Then,

$$V_{,i} = V_{,x_i} = \sum_j y_{j,x_i} V_{,y_j} = \sum_j u_{ij} V_{,y_j}$$

so that

$$V_{,y_j} = \sum_i u_{ij} V_{,i}$$

Thus the generalized Poisson equation assumes the form

$$\sum_k \lambda_k V_{,y_k y_k} = -\rho$$

Define

$$z_k = \frac{y_k}{\sqrt{\lambda_k}}$$

Then, the above equation becomes

$$\sum_k V_{,z_k z_k} = -\rho$$

which is the standard Poisson equation and can be solved by means of the standard Green's function $1/|z - z'|$. Note that

$$\mathbf{z} = \Lambda^{-1/2}\mathbf{y} = \Lambda^{-1/2}\mathbf{U}^T\mathbf{x}$$

or equivalently,

$$\mathbf{x} = \mathbf{U}\lambda^{1/2}\mathbf{z}$$

[97] Maxwell's equations for sinusoidal time variations in a medium consisting of anisotropic dielectric with 3×3 permittivity matrix ϵ can be written as

$$\nabla \times \mathbf{E} = -j\omega\mu.\mathbf{H}, \nabla \times \mathbf{H} = j\omega\epsilon.\mathbf{E}$$

which in terms of components reads

$$E_{z,y} - E_{y,z} = -j\omega\mu H_x, E_{x,z} - E_{z,x} = -j\omega\mu H_y, E_{y,x} - E_{x,y} = -j\omega\mu H_z,$$

$$H_{z,y} - H_{y,z} = j\omega(\epsilon_{xx}E_x + \epsilon_{xy}E_y + \epsilon_{xz}E_z),$$

$$H_{x,z} - H_{z,x} = j\omega(\epsilon_{yx}E_x + \epsilon_{yy}E_y + \epsilon_{yz}E_z)$$

$$H_{y,x} - H_{x,y} = j\omega(\epsilon_{zx}E_x + \epsilon_{zy}E_y + \epsilon_{zz}E_z)$$

Assume that the z dependence of the fields is $exp(-\gamma z)$ with γ a complex number and then using the x and y components of the two Maxwell equations express E_x, E_y, H_x, H_y in terms of $E_z, H_z, E_{z,x}, E_{z,y}, H_{z,x}, H_{z,y}$. Then substitute these expressions into the z components of the two Maxwell equations and hence derive a pair of coupled linear second order partial differential equations for E_z, H_z. Explain how approximate solutions to these equations can then be obtained by the method of moments.

[98] Let $\alpha > 0$. Determine the Green's function $G(x, x')$ that satisfies

$$G_{,xx}(x, x') - \alpha^2 G(x, x') = \delta(x - x'), x, x' \in \mathbb{R}$$

with the boundary condition $G(x, x') \to 0$ as $|x| \to \infty$. Use this Green's function to solve the differential equation

$$f''(x) - \alpha^2 f(x) = g(x), x \in \mathbb{R}$$

where g is known.

hint: Solve for $G(x, x')$ in the regions $x > x'$ and $x < x'$ and then use the fact that

$$G_{,x}(x'+, x') - G_{,x}(x'-, x') = 1, G(x'+, x') - G(x'-, x') = 0$$

as follows from the differential equation satisfied by G.

[99] Waveguide with random excitation. Consider the TE modes propagation inside a rectangular waveguide with dimensions a, b. $E_z(x, y)$ at frequency ω satisfies the Helmholtz equation

$$E_{z,xx} + E_{z,yy} + (\omega^2 \mu \epsilon + \gamma^2)E_z = 0$$

where γ is the propagation constant along the z direction. Using the boundary conditions that E_z vanishes at $x = 0, a$ and at $y = 0, b$, we get

$$E_z(x, y) = C.sin(m\pi x/a)sin(n\pi y/b)$$

where m, n are positive integers

$$\omega^2 \mu \epsilon + \gamma^2 = h_{mn}^2 = (\pi/a)^2 + (n\pi/b)^2$$

and hence the general solution can be expressed as

$$E_z(x, y, z, t) = \int_{\mathbb{R}} \sum_{n,m \geq 1} C_{mn}(\omega)sin(m\pi x/a)sin(n\pi y/b)exp(-z\sqrt{h_{mn^2} - \omega^2})exp(j\omega t)d\omega$$

The excitation is provided at $z = 0$ and is given by

$$E_z(x, y, 0, t) = \psi(x, y, t)$$

Thus,

$$C_{mn}(\omega) = \frac{2}{\pi ab} \int_{[0,a] \times [0,b] \times \mathbb{R}} \psi(x, y, t)sin(m\pi x/a)sin(n\pi y/b)exp(-j\omega t)dxdydt$$

Thus corresponding to a given excitation at $z = 0$, we can obtain expressions for the electromagnetic field at any point inside the waveguide. If the excitation $\psi(x, y, t)$ is a Gaussian random field, with prescribed mean and autocorrelation function, then the the electromagnetic field at any point inside the waveguide is also a Gaussian random field and its mean and autocorrelation can be obtained using the above expression.

[100] Propagation of electromagnetic waves in an anisotropic field dependent dielectric.

The permittivity is a 3×3 matrix with each entry being a function of the electric field. Formulate the wave equation and obtain approximate solutions using perturbation theory. Specifically, the permittivity tensor can be expressed as

$$\epsilon(\mathbf{E}) = \begin{pmatrix} \epsilon_{xx}(\mathbf{E}) & \epsilon_{xy}(\mathbf{E}) & \epsilon_{zz}(\mathbf{E}) \\ \epsilon_{yx}(\mathbf{E}) & \epsilon_{yy}(\mathbf{E}) & \epsilon_{yz}(\mathbf{E}) \\ \epsilon_{zx}(\mathbf{E}) & \epsilon_{zy}(\mathbf{E}) & \epsilon_{zz}(\mathbf{E}) \end{pmatrix}$$

The field equations are

$$div(\epsilon(\mathbf{E})\mathbf{E}) = 0,$$

$$\nabla \times \mathbf{E} = -\mu \mathbf{H}_{,t}, \nabla \times \mathbf{H} = (\epsilon(\mathbf{E})\mathbf{E})_{,t},$$

$$div\mathbf{H} = 0$$

Note that

$$\epsilon(\mathbf{E}).\mathbf{E} = (\epsilon_{xx}E_x + \epsilon_{xy}E_y + \epsilon_{xz}E_z, \epsilon_{yx}E_x + \epsilon_{yy}E_y + \epsilon_{yz}E_z,$$

$$\epsilon_{zx}E_x + \epsilon_{zy}E_y + \epsilon_{zz}E_z)^T$$

For $\alpha, \beta = x, y, z$, we assume that $\epsilon_{\alpha\beta}(\mathbf{E})$ can be expanded in a Taylor series as

$$\epsilon_{\alpha\beta}(\mathbf{E}) = \sum_{n_x,n_y,n_z=0}^{\infty} \epsilon_{\alpha\beta,n_x n_y n_z} E_x^{n_x} E_y^{n_y} E_z^{n_z}$$

We assume that $\epsilon_{\alpha\beta,n_x n_y n_z}$ is of the order $n_x + n_y + n_z$ of smallness and apply perturbation theory to obtain a series solution to the problem.

[101] Abstract:We consider Maxwell's equations inside a dielectric whose susceptibility is a function of the electric field. From these equations, we derive the nonlinear wave propagation equation inside the dielectric. The nonlinear operator in the resulting equation is decomposed into a sum of a linear operator and a nonlinear operator. By introduction of a homotopic parameter p, we formulate a parameteric nonlinear equation which when $p = 0$, reduces to the linear equation and when $p = 1$ to the nonlinear equation. The solution to this equation is expressed in powers of p with the the successive terms being determined as solutions to linear differential equations. A comparison with other methods is presented.

Problem Formulation

The susceptibility of the medium in which the electromagnetic waves are propagating is $\chi(\mathbf{E})$, or to be more explicit, $\chi(E_x, E_y, E_z)$. The dielectric constant now becomes a nonlinear function of the electric field given by

$$\epsilon(\mathbf{E}) = \epsilon_0(1 + \chi(\mathbf{E}))$$

We usually assume that $\chi(E_x, E_y, E_z)$ has a convergent Taylor series expansion in powers of the components of the electric field provided that the field magnitude $E = (E_x^2 + E_y^2 + E_z^2)^{1/2}$ is not too large. Such a Taylor expansion can be cast as

$$\chi(\mathbf{E}) = \sum_{n_x,n_y,n_z=0}^{\infty} \chi_{n_x,n_y,n_z} E_x^{n_x} E_y^{n_y} E_z^{n_z}$$

where $\{\chi_{n_x,n_y,n_z}\}$ are constants. We may assume that $\chi_{000} = 0$ so that upto quadratic order in the fields, we have

$$\chi(\mathbf{E}) \approx \chi_x E_x + \chi_y E_y + \chi_z E_z + \frac{1}{2}(\chi_{xx}E_x^2 + \chi_{yy}E_y^2 + \chi_{zz}E_z^2$$

$$+2\chi_{xy}E_x E_y + 2\chi_{yz}E_y E_z + 2\chi_{xz}E_x E_z)$$

where for simplicity of notation

$$\chi_x = \chi_{100}, \chi_y = \chi_{010}, \chi_z = \chi_{001},$$

$$\chi_{xx} = 2\chi_{200}, \chi_{yy} = 2\chi_{020}, \chi_{zz} = 2\chi_{002},$$

$$\chi_{xy} = \chi_{110}, \chi_{yz} = \chi_{011}, \chi_{xz} = \chi_{101}$$

Maxwell's equations that describe the propagation of electromagnetic waves inside such a medium in the absence of external charge and current densities are

$$\nabla.(\epsilon(\mathbf{E})\mathbf{E}) = 0, \nabla \times \mathbf{E} = -\mu_0 \mathbf{H}_{,t},$$

$$\nabla \times \mathbf{H} = (\epsilon(\mathbf{E})\mathbf{E})_{,t}, \nabla.\mathbf{H} = 0$$

The first Maxwell equation gives

$$(\nabla \epsilon, \mathbf{E}) + \epsilon \nabla.\mathbf{E} = 0$$

so that

$$\nabla.\mathbf{E} = -((\nabla \epsilon)/\epsilon, \mathbf{E}) \approx -(\nabla \chi, \mathbf{E})$$

where we have used

$$(\nabla \epsilon)/\epsilon = \nabla(log(\epsilon)) = \nabla log(1 + \chi) \approx \nabla \chi$$

Taking the curl of the second Maxwell equation and using the third and first gives us

$$\nabla^2 \mathbf{E} - \epsilon_0 \mu_0((1 + \chi(\mathbf{E}))\mathbf{E})_{,tt} + \nabla(\nabla \chi(\mathbf{E}), \mathbf{E}) = 0$$

Setting $\mu_0 \epsilon_0 = 1/c^2$, we can write this equation as

$$\nabla^2 \mathbf{E} - \frac{1}{c^2}\mathbf{E}_{,tt} + \nabla(\nabla \chi(\mathbf{E}), \mathbf{E}) - \frac{1}{c^2}(\chi(\mathbf{E})\mathbf{E})_{,tt} = 0$$

In accordance with the homotopy method for solving differential equations, we now consider the modified equation

$$p(\nabla^2 \mathbf{E} - \frac{1}{c^2}\mathbf{E}_{,tt} + \nabla(\nabla \chi(\mathbf{E}), \mathbf{E}) - \frac{1}{c^2}(\chi(\mathbf{E})\mathbf{E})_{,tt})$$

$$+(1-p)(-\frac{1}{c^2}\mathbf{E}_{,tt} + \frac{1}{c^2}\mathbf{E}_{0,tt}) = 0$$

where $0 \le p \le 1$ and $\mathbf{E}_0$ is a known field. This modified equation reduces to the above nonlinear wave equation when $p = 1$ and to the linear differential equation

$$\mathbf{E}_{,tt} = \mathbf{E}_{0,tt}$$

when $p = 0$. The modified equation can be written as

$$p(\nabla^2 \mathbf{E} - \frac{1}{c^2}\mathbf{E}_{0,tt} + \nabla(\nabla \chi(\mathbf{E}), \mathbf{E}) - \frac{1}{c^2}(\chi(\mathbf{E})\mathbf{E})_{,tt})$$

$$+\frac{1}{c^2}(\mathbf{E}_{0,tt} - \mathbf{E}_{,tt}) = 0$$

We expand the solution in powers of p

$$\mathbf{E} = \sum_{m=0}^{\infty} p^m \mathbf{E}_m$$

Equating the coefficient of p^0 gives us a trivial equation, ie the $\mathbf{E}_{0,tt} = \mathbf{E}_{0,tt}$. Equating the coefficient of p to zero gives

$$\frac{1}{c^2}\mathbf{E}_{1,tt}$$

$$= (\nabla^2\mathbf{E}_0 - \frac{1}{c^2}\mathbf{E}_{0,tt} + \nabla(\nabla\chi(\mathbf{E}_0), \mathbf{E}_0) - \frac{1}{c^2}(\chi(\mathbf{E}_0)\mathbf{E}_0)_{,tt})$$

In general, equating the coefficient of p^m gives us

$$\frac{1}{c^2}\mathbf{E}_{m,tt} = \Phi_{m-1}, m \geq 2$$

where Φ_{m-1} is the coefficient of p^{m-1} in the expansion of

$$(\nabla^2\mathbf{E} + \nabla(\nabla\chi(\mathbf{E}), \mathbf{E}) - \frac{1}{c^2}(\chi(\mathbf{E})\mathbf{E})_{,tt})$$

It is clear that

$$\Phi_{m-1} = \nabla^2\mathbf{E}_{m-1} + \Psi_{m-1}$$

where Ψ_{m-1} is the coefficient of p^{m-1} in the expansion of

$$\nabla(\nabla\chi(\sum_{k\geq 0} p^k\mathbf{E}_k), \sum_{k\geq 0} p^k\mathbf{E}_k) - \frac{1}{c^2}(\chi(\sum_{k\geq 0} p^k\mathbf{E}_k)\sum_{k\geq 0} p^k\mathbf{E}_k)_{,tt}$$

The case of nonlinear permittivity and permeability. Consider the case when the permittivity is a nonlinear function of the electric field and the permeability is a nonlinear function of the magnetic field, ie, the permittivity and permeability are respectively given by

$$\epsilon(\mathbf{E}) = \epsilon_0(1 + \chi(\mathbf{E})), \mu(\mathbf{H}) = \mu_0(1 + \chi_m(\mathbf{H}))$$

The Maxwell equations in such a medium in the absence of external charges and currents are

$$(\nabla, \epsilon.\mathbf{E}) = 0, \nabla \times \mathbf{E} = -(\mu.\mathbf{H})_{,t},$$

$$\nabla \times \mathbf{H} = (\epsilon.\mathbf{E})_{,t}, (\nabla, \mu.\mathbf{H}) = 0$$

The first one gives for small χ,

$$\nabla.\mathbf{E} = -(\nabla\chi, \mathbf{E})$$

and likewise, the last gives for small χ_m,

$$\nabla.\mathbf{H} = -(\nabla\chi_m, \mathbf{H})$$

The Taking the curl of the second Maxwell equation gives

$$\nabla(\nabla.\mathbf{E}) - \nabla^2\mathbf{E} = -\nabla \times (\mu\mathbf{H})_{,t} = -(\nabla\mu \times \mathbf{H})_{,t} - (\mu\nabla \times \mathbf{H})_{,t}$$

Substituting for $\nabla \times \mathbf{H}$ from the third Maxwell equation and for $\nabla.\mathbf{E}$ from the above expression gives

$$\nabla^2 \mathbf{E} + \nabla(\nabla\chi, \mathbf{E}) - (\nabla\mu \times \mathbf{H})_{,t} - (\mu(\epsilon\mathbf{E})_{,t})_{,t} = \mathbf{0}$$

Noting that

$$\nabla\mu = \mu_0 \nabla\chi_m,$$

and

$$(\epsilon\mathbf{E})_{,t} = \epsilon_0((1+\chi)\mathbf{E})_{,t} = \epsilon_0\mathbf{E}_{,t} + \epsilon_0(\chi.\mathbf{E})_{,t}$$

so that

$$(\mu(\epsilon\mathbf{E})_{,t})_{,t}$$

$$= \mu_{,t}(\epsilon_0\mathbf{E}_{,t} + \epsilon_0(\chi.\mathbf{E})_{,t}) + \mu(\epsilon_0\mathbf{E}_{,t} + \epsilon_0(\chi.\mathbf{E})_{,t})_{,t}$$

$$= \mu_0\epsilon_0\chi_{m,t}\mathbf{E}_{,t} + \mu_0\epsilon_0\chi_{m,t}(\chi.\mathbf{E})_{,t}$$

$$+ \mu_0\epsilon_0\mathbf{E}_{,tt} + \mu_0\epsilon_0\chi_m\mathbf{E}_{,tt} + \mu_0\epsilon_0\chi_m(\chi\mathbf{E})_{,tt}$$

we can write the above equation as

$$\nabla^2 \mathbf{E} - \mu_0\epsilon_0\mathbf{E}_{,tt} + \nabla(\nabla\chi, \mathbf{E})$$

$$- \mu_0\epsilon_0\chi_{m,t}\mathbf{E}_{,t} - \mu_0\epsilon_0\chi_{m,t}(\chi.\mathbf{E})_{,t}$$

$$- \mu_0\epsilon_0\chi_m\mathbf{E}_{,tt} - \mu_0\epsilon_0\chi_m(\chi\mathbf{E})_{,tt} - \mu_0(\nabla\chi_m \times \mathbf{H})_{,t} = \mathbf{0}$$

This is to be viewed as a modified wave equation for the electric field with the modification incorporating nonlinear terms and coupling with the magnetic field.

[102] The amplitude pattern produced by an N-element array of sensors has the form

$$A(\theta, \phi) = \sum_{n=1}^{N} I_n exp(jk(x_n cos(\phi)sin(\theta) + y_n sin(\phi)sin(\theta) + z_n cos(\theta)))$$

where I_n is the excitation of the nth element and (x_n, y_n, z_n) is the position of the nth sensor. Suppose that (x_n, y_n, z_n) are random vectors with characteristic function

$$\Phi_n(t_1, t_2, t_3) = \mathbb{E}exp(j(t_1 x_n + t_2 y_n + t_3 z_n))$$

Then, the average amplitude pattern of the array is given by

$$\mathbb{E}(A(\theta, \phi)) = \sum_{n=1}^{N} I_n \Phi_n(k.cos(\phi)sin(\theta), k.sin(\phi)sin(\theta), k.cos(\theta))$$

Assume further that the random vectors $(x_n, y_n, z_n), (x_m, y_m, z_m)$ for $n \neq m$ have joint characteristic function

$$\Phi_{n,m}(t_1, t_2, t_3, s_1, s_2, s_3) = \mathbb{E}exp(j(t_1 x_n + t_2 y_n + t_3 z_n + s_1 x_m + s_2 y_m + s_3 z_m))$$

Then the average power pattern is given by

$$P(\theta, \phi) = \mathbb{E}(|A(\theta, \phi)|^2)$$

$$= \sum_n |I_n|^2 + \sum_{n \neq m} \Psi_{n,m}(\theta, \phi)$$

where

$$\Psi_{n,m}(\theta, \phi) = \Phi_{n,m}(k.cos(\phi)sin(\theta), k.sin(\phi)sin(\theta), k.cos(\theta), -k.cos(\phi)sin(\theta), -k.sin(\phi)sin(\theta), -k.cos(\theta))$$

More generally, the amplitude correlation function is given by

$$\mathbb{E}(A(\theta_1, \phi_1).\bar{A}(\theta_2, \phi_2))$$

$$= \sum_n |I_n|^2.\Phi_n(k.(cos(\phi_1)sin(\theta_1)-cos(\phi_2)sin(\theta_2)), k.(sin(\phi_1)sin(\theta_1)-sin(\phi_2)sin(\theta_2)), k.(cos(\theta_1)-cos(\theta_2)))$$

$$+ \sum_{n \neq m} I_n \bar{I}_m.\Phi_{n,m}(k.cos(\phi_1)sin(\theta_1), k.sin(\phi_1)sin(\theta_1), k.cos(\theta_1),$$

$$-k.cos(\phi_2)sin(\theta_2), -k.sin(\phi_2)sin(\theta_2), -k.cos(\theta_2))$$

[103] The equations for electromagnetic waves propagating in a plasma. Consider a fluid consisting of ions all having the same mass m and same charge Q. Let $\mathbf{E}(t, \mathbf{r})$ and $\mathbf{B}(t, \mathbf{r})$ be the electric and magnetic fields inside this fluid. Let $f(t, \mathbf{r}, \mathbf{v})$ be the number of ions per unit volume in phase space, ie, the number of particles at time t having positions in the set $A_1 \subset \mathbb{R}^3$ and velocities in the set $A_2 \subset \mathbb{R}^3$ is given by

$$N_t(A_1 \times A_2) = \int_{A_1 \times A_2} f(t, \mathbf{r}, \mathbf{v})d^3r d^3v$$

Boltzmann's equation for such a fluid has the form

$$f_{,t}(t, \mathbf{r}, \mathbf{v}) + (\mathbf{v}, \nabla_r)f(t, \mathbf{r}, \mathbf{v}) + \frac{1}{m}(\mathbf{F}(t, \mathbf{r}, \mathbf{v}), \nabla_v)f(t, \mathbf{r}, \mathbf{v}) = I(f)$$

where $I(f)$ is the Boltzmann collision integral. In the relaxation time approximation, $I(f)$ is replaced by $\tau_v^{-1}(f_0(\mathbf{r}, \mathbf{v}) - f(t, \mathbf{r}, \mathbf{v}))$, where τ_v is the relaxation time and f_0 is the equilibrium density. $\mathbf{F}(t, \mathbf{r}, \mathbf{v})$ is the force at time t on an ion located at $\mathbf{r}$ and moving with velocity $\mathbf{v}$. It is given by

$$\mathbf{F}(t, \mathbf{r}, \mathbf{v}) = \frac{Q}{m}(\mathbf{E}(t, \mathbf{r}) + \mathbf{v} \times \mathbf{B}(t, \mathbf{r}))$$

The charge density in space is given by

$$\rho(t, \mathbf{r}) = Q \int f(t, \mathbf{r}, \mathbf{v})d^3v$$

and the current density by

$$\mathbf{J}(t,\mathbf{r}) = Q \int \mathbf{v} f(t,\mathbf{r},\mathbf{v}) d^3 v$$

The Maxwell equations for the electric and magnetic fields $\mathbf{E}, \mathbf{B}$ are to be solved taking these as the sources. Thus the problem of describing an electromagnetic wave propagating inside this charged ion fluid involves solving the combined Boltzmann and Maxwell equations. Let $V(t,\mathbf{r})$ and $\mathbf{A}(t,\mathbf{r})$ denote respectively the electric potential and magnetic vector potential. Then the electric and magnetic fields are given by

$$\mathbf{E} = -\nabla V - \mathbf{A}_{,t}, \mathbf{B} = \nabla \times \mathbf{A}$$

and $V, \mathbf{A}$ satisfy the wave equations with source, ie,

$$\nabla^2 V(t,\mathbf{r}) - c^{-2} V_{,tt}(t,\mathbf{r}) = -\rho(t,\mathbf{r})/\epsilon_0$$

$$\nabla^2 \mathbf{A}(t,\mathbf{r}) - c^{-2} \mathbf{A}_{,tt} = -\mu_0 \mathbf{J}(t,\mathbf{r})$$

These are equivalent to the four Maxwell equations. The solution is given by the retarded potentials

$$V(t,\mathbf{r}) = \frac{1}{4\pi\epsilon_0} \int \frac{\rho(t - |\mathbf{r} - \mathbf{r}'|/c, \mathbf{r}')}{|\mathbf{r} - \mathbf{r}'|} d^3 r'$$

$$\mathbf{A}(t,\mathbf{r}) = \frac{\mu_0}{4\pi} \int \frac{\mathbf{J}(t - |\mathbf{r} - \mathbf{r}'|/c, \mathbf{r}')}{|\mathbf{r} - \mathbf{r}'|} d^3 r'$$

The electric and magnetic fields are computed from these potentials by differentiation thereby enabling the Lorentz force $\mathbf{F}$ to be expressed in terms of the particle distribution function f.

[104] Consider a semiconductor in which the particles are electrons and holes. Let $f(t,\mathbf{r},\mathbf{k})$ be the probability that a state at the phase point $(\mathbf{r},\mathbf{k})$ in position momentum space is occupied by an electron at time t. Then $1 - f(t,\mathbf{r},\mathbf{k})$ is the probability that a state at the phase point $(\mathbf{r},\mathbf{k})$ in momentum space is not occupied. The number of states per unit volume centred at the position $\mathbf{r}$ within the momentum volume $d^3 k$ centred at momentum $\mathbf{k}$ is denoted by $d\tau_{\mathbf{k}}$. This is proportional to $d^3 k$. Let $w(\mathbf{k},\mathbf{k}')d\tau_{\mathbf{k}'}$ be the probability per unit time that an electron having momentum $\mathbf{k}$ following a collision, has momentum in the volume $d\tau_{\mathbf{k}'}$. Then the rate of increase of the number density $f(t,\mathbf{r},\mathbf{k})$ of electrons due to collisions (the collision integral) is given by

$$I(f)(t,\mathbf{r},\mathbf{k}) = \int (w(\mathbf{k}',\mathbf{k})f(t,\mathbf{r},\mathbf{k}')(1 - f(t,\mathbf{r},\mathbf{k}) - w(\mathbf{k},\mathbf{k}')f(t,\mathbf{r},\mathbf{k})(1 - f(t,\mathbf{r},\mathbf{k}'))d\tau_{\mathbf{k}'}$$

If we assume that $w(\mathbf{k},\mathbf{k}') = w(\mathbf{k}',\mathbf{k})$, then we get from the above

$$I(f)(t,\mathbf{r},\mathbf{k}) = \int (w(\mathbf{k}',\mathbf{k})(f(t,\mathbf{r},\mathbf{k}') - f(t,\mathbf{r},\mathbf{k}))d\tau_{\mathbf{k}'}$$

The Boltzmann equation can now be written for a force field $\mathbf{F}(t,\mathbf{r},\mathbf{k})$ as

$$f_{,t}(t,\mathbf{r},\mathbf{k}) + m^{-1}(\mathbf{k},\nabla_r)f(t,\mathbf{r},\mathbf{k}) + (\mathbf{F}(t,\mathbf{r},\mathbf{k}),\nabla_k)f(t,\mathbf{r},\mathbf{k}) = I(f)(t,\mathbf{r},\mathbf{k})$$

[105] Steady state version of the Boltzmann kinetic transport equation for a charged fluid consisting of electrons is given by

$$(\mathbf{v}, \nabla_r)f(\mathbf{r}, \mathbf{k}) - e(\mathbf{E}(\mathbf{r}) + \mathbf{v} \times \mathbf{B}(\mathbf{r}), \nabla_k)f(\mathbf{r}, \mathbf{k}) = \frac{f_0(\mathbf{r}, \mathbf{k}) - f(\mathbf{r}, \mathbf{k})}{\tau(\mathbf{k})}$$

where the velocity $\mathbf{v}$ is calculated using the equation

$$\mathbf{v} = \nabla_k E(\mathbf{k})$$

with $E(\mathbf{k})$ being the energy of the electron as a function of the momentum $\mathbf{k}$. $\tau(\mathbf{k})$ is the relaxation time. The function $E(\mathbf{k})$ is determined from quantum mechanics, ie, by solving the Schrodinger wave equation. f_0 is the equilibrium distribution function given by

$$f_0(\mathbf{r}, \mathbf{k}) = \frac{1}{exp((E(\mathbf{k}) - F(\mathbf{r}))/kT(\mathbf{r})) + 1}$$

with F being the Fermi level and T the temperature. These are in general, functions of the position. Set

$$f(\mathbf{r}, \mathbf{k}) = f_0(\mathbf{r}, \mathbf{k}) + f_1(\mathbf{r}, \mathbf{k})$$

where f_1 is the perturbation to the equilibrium distribution function produced by the applied electromagnetic field. Substituting this into the above equation gives

$$-f_1(\mathbf{r}, \mathbf{k})/\tau(\mathbf{k}) = (\mathbf{v}, \nabla_r)f_0(\mathbf{r}, \mathbf{k}) - e(\mathbf{E} + \mathbf{v} \times \mathbf{B}, \nabla_k)f_0(\mathbf{r}, \mathbf{k})$$

$$+(\mathbf{v}, \nabla_r)f_1(\mathbf{r}, \mathbf{k}) - e(\mathbf{E} + \mathbf{v} \times \mathbf{B}, \nabla_k)f_1(\mathbf{r}, \mathbf{k})$$

Now,

$$\nabla_k f_0 = -(\nabla_k E/kT)(exp((E - F)/kT) + 1)^{-2} = -(\mathbf{v}/kT)(exp((E - F)/kT) + 1)^{-2}$$
$$\nabla_r f_0 = (\nabla_r(F/kT))(exp((E - F)/kT) + 1)^{-2}$$

and

$$\nabla_r(F/T) = T^{-1}\nabla_r F - \frac{F}{T^2}\nabla_r T$$

[106] The susceptibility of a medium in which electromagnetic waves are propagating is field dependent and has the form

$$\chi(\mathbf{E}) = b.|\mathbf{E}|^2$$

The electric field in such a medium satisfies a nonlinear wave equation

$$\nabla^2\mathbf{E} - \mu_0\epsilon_0\mathbf{E}_{,tt} + \nabla(\nabla\chi, \mathbf{E}) - \mu_0\epsilon_0(\chi(\mathbf{E})\mathbf{E})_{,tt} = 0$$

Assume that the electric field has only a z component given by

$$\mathbf{E}(x, y, z, t) = Re(exp(i(kz - \omega t))\psi(x, y, z))\hat{z}$$

where ψ is a complex function. Assume further that $k^2 >> |\psi_{,zz}|/|\psi|$. We have

$$\nabla^2 \mathbf{E} \approx Re((2ik\psi_{,z} - k^2\psi + \psi_{,xx} + \psi_{,yy})exp(i(kz - \omega t)))\hat{z}$$

where the term $\psi_{,zz}$ has been neglected in comparison with $k\psi_{,z}$ and $k^2\psi$.

$$\mathbf{E}_{,tt} = -Re(\omega^2\psi.exp(i(kz - \omega t)))\hat{z}$$

$$\chi(\mathbf{E}) = b|\mathbf{E}|^2 == \frac{b}{4}(2|\psi|^2 + 2Re(\psi^2.exp(i(2kz - 2\omega t))))$$

The time averaged value of $\chi(\mathbf{E})$ is thus, $\frac{b}{2}|\psi|^2$. Also

$$(\chi(\mathbf{E})\mathbf{E})_{,tt} \approx -\frac{b\omega^2}{2}|\psi|^2\psi.exp(i(kz - \omega t))\hat{z}$$

provided that we replace χ by its time average. Also,

$$\nabla\chi \approx \frac{b}{2}\nabla|\psi|^2$$

$$(\nabla\chi, \mathbf{E}) \approx \frac{b}{2}(|\psi|^2)_{,z}Re(\psi.exp(i(kz - \omega t)))$$

$$(\nabla\chi, \mathbf{E})_{,z} \approx \frac{b}{2}(|\psi|^2)_{,z}Re(ik.\psi.(exp(i(kz - \omega t))))$$

Making these substitutions in the z component of the nonlinear wave equation gives

$$2ik\psi_{,z} - k^2\psi + \psi_{,xx} + \psi_{,yy} + (\omega^2/c^2)\psi + \frac{ikb}{2}(|\psi|^2)_{,z}\psi - \frac{1}{c^2}(-b\omega^2/2)|\psi|^2\psi = 0$$

Taking $k = \omega/c$, this reduces to

$$2ik\psi_{,z} + \psi_{,xx} + \psi_{,yy} + (ikb/2)(|\psi|^2)_{,z}\psi - (bk^2/2)|\psi|^2\psi = 0$$

or

$$2ik\psi_{,z} + \psi_{,xx} + \psi_{,yy} - V(\psi)\psi = 0$$

where

$$V(\psi) = (bk^2/2)|\psi|^2 - (ikb/2)(|\psi|^2)_{,z}$$

The nonlinear wave equation for the electric field after making approximations thus becomes a nonlinear Schrodinger equation. Assume that ψ is a function of only x, z. We can apply the homotopy method for solving the above equation. Let $p \in [0, 1]$ and consider the pde

$$(1 - p)(2ik\psi_{,z} - 2ik\psi_{0,z}) + p(2ik\psi_{,z} + \psi_{,xx} - V(\psi)\psi) = 0$$

When $p = 0$, it reduces to

$$\psi_{,z} = \psi_{0,z}$$

which has solution

$$\psi(x, z) = \psi_0(x, z) + f(x)$$

where f is any smooth function of x. We may assume that f has been absorbed into ψ_0 and write

$$\psi(x,z) = \psi_0(x,z)$$

When $p = 1$, it reduces to the nonlinear Schrodinger equation. For general p, we can expand the above to get

$$2ik(\psi_{,z} - \psi_{0,z}) + p(\psi_{,xx} - V(\psi)\psi + 2ik\psi_{0,z}) = 0$$

We write

$$\psi(x,z) = \sum_{m=0}^{\infty} p^m \psi_m(x,z)$$

Substituting this into the equation and equating coefficients of p^0 gives the trivial equation $\psi_{0,z} = \psi_{0,z}$ or equivalently, $\psi_0(x,z) = \psi_0(x,z)$, provided that we take the arbitrary additive function of x as zero. This shows that our choice of ψ_0 is indeed the zeroth order approximant. Equating coefficients of p gives

$$2ik\psi_{1,z} + (2ik\psi_{0,z} + \psi_{0,xx} - V(\psi_0)\psi_0) = 0$$

so that

$$2ik\psi_1(x,z) = -\int_0^z (2ik\psi_{0,z}(x,\xi) + \psi_{0,xx}(x,\xi) - (V(\psi_0)\psi_0)(x,\xi))d\xi + g_1(x)$$

where g is an arbitrary function of x. We next determine ψ_2 for which we must equate the coefficients of p^2. Note that

$$V(\psi)\psi = (bk^2/2)|\psi_0 + p\psi_1|^2 - (ikb/2)(|\psi_0 + p\psi_1|^2)_{,z} + O(p^2)$$

So the coefficient of p in $V(\psi)\psi$ is given by

$$(V(\psi)\psi)_1 = bk^2.Re(\bar{\psi}_0\psi_1) - ikb.Re(\bar{\psi}_0\psi_1)_{,z}$$

Thus

$$2ik\psi_{2,z} + (\psi_{1,xx} - bk^2.Re(\bar{\psi}_0\psi_1) + ikb.Re(\bar{\psi}_0\psi_1)_{,z}) = 0$$

and we get

$$2ik\psi_2(x,z) = -\int_0^z (\psi_{1,xx}(x,\xi) - bk^2.Re(\bar{\psi}_0(x,\xi).\psi_1(x,\xi)) + ikb.Re(\bar{\psi}_0(x,\xi)\psi_1(x,\xi))_{,xi}d\xi + g_2(x)$$

where g_2 is an arbitrary function of x.

[107] Derive the nonlinear wave equations satisfied by the electric and megnetic fields in a medium in which the permittivity and permeability are anisotropic and each component of the permittivity and permeability tensors depend on the electric and magnetic fields, ie, the permittivity tensor has the form

$$\epsilon(\mathbf{E}, \mathbf{H}) = \begin{pmatrix} \epsilon_{xx}(\mathbf{E}, \mathbf{H}) & \epsilon_{xy}(\mathbf{E}, \mathbf{H}) & \epsilon_{xz}(\mathbf{E}, \mathbf{H}) \\ \epsilon_{yx}(\mathbf{E}, \mathbf{H}) & \epsilon_{yy}(\mathbf{E}, \mathbf{H}) & \epsilon_{yz}(\mathbf{E}, \mathbf{H}) \\ \epsilon_{zx}(\mathbf{E}, \mathbf{H}) & \epsilon_{zy}(\mathbf{E}, \mathbf{H}) & \epsilon_{zz}(\mathbf{E}, \mathbf{H}) \end{pmatrix}$$

and likewise for the permeability tensor $\mu(\mathbf{E}, \mathbf{H})$.

[108] Let $V(\mathbf{r}, t)$ be the electrostatic potential and $\mathbf{A}(\mathbf{r}, t)$ the magnetic vector potential. Assume that these are random functions of space-time owing to the charge and current densities being random fields. Then compute the correlations in the electric and magnetic fields.

hint:

$$E_i = -V_{,i} - A_{i,t}, i = 1, 2, 3$$

where $V_{,i} = \frac{\partial V}{\partial x^i}, i = 1, 2, 3$ and $V_{,t} = \frac{\partial V}{\partial t}$ and likewise for A_i. Note that $x^1 = x, x^2 = y, x^3 = z$. The electric field is

$$\mathbf{E} = E_1 \hat{x} + E_2 \hat{y} + E_3 \hat{z}$$

or denoting $\mathbf{e}_1 = \hat{x}, \mathbf{e}_2 = \hat{y}, \mathbf{e}_3 = \hat{z}$, we can write

$$\mathbf{E} = \sum_{i=1}^{3} E_i \mathbf{e}_i$$

Likewise, the magnetic field components are

$$B_i = \sum_{j,k=1}^{3} \epsilon_{ijk} A_{k,j}, i = 1, 2, 3$$

where ϵ_{ijk} is antisymmetric with $\epsilon_{123} = 1$. Thus,

$$B_1 = A_{3,2} - A_{2,3}, B_2 = A_{1,3} - A_{3,1}, B_3 = A_{2,1} - A_{1,2}$$

Define the potential correlations

$$R_{VV}(\mathbf{r}_1, t_1, \mathbf{r}_2, t_2) = \mathbb{E}(V(\mathbf{r}_1, t_1) . V(\mathbf{r}_2, t_2)),$$

$$R_{ij}(\mathbf{r}_1, t_1, \mathbf{r}_2, t_2) = \mathbb{E}(A_i(\mathbf{r}_1, t_1) A_j(\mathbf{r}_2, t_2)), i, j = 1, 2, 3,$$

$$R_{Vj}(\mathbf{r}_1, t_1, \mathbf{r}_2, t_2) = \mathbb{E}(V(\mathbf{r}_1, t_1) A_j(\mathbf{r}_2, t_2)), j = 1, 2, 3$$

Then,

$$\mathbb{E}(E_i(\mathbf{r}_1, t_1) E_j(\mathbf{r}_2, t_2)) = \mathbb{E}((V_{,i}(\mathbf{r}_1, t_1) + A_{i,t}(\mathbf{r}_1, t_1)).(V_{,j}(\mathbf{r}_2, t_2) + A_{j,t}(\mathbf{r}_2, t_2)))$$

$$= \frac{\partial^2 R_{VV}(\mathbf{r}_1, t_1, \mathbf{r}_2, t_2)}{\partial x_{1i} \partial x_{2j}}$$

$$+ \frac{\partial^2 R_{Vj}(\mathbf{r}_1, t_1, \mathbf{r}_2, t_2)}{\partial x_{1i} \partial t_2}$$

$$+ \frac{\partial^2 R_{Vi}(\mathbf{r}_2, t_2, \mathbf{r}_1, t_1)}{\partial x_{2j} \partial t_1}$$

$$+ \frac{\partial^2 R_{ij}(\mathbf{r}_1, t_1, \mathbf{r}_2, t_2)}{\partial t_1 \partial t_2}$$

Along similar lines, we can compute $\mathbb{E}(B_i(\mathbf{r}_1, t_1) . B_j(\mathbf{r}_2, t_2))$ and $\mathbb{E}(E_i(\mathbf{r}_1, t_1) . B_j(\mathbf{r}_2, t_2))$.

[109] Assume that the magnetic field phasor at frequency ω in the region $0 \leq x \leq a, 0 \leq y \leq b, z = 0$ is given by $H_x(x,y)\hat{x} + H_y(x,y)\hat{y}$. The equivalent surface current density phasor in this region is then

$$\mathbf{J}_S(x,y) = \hat{z} \times (H_x(x,y)\hat{x} + H_y(x,y)\hat{y}) = H_x(x,y)\hat{y} - H_y(x,y)\hat{x}$$

The magnetic vector potential phasor in the far field region is then approximately

$$\mathbf{A}(r,\theta,\phi)) = \frac{\mu.exp(-jkr)}{4\pi r} \int_0^a \int_0^b \mathbf{J}_S(\xi,\eta)exp(j(\xi.cos(\phi)sin(\theta) + \eta.sin(\phi)sin(\theta)))d\xi d\eta$$

The problem is to calculate the electromagnetic fields in the far field region and hence determine the power radiated by this aperture in the far field region.

[110] Let $I(z), 0 \leq z \leq L$ be the current phasor in a linear wire antenna of length L. Then the far field magnetic vector potential is given by

$$\mathbf{A}(r,\theta,\phi) = \hat{z}\frac{\mu.exp(-jkr)}{4\pi r} \int_0^L I(\xi)exp(jk\xi.cos(\theta))d\xi$$

The far field electric field is given by

$$\mathbf{E} = j\omega.A_z sin(\theta)\hat{\theta}$$

Assume that $I(\xi)$ is a random field. Then for the higher order electric field correlations, we have

$$< \Pi_{k=1}^m E_\theta(r_k,\theta_k,\phi_k)\Pi_{p=m+1}^q \bar{E}_\theta(r_p,\theta_p,\phi_p) >$$

$$= (j\omega)^q(-1)^{q-m}(\Pi_{k=1}^q sin(\theta_k))(\mu/4\pi)^q(exp(-jk(\sum_{l=1}^m r_l - \sum_{p=m+1}^q r_p))/\Pi_{p=1}^q r_p)$$

$$\int_{[0,L]^q} < \Pi_{k=1}^m I(\xi_k)\Pi_{p=m+1}^q \bar{I}(\xi_p) > exp(jk(\sum_{l=1}^m \xi_l cos(\theta_l) - \sum_{p=m+1}^q \xi_p cos(\theta_p)))\Pi_{p=1}^q dx i_p$$

[112] Nonlinear wave equation from an action principle. Let

$$S[\phi] = \frac{1}{2}\int (\phi_{,t})^2 d^3rdt - \frac{1}{2}\int |\nabla\phi|^2 d^3rdt - \int V(\phi)d^3rdt$$

We compute its variation using the assumption that ϕ vanishes as $|t| \to \infty$ or as $r \to \infty$.

$$\delta S[\phi] = \int \phi_{,t}\delta\phi_{,t}d^3rdt - \int (\nabla\phi, \nabla\delta\phi)d^3rdt - \int V'(\phi)\delta\phi d^3rdt$$

$$= -\int \phi_{,tt}\delta\phi d^3rdt + \int (\nabla^2\phi)\delta\phi d^3rdt - \int V'(\phi)\delta\phi d^3rdt$$

Setting this to zero for all variations $\delta\phi$ gives us the nonlinear wave equation

$$\phi_{,tt} - \nabla^2\phi + V'(\phi) = 0$$

[113] Derive the Lorentz transformation on $\mathbb{R}^2$ from the assumption that it is linear in (t, x) and preserves the bilinear form $c^2t^2 - x^2$.

[114] Show that if f, g are two differentiable functions on $\mathbb{R}^3$, then

$$\nabla^2(fg) = f\nabla^2 g + g\nabla^2 f + 2(\nabla f, \nabla g)$$

[115] (a) Derive the partial differential equation satisfied by the function $u(x, y, z)$ from the action principle $\delta S[u] = 0$ where

$$S[u] = \int D(u)|\nabla u|^2 d^3r$$

where $D : \mathbb{R} \to \mathbb{R}$ is a smooth function.

(b) Derive the partial differential equation satisfied by $u(t, x, y, z)$ from the action principle $\delta S[u] = 0$, where

$$S[u] = \int (\partial_t u)^2 - D(u)|\nabla u|^2)d^4x$$

Chapter 6

Special and General Theories of Relativity

Introduction: This chapter discusses notions like covariant derivative of a tensor field, Maxwell's equations in the gravitational field of a blackhole, the torsion and curvature tensor associated with a metric and with a connection on the manifold, cosmological models, the tetrad formalism of general relativity, fluid dynamics in curved space-time, mechanical systems in special relativity and the geodesic equation on a Riemannian manifold.

[1] General relativity and gravitation:

1.Abstract : This paper deals with the formulation of Maxwell's equations in a cavity resonator in the presence of the gravitational field produced by a blackhole. The metric of space-time due to the blackhole is the Schwarzchild metric. Conventionally, this is expressed in spherical polar coordinates. In order to adapt this metric to our problem, we have considered this metric in a small region close to the blackhole and expressed this metric in a cartesian system locally. Suppose (X, Y, Z) are the cartesian coordinates of the centre of our system relative to the centre of the blackhole. Then the metric is expressed in terms of differentials dx, dy, dz where $(X + x, Y + y, Z + z)$ are the cartesian coordinates of out point relative to the centre of the blackhole. The metric is then linearized in the mass of the blackhole yielding a local metric having the form of the conventional Minkowski metric of flat space-time plus a small perturbation about the Minkowski metric due to the presence of the gravitational field of the blackhole. The Maxwell equations are then written down for this metric. The model assumed is a box (cavity) with centre at (X, Y, Z) and the point $(X + x, Y + y, Z + z)$ varies inside the box. The Maxwell equations are expressed in terms of the electromagnetic four potential yielding after applying the Lorentz gauge condition, a perturbed wave equation. This equation has the form $(L_0 + mL_1)(A^\mu) = 0$ where L_0 is the standard D'Alembertian wave operator and L_1 is the perturbation to this wave operator caused by the blackhole. m is the mass of the blackhole. We then assume sinusoidal dependence on time and reduce this wave equation to a modified Helmholtz equation which is an eigenequation for the frequencies. Vanishing boundary conditions for the potential on the faces of the box are assumed and then by applying the method of moments which involves expanding the potentials inside the box as linear combinations of test functions, we obtain a matrix generalized eigenvalue problem for the frequencies of oscillation. The solution for the eigen-frequencies can then be obtained using standard perturbtation theory of matrix generalized eigenvalue problems. The method of perturbation theory required here involves the following: Let A_0, B_0, A_1, B_1 be square matrices of the same size and let ϵ be a small parameter. Then given the solutions to the generalized eigenvalue problem $(A_0 - \lambda.B_0)x = 0$, determine as a power series in ϵ the solutions to the perturbed generalized eigenvalue problem $(A_0 + \epsilon A_1 - \lambda(B_0 + \epsilon B_1))x = 0$. The generalized eigenvalues λ and eigenvectors x are to be expressed as power series in the parameter ϵ.

[2]. Explain how you would compute the covariant derivative of a vector field along a curve in terms of the connection.

hint: Let the curve be given in coordinates by $t \to x^i(t)$ and let Γ^i_{jk} be the connection coefficients. Then if v^i is a vector field, its covariant partial derivative with respect to x^j is given by

$$v^i_{:j} = v^i_{,j} + \Gamma^i_{jk} v^k$$

It follows that along the curve, the covariant derivative of v^i is given by

$$Dv^i/Dt = (dx^j/dt)v^i_{:j} = (dx^j/dt)v^i_{,j} + (dx^j/dt)\Gamma^i_{jk} v^k = dv^i/dt + \Gamma^i_{jk} v^k (dx^j/dt)$$

[3]. Derive the transformation property of the connection coefficient using the fact that the covariant derivative of a vector field is a tensor field.

hint: Let v^i be a vector field and Γ^i_{jk} the connection coefficients in the coordinate system (x^i). Then define

$$v^i_{:j} = v^i_{,j} + \Gamma^i_{jk} v^k$$

The transformation properties of (v^i) and $(v^i_{:j})$ are

$$\bar{v}^i = \frac{\partial \bar{x}^i}{\partial x^j} v^j$$

and

$$\bar{v}^i_{:j} = v^k_{:m} \frac{\partial \bar{x}^i}{\partial x^k} \frac{\partial x^m}{\partial \bar{x}^j}$$

Using these two equations derive the transformation law of Γ^i_{jk}. Note that

$$\bar{v}^i_{:j} = \bar{v}^i_{,j} + \bar{\Gamma}^i_{jk} \bar{v}^k$$

[4]. Find the covariant derivative in components if the connection has a vanishing torsion and the metric is invariant under the connection.

hint: Let $(X, Y) \to \nabla_X Y$ denote the connection. The torsion $T(X, Y)$ is defined by

$$T(X, Y) = \nabla_X Y - \nabla_Y X - [X, Y]$$

The metric is a symmetric bilinear form $(X, Y) \to g(X, Y)$ and the condition for its covariant derivative to vanish is that

$$0 = (\nabla_X g)(Y, Z) = X(g(Y, Z)) - g(\nabla_X Y, Z) - g(Y, \nabla_X Z)$$

Now consider a local coordinate system x^i and define the local basis $e_i = \frac{\partial}{\partial x^i}$ for the tangent space. We set

$$\nabla_{e_i}(e_j) = \Gamma_{ij}^k e_k$$

Then $[e_i, e_j] = 0$ and the condition for $T(e_i, e_j)$ to vanish is clearly that

$$\Gamma_{ij}^k = \Gamma_{ji}^k$$

ie, the connection coefficients are symmetric. The condition for $\nabla_{e_i} g = 0$ can now be written as

$$e_i(g(e_j, e_k)) - g(\nabla_{e_i}(e_j), e_k) - g(e_j, \nabla_{e_i} e_k) = 0$$

or

$$g_{jk,i} - g(\Gamma_{ij}^m e_m, e_k) - g(e_j, \Gamma_{ik}^m e_m) = 0$$

or

$$g_{jk,i} - g_{mk}\Gamma_{ij}^m - g_{jm}\Gamma_{ik}^m = 0$$

or

$$\Gamma_{kij} + \Gamma_{jik} = g_{jk,i}$$

or using the symmetry of the connection coefficients,

$$g_{jk,i} = \Gamma_{kji} + \Gamma_{jki}$$

This equation can be solved for the connection coefficients giving

$$\Gamma_{ijk} = \frac{1}{2}(g_{ij,k} + g_{ik,j} - g_{jk,i})$$

[5]. Consider a metric of the form

$$d\tau^2 = A(r,t)dt^2 - B(r,t)dr^2 - r^2 d\theta^2 - r^2 sin^2(\theta)d\phi^2$$

Formulate the Maxwell equations in this metric and obtain approximate solutions corresponding to the propagation of waves.

hint: For this metric,

$$g_{00} = A, g_{11} = -B, g_{22} = -r^2, g_{33} = -r^2 sin^2(\theta)$$

where

$$x^0 = t, x^1 = r, x^2 = \theta, x^3 = \phi$$

The four contravariant potentials are defined

$$A^0 = V, A^1 = A_r, A^2 = A_\theta, A^3 = A_\phi$$

We then compute the covariant components of the same:

$$A_0 = g_{00}A^0 = A.V, A_1 = g_{11}A^1 = -BA_r, A_2 = g_{22}A^2 = -r^2 A_\theta, A_3 = g_{33}A^3 = -r^2 sin^2(\theta)A_\phi$$

We then compute the components of the antisymmetric field tensor:

$$F_{ij} = A_{j,i} - A_{i,j}$$

and then formulate the Maxwell equations in free space as

$$((g^{ia}g^{jb}F_{ab}\sqrt{-g})_{,j} = 0$$

where

$$g = -r^4 sin^2(\theta)AB$$

so that

$$\sqrt{-g} = r^2 sin(\theta)\sqrt{AB}$$

[6].Explain the tetrad formulation of the equations of general relativity.

hint: Suppose g_{ij} is the metric tensor. Then we choose contravariant vectors e_a^i such that

$$g_{ij}e_a^i e_b^j = \eta_{ab}$$

where η_{ab} are constant scalars. We can equivalently express the metric as

$$g_{ij} = \eta_{ab}e_i^a e_j^b$$

where

$$e_i^a e_a^j = \delta_i^j$$

If A^i is a contravariant vector, then its tetrad form is the set of scalars

$$A_a = A^i e_{ai}, e_{ai} = \eta_{ab}e_i^a$$

If T^{ij} is a tensor, its tetrad form is the set of scalars

$$T_{ab} = T^{ij}e_{ai}e_{bj}$$

We have

$$A_{,b}^a = e_b^i A_{,i}^a = e_b^i A_{:i}^a = e_b^i (A^j e_j^a)_{:i} = e_b^i e_{j:i}^a A^j + e_b^i e_j^a A_{:i}^j$$

since A^a is a scalar. This formula gives the directional derivative in the tetrad formalism in terms of the covariant derivative.

Reference:Chandraskehar, The mathematical theory of blackholes.

[7]. Let M be a differentiable manifold and ∇ a connection on this manifold. For vector fields X, Y, Z, define a vector field $R(X, Y, Z)$ by

$$R(X, Y, Z) = \nabla_X \nabla_Y Z - \nabla_Y \nabla_X Z - \nabla_{[X,Y]} Z$$

Evaluate R in a local coordinate chart, ie, by taking $X = \frac{\partial}{\partial x^i}, Y = \frac{\partial}{\partial x^j}$ and $Z = \frac{\partial}{\partial x^k}$. Justify the term curvature tensor for R by computing the parallel displacement of a vector around an infinitesimal loop on the manifold.

[8]. Formulate the equations of motion of a fluid in general relativity starting with the energy momentum tensor of matter

$$T^{ij} = (\rho + p)v^i v^j + pg^{ij}$$

[9].Prove that if A^i is a contravariant vector field and g, the determinant of the metric, then the covariant divergence of A^i is given by

$$A^i_{:i} g^{1/2} = (A^i g^{1/2})_{,i}$$

[10]. Formulate the Einstein field equations for the metric

$$d\tau^2 = A(r, \theta, t)dt^2 - B(r, \theta, t)dr^2 - r^2 d\theta^2 - D(r, \theta, t)(d\phi - \omega(r, \theta, t)dt)^2$$

[11].Show that the metric on a spherical surface of radius R is given by

$$ds^2 = R^2(d\theta^2 + sin^2(\theta)d\phi^2)$$

Hence obtain the equations for infinitesimal parallel displacement on a spherical surface.

[12]. State and prove the quotient theorem for tensor transformation laws.

hint: Let T^{ij} be such that for every covariant vector A_i, $B^i = T^{ij}A_j$ is a contravariant vector. Then T^{ij} is a tensor. More generally, suppose $T^{i_1 \cdots i_p}_{j_1 \cdots j_q}$ is such that for every set of r covariant vectors $A_{1i}, ..., A_{ri}$, with $r \leq p$, $T^{i_1 \cdots i_p}_{j_1 \cdots j_q} A_{1i_1} ... A_{ri_r}$ is a tensor. Then, $T^{i_1 \cdots i_p}_{j_1 \cdots j_q}$ is a tensor. This can be further generalized.

[13]. Develop a numerical package for simulating the Maxwell equations in curved space-time in the absence of charges and currents.

hint: Let A^i be the contravariant electromagnetic four potential and let g_{ij} be the space-time metric. Then the covariant electromangetic four potential has components $A_i = g_{ij}A^j$ and the covariant antisymmetric field tensor is given by

$$F_{ij} = A_{j,i} - A_{i,j}$$

The contravariant antisymmetric field tensor is then given by

$$F^{ij} = g^{ia} g^{jb} F_{ab}$$

where the contravariant metric tensor (g^{ij}) is the inverse of the covariant metric tensor (g_{ij}). The Maxwell field equations in the absence of charges and currents are then given by

$$(F^{ij} \sqrt{-g})_{,j} = 0$$

where $g = det((g_{ij}))$. This set of partial differential equations must be combined with the Lorentz gauge condition $(A^i \sqrt{-g})_{,i} = 0$ to obtain the modified wave equation satisfied by A^i. Approximate solutions are then obtained using perturbation theory for partial differential equations. Note that

$$F^{ij} = g^{ia} g^{jb} (A_{b,a} - A_{a,b}) = g^{ia} g^{jb} ((g_{bc} A^c)_{,a} - (g_{ac} A^c)_{,b})$$

from which it follows that $(F^{ij} \sqrt{-g})_{,j}$ can be put in the form

$$C_a^{ibc} A_{,bc}^a + D_a^{ib} A_{,b}^a + E_a^i A^a = 0$$

where C_a^{ibc}, D_a^{ib} and E_a^i are all functions of the space-time location.

[14]. Develop a numerical package for simulating the geodesic motion in a the curved space-time generated by a rotating blackhole. The metric has the general form

$$d\tau^2 = a(r,\theta)dt^2 - b(r,\theta)dr^2 - c(r,\theta)(d\phi - \omega(r,\theta)dt)^2 - f(r,\theta)d\theta^2$$

The first step in simulating geodesic motion is to compute the components of the connection. For that we first need to determine the components of the metric tensor.

[15]. Develop numerical packages for computing the Riemann Christoffel curvature tensor and the Ricci tensor for various kinds of connections defined on a differentiable manifold. Given a differentiable manifold $\mathcal{M}$, let us associate with each pair of vector fields X, Y another vector field $\nabla_X Y$ such that

$$\nabla_X (fY) = X(f)Y + f \nabla_X Y$$

for all differentiable functions f and

$$\nabla_{fX} Y = f \nabla_X Y$$

for all differentiable functions f. X, Y are arbitrary vector fields on $\mathcal{M}$. Then given a local coordinate system (x^i), we define the local basis for TM_x as $e_i = \frac{\partial}{\partial x^i}$. Then put

$$\nabla_{e_j} e_i = \Gamma_{ji}^k e_k$$

where Γ_{ji}^k are differentiable functions on $\mathcal{M}$. The Riemann Christoffel curvature tensor is a map that associates with each triplet of vector fields X, Y, Z, a vector field

$$R(X,Y)Z = \nabla_X \nabla_Y Z - \nabla_Y \nabla_X Z - \nabla_{[X,Y]} Z$$

or equivalently,

$$R(X,Y) = \nabla_X \nabla_Y - \nabla_Y \nabla_X - \nabla_{[X,Y]}$$

The components of the tensor R must be computed in the given local coordinate basis. Note that if ω is a one form on the manifold $\mathcal{M}$ and X, Y, Z are three vector fields, then $\omega(R(X,Y)Z)$ is a scalar field. In terms of local coordinates, we have $X = X^i e_i, Y = Y^i e_i, Z = Z^i e_i, \omega = \omega_i e^i, e^i = dx^i$, so that

$$\omega(R(X,Y)Z) = R^i_{jkl}\omega_i X^j Y^k Z^l$$

The components R^i_{jkl} of the tensor R must be computed.

[16]. Develop numerical packages for solving the Klein-Gordon equation in curved space-time defined by a metric. The Klein-Gordon equation is

$$(g^{ij}V_{,i}\sqrt{-g})_{,j} + m^2 V = 0$$

This can be expressed as

$$g^{ij}\sqrt{-g}V_{,ij} + (g^{ij}\sqrt{-g})_{,j}V_{,i} + m^2 V = 0$$

This equation can be derived from an action principle and then action can be discretized and a finite element implementation of this action principle can be carried out.

[17]. Develop numerical algorithms for solving the geodesic equation given the space-time metric.

hint: Let g_{ij} be the metric. Then the geodesic equations are

$$\frac{d^2 x^i}{ds^2} + \Gamma^i_{jk}\frac{dx^j}{ds}\frac{dx^k}{ds} = 0$$

These differential equations must be discretized with respect to the s variable transforming them thereby into nonlinear difference equations.

[18]. Using MATLAB implement the geodesic equations corresponding to the space-time metric

$$d\tau^2 = A(r,t)dt^2 - B(r,t)dr^2 - r^2(d\theta^2 + sin^2(\theta)d\phi^2)$$

Specialize to the case of null geodesics. What do the null radial geodesics look like.

hint:The null radial geodesics satisfy the differential equation

$$A(r,t)dt^2 - B(r,t)dr^2 = 0$$

or

$$dr/dt = \pm(A(r,t)/B(r,t))^{1/2}$$

[19]. Define an isometry of space time for a specified metric. Derive the differential equations satisfied by an infinitesimal isometry.

hint: Suppose $g_{\mu\nu}(x)$ is the given metric of space-time. A transformation $x^\mu \to \bar{x}^\mu$ is said to be an isometry if the metric is form invariant under this transformation, ie, if

$$g_{\mu\nu}(\bar{x}) \frac{\partial \bar{x}^\mu}{\partial x^\alpha} \frac{\partial \bar{x}^\nu}{\partial x^\beta} = g_{\alpha\beta}(x)$$

Suppose v^μ is a four vector at the point x of the Riemannian manifold having metric $g_{\mu\nu}(x)$. Then, this vector transforms to the vector $\bar{v}^\mu$ at $\bar{x}$ under the coordinate transformation $x \to \bar{x}$ where

$$\bar{v}^\mu = v^\alpha \frac{\partial \bar{x}^\mu}{\partial x^\alpha}$$

The condition of isometry implies that the length of the vector v^μ defined by $g_{\mu\nu}(x) v^\mu v^\nu$ is invariant under the given coordinate transformation, i.e.,

$$g_{\mu\nu}(x) v^\mu v^\nu = g_{\mu\nu}(\bar{x}) \bar{v}^\mu \bar{v}^{\nu|nu}$$

Substituting for $\bar{v}^\mu$, we get the condition for the coordinate transformation $x \to \bar{x}$ to be an isometry as

$$g_{\mu\nu}(\bar{x}) \frac{\partial \bar{x}^\mu}{\partial x^\alpha} \frac{\partial \bar{x}^\nu}{\partial x^\beta} = g_{\alpha\beta}(x)$$

which is the above stated condition for isometry. Thus an isometry of a Riemannian manifold can either be defined as a smooth coordinate transformation that leaves the metric form invariant or equivalently, as a smooth coordinate transformation that leaves the length of vectors invariant. Consider now the infinitesimal isometry

$$x^\mu \to \bar{x}^\mu = x^\mu + \epsilon.\xi^\mu(x)$$

The condition for the metric to be form invariant under this infinitesimal transformation is given by

$$g_{\alpha\beta}(x) = g_{\mu\nu}(\bar{x}) \bar{x}^\mu_{,\alpha} \bar{x}^\nu_{,\beta}$$

or

$$g_{\alpha\beta}(x) = g_{\mu\nu}(x + \epsilon.xi)(\delta^\mu_\alpha + \epsilon.\xi^\mu_{,\alpha})(\delta^\nu_\beta + \epsilon.\xi^\nu_{,\beta})$$

Retaining terms only upto $O(\epsilon)$, this equation gives

$$g_{\mu\nu}(x)\xi^\mu_{,\alpha}\delta^\nu_\beta + g_{\mu\nu}(x)\delta^\mu_\alpha \xi^\nu_{,\beta} + g_{\mu\nu,\rho}\xi^\rho \delta^\mu_\alpha \delta^\nu_\beta = 0$$

or

$$g_{\mu\beta}\xi^\mu_{,\alpha} + g_{\alpha\mu}\xi^\mu_{,\beta} + g_{\alpha\beta,\rho}\xi^\rho = 0$$

This can be expressed in the covariant form

$$\xi_{\mu:\nu} + \xi_{\nu:\mu} = 0$$

A vector field $\xi^\mu(x)$ that satisfies this condition is called a Killing vector field.

[20]. Geodesic deviation. Let $x^\mu(\tau)$ and $x^\mu(\tau) + \delta x^\mu(\tau)$ be two neighbouring geodesics. We want to derive a differential equation satisfied by the geodesic deviation $\delta x^\mu(\tau)$.

$$\frac{d^2 x^\mu}{d\tau^2} + \Gamma^\mu_{\alpha\beta}(x) \frac{dx^\alpha}{d\tau} \frac{dx^\beta}{d\tau} = 0$$

$$\frac{d^2(x^\mu + \delta x^\mu)}{d\tau^2} + \Gamma^\mu_{\alpha\beta}(x + \delta x)\frac{d(x^\alpha + \delta x^\alpha)}{d\tau}\frac{d(x^\beta + \delta x^\beta)}{d\tau} = 0$$

Retaining terms only upto linear in δx, we get from these two equations

$$\frac{d^2\delta x^\mu}{d\tau^2} + 2\Gamma^\mu_{\alpha\beta}(x)\frac{dx^\alpha}{d\tau}\frac{d\delta x^\beta}{d\tau} + \Gamma^\mu_{\alpha\beta,\rho}(x)\frac{dx^\alpha}{d\tau}\frac{dx^\beta}{d\tau}\delta x^\rho = 0$$

This is the differential equation for geodesic deviation. In can be brought to the standard form involving the Riemann-Christoffel curvature tensor. The implementation of this differential equation using MATLAB involves discretization of the proper time.

Reference:Weinberg, Gravitation and Cosmology, Principles and Applications of the General Theory of Relativity.

[21]. Consider the metric in $\mathbb{R}^2$ given by

$$ds^2 = A(x,y)dx^2 + 2B(x,y)dxdy + C(x,y)dy^2$$

Show how this metric can be transformed into a form in which the cross term $dxdy$ does not appear.

hint: Consider $x = u, y = v + f(u,v)$. Then,

$$dx = du, dy = (1 + f_{,v})dv + f_{,u}du$$

so that

$$ds^2 = Adu^2 + 2Bdu(f_{,u}du + (1 + f_{,v})dv) + C((1 + f_{,v})dv + f_{,u}du)^2$$

To make the term involving $dudv$ zero, we must choose f so that

$$B + Cf_{,u} = 0$$

or equivalently,

$$f_{,u}(u,v) = -\frac{B(u, v + f(u,v))}{C(u, v + f(u,v))}$$

[22]. Suppose a spherical distribution of matter has density $\rho(r)$ a function of the radial distance r from the origin. Assume that the matter forms a gas with equation of state $p = p(\rho) = K\rho^\gamma$. The total mass inside a spherical shell of radius r is given by

$$M(r) = 4\pi \int_0^r r^2\rho(r)dr$$

The inward radial gravitational force on a unit test mass at a distance r from the origin is then

$$F_g(r) = GM(r)/r^2$$

The total inward gravitational force on the shell $[r, r + dr]$ is then

$$F_g(r)4\pi r^2\rho(r)dr = GM(r).4\pi\rho(r)dr$$

and the outward pressure force on the same is

$$-(dp/dr)4\pi r^2 dr$$

For equilibrium, the gravitational force and pressure force must match giving

$$GM(r)\rho(r) = -(dp/dr).r^2$$

Now observe that

$$\rho(r) = M'(r)/4\pi r^2$$

and

$$dp/dr = K\gamma.\rho^{\gamma-1}\rho'(r) = K\gamma(M'/4\pi r^2)^{\gamma-1}(M'/4\pi r^2)'$$

A differential equation for $M(r)$ can be derived from these equations. Thus,

$$GMM'/4\pi r^2 = -K\gamma(M'/4\pi r^2)^{\gamma-1}(M'/4\pi r^2)'$$

This differential equation is the starting point for deriving the Chandrasekhar limit.

[23]. Consider the metric on the surface of a sphere S of radius R. Introducing standard polar coordinates (θ, ϕ) on this surface, the metric is given by

$$d\tau^2 = R^2(d\theta^2 + sin^2(\theta)d\phi^2)$$

The geodesics are obtained from the variational principle

$$\delta \int_a^b (d\theta^2 + sin^2(\theta)d\phi^2)^{1/2} = 0$$

This results in the Euler-Lagrange equations

$$\frac{d}{dt}\frac{\partial L}{\partial \theta'} = \frac{\partial L}{\partial \theta},$$

$$\frac{d}{dt}\frac{\partial L}{\partial \phi'} = \frac{\partial L}{\partial \phi} = 0$$

where

$$L = L(\theta, \theta', \phi') = (sin^2(\theta)\theta'^2 + \phi'^2)^{1/2}$$

with primes denoting derivative with respect to t. Write down these differential equations and show that a first integral of these equations is given by

$$(d\theta/d\tau)^2 + sin^2(\theta)(d\phi/d\tau)^2 = 1/R^2$$

Obtain using these equations a differential equation for θ as a function of ϕ and by integrating this show, that the geodesics are great circles.

[24] General relativistic analysis of antennas. An antenna is placed close to a gravitating object and the aim is to formulate the Maxwell equations in the curved space-time generated by the gravitating object

and determine perturbation theoretic expansions for the electromagnetic fields. Let $A^\mu(x), \mu = 0, 1, 2, 3$ be the contravariant electromagnetic four potential and $g_{\mu\nu}(x)$ the metric tensor of our curved space-time. Here, the notation $x = (x^\mu)$ are the space-time coordinates of an event. The antisymmetric electromagnetic tensor field is given by

$$F_{\mu\nu} = A_{\nu:\mu} - A_{\mu:\nu} = A_{\nu,\mu} - A_{\mu,\nu}, A_\mu = g_{\mu\nu} A^\nu$$

Here, $A_{\nu:\mu}$ stands for the covariant partial derivative of the covariant electromagnetic four potential A_ν. With J^μ denoting the four current density, Maxwell's equations are

$$(F^{\mu\nu}\sqrt{-g})_{,\nu} = \mu_0 J^\mu \sqrt{-g}$$

where $g = det(g_{\mu\nu})$. These equations must be solved using perturbation series.

[25] Motion of charges in an electromagnetic and gravitational field in the general relativistic formulation. The presence of the gravitational field shows up in the form of the Christoffel symbols that appear in the geodesic equation while the presence of the electromagnetic field shows up in the form of the Lorentz force. The form of the equations is

$$\frac{d^2 x^\mu}{d\tau^2} + \Gamma^\mu_{\alpha\beta}(x) \frac{dx^\alpha}{d\tau} \frac{dx^\beta}{d\tau} = K e F^\mu_\alpha \frac{dx^\alpha}{d\tau}$$

where e is the charge and K is a numerical constant which can be determined from special relativistic considerations. For example, take $A^0 = V, A^1 = A_x, A^2 = A_y, A^3 = A_z, x^0 = t, x^1 = x, x^2 = y, x^3 = z$. Then $(g_{\mu\nu}) = diag[1, -1, -1, -1]$. Then,

$$F_{01} = A_{1,0} - A_{0,1} = -A_{x,t} - V_{,x} = E_x$$

etc and $F^1_0 = -F_{10} = F_{01} = E_x$ so that $K = 1$.

[26] Let e^i_a be the components of four contravariant vector fields. The vector fields can thus be written as

$$\mathbf{e}_a = e^i_a \frac{\partial}{\partial x^i}$$

Suppose that these vector fields are chosen so that

$$\eta_{ab} = g_{ij} e^i_a e^j_b$$

are constants. In other words, the vectors e^i_a define a basis for a locally inertial reference frame. Such vectors are called a tetrad basis. Suppose A_i is a covariant vector field. Then $A_{(a)} = A_i e^i_a$ are scalar fields. Develop the calculus of covariant derivatives relative to a tetrad basis. For example, the directional derivative can be computed relative to the tetrad basis as

$$A_{(a),(b)} = e^i_b A_{(a),i} = e^i_b (A_j e^j_a)_{,i} = e^i_b (A_j e^j_a)_{:i}$$

$$= e^i_b (A_{j:i} e^j_a + A_j e^j_{a:i}) = A_{j:i} e^i_b e^j_a + A_{(c)} e^c_j e^i_b e^j_{a:i}$$

Note that

$$e^i_a e_{bi} = \eta_{ab}, \ e^i_a e^a_j = \delta^i_j$$

[27] Formulate the discretized form of Einstein's field equations of gravitation.

hint: Express the Ricci tensor explicitly in terms of the metric tensor and its first and second order partial derivatives with respect to the space-time coordinates. Then choose an appropriate discretization step size and discretize the field equations $R_{\mu\nu} = 0$ in free space. In the presence of matter and radiation, the energy-momentum tensor $T_{\mu\nu}$ must in addition be taken into account and the field equations are $R_{\mu\nu} - \frac{1}{2}R.g_{\mu\nu} = k.T_{\mu\nu}$. These must be discretized appropriately.

[28] Starting from the definition of parallel displacement of a vector on a p dimensional surface immersed in N dimensional Euclidean space, derive the formulae for the connection coefficients in terms of the metric tensor.

[29] Compute using the spherical polar coordinates (θ, ϕ) on the two dimensional unit sphere, the components of the Riemann-Christoffel curvature tensor, the Ricci tensor and the scalar curvature.

[30] From the transformation law of a scalar and a contravariant vector under a diffeomorphism of a Riemannian manifold, deduce the transformation laws of a covariant vector and of an arbitrary (p, q) tensor. Note that a contravariant vector is a $(1, 0)$ tensor while a covariant vector is a $(0, 1)$ tensor.

hint: A (p, q) tensor can be expressed as a linear combination of the outer product of p contravariant and q covariant vectors.

[31] Starting from the Lagrangian for a relativistic charged particle moving in an electromagnetic field, derive the Hamiltonian for the particle.

hint: The Lagrangian is given by

$$L(\mathbf{r}, \mathbf{v}, t) = -m_0 c^2 \sqrt{1 - v^2/c^2} - q(V(t, \mathbf{r}) - (\mathbf{v}, \mathbf{A}(t, \mathbf{r})))$$

and the canonical momenta are given by

$$\mathbf{p} = \frac{\partial L}{\partial \mathbf{v}} = \frac{m_0 \mathbf{v}}{\sqrt{1 - v^2/c^2}} + e\mathbf{A}$$

[32] Let $\mathcal{M}$ be a Riemannian manifold with metric g, ie, for any $p \in \mathcal{M}$, $g_p : T\mathcal{M}_p \times T\mathcal{M}_p \to \mathbb{R}$ is bilinear and positive definite. Let $\gamma(t), 0 \leq t \leq 1$ be a curve in $\mathcal{M}$ with $\gamma(0) = p, \gamma(1) = q$. The length of this curve is defined by $L(\gamma) = \int_0^1 (g_{\gamma(t)}(\gamma'(t), \gamma'(t)))^{1/2} dt$. We define $d(p, q)$ to be the infimum of $L(\gamma)$ where γ varies over all curves in $\mathcal{M}$ joining p and q. Under what conditions does there exist a curve γ_{pq} joining p and q such that $L(\gamma_{pq}) = d(p, q)$. If there exists such a curve, then γ_{pq} is a geodesic, ie,

the tangent vectors on γ_{pq} are obtained by parallel displacement of the tangent vector to the same curve at p with respect to the Riemannian connection. The geodesic curve satisfies a system of second order differential equations obtained by applying the Euler-Lagrange variational method to the Lagrangian $(g_{\gamma(t)}(\gamma'(t), \gamma'(t))))^{1/2}$.

[33] Suppose $\mathcal{M}$ is a differentiable manifold with a connection ∇. Let X be a vector field on $\mathcal{M}$ and let $\gamma(t)$ be a curve in $\mathcal{M}$. Suppose Y is any smooth vector field on $\mathcal{M}$ such that $Y_{\gamma(t)}$ that coincides with $\gamma'(t)$. We say that X is parallel on γ if $(\nabla_Y X)_{\gamma(t)} = 0$ for all t. We want to show that this definition of parallelism of the vector field X along γ is independent of the choice of Y and also of the values of X at points that do not fall on the curve γ. Indeed, relative to a local coordinate chart (x^i), we have $X(x) = X^i(x)\frac{\partial}{\partial x^i}$ and $Y(x) = Y^i\frac{\partial}{\partial x^i}$. Let $x^i(\gamma(t)) = \gamma^i(t)$. Then

$$0 = (\nabla_Y X)_{\gamma(t)} = \nabla_{Y^j \partial_j}(X^i \partial_i)_{\gamma(t)}$$

$$= Y^j(\gamma(t))\nabla_{\partial_j}(X^i \partial_i)_{\gamma(t)}$$

$$= Y^j(\gamma(t))(X^i_{,j}\partial_i + X^i \nabla_{\partial_j}\partial_i)_{\gamma(t)}$$

$$= Y^j(\gamma(t))((X^k_{,j} + X^i \Gamma^k_{ji})\partial_k)_{\gamma(t)}$$

$$= (\gamma^{j'}(t)X^k_{,j}(\gamma(t)) + \gamma^{j'}(t)X^i(\gamma(t))\Gamma^k_{ji}(\gamma(t)))\partial_k$$

so that the condition for parallelism reads

$$\frac{dX^k(\gamma(t))}{dt} + \Gamma^k_{ji}(\gamma(t))\gamma^{j'}(t)X^i(\gamma(t)) = 0$$

and it is clear that this condition depends only on the values of the vector field X on the curve γ. Here,

$$\partial_i = \frac{\partial}{\partial x^i}, \nabla_{\partial_i}\partial_j = \Gamma^k_{ij}\partial_k$$

[34] Determine all the linear transformations on $\mathbb{R}^4$ that preserve the quadratic form

$$Q(x^\mu) = c^2 t^2 - x^2 - y^2 - z^2$$

where $x^0 = t, x^1 = x, x^2 = y, x^3 = z$.

[35] Let $T^{\mu\nu}$ be a symmetric tensor field. Compute the covariant divergence $T^{\mu\nu}_{:\nu}$ in terms of ordinary partial derivatives.

hint: Suppose $T^{\mu\nu} = A^\mu B^\nu$ where A^μ and B^μ are four vectors. Then,

$$T^{\mu\nu}_{:\nu} = A^\mu_{:\nu}B^\nu + A^\mu B^\nu_{:\nu}$$

$$= (A^\mu_{,\nu} + \Gamma^\mu_{\sigma\nu}A^\sigma)B^\nu + A^\mu(B^\nu_{,\nu} + \Gamma^\nu_{\sigma\nu}B^\sigma)$$

$$= T^{\mu\nu}_{,\nu} + \Gamma^{\mu}_{\sigma\nu}T^{\sigma\nu} + \Gamma^{\nu}_{\sigma\nu}T^{\mu\sigma}$$

By linearity of the covariant derivative, this formula holds for any $(2,0)$ tensor $T^{\mu\nu}$. Now suppose $T^{\mu\nu} = T^{\nu\mu}$, ie, $T^{\mu\nu}$ is a symmetric tensor. Then, the above gives

$$T^{\mu\nu}_{:\nu} = T^{\mu\nu}_{,\nu} + (g)^{-1/2}(g^{1/2})_{,\sigma}T^{\mu\sigma} + \Gamma^{\mu}_{\sigma\nu}T^{\sigma\nu}$$

or

$$g^{1/2}T^{\mu\nu}_{:\nu} = (T^{\mu\nu}\sqrt{g})_{,\nu} + g^{1/2}\Gamma^{\mu}_{\sigma\nu}T^{\sigma\nu}$$

Note that if $g < 0$, then $\sqrt{g}$ should be replaced by $-g$.

[36] Let M be a Riemannian manifold with metric g. Assume that ∇ is a connection on the manifold such that the torsion of this connection is zero and $\nabla_X g = 0$ for all vector fields X on the manifold. Deduce from these conditions an expression for the connection in a local coordinate basis in terms of the metric tensor components.

hint. Let x^i be local coordinates and $\partial_i = \frac{\partial}{\partial x^i}$. The torsion is zero implies

$$\nabla_{\partial_i}\partial_j - \nabla_{\partial_j}\partial_i = 0$$

and hence setting

$$\nabla_{\partial_i}\partial_j = \Gamma^k_{ij}\partial_k$$

and hence,

$$\Gamma^k_{ij} = \Gamma^k_{ji}$$

The condition $\nabla_X g = 0$ implies

$$X(g(Y,Z)) = g(\nabla_X Y, Z) + g(Y, \nabla_X Z)$$

for all vector fields X, Y, Z. Put $X = \partial_i, Y = \partial_j, Z = \partial_k$ and deduce that

$$g_{jk,i} = g(\Gamma^m_{ij}\partial_m, \partial_k) + g(\partial_j, \Gamma^m_{ik}\partial_m)$$

$$= \Gamma^m_{ij}g_{mk} + \Gamma^m_{ik}g_{jm} = \Gamma_{kij} + \Gamma_{jik}$$

where by definition

$$\Gamma_{ijk} = g_{im}\Gamma^m_{jk}$$

From these equations derive the expression

$$\Gamma_{ijk} = \frac{1}{2}(g_{ij,k} + g_{ik,j} - g_{jk,i})$$

[37] Let V be the electrostatic potential and $\mathbf{A}$ the magnetic vector potential. Choosing space-time coordinates $x^0 = t, x^1 = x, x^2 = y, x^3 = z$ and introducing the contravariant electromagnetic four

potential A^μ so that $A^0 = V, A^1 = A_x, A^2 = A_y, A^3 = A_z$, and letting $g_{\mu\nu}$ be the space-time metric, we introduce the covariant electromagnetic four potential A_μ so that

$$A_\mu = g_{\mu\nu} A^\nu$$

Then derive the expressions for the components of the contravariant electromagnetic antisymmetric field tensor

$$F^{\mu\nu} = g^{\mu\alpha} g^{\nu\beta} F_{\alpha\beta}, \; F_{\alpha\beta} = A_{\beta,\alpha} - A_{\alpha,\beta}$$

where $((g^{\mu\nu}))$ is the inverse of $((g_{\mu\nu}))$, ie,

$$g^{\mu\alpha} g_{\alpha\nu} = \delta^\mu_\nu$$

[38] What do you understand by parallel displacement of a vector on a Riemannian manifold. Determine the exact equations for this when the manifold is the surface of a sphere in $\mathbb{R}^3$ and more generally when the surface has the form $z = f(x, y)$.

hint:First determine in terms of the metric the Christoffel symbols for an infinitesimal parallel displacement on a p-dimensional surface imbedded in n-dimensional Euclidean space. Then use the fact that the metric on the spherical surface of radius R is

$$ds^2 = R^2(d\theta^2 + sin^2(\theta)d\phi^2)$$

and the metric on the surface $z = f(x, y)$ is

$$ds^2 = dx^2 + dy^2 + (f_x(x, y)dx + f_y(x, y)dy)^2$$

where

$$f_x = \frac{\partial f}{\partial x}, f_y = \frac{\partial f}{\partial y}$$

[39] Determine the Riemann-Christoffel curvature tensor on the the two dimensional surface $z = f(x, y)$ immersed in $\mathbb{R}^3$.

[40] Fluid dynamical equations in curved space-time. Let the metric be $g_{\mu\nu}(x)$ where $x = (x^\mu)_{0 \le \mu \le 3}$. The proper time interval $d\tau$ is given by

$$d\tau^2 = g_{\mu\nu} dx^\mu dx^\nu$$

Let x^μ denote the space-time coordinates of a fluid particle. Then its four velocity is

$$v^\mu = \frac{dx^\mu}{d\tau}, \mu = 0, 1, 2, 3$$

Let p be the pressure and ρ the density of the fluid. Then the energy-momentum tensor is

$$T^{\mu\nu} = \rho . v^\mu v^\nu - p g^{\mu\nu}$$

Assuming that the fluid is non-viscous, the energy momentum conservation equation reads

$$T^{\mu\nu}_{:\nu} = 0$$

where : means covariant derivative. Using the mass conservation equation

$$(\rho v^{\mu})_{:\mu} = 0$$

we get

$$\rho v^{\nu} v^{\mu}_{:\nu} = g^{\mu\nu} p_{,\nu}$$

In flat space time, we have

$$(g_{\mu\nu}) = diag[1, -1, -1, -1], (g^{\mu\nu}) = diag[1, -1, -1, -1]$$

and the above equation gives

$$\rho(v^0 v^r_{,0} + \sum_{s=1}^{3} v^s v^r_{,s}) = -p_{,r}, r = 1, 2, 3$$

and for low velocities

$$v^0 \approx 1$$

so that we get the usual non-relativistic Navier-Stokes equation. In the general case, we have

$$\rho(v^{\nu} v^{\mu}_{,\nu} + \Gamma^{\mu}_{\nu\sigma} v^{\nu} v^{\sigma}) = g^{\mu\nu} p_{,\nu}$$

Using the equation

$$g_{\mu\nu} v^{\mu} v^{\nu} = 1$$

we can solve for v^0 in terms of $v^r, r = 1, 2, 3$. Assuming that this has been done, we can express the above general relativistic Navier-Stokes equation in terms of $v^r, r = 1, 2, 3, p$. The above equations are thus essentially three in number. The fourth equation is the mass conservation equation which can be expressed as

$$(\rho v^{\mu} \sqrt{-g})_{,\mu} = 0$$

If the fluid is incompressible, then ρ is a constant and this equation reduces to

$$(v^{\mu} \sqrt{-g})_{,\mu} = 0$$

We thus have four equations in all for the four variables $v^r, r = 1, 2, 3, p$.

[41] Kerr metric corresponding to a rotating blackhole has the form

$$d\tau^2 = E(t, r, \theta)dt^2 - F(t, r, \theta)dr^2 - G(t, r, \theta)d\theta^2 - C(t, r, \theta)(d\phi - \omega(t, r, \theta)dt)^2$$

The metric coefficients E, F, G, C, ω are determined from the Einstein field equations. Suppose we are given these coefficients and we wish to solve Maxwell's equations in this metric in the absence of charges and currents. Then we would be able to describe the propagation of electromagnetic waves in the vicinity of a rotating blackhole. Let

$$(A^\mu) = (A^0, A^1, A^2, A^3) = (V, A_r, A_\theta, A_\phi)$$

be the contravariant electromagnetic four potential. The space-time coordinates are

$$(x^\mu) = (x^0, x^1, x^2, x^3) = (t, r, \theta, \phi)$$

Thus,

$$g_{00} = E - C\omega^2, g_{11} = -F, g_{22} = -G, g_{33} = -C, g_{13} = g_{31} = C\omega$$

We compute the covariant components of the electromagnetic four potential.

$$A_0 = g_{00}A^0 = (E - c\omega^2)V, A_1 = g_{11}A^1 + g_{13}A^3 = -FA_r + c\omega A_\phi,$$

$$A_2 = g_{22}A^2 = -GA_\theta, A_3 = g_{33}A^3 + g_{31}A^1 = -CA_\phi + C\omega A_r$$

We next compute the covariant components of the antisymmetric electromagnetic field tensor.

[42] Obtain the components of the contravariant antisymmetric electromagnetic field tensor in a homogeneous and isotropic universe described by the Robertson-Walker metric.

hint: The equation of a sphere in $\mathbb{R}^4$ is given by

$$x^2 + y^2 + z^2 + u^2 = R^2$$

The spatial line element on such a surface is

$$dl^2 = dx^2 + dy^2 + dz^2 + du^2 = dr^2 + r^2 d\theta^2 + r^2 sin^2(\theta)d\phi^2 + (d\sqrt{R^2 - r^2})^2$$

$$= \frac{R^2}{R^2 - r^2}dr^2 + r^2 d\theta^2 + r^2 sin^2(\theta)d\phi^2$$

where the substitution

$$x = r.cos(\phi)sin(\theta), y = r.sin(\phi)sin(\theta), z = r.cos(\theta)$$

has been made. The complete space-time metric of our universe thus has the form

$$d\tau^2 = dt^2 - dl^2 = dt^2 - \frac{R^2(t)}{R^2(t) - r^2}dr^2 - r^2 d\theta^2 - r^2 sin^2(\theta)d\phi^2$$

where we have now allowed R to be a function of time. In the coordinate system

$$x^0 = t, x^1 = r, x^2 = \theta, x^3 = \phi$$

the components of the metric tensor are given by

$$g_{00} = 1, g_{11} = -\frac{R^2(t)}{R^2(t) - r^2}, g_{22} = -r^2, g_{33} = -r^2 sin^2(\theta)$$

It can be verified that the coordinate system chosen is comoving, ie, $r = c_1, \theta = c_2, \phi = c_3$ where c_1, c_2, c_3 are constants is a geodesic for this metric. Let the contravariant four potential be

$$(A^\mu) = (V, A_r, A_\theta, A_\phi)$$

Then,

$$A_0 = g_{00}A^0 = V, A_1 = g_{11}A^1 = -\frac{R^2(t)}{R^2(t) - r^2}A_r, A_2 = g_{22}A^2 = -r^2 A_\theta, A_3 = g_{33}A^3 = -r^2 sin^2(\theta)A_\phi$$

From these equations obtain the covariant antisymmetric field tensor

$$F_{\mu\nu} = A_{\nu,\mu} - A_{\mu,\nu}$$

and then raise the indices using the contravariant metric tensor $((g^{\mu\nu})) = ((g_{\mu\nu}))^{-1}$, ie,

$$g^{00} = 1, g^{11} = -(1 - r^2/R^2(t)), g^{22} = -1/r^2, g^{33} = -\frac{1}{r^2 sin^2(\theta)}$$

[43] Harmonic oscillator in special relativity. The equation of motion is given by

$$\frac{d}{dt}\frac{v(t)}{\sqrt{1 - v^2(t)/c^2}} + \omega^2 x(t) = 0, v(t) = x'(t)$$

This leads to the energy integral

$$\frac{c^2}{\sqrt{1 - x'2/c^2}} + \omega^2 x^2/2 = E$$

where E is a constant. This is a first order nonlinear differential equation for $x(t)$. It can be expressed as

$$c^4(E - \omega^2 x^2/2)^{-2} = (1 - x'^2/c^2)$$

or

$$x' = \pm c(1 - c^4(E - m\omega^2 x^2/2)^{-2})^{1/2}$$

This equation can be solved in series form using perturbation theory.

[44] Consider the surface of a four dimensional sphere of radius R defined by the equation

$$x^2 + y^2 + z^2 + u^2 = R^2$$

What metric is induced on this surface by the Euclidean metric on $\mathbb{R}^4$, ie, by the metric

$$(dl)^2 = (dx)^2 + (dy)^2 + (dz)^2 + (du)^2$$

[45] Formulate the Maxwell equations in n dimensional space using tensor calculus.

hint: Start with a contravariant potential $A^i, i = 0, 1, 2, ..., n$ and let the standard metric be given by

$$(d\tau)^2 = (dx^0)^2 - \sum_{r=1}^{n}(dx^r)^2 = \sum_{i,j=0}^{n} g_{ij}dx^i dx^j$$

The covariant components of the potential are given by

$$A_i = \sum_{j=0}^{n} g_{ij}A^j$$

Then the antisymmetric field tensor is

$$F_{ij} = A_{j,i} - A_{i,j}$$

Define the contravariant current density $J^i, i = 0, 1, ..., n$ so that

$$\sum_{j=0}^{n} F^{ij}_{,j} = \alpha . J^i, i = 0, 1, ..., n$$

where α is a constant.

[46] Let $\tau_{p,q}$ denote the parallel translation operator from the point p to the point q along the geodesic joining p and q on a Riemannian manifold M. Thus, $\tau_{p,q} : TM_p \to TM_q$ is a linear operator. Compute the operator $\tau_{p,q}$ for the two dimensional sphere in $\mathbb{R}^3$.

hint: If Γ^i_{jk} are the Christoffel symbols relative to a local coordinate chart, $\gamma^i(s), s \in [a, b]$ the geodesic curve in M relative to this local coordinate chart so that $\gamma(a) = p, \gamma(b) = q$ and ξ^i are the components of a vector in TM_p, then we solve

$$\frac{dX^i(s)}{ds} + \Gamma^i_{jk}(\gamma(s))\gamma^{j'}(s)X^k(s) = 0, a \leq s \leq b$$

with the initial condition

$$X^i(a) = \xi^i$$

Then the components of $\tau_{p,q}(\xi)$ are $X^i(b)$. This is possible provided p, q are inside the same local coordinate neighbourhood, ie, provided (U, ϕ) is a chart with $p, q \in U$. If not, then we must use the method of patching local overlapping neighbourhoods between p and q.

Reference:S.Helgason, Differential Geometry, Lie Groups and Symmetric Spaces.

Chapter 7

Group Representations Theory

This chapter introduces representation theory of Lie groups and Lie algebras including important concepts like the exponential map, root space decomposition and the Schur Lemmas.

[1]. Explain the notion of an irreducible representation of a group and of the group algebra.

hint:Take a look at Hermann Weyl's book, "Group theory and quantum mechanics". Suppose G is a group and let $\pi : G \to GL(V)$ be a representation of G in a vector space V. Here $GL(V)$ is the group of all non-singular transformations on V. By representation, we mean that π is a homomorphism. Then a subspace W of V is invariant if $\pi(g)(W) \subset W$ for all $g \in G$. If the only invariant subspaces of V are $\{0\}$ and V, we say that the representation π is irreducible. Suppose G is a finite group and $\mathcal{A}(G)$ is the set of all functions $f : G \to \mathbb{C}$. For $f \in \mathcal{A}(G)$ define

$$\pi(f) = \sum_{g \in G} f(g)\pi(g)$$

Here, π is a representation of G. Then it is clear that for $f_1, f_2 \in \mathcal{A}(G)$, we have

$$\pi(f_1)\pi(f_2) = \sum_{g_1, g_2 \in G} f_1(g_1)f_2(g_2)\pi_1(g_1 g_2)$$

$$= \sum_{g_1, g \in G} f_1(g_1)f_2(g_1^{-1}g)\pi(g) = \pi(f_1 * f_2)$$

where

$$f_1 * f_2(g) = \sum_{g_1 \in G} f_1(g_1)f_2(g_1^{-1}g)$$

is the group convolution of f_1 and f_2. Thus if the group algebra multiplication operation is convolution, then π becomes a representation of the group algebra.

[2]. State Schur's lemma for intertwining operators between two irreducible representations of G.

hint: Let π_1, π_2 be two irreducible representations of a group G in two vector spaces V_1, V_2. Assume that $T : V_1 \to V_2$ is a linear operator such that $T\pi_1(g) = \pi_2(g)T$ for all $g \in G$. Then it is clear that $\mathcal{N}(T)$ is invariant under π_1 and $\mathcal{R}(T)$ is invariant under π_2. It follows from the irreducibility of π_1 and π_2 that $\mathcal{N}(T)$ is either V_1 or $\{0\}$ and so is $\mathcal{R}(T)$. Assume that $T \neq 0$. Then $\mathcal{N}(T) = \{0\}$ and $\mathcal{R}(T) = V_2$. It follows that T is an isomorphism between the spaces V_1 and V_2 and hence π_1 and π_2 are equivalent. In case π_1 and π_2 are inequivalent, it is clear that $T = 0$.

[3]. Let G be a group acting transitively on a set M. Then from the left invariant Haar measure on G, construct an invariant measure on M.

hint: Choose a $x_0 \in M$ and consider the map $\tau : G \to M$ defined by $\tau(g) = gx_0$.

[4]. What do you understand by an analytic representation of a Lie group. Give examples.

Hint: for every vector ψ in the representation space, the map $x \to \pi(x)\psi$ should be an analytic map.

[5]. Explain how you would introduce coordinates on a manifold to make it into a topological manifold. Explain using this, the notion of a differentiable manifold.

[6]. What do you understand by a conjugation on a Lie algebra. Using this tool, explain the notion of compact and noncompact roots.

[7]. Derive the Schur orthogonality relations.

hint:Let π_1 be an irreducible representation of a compact group G in a finite dimensional vector space V_1 and π_2 another irreducible representation of G in another vector space V_2. Let $A : V_1 \to V_2$ be a linear transformation and consider

$$T = \int_G \pi_2(g) A \pi_1(g^{-1}) dg$$

where dg is the Haar measure on G. Then $T : V_1 \to V_2$ and

$$\pi_2(h)T = \int \pi_2(hg) A \pi_1(g^{-1}) dg = \int \pi_2(g) A \pi_1(g^{-1}h) dg = T\pi_1(h), h \in G$$

In other words, T intertwines the representations π_1 and π_2.

[8]. Suppose G is a Lie group and $\}$ is its Lie algebra. Let π be a representation of G in a vector space V and for $X \in \}$, define

$$\tilde{\pi}(X) = \frac{d}{dt}\pi(exp(tX))|_{t=0}$$

Then show that

$$\tilde{\pi}([X,Y]) = [\tilde{\pi}(X), \tilde{\pi}(Y)]$$

Specialize this idea to the three dimensional rotation group $SO(3)$. Write down the general form of any element in its Lie algebra and obtain a basis for this. Then consider a representation of $SO(3)$ for example the tensor representations, and calculate the elements of the corresponding Lie algebra.

Note: The Lie algebra of $SO(3)$ consists of all 3×3 real skew symmetric matrices. This Lie algebra has dimension three, and the standard basis for this Lie algebra is $iL_a, a = 1, 2, 3$ where $L_a, a = 1, 2, 3$ are the angular momentum operators. The commutation relations between these generators can be evaluated easily.

[9]. Suppose } is a semisimple Lie algebra and ⟨ is a Cartan subalgebra. Let P denote the set of positive roots with respect to this Cartan subalgebra. The set of all roots is therefore $\Delta = P \cup (-P)$. We have the root space decomposition

$$ \} = \langle + \bigoplus_{\alpha \in P} (\}_\alpha \oplus \}_{-\alpha}) $$

and for $H \in \langle$, we have

$$ exp(ad(H))(H') = 0, \quad exp(ad(H))(X_\alpha) = exp(\alpha(H))X_\alpha, \alpha \in \Delta $$

[10]. Prove the Leibniz rule for multiple differentials of the product of two functions.

hint: Let f, g be two functions and let D_f be the derivative acting on f and D_g the derivative acting on g. Thus,

$$ D = D_f + D_g $$

For example,

$$ D(fg) = (Df)g + fDg = (D_f + D_g)(fg) $$

Note that D_f and D_g commute. Thus

$$ D^n = (D_f + D_g)^n = \sum_{m=0}^{n} \binom{n}{m} D_f^m D_g^{n-m} $$

and hence

$$ D^n(fg) = \sum_{m=0}^{n} \binom{n}{m} D^m f D^{n-m} g $$

[11]. Let } be a semisimple Lie algebra and ⟨ a Cartan subalgebra of }. Let $\sqcup_0$, $\surd_0$ be as usual, ie, $\}_0 = \sqcup_0 \oplus \surd_0$ where $\}_0$ is a real form of }. Note that if $\sqcap$ is a compact real form of }, then $\sqcup_0 = \sqcap \cap \}_0$ and $\surd_0 = i\sqcap \cap \surd_0$. Show that if $\langle_0$ is a Cartan subalgebra of $\}_0$, then

$$ \langle_0 = \langle_0 \cap \sqcup_0 \oplus \langle_0 \cap \surd_0 $$

Hint:Make use of the conjugation $\tilde{\theta}$ of $\}$ with respect to $\sqcap$ or equivalently with respect to $\sqcup_0 + i \sqrt{}_0$.

Suppose $\langle_0 \subset \sqcup_0$. Then let α be a root and X_α a corresponding root vector. For $H = H_1 + H_2$ with $H_1 \in \sqcup_0, H_2 \in \sqrt{}_0$, we have

$$[H_1 + H_2, X_\alpha] = (\alpha(H_1) + \alpha(H_2))X_\alpha$$

and on applying the conjugation θ here, we get

$$[H_1 - H_2, \tilde{\theta}(X_\alpha)] = (-\alpha(H_1) + \alpha(H_2))\tilde{\theta}(X_\alpha)$$

since $\alpha(H_1)$ is pure imaginary while $\alpha(H_2)$ is real. Taking $H_2 = 0$, we get

$$[H_1, X_\alpha] = \alpha(H_1)X_\alpha, [H_1, \tilde{\theta}(X_\alpha)] = -\alpha(H_1)\tilde{\theta}(X_\alpha)$$

and taking $H_1 = 0$, we get

$$[H_2, X_\alpha] = \alpha(H_2)X_\alpha, [H_2, \tilde{\theta}(X_\alpha)] = -\alpha(H_2)\tilde{\theta}(X_\alpha)$$

Thus, $\tilde{\theta}(X_\alpha)$ is proportional to $X_{-\alpha}$. Now, let $\sqcup$ be the complexification of $\sqcup_0$ and $\sqrt{}$ the complexification of $\sqrt{}_0$ and let θ denote the conjugation that takes $X + Y$ to $X - Y$ where $X \in \sqcup$ and $Y \in \sqrt{}$. Then for $H \in \langle$ and α a root, we have

$$[H, X_\alpha] = \alpha(H)X_\alpha$$

Assume that $\langle_0 \subset \sqcup_0$. Then we have $\langle \subset \sqcup$ and hence $\theta(H) = H$. it follows that

$$[H, \theta(X_\alpha)] = \alpha(H)\theta(X_\alpha)$$

Note that if c is any complex number and $X \in \sqcup$, then $cX \in \sqcup$ and hence $\theta(cX) = cX$. On the other hand, if $Y \in \sqrt{}$, then $cY \in \sqrt{}$ and hence $\theta(cY) = -cY$. This gives $\theta(c(X + Y)) = c(X - Y) = c\theta(X + Y)$, ie, for every $Z \in \}$ and $c \in \mathbb{C}$, we have $\theta(c.Z) = c\theta(Z)$. It follows that

$$[H, X_\alpha \pm \theta(X_\alpha)] = \alpha(H)(X_\alpha \pm \theta(X_\alpha))$$

and since $X_\alpha + \theta(X_\alpha) \in \sqcup$ while $X_\alpha - \theta(X_\alpha) \in \sqrt{}$, it follows that we can choose the root vectors, such that each root vector belongs either to $\sqcup$ or to $\sqrt{}$. Note that the above argument and the fact that the root spaces are one dimensional implies that for each α, exactly one of $X_\alpha \pm \theta(X_\alpha)$ is zero. In other words, if α is a root, then X_α is either in $\sqcup$ or in $\sqrt{}$. In the former case, we say that α is a compact root and in the latter case, we say that α is a non-compact root.

[12]. Definition of totally positive roots:Let $\}$ be a Semisimple Lie algebra and $\langle$ a Cartan subalgebra. We assume that $\}_0$ is a real form of $\}$ and that $\langle_0$ is a Cartan subalgebra of $\}_0$ and its complexification equals $\langle$. Let $\approx_0$ and $\sqrt{}_0$ be as usual and assume that $\langle_0 \subset \approx_0$. It follows that $\langle \subset \approx$. Let χ denote the universal enveloping algebra of $\approx$, the complexification of $\approx_0$. Then let σ denote the adjoint representation of χ in $\}$. Let P be a positive system of roots relative to $\langle$. Let α be a root and X_α a corresponding root

vector. Suppose $\sigma(\chi)X_\alpha$ is contained in a sum of positive root spaces, then we say that α is a totally positive root. We want to show that a totally positive root is also positive. Indeed, suppose α is a totally positive root. Then $\sigma(\chi)X_\alpha \subset \sum_{\beta \in P'} \}_\beta$ where $P' \subset P$. It follows in particular that $X_\alpha \subset \sum_{\beta \in P'} \}_\beta$ and hence from the linear independence of root spaces, it follows that $\alpha \in P'$, proving that α is a positive root. We want to show next that a totally positive root is necessarily non-compact. Suppose α is a totally positive root such that $X_\alpha in \approx$, ie, α is compact.

[13]. What do you understand by the adjoint representation of a Lie group in its Lie algebra ?

ans:Let G be a Lie a group and $\}$ its Lie algebra. Then for each $g \in \mathcal{G}$ and $X \in \}$, we define $Ad(g)(X) \in \}$ so that $Ad(g)$ is the differential of the map $x \to gxg^{-1}$ from G into G at e. In case G is a matrix Lie group, $\}$ becomes a matrix Lie algebra and then $Ad(g)(X) = gXg^{-1}$. Note that $Ad(g)$ is a linear transformation in $\}$ and $Ad(gh) = Ad(g)Ad(h)$ for all $g, h \in G$.

[14]. What do you understand by Haar measure on a non-Abelian group. What is the modular function. What do you understand by the statement "A compact group is unimodular".

[15]. Let $T : \mathbb{R}^n \to \mathbb{R}^n$ and $S : \mathbb{R}^n \to \mathbb{R}^n$ be differentiable map. Let $J_T(x)$ be the Jacobian determinant of T and $J_S(x)$ the Jacobian determinant of S. Justify the statement $J_{ST}(x) = J_S(Tx).J_T(x)$. Writing this in cocyle form by setting $J_T(x) = J(T,x)$, we have $J(ST,x) = J(S,Tx).J(T,x)$.

[16]. Let G be a finite group and $T : G \to GL(V)$ a representation of G in a vector space V. Assume that V is equipped with an inner product $< .,. >$. Consider another inner product $(.,.)$ on V defined by

$$(x,y) = \sum_{g \in G} < T(g)x, T(g)y >$$

Then show that this new inner product is G-invariant. In other words, show that the operator $T(g)$ is unitary with respect to the inner product $(.,.)$.

[17].Let V be a vector space over a field $\mathbb{F}$ and let $\otimes$ be a tensor product on V. Let G be a group and $T : G \to GL(V)$ be a representation of G in V and consider the tensor representation $g \to T(g)^{\otimes m}$ of G into $V^{\otimes m}$. Show that if χ_T is the character of the representation T, then χ_T^m is the character of this tensor representation. Deduce that the tensor representation is not necessarily irreducible when T is.

[18]. Determine the structure constants of the Lorentz group in an appropriate coordinate system.

hint: The infinitesimal Lorentz transformation on $\mathbb{R}^4$ has the form

$$x^\mu \to x^\mu + \epsilon^\mu_\nu x^\nu$$

Let $(g_{\mu\nu})$ denote the standard metric tensor, ie,

$$(g_{\mu\nu}) = diag[1, -1, -1, -1]$$

Then let $\epsilon_{\mu\nu} = g_{\mu\alpha}\epsilon_\nu^\alpha$. Show that

$$\epsilon_{\mu\nu} = -\epsilon_{\nu\mu}$$

Use this equation to obtain explicit formulae for the 4×4 matrix generators of the Lorentz group. The commutation relations for the generators may be evaluated explicitly from these expressions.

[19].Let $\}$ be a Lie algebra and let $X_1, ..., X_m$ be a set of generators of this Lie algebra. Let $c(mnp)$ be the structure constants of this Lie algebra relative to this basis, ie,

$$[X_m, X_k] = \sum_p c(mkp)X_p$$

Then use the Jacobi identity to deduce a quadratic relation for the coefficients $c(mkp)$.

hint:Make use of the Jacobi identity in the form

$$[[X_m, X_k], X_p] + [[X_k, X_p], X_m] + [[X_p, X_m], X_k] = 0$$

[20].Let $\}$ be a Lie algebra and let $\sqcap$ be a compact real form of this Lie algebra, ie, $\sqcap$ is the Lie algebra of a compact Lie subgroup of G where $\}$ is the Lie algebra of G and $\} = \sqcap + i.\sqcap$. Let $\approx_0 = \sqcap \cap \}_0$ and $\sqrt{}_0 = i\sqcap \cap \}_0$. Here, $\}_0$ is a real form of $\}$. We see that

$$\approx_0 + i.\sqrt{}_0 = \sqcap \cap (\}_0 + i.\}_0) = \sqcap \cap \} = \sqcap$$

$\tilde{\theta}$ denotes conjugation with respect to $\sqcap$. It follows that $\tilde{\theta}$ is conjugation with respect to $\approx_0 + i\sqrt{}_0$. Let $\approx$ be the complexification of $\approx_0$ and $\sqrt{}$ the complexification of $\sqrt{}_0$. We then have $\} = \approx + \sqrt{}$ as a direct sum. Let η denote conjugation with respect to $\}_0$, ie, $\eta(X + iY) = X - iY$ for $X, Y \in \}_0$. We then have on setting $\theta = \tilde{\theta}o\eta$ that $\theta(X + Y) = X - Y$ for all $X \in \approx$ and $Y \in \sqrt{}$. Assume that $\langle_0$ is a Cartan subalgebra of $\}_0$ such that $\langle_0 \subset \approx_0$ Consider $\langle = \langle_0 + i.\langle_0$, the complexification of $\langle_0$. Let α be a root of $\}$ with respect to the Cartan subalgebra $\langle$. Then we have for $H = H_1 + iH_2$ with $H_1, H_2 \in \langle_0$ and X_α a root vector that

$$[H, X_\alpha] = [H_1 + iH_2, X_\alpha] = \alpha(H)X_\alpha = (\alpha(H_1) + i\alpha(H_2))X_\alpha$$

Now $\alpha(H_1) and \alpha(H_2)$ are pure imaginary since $\langle_0 \subset \approx_0$ and $\approx_0$ is a compact Lie algebra. It follows that

$$[H_1 - iH_2, \tilde{\theta}(X_\alpha)] = \theta([H_1 + iH_2, X_\alpha]) = \tilde{\theta}((\alpha(H_1) + i\alpha(H_2))X_\alpha)$$

$$= (-\alpha(H_1) + i\alpha(H_2))\tilde{\theta}(X_\alpha)$$

since $\alpha(H_i), i = 1, 2$ are pure imaginary. It follows on replacing H_2 by $-H_2$ that

$$[H_1 + iH_2, \tilde{\theta}(X_\alpha)] = -\alpha(H_1 + iH_2)\tilde{\theta}(X_\alpha)$$

which proves that

$$\tilde{\theta}(X_\alpha) = c_\alpha X_{-\alpha}$$

where c_α is some complex constant. We get on again operating with $\tilde{\theta}$ that

$$X_\alpha = c_\alpha c_{-\alpha} X_\alpha$$

so that

$$c_\alpha c_{-\alpha} = 1$$

Assume now that $\langle_0 \subset \approx_0$ does not necessarily hold but that $\langle$ is invariant under θ, ie, $\theta(\langle) = \langle$. Consider now the direct sum decomposition $\langle = \langle_t + \langle_p$ where $\langle_t = \langle \cap \approx$ and $\langle_p = \langle \cap \sqrt{}$. This equation can be justified as follows. Let $X \in \langle$. Then $X = X_1 + X_2$ where $X_1 \in \approx$ and $X_2 \in \sqrt{}$. It follows that $\theta(X) = X_1 - X_2$ and hence $X_1 = \frac{1}{2}(X + \theta(X))$ and $X_2 = \frac{1}{2}(X - \theta(X))$. But then it is clear that $X_1, X_2 \in$ *mathcalh* since $\langle$ is invariant under θ. Now consider the equation

$$[H, X_\alpha] = \alpha(H) X_\alpha$$

Write $H = H_1 + H_2$ where $H_1 \in \approx$ and $H_2 \in \sqrt{}$. Then we get on applying θ to both sides,

$$[H_1 - H_2, \theta(X_\alpha)] = \alpha(H)\theta(X_\alpha)$$

Suppose $\langle \subset \approx$. It then follows that $H_2 = 0$ and hence $\theta(X_\alpha) = a_\alpha X_\alpha$ for some complex constant a_α. Since $\theta^2 = 1$, we get

$$X_\alpha = a_\alpha X_{-\alpha}$$

and hence $a_\alpha = \pm 1$. Note that we are making use of the fact that if c is any complex number and $X \in \}$, then $\theta(a.X) = a\theta(X)$. Indeed, write $X = X_1 + X_2$ where $X_1 \in \sqcup$ and $X_2 \in \sqrt{}$. Then $cX_1 \in \approx$ and $cX_2 \in \sqrt{}$ so that $\theta(cX_1) = cX_1$ and $\theta(cX_2) = -cX_2$ which implies

$$\theta(c.X) = c\theta(X)$$

[21]. Let $\}, \approx_0, \sqrt{}_0, \approx, \sqrt{}, \langle_0, \langle$ be as before. Consider $\langle^* = \langle_{p_0} + i\langle_{t_0}$. Choose a basis $H_1, ..., H_l$ for $\langle_0$ so that $H_1, ..., H_m$ is a basis for $\langle_{p_0}$ and $H_{m+1}, ..., H_l$ is a basis for $\langle_{t_0}$. Define $H_k^* = H_k, 1 \le k \le m$ and $H_k^* = iH_k$ if $m + 1 \le k \le l$. Then it is clear that $H_1^*, ..., H_l^*$ is a basis for $\langle^*$. For $H \in \langle$ we can write $H = \sum_{j=1}^{l} c_j H_j^*$. If λ is a linear functional on $\langle$, we have $\lambda(H) = B(H_\lambda, H)$ for some unique $H_\lambda \in \langle$. We say that λ is real if $H_\lambda \in \langle^*$. In this case, we can write $H_\lambda = H_1 + iH_2$ where $H_1 \in \langle_{p_0}$ and $H_2 \in \langle_{t_0}$. Now let $H' \in \langle^*$. Then, $H' = H_1' + iH_2'$ for some $H_1' \in \langle_{p_0}$ and $H_2' \in \langle_{t_0}$. Then

$$\lambda(H) = B(H_\lambda, H) = B(H_1 + iH_2, H_1' + iH_2') = B(H_1, H_1') - B(H_2, H_2')$$

Now $ad(H_1)$ and $ad(H_1')$ have all eigenvalues real for the following reason: $\sqrt{}_0 = i \sqcap \cap \}_0$ where $\sqcap$ is a compact real form of $\}$. If U is the Lie group corresponding to the Lie algebra $\sqcap$, then U is compact and hence $Ad(g)$ can be made unitary with respect to an averaged inner product for all $g \in G$. This means that $Ad(g)$ has all its eigenvalues on the unit circle whence $ad(X)$ has all eigenvalues pure imaginary for all $X \in \sqcap$. It follows that if $X \in \rangle u$ then $ad(X)$ has all eigenvalues real. Now, $B(H_1, H_1') = Tr(ad(H_1).ad(H_1'))$. If we assumes $ad(H_1), ad(H_1')$ to be semisimple, then both are diagonal relative to an appropriate basis in both have all eigenvalues real. Hence $B(H_1, H_1')$ is real. Likewise $B(H_2, H_2')$ is also real. Thus, $\lambda(H)$ is real for almost all $H \in \langle^*$ and this justifies the definition of reality of a linear functional λ on $\langle$. Suppose that $H \in \langle^*$. We can write $H = \sum_{i=1}^{l} c_i H_i^*$ for some $c_1, ..., c_l \in \mathbb{R}$. Let j be the smallest index k for which $c_k \neq 0$. Then we say that $H > 0$ if $c_j > 0$. This is an order on $\langle^*$. For real λ, μ, we say that $\lambda > \mu$ if $H_{\lambda-\mu} = H_\lambda - H_\mu > 0$. This puts an order on real linear functionals λ on $\langle$. Suppose α is a root. Then $\alpha(H)$ is pure imaginary for all $H \in \langle_{t_0}$ (since $Ad(exp(H))(X_\alpha) = exp(\alpha(H))X_\alpha$ and $Ad(exp(H))$ has pure imaginary eigenvalues for all $H \in \approx_0$). It follows that $\alpha(H)$ is real for all $H \in \langle_{p_0}$ and hence $\alpha(H)$ is real for all $H \in \langle_{p_0} + i\langle_{t_0} = \langle^*$. This proves that every root is a real linear functional. Now let P be the positive system of roots defined by the above order relation. Let P_+ denote the set of all roots $\alpha \in P$ for which $\theta\alpha \neq \alpha$ and P_- the set of all roots $\alpha \in P$ for which $\theta\alpha = \alpha$, where $\theta(X + Y) = X - Y$ for all $X \in \approx$ and $Y \in \sqrt{}$. Note that if λ is a linear functional on $\langle$, then $\theta\lambda(H) = \lambda(\theta(H))$ by definition.

It is clear that P is the disjoint union of P_+ and P_-. Now suppose $\alpha \in P_+$. Then we claim that $\theta\alpha < 0$. $\alpha \in P_+$ implies $\theta\alpha \neq \alpha$ implies $H_{\theta\alpha} \neq H_\alpha$. Now

$$B(H_{\theta\alpha}, H) = \theta\alpha(H) = \alpha(\theta H) = B(H_\alpha, \theta(H)) = B(\theta(H_\alpha), H), H \in \langle$$

and therefore

$$H_{\theta\alpha} = \theta(H_\alpha)$$

Thus, $\alpha \in P_+$ implies

$$\theta(H_\alpha) \neq H_\alpha$$

It follows that $H_\alpha \notin \rangle h_{\approx_0}$. Hence H_α has a component in $\langle_{p_0}$. Since $H_\alpha > 0$, we can write

$$H_\alpha = \sum_{j=r}^{l} c_j H_j^*$$

where $r \leq m$ and $c_r > 0$. It follows that

$$\theta(H_\alpha) = -c_r H_r^* + \sum_{j>r} c_j H_j^*$$

and hence $\theta(H_\alpha) < 0$, or equivalently, $-H_{\theta\alpha} > 0$ implying that $\theta\alpha < 0$.

Remark:This result has been proved by Iwasawa and has been used by HarishChandra in his papers on representations of semisimple Lie groups.

[22].Suppose $\}$ is a semisimple Lie algebra and $\backslash_+$ is the set of positive root vectors and $\backslash_-$ is the set of negative root vectors relative to a Cartan subalgebra $\langle$. We have the direct sum decomposition

$$\} = \langle + \backslash_+ + \backslash_-$$

Let $\sqcap$ be a compact real form of $\}$ and let $\}_0$ be a real form of $\}$. Consider $\approx_0 = \sqcap \cap \}_0$ and $\sqrt{}_0 = i \sqcap \cap \}_0$. Then show that $\approx_0 + i\sqrt{}_0 = \sqcap$. Let $\langle_0 = \langle \cap \}_0$. Let $\approx = \approx_0 + i\approx_0$ and $\sqrt{} = \sqrt{}_0 + i\sqrt{}_0$. Then prove the direct sum decomposition

$$\} = \approx + \sqrt{}$$

Now let $\theta : \} \to \}$ be defined so that $\theta(X + Y) = X - Y$ for $X \in \approx, Y \in \sqrt{}$. Then, show that θ is the composition of the two conjugations on $\}$, one relative to $\sqcap$ and the other relative to $\}_0$.

[23]. Let $x^0 = t, x^1 = x, x^2 = y, x^3 = z$. $(g_{\mu\nu}) = diag[1, -1, -1, -1]$. The matrix of an infinitesimal Lorentz transformation is antisymmetric after lowering an index to make it covariant:

$$\epsilon_{\mu\nu} = -\epsilon_{\nu\mu}$$

The actual matrix of the corresponding infinitesimal Lorentz transformation is

$$(\epsilon^\mu_\nu),$$

where

$$\epsilon^\mu_\nu = g^{\mu\alpha}\epsilon_{\alpha\nu}$$

We can write this matrix as

$$(\epsilon^\mu_\nu) = \epsilon_{01} J^{01} + \epsilon_{02} J^{02} + \epsilon_{03} J^{03} + \epsilon_{12} J^{12} + \epsilon_{23} J^{23} + \epsilon_{31} J^{31}$$

$$= \epsilon^0_1 J^{01} + \epsilon^0_2 J^{02} + \epsilon^0_3 J^{03} - \epsilon^1_2 J^{12} - \epsilon^2_3 J^{23} - \epsilon^3_1 J^{31}$$

so that

$$J^{01} = \begin{pmatrix} 0 & 1 & 0 & 0 \\ 1 & 0 & 0 & 0 \\ 0 & 0 & 0 & 0 \\ 0 & 0 & 0 & 0 \end{pmatrix}$$

$$J^{02} = \begin{pmatrix} 0 & 0 & 1 & 0 \\ 0 & 0 & 0 & 0 \\ 1 & 0 & 0 & 0 \\ 0 & 0 & 0 & 0 \end{pmatrix}$$

$$J^{03} = \begin{pmatrix} 0 & 0 & 0 & 1 \\ 0 & 0 & 0 & 0 \\ 0 & 0 & 0 & 0 \\ 1 & 0 & 0 & 0 \end{pmatrix}$$

$$J^{23} = \begin{pmatrix} 0 & 0 & 0 & 0 \\ 0 & 0 & 0 & 0 \\ 0 & 0 & 0 & -1 \\ 0 & 0 & 1 & 0 \end{pmatrix}$$

$$J^{31} = \begin{pmatrix} 0 & 0 & 0 & 0 \\ 0 & 0 & 0 & 1 \\ 0 & 0 & 0 & 0 \\ 0 & -1 & 0 & 0 \end{pmatrix}$$

$$J^{12} = \begin{pmatrix} 0 & 0 & 0 & 0 \\ 0 & 0 & -1 & 0 \\ 0 & 1 & 0 & 0 \\ 0 & 0 & 0 & 0 \end{pmatrix}$$

From these equations, it follows that

$$[J_1, J_2] = J_3, [J_2, J_3] = J_1, [J_3, J_1] = J_2,$$

$$[K_1, K_2] = -J_3, [K_2, K_3] = -J_1, [K_3, K_1] = -J_2,$$

$$[J_1, K_2] = K_3, [J_2, K_3] = K_1, [J_3, K_1] = K_2$$

where

$$K_i = J^{0i}, i = 1, 2, 3, J_1 = J^{23}, J_2 = J^{31}, J_3 = J^{12}$$

Also

$$[J_a, K_a] = 0, a = 1, 2, 3$$

Note that $\{K_1, K_2, K_3\}$ are the generators of boosts while $\{J_1, J_2, J_3\}$ are the generators of rotations. Define

$$A_a = iJ_a + K_a, B_a = iJ_a - K_a$$

Note that since $J_a, a = 1, 2, 3$ are real and antisymmetric while K_a are real and symmetric, iJ_a and K_a are both Hermitian and hence $A_a^* = B_a$. Since

$$[J_a, J_b] = \epsilon_{abc} J_c, [K_a, K_b] = -\epsilon_{abc} J_c, [J_a, K_b] = \epsilon_{abc} K_c$$

it follows that

$$[A_a, A_b] = -[J_a, J_b] + i[J_a, K_b] + i[K_a, J_b] + [K_a, K_b]$$

$$= -\epsilon_{abc} J_c + i\epsilon_{abc} K_c + i\epsilon_{abc} K_c - \epsilon_{abc} J_c = 2\epsilon_{abc}(-J_c + iK_c)$$

$$= 2i\epsilon_{abc}(iJ_c + K_c) = 2i\epsilon_{abc} A_c$$

Likewise,

$$[B_a, B_b] = [iJ_a - K_a, iJ_b - K_b] = -[J_a, J_b] - i[J_a, K_b] - i[K_a, J_b] + [K_a, K_b]$$

$$= -\epsilon_{abc} J_c - i\epsilon_{abc} K_c - i\epsilon_{abc} K_c - \epsilon_{abc} K_c$$

$$= -2\epsilon_{abc}(J_c + iK_c) = 2i\epsilon_{abc}(iJ_c - K_c) = 2i\epsilon_{abc} B_c$$

Finally,

$$[A_a, B_b] = [iJ_a + K_a, iJ_b - K_b] = -[J_a, J_b] - i[J_a, K_b] + i[K_a, J_b] - [K_a, K_b]$$

$$-\epsilon_{abc} J_c - i\epsilon_{abc} K_c + i\epsilon_{abc} K_c + \epsilon_{abc} J_c = 0$$

If we replace A_a and B_a respectively by $A_a/\sqrt{2}$ and $B_a/\sqrt{2}$ respectively, then we get the commutation relations

$$[A_a, A_b] = i\epsilon_{abc}A_c, [B_a, B_b] = i\epsilon_{abc}B_c, [A_a, B_b] = 0$$

In other words, the matrices $\{A_a, a = 1, 2, 3\}$ form a basis for a Lie algebra isomorphic to $so(3)$ and so do the matrices $\{B_a, a = 1, 2, 3\}$. Moreover, the matrices $\{A_1, A_2, A_3, B_a, B_2, B_3\}$ form a basis for a Lie algebra isomorphic to the tensor product of $so(3)$ with itself. Note the way in which we define the tensor product of two Lie algebras. If $\}_i, i = 1, 2$ are two Lie algebras and $\otimes$ is a tensor product between the the two Lie algebras, then the set of all elements of the form $X \otimes I + I \otimes Y$ where $X \in \}_1$ and $Y \in \}_2$ forms a Lie algebra if the Lie algebras $\}_i$ are matrix Lie algebras. In general, an element of $\}_1 \otimes \}_2$ is given by (X, Y) where $X \in \}_1, Y \in \}_2$ and the Lie bracket is

$$[(X_1, Y_1), (X_2, Y_2)] = ([X_1, X_2], [Y_1, Y_2])$$

In the above case, we identify the generator A_a with $A_a \otimes I$ and B_a with $I \otimes B_a$ and then the commutation relations follow. Suppose G is a matrix Lie group and $\}$ is its Lie algebra. Let $\otimes$ be a tensor product of G with itself. For $X, Y \in \}$ and $t \in \mathbb{R}$, we consider the element of $\} \otimes \}$ defined by

$$\frac{d}{dt}exp(tX) \otimes exp(tY)|_{t=0} = X \otimes I + I \otimes Y$$

Note that

$$[X \otimes I + I \otimes Y, X' \otimes I + I \otimes Y']$$

$$= (X \otimes I + I \otimes Y)(X' \otimes I + I \otimes Y') - (X' \otimes I + I \otimes Y')(X \otimes I + I \otimes Y)$$

$$= XX' \otimes I + X \otimes Y' + X' \otimes Y + I \otimes YY' - X'X \otimes I - X' \otimes Y - X \otimes Y' - I \otimes YY'$$

$$= (XX' - X'X) \otimes I + I \otimes (YY' - Y'Y) = [X, X'] \otimes I + I \otimes [Y, Y']$$

Thus, if we define

$$[X, Y] = XY - YX$$

and

$$(X, Y) = X \otimes I + I \otimes Y$$

then we have

$$[(X, Y), (X', Y')] = ([X, X'], [Y, Y'])$$

For more details on tensor product, see the book "Lie groups, Lie algebras and their Representations" by V.S.Varadarajan.

[24].Clebsch-Gordon coefficients:Let $\mathbf{J}_1, \mathbf{J}_2$ be two sets of mutually commuting angular momentum vector operators. Let $|j_1, m_1, j_2, m_2 >$ or for short $|m_1, m_2 >$ be a state in which $\mathbf{J}_1^2 = \sum_{a=1}^{3} J_{1a}^2$, $\mathbf{J}_2^2 = \sum_a J_{2a}^2$, J_{13}, J_{23} respectively have eigenvalues $j_1(j_1 + 1), j_2(j_2 + 1), m_1, m_2$. It is known that if $\mathbf{J} = \mathbf{J}_1 + \mathbf{J}_2$, then the possible eigenvalues of $\mathbf{J}^2$ in a state in which $\mathbf{J}_1^2$ and $\mathbf{J}_2^2$ respectively have eigenvalues $j_1(j_1 + 1)$ and $j_2(j_2 + 1)$ are $j(j + 1)$ where $|j_1 - j_2| \le j \le j_1 + j_2$. Fix j_1, j_2 and consider the state $|j, m >$ in which $\mathbf{J}^2, J_3$ respectively have eigenvalues $j(j + 1)$ and m. Then $\{|m_1, m_2 >: |m_i| \le j_i, i = 1, 2, \}$ is

an orthonormal basis for the set of all states and so is $\{|j,m> : |j_1 - j_2| \leq j \leq j_1 + j_2, |m| \leq j\}$. The number of elements in both the sets tally since

$$\sum_{j=|j_1-j_2|}^{j_1+j_2} (2j+1) = (2j_1+1)(2j_2+1)$$

Thus, we can write

$$|j,m> = \sum_{|m_1|\leq j_1, |m_2|\leq j_2} C(j,m|m_1,m_2)|m_1,m_2>$$

where the matrix $((C(j,m|m_1,m_2)))$ is unitary. We thus have

$$|m_1,m_2> = \sum_{j=|j_1-j_2|}^{j_1+j_2} \sum_{m=-j}^{j} \bar{C}(j,m|m_1,m_2)|j,m>$$

The fundamental problem is to determine the Clebsch-Gordon coefficients $C(j,m|m_1,m_2)$. It is clear that $C(j,m|m_1,m_2)$ is zero unless $m = m_1 + m_2$. We have

$$C(j,m|m_1,m_2) = <m_1,m_2|j,m>$$

Now, consider the operators

$$J_+ = J_{1+} + J_{2+} = J_{11} + iJ_{12} + J_{21} + iJ_{22}$$

and

$$J_-$$

[25]. Show that a semisimple Lie group is unimodular, ie, the left and right Haar measures conicide.

ans: Let μ_l be the left Haar measure and μ_r the right Haar measure. Consider

$$\Delta(g) = d\mu_l(g)/d\mu_r$$

We have

$$\int f(g)\Delta(g)d\mu_r(g) = \int f(g)d\mu_l(g)$$

and hence,

$$\int f(g)\Delta(gh)d\mu_r(g) = \int f(gh^{-1})\Delta(g)d\mu_r(g) = \int f(gh^{-1})d\mu_l(g) = \int f(gh^{-1})\Delta(g)d\mu_r(g)$$

We wish to prove that $\Delta(gh) = \Delta(g)\Delta(h), g,h \in G$. We have for any Borel set E,

$$\mu_l(Eg) = \Delta(g)\mu_l(E)$$

by definition so that

$$\int f(x)d\mu_l(xg) = \Delta(g) \int f(x)d\mu_l(x)$$

or

$$\int f(xg^{-1})d\mu_l(x) = \Delta(g) \int f(x)d\mu_l(x)$$

or

$$\int f(xg^{-1})\psi(x)d\mu_r(x) = \Delta(g) \int f(x)\psi(x)d\mu_r(x)$$

where

$$\psi(x) = d\mu_l(x)/d\mu_r$$

It follows that

$$\int f(x)\psi(xg)d\mu_r(x) = \Delta(g) \int f(x)\psi(x)d\mu_r(x)$$

and since f is arbitrary, it follows that

$$\psi(xg) = \Delta(g)\psi(x)$$

and hence normalizing so that $\psi(e) = 1$, we get

$$\psi(g) = \Delta(g)$$

Also

$$\Delta(gh)\mu_l(E) = \mu_l(Egh) = \Delta(h)\mu_l(Eg) = \Delta(h)\Delta(g)\mu_l(E)$$

so that

$$\Delta(gh) = \Delta(g)\Delta(h), g, h \in G$$

ie, $\Delta : G \to \mathbb{R}_+$ is a homomorphism. Suppose G_1, G_2 are Lie groups and $T : G_1 \to G_2$ is a homomorphism. Then $dT : \}_1 \to \}_2$ is a Lie algebra homomorphism where $\}_i$ is the Lie algebra of G_i. Hence, $d\Delta : \} \to \mathbb{R}$ is a non-trivial Lie algebra homomorphism. Let $\langle$ denote its kernel. Then, $\}/\langle$ is isomorphic to $\mathcal{R}$ and hence $\}/\langle$ is solvable which means that $\}$ has a solvable ideal and hence cannot be semisimple.

[26] Let $\}, \}_0, \sqcap, \approx_0, \sqrt{}_0, \approx, \sqrt{}$ be as usual and let $\langle$ be a Cartan subalgebra of $\}$ and $\langle_0 = \langle \cap \}_0$. Assume that $\langle$ is invariant under $\tilde{\theta}$. Here η is the conjugation with respect to $\}_0$, $\tilde{\theta}$ is the conjugation with respect to $\sqcap = \approx_0 + i \sqrt{}_0$ and $\theta = \tilde{o}\eta$. We assume that $\theta(\langle) = \langle$ and hence $\tilde{(}\langle_0) = \langle_0$. Let $\langle_p = \langle \cap \sqrt{}$ and let β be a root that takes real values on $\langle_0$. Then we claim that $H_\beta \in \langle_{p0}$. For we have for any $H \in \langle_0, H = H_1 + H_2$ where $H_1 \in \langle_{t0}$ and $H_2 \in \langle_{p0}$ and then

$$B(H_\beta, H) = \beta(H) = \beta(H_1) + \beta(H_2)$$

$\beta(H_1)$ is pure imaginary while $\beta(H_2)$ is real. By hypothesis, $\beta(H_1) = 0$ and hence

$$B(H_\beta, H - H_2) = 0$$

or

$$B(H_\beta, H_1) = 0$$

Since $H_1 \in \approx_0$ is arbitrary and $\sqrt{}_0$ is the orthogonal complement of $\approx_0$, it necessarily follows that $H_\beta \in \sqrt{}_0$ and hence $H_\beta \in \langle_{p_0}$. We denote by Q_+, the set of all roots that take only real values on $\langle_{p_0}$. Let P_+ denote the set of all roots that do not vanish identically on $\langle_p$. Then we claim that $Q_+ \subset P_+$. To see this, let $\beta \in Q_+$. Then $\beta(\langle_{t_0})$ is pure imaginary and hence $\beta(\langle_{t_0}) = 0$ since β is real on $\langle_0$ and $\langle_{t_0} \subset \langle_0$. It follows that if β vanished identically on $\langle_{p_0}$, then β would vanish identically on $\langle_0$ which is false. Hence $\beta \in P_+$. Let Suppose β is a root that falls in the complement of P_+. Then β vanishes on $\langle_p$. Then we claim that X_β is either in $\approx$ or in $\sqrt{}$. Indeed, for $H_1 \in \langle_t$ and $H_2 \in \langle_p$, we have

$$[H_1 + H_2, X_\beta] = \beta(H_1)X_\beta$$

and applying the automorphism θ to both sides gives us

$$[H_1 - H_2, \theta(X_\beta)] = \beta(H_1)\theta(X_\beta)$$

It follows on replacing H_2 by $-H_2$ in this equation that for $H = H_1 + H_2$, we have

$$[H, X_\beta] = \beta(H)X_\beta, [H, \theta(X_\beta)] = \beta(H)\theta(X_\beta)$$

Since the root spaces are one dimensional, we must therefore have

$$\theta(X_\beta) = c.X_\beta$$

For some complex scalar c. Again applying θ to both sides gives

$$X_\beta = c^2 X_\beta$$

and hence $c = \pm 1$. Thus $\theta(X_\beta) = \pm X_\beta$ and hence X_β is either in $\approx$ or in $\sqrt{}$. Thus we can write

$$P_+^c = P_0 \cup P_-$$

where P_0 is the set of roots β for which $X_\beta \in \sqrt{}$ while P_- is the set of roots for which $X_\beta \in t$.

Let H, X, Y be three elements satisfying the commutation relations

$$[H, X] = 2X, [H, Y] = -2Y, [X, Y] = H$$

Then

$$\mathfrak{l}_1 = \mathbb{R}H + \mathbb{R}X + \mathbb{R}Y$$

is a real Lie algebra. Let

$$\mathfrak{l}_2 = \mathbb{R}.iH + \mathbb{R}.i(X - Y) + \mathbb{R}.(X + Y)$$

This is also a real Lie algebra in view of the commutation relations

$$[iH, i(X - Y)] = -[H, X - Y] = -[H, X] + [H, Y] = -2X - 2Y = -2(X + Y)$$

$$[iH, X + Y] = i[H, X] + i[H, Y] = 2iX - 2iY = 2i(X - Y)$$

$$[i(X - Y), X + Y] = i[X, Y] - i[Y, X] = 2i[X, Y] = 2iH$$

Finally, set

$$l_3 = \mathbb{R}.iH + \mathbb{R}.(X - Y) + \mathbb{R}.i(X + Y)$$

This is also a real Lie algebra since

$$[iH, X - Y] = i[H, X] - i[H, Y] = 2iX + 2iY = 2i(X + Y),$$

$$[iH, i(X + Y)] = -[H, X + Y] = -[H, X] - [H, Y] = -2X + 2Y = -2(X - Y),$$

$$[X - Y, i(X + Y)] = i[X, Y] - i[Y, X] = 2i[X, Y] = 2iH$$

We shall evaluate the Cartan-Killing forms of these Lie algebras. Set

$$\xi(t, x, y) = tH + xX + yY, t, x, y \in \mathbb{R}$$

Then,

$$\Omega(\xi) = Tr(ad(\xi)^2) = Tr(ad(tH + xX + yY)^2)$$

Now,

$$ad(\xi)(H) = x[X, H] + y[Y, H] = -2xX + 2yY, ad(\xi)^2(H)$$

$$= [tH + xX + yY, -2xX + 2yY] = -2tx[H, X] + 2ty[H, Y] + 8xy[X, Y]$$

$$= -4txX - 4tyY + 8xyH$$

$$ad(\xi)(X) = [tH + xX + yY, X] = t[H, X] + y[Y, X] = 2tX - yH$$

$$ad(\xi)^2(X) = [tH + xX + yY, 2tX - yH]$$

$$= 2t^2[H, X] - xy[X, H] + 2ty[Y, X] - y^2[Y, H] = 4t^2X + 2xyX - 2tyH + 2y^2Y$$

$$ad(\xi)(Y) = [tH + xX + yY, Y] = t[H, Y] + x[X, Y] = -2tY + xH$$

$$ad(\xi)^2(Y) = [tH + xX + yY, -2tY + xH]$$

$$= -2t^2[H, Y] - 2tx[X, Y] + x^2[X, H] + xy[Y, H] = 4t^2Y - 2txH - 2x^2X + 2xyY$$

This gives

$$\Omega(\xi) = Tr(ad(\xi)^2) = 8xy + 4t^2 + 2xy + 4t^2 + 2xy = 12xy + 8t^2$$

Note that t, x, y vary over $\mathbb{R}$. It is clear that Ω is not negative definite and hence, the Lie algebra l_1 is noncompact. Note that a general element in l_2 can be written as

$$\eta = t(iH) + x(i(X - Y)) + y(X + Y) = itH + (y + ix)X + (y - ix)Y, t, x, y \in \mathbb{R}$$

and hence from the above formula,

$$\Omega(\eta) = 12(y^2 + x^2) - 8t^2$$

This is again not negative definite and hence l_2 is noncompact. Finally, any element of l_3 can be expressed as

$$\zeta = tiH + x(X - Y) + y(i(X + Y)) = itH + (x + iy)X + (iy - x)Y$$

and hence

$$\Omega(\zeta) = -12(x^2 + y^2) - 8t^2$$

and clearly, this is negative definite proving that l_3 is compact. Thus, l_1, l_2 are noncompact real forms of the Lie algebra $\mathbb{C}.H + \mathbb{C}.X + \mathbb{C}.Y$ while l_3 is a compact real form of the same.

[27]. $X_1, ..., X_n$ form a basis for the Lie algebra } which is assumed to be semisimple. Define $g_{ij} = B(X_i, X_j)$ and let (g^{ij}) denote the inverse of the matrix (g_{ij}). Then $\omega = \sum_{i,j} g^{ij} X_i X_j$ is in the centre of the universal enveloping algebra $\mathcal{B}$ of }. To show this, we must show that ω commutes with every element X_k of the basis. The structure constants are $c(mpq)$ and are defined by

$$[X_m, X_p] = \sum_q c(mpq) X_q$$

Then,

$$[\omega, X_m] = \sum_{i,j} g^{ij}([X_i, X_m]X_j + X_i[X_j, X_m]) = \sum_{i,j,p} g^{ij}(c(imp)X_p X_j + c(jmp)X_i X_p)$$

$$= \sum_{i,j,p} (g^{pj}c(pmi) + g^{ip}c(pmj))X_i X_j$$

Now,

$$\sum_p g_{mp}c(ijp) = B(\sum_p c(ijp)X_p, X_m) = B([X_i, X_j], X_m)$$

$$= B(\sum_p c(ijp)X_p, X_m) = \sum_p g_{mp}c(ijp) = B(X_i, [X_j, X_m]) = B(X_i, \sum_p c(jmp)X_p)$$

$$= \sum_p g_{ip}c(jmp)$$

Raising the index m gives us

$$c(ijm) = \sum_{p,q} g_{ip}g^{mq}c(jqp)$$

Raising the index i gives us

$$\sum_p g^{ip}c(pjm) = \sum_q g^{mq}c(jqi) = -\sum_q g^{mq}c(qji)$$

which proves the claim.

[28]. Problems in Linear algebra and group representation theory:

1.Obtain a basis for the Lie algebra $sl(2, \mathbb{C})$ of all 2×2 complex matrices having trace zero. Show that any element of the form $exp(A)$ where $A \in sl(2, \mathbb{C})$ has unit determinant. Determine a Cartan subalgebra ⟨ of $sl(2, \mathbb{C})$. Classify all the finite dimensional irreducible representations of $sl(2, \mathbb{C})$.

2.What do you understand by an irreducible representation of a Lie group and an irreducible representation of a Lie algebra ?

3.What do you understand by the term "roots of a Cartan algebra" in a semisimple Lie algebra ?

4.If S,T are two inequivalent irreducible representations of a group G in the vector spaces V,W respectively and $A : V \to W$ is a linear operator such that $S(g)A = AT(g)$ for all $g \in G$, then prove that $A = 0$. On the other hand suppose $A : V \to V$ is a linear operator such that $AT(g) = T(g)A$ for all $g \in G$, then prove that $A = c.I$ for some complex scalar c. For the latter part, assume that V is a complex vector space and hence A has an eigenvalue.

5.What is the analogue of the convolution property of Fourier transforms for a group.

hint:Let π be an irreducible representation of G assumed to be compact and for $f_1, f_2 \in L^2(G)$, define

$$(f_1 * f_2)(g) = \int_G f_1(h^{-1}g)f_2(h)dh$$

Its Fourier transform at π is given by

$$\pi(f_1 * f_2) = \int (f_1 * f_2)(g)\pi^*(g)dg$$

Show that this equals $\pi(f_1)\pi(f_2)$. Note that $\pi(f_1)\pi(f_2) \neq \pi(f_2)\pi(f_1)$ in general.

6. Suppose V is a vector space on which a group G acts linearly. For $X,Y \in V$ and $f : V \to \mathbb{C}$, define $\partial(X)f(Y) = \frac{d}{dt}f(Y + tX)|_{t=0}$. Then define $\tau_x f(Z) = f(x^{-1}Z)$. Then

$$tau_x\partial(X)\tau_x^{-1}f(Y) = (\partial(X)\tau_x^{-1}f)(x^{-1}Y) = \frac{d}{dt}(\tau_x^{-1}f)(x^{-1}Y + tX)|_{t=0}$$

$$= \frac{d}{dt}f(Y + txX)|_{t=0} = \partial(xX)f(Y)$$

Thus,

$$\tau_x\partial(X)\tau_x^{-1} = \partial(xX)$$

7. Suppose $\}$ is a Lie algebra of a Lie group G and consider $f : \} \to \mathbb{C}$. Define for $g \in G, X \in \}$, $f(g : X) = f(Ad(g)X)$. Then determine an expression for $\frac{d}{dt}f(g.exp(tY) : X)|_{t=0}$ and also for $\frac{d}{dt}f(g : X + tY)|_{t=0}$ where $Y \in \}$.

hint:

$$f(g.exp(tY) : X) = f(g.exp(tY).X.exp(-tY).g^{-1}) = f(gXg^{-1} + tg[X,Y]g^{-1}) + o(t)$$

so that

$$\frac{d}{dt}f(g.exp(tY) : X)|_{t=0} = f(g : X; \partial([X,Y]))$$

8.What do you understand by an invariant distribution on a Lie algebra.

9.What do you understand by an invariant differential operator on a Lie group and on a Lie algebra.

10. Let G be a Lie group and $\}$ its Lie algebra. Let $f : \} \to \mathbb{C}$ be a smooth function. For $X,Y \in \}$, obtain a formula for $\frac{d}{dt}f(Ad(exp(tX))Y)|_{t=0}$.

hint:

$$Ad(exp(tX))Y = Y + t[X,Y] + o(t)$$

and hence

$$\frac{d}{dt}f(Ad(exp(tX))Y)|_{t=0} = (\partial([X,Y])f)(Y)$$

where for $Z \in \}$, we define

$$(\partial(Z)f)(Y) = \frac{d}{dt}f(Y + tZ)|_{t=0}$$

11.Suppose V is a vector space over $\mathbb{R}$ and $B : V \times V \to \mathbb{R}$ is a non-degenerate bilinear form on V. For $X \in V$, define $f_X \in V^*$ so that $f_X(Y) = B(X,Y)$ for $Y \in V$. V^* is the space of linear functionals on V. Let $S(V)$ be the symmetric algebra on V. Thus, if $\otimes$ is the tensor product on V and $X_1, ..., X_n \in V$, then, $X_1...X_n$ stands for the element in $S(V)$ given by $\frac{1}{n!}\sum_{\sigma \in S_n} X_{\sigma 1} \otimes ... \otimes X_{\sigma n}$. For $X_i \in V, i = 1, 2, ..., n$ and $a = X_1...X_n$, define $\partial(a)f(X) = \partial(X_1)...\partial(X_n)f(X)$ and extend the definition of $a \to \partial(a)$ by linearity to the whole of $S(V)$. Thens show that $\partial(ab) = \partial(a)\partial(b) = \partial(b)\partial(a)$, where $a, b \in S(V)$ and $ab = ba \in S(V)$ is defined by symmetrizing the tensor product of a and b.

12. Let $G = SL(2, \mathbb{R})$, the group of all 2×2 real matrices having determinant unity and let $\}sl(2, \mathbb{R})$ denote its Lie algebra. Show that $sl(2, \mathbb{R})$ is the set of all 2×2 real matrices having trace zero. Show that if

$$H = \begin{pmatrix} 1 & 0 \\ 0 & -1 \end{pmatrix}, X = \begin{pmatrix} 0 & 1 \\ 0 & 0 \end{pmatrix},$$

$$Y = \begin{pmatrix} 0 & 0 \\ 1 & 0 \end{pmatrix}$$

then $\{H, X, Y\}$ is a basis for $\}$. Show that $\dashv = \mathbb{R}.H$ and $\lfloor = \mathbb{R}.(X - Y)$ are two non-conjugate Cartan subalgebras of $\}$. What are the subgroups corresponding to $\dashv$ and $\lfloor$. Show that if $\}_C$ is the complexification of $\}$, ie, $\}_C = sl(2, \mathbb{C})$, then $\dashv_C = \mathbb{C}.H$ and $\lfloor_C = \mathbb{C}.(X - Y)$ are conjugate Cartan subalgebras of $\}_C$.

13. State and prove the QR decomposition for nonsingular square matrices over $\mathbb{C}$.

hint:Let A be a $n \times n$ nonsingular matrix. Set $A = [a_1, ..., a_n]$ where a_i is the ith column of A. Then using the Gram-Schmidt orthonormalization process, construct an orthonormal basis $\{e_1, ..., e_n\}$ for $\mathbb{C}^n$ such that

$$span\{e_1, ..., e_i\} = span\{a_1, ..., a_i\}, i = 1, 2, ..., n$$

It follows that

$$a_i = \sum_{j=1}^{i} r_{ij}e_j, i = 1, 2, ..., n$$

where $r_{ij}, 1 \leq j \leq i \leq n$ are complex scalars. From this, the QR decomposition can be obtained, i.e, $A = QR$ where Q is an $n \times n$ unitary matrix and R is an $n \times n$ upper triangular matrix.

14. Let T be a linear operator in a finite dimensional vector space V over $\mathbb{C}$. For any eigenvalue λ of T, define the subspace W_λ to consist of all $x \in V$ for which $(T - \lambda.I)^m x = 0$ for some $m = 1, 2, ...$ Show that if S is a linear operator in V that commutes with T, then W_λ is invariant under S.

15. What do you understand by simultaneous diagonability of a family of matrices. Show that if we have a commuting family of diagonable matrices, then this family is simultaneously diagonable.

16. Let W be a subspace of a vector space V and let T be a linear operator in V that leaves W invariant. Then let α be a vector in V that is not contained in W. Let $s_T(\alpha, W)$ be the monic polynomial f of least degree for which $f(T)\alpha \in W$. Then show that if $I_T(\alpha, W)$ is the set of all polynomials g such that $g(T)\alpha \in W$, then $I_T(\alpha, W)$ is an ideal in $\mathbb{F}[x]$ and that this ideal is generated by $s_T(\alpha, W)$.

17. Let V be a vector space and let π be a representation of a group G in V. Suppose π is irreducible. Then show that given any nonzero $x \in V$, $V = span\{\pi(g)x : g \in G\}$.

18. What do you understand by an ideal in a Lie algebra }. What is a solvable ideal. When is the Lie algebra } said to be semisimple. What is a nilpotent Lie algebra.

19. Suppose that a matrix A is expressed in Jordan canonical form, ie, as a direct sum of Jordan blocks. Then give a formula for $exp(tA)$ and state the importance of this result in solving linear differential equations of the form

$$\frac{d\mathbf{x}(t)}{dt} = \mathbf{A}\mathbf{x}(t) + \mathbf{u}(t)$$

20. Give an example of three 2×2 complex matrices X, Y, Z that satisfy the equations

$$X^2 = Y^2 = Z^2 = I, XY + YX = YZ + ZY = ZX + XZ = 0, XY = Z, YZ = X, ZX = Y$$

21. Give an example of four 4×4 matrices $\gamma^\mu, \mu = 0, 1, 2, 3$ that satisfy the anticommutation relations

$$\gamma^\mu \gamma^\nu + \gamma^\nu \gamma^\mu = 2g^{\mu\nu}$$

where $g^{\mu\nu}$ is the standard metric tensor of special relativity, ie,

$$((g^{\mu\nu})) = diag(1, -1, -1, -1)$$

or in terms of components,

$$g^{00} = 1, g^{rr} = -1, r = 1, 2, 3, g^{0r} = g^{r0} = 0, r = 1, 2, 3,$$

$$g^{rs} = 0, r, s = 1, 2, 3, r \neq s$$

22. Let S, T be representations of a compact group G in finite dimensional vector spaces V, W respectively. Let $X : V \to W$ be a linear operator and define

$$A = \int_G S(g^{-1})XT(g)dg$$

Show that

$$S(h)A = \int_G S(hg^{-1})XT(g)dg = \int_G S(g^{-1})XT(gh)dg = AT(h), h \in G$$

Deduce the Schur orthogonality relations for the matrix elements of irreducible representations of a compact group.

23. If T is an irreducible representation of a compact group G, and if χ denotes the character of T, then show that $\int_G |\chi(g)|^2 dg = 1$. Conversely show that $\int_G |\chi(g)|^2 dg = 1$ implies that T is irreducible.

hint:Use the Schur orthogonality relation to deduce orthogonality relations for the characters and the fact that every representation of G can be decomposed into a direct sum of irreducibles. Specifically, show that if T is a representation of G and

$$T = \bigoplus_i m_i \pi_i$$

where m_i are positive integers and $\pi_i, i = 1, 2, \ldots$ are inequivalent irreducible representations, then

$$\chi = \sum_i m_i \chi_i$$

where χ is the character of T and χ_i is the character of π_i.

[29].Let $\}$ be a semisimple Lie algebra and $\langle$ a Cartan subalgebra. Let $\approx$ and $\sqrt{\ }$ be as usual, ie, $\sqcap$ is a compact real form of $\}$ and $\sqcap \cap \}_0 = \approx_0$ and $i \sqcap \cap \}_0 = \sqrt{\ }_0$ where $\}_0$ is a real form of $\}$. Then $\approx$ and $\sqrt{\ }$ are respectively the complexifications of $\approx_0$ and $\sqrt{\ }_0$. Let $\dashv$ be a maximal Abelian subspace of $\sqrt{\ }$. Then let the Cartan subalgebra $\langle$ can be chosen so that $\langle \cap \sqrt{\ } = \dashv$. Let $\updownarrow$ denote the centralizer of $\dashv$ in $\}$. Then $\updownarrow$ can be expressed as the direct sum of $\langle$ and the root spaces $\}_\alpha$ corresponding to those roots α which vanish on $\dashv$. We have the Iwasawa decomposition:$\} = \approx + \dashv + \backslash$ where $\backslash$ is the direct sum of the root spaces $\}_\alpha$ for which α is a positive root that does not vanish identically on $\dashv$. Let K denote the analytic subgroup corresponding to $\approx$ and A the analytic subgroup corresponding to $\dashv$ and N the analytic subgroup corresponding to $\backslash$. Then we have the corresponding decomposition $G = KAN$. Every element $x \in G$ can be uniquely expressed as $x = kan$ with $k \in K, a \in A, n \in N$. We want to compute the Haar measure dx on G in terms of dk, da, dn. Note that $Ad(a)$ on $\}$ has unit determinant and hence $Ad(a)$ restricted to $\approx$ will have determinant equal to the reciprocal of the determinant of $Ad(a)$ restricted to $\dashv + \backslash$. However $Ad(a)(X) = X$ for $a \in A$ and $X \in \dashv$. It follows that $Ad(a)$ restricted to $\approx$ will have determinant equal to the reciprocal of the determinant of $Ad(a)$ restricted to $\backslash$. Now let $X \in K$. Then

$$k.exp(tX)an = kann^{-1}a^{-1}exp(tX)an$$

and hence the differential of $(k, a, n) \to kan$ with respect to k is given by $X \to n^{-1}a^{-1}Xan$ and by the above argument, it has determinant equal to the reciprocal of $Ad(a^{-1})$ restricted to $\backslash$. Now let $a = exp(H), H \in \dashv \subset \langle$. Then, let α be a root that is not identically zero on $\dashv$ and let X_α be a corresponding root vector. Then, $Ad(a)X_\alpha = exp(ad(H))(X_\alpha) = exp(\alpha(H))X_\alpha$ and hence $Ad(a)$

restricted to \ has determinant equal to the $exp(\sum_{\alpha \in P_+} \alpha(H))$ where P_+ is the set of positive roots that do not vanish identically on ⊣. We set $\sigma(H) = \frac{1}{2}\sum_{\alpha \in P_+} \alpha(H)$ for $H \in$ ⊣. Then, we get from the above that $Ad(a^{-1})$ restricted to K has determinant equal to $exp(2\sigma(log(a)))$ and hence we get the desired formula

$$dx = exp(2\sigma(loga))dkdadn$$

(Ref:Collected papers of HarishChandra). We now consider the differential of the map $(x,a) \to xax^{-1}$ from $G \times A \to G$ with respect to x. First let x be perturbed to $x.exp(tX)$ where $X \in$ \. Then, xax^{-1} gets perturbed to $x.exp(tX)aexp(-tX)x^{-1}$ and is derivative with respect to t at $t = 0$ is given by $x[X,a]x^{-1}$. Write $X = X_\alpha$ and $a = exp(H)$. Then $XH - HX = -\alpha(H)X$ or $XH = (H - \alpha(H))X$ giving $XH^n = (H - \alpha(H))^n X$ so that $X.exp(H) = exp(H - \alpha(H))X$ so that $[X,a] = [X, exp(H)] = exp(H)(exp(-\alpha(H)) - 1)X = a(exp(-\alpha(H)) - 1)X$. Thus the differential of xax^{-1} with respect to x at X is given by

$$xax^{-1}xXx^{-1}(exp(-\alpha(H)) - 1)$$

[30]. Suppose $(\Omega, \mathcal{F}, \mu)$ is a measure space. Define $\mathcal{H} = L^2(\Omega, \mathcal{F}, \mu)$ the Hilbert space with inner product $<u,v> = \int_\Omega u\bar{v}d\mu$. Define $U_T : \mathcal{H} \to \mathcal{H}$ so that

$$U_T f(x) = (\frac{d\mu oT}{d\mu}(x))^{1/2} f(Tx)$$

where T is a measure preserving transformation on Ω. Assume here that $\mu oT << \mu$, ie, μoT is absolutely continuous with respect to μ. Show that U_T is an isometry, ie,

$$< U_T f, U_T g > = < f, g >, f, g \in \mathcal{H}$$

[31]. What do you understand by induced representation.

hint: G is a group and H is a subgroup of G. Let σ be a unitary representation of H in a Hilbert space $\mathcal{H}$ and let V_1 be the set of all functions $f : G \to \mathcal{H}$ such that $f(gh) = \sigma(h)^{-1}f(g)$ for all $g \in G$ and $h \in H$. Then obviously $\| f(gh) \|^2 = \| f(g) \|^2, g \in G, h \in H$. Hence if μ is the invariant measure on G/H induced by the left invariant measure on G, we can define $\int_{G/H} \| f \|^2 d\mu$ and we require this to be finite. Let V be the subset of V_1 consisting of such functions. Then V_1 is a Hilbert space and we can define a representation π of G in this space as $\pi(g)f(x) = f(g^{-1}x), g, x \in G$. This is the representation of G in V induced by the representation σ of H in $\mathcal{H}$. Consistency of this definition follows from $f(g^{-1}xh) = \sigma(h)^{-1}f(g^{-1}x)$ for $g, x \in G, h \in H$.

[32]. What is a cocycle.

hint: G is a group acting on a set X. Consider a map $\sigma : G \times M \to G_1$ where G_1 is another group. Suppose σ satisfies $\sigma(g_2 g_1, x) = \sigma * g_2, g_1 x)\sigma(g_1, x)$ for $g_1, g_2 \in G, x \in X$. Then, σ is called a cocycle.

[33]. Show how cocycles can be used to construct induced representations.

[34]. Define the adjoint representation of a Lie algebra in terms of the adjoint representation of the corresponding Lie group.

hint: Let G be a Lie group with Lie algebra } . For $g \in G$, define $Ad(g) : G \to G$ so that $Ad(g)(h) = ghg^{-1}$. Then $A(g)$ maps e to e and hence, $dAd(g)$ maps TG_e into TG_e, ie, $dAd(g) : \} \to \}$. We use the same notation $Ad(g)$ for $dAd(g)$. If G is a matrix group then $Ad(g)(X) = gXg^{-1}$ for $X \in \}$, where } is the set of all matrices of the form $log(A)$ with $A \in \}$. For example, if $G = GL(n,\mathbb{R})$, the group of all nonsingular real $n \times n$ matrices, then $\} = M_n(\mathbb{R})$, the set of all $n \times n$ real matrices. Then Given an $X \in M_n(\mathbb{R})$, the corresponding left invariant vector field on G is given by $X(g) = gX, g \in G$ or equivalently as a differential operator,

$$X(g) = \sum_{i,j}(gX)_{ij}\frac{\partial}{\partial g_{ij}}$$

[35]. Suppose M is a Riemannian manifold with metric g_{ij} relative to a local coordinate system. Choose $p \in M$ and a vector $X \in TM_p$. Define $\xi(t,p) = exp(tX)(p)$ as the solution to the differential equations

$$d\xi^i(t,p)/dt = v^i(t,p), dv^i(t,p)/dt + \Gamma^i_{jk}v^j(t,p)v^k(t,p) = 0, t \in \mathbb{R}$$

with the initial conditions

$$\xi^i(0,p) = p, v^i(0,p) = X^i$$

Thus, $t \to exp(tX)(p)$ is a geodesic on M that passes through p and has a tangent X at p. Show that the exponential map defined in this way is the standard exponential map for Lie groups, ie, geodesics on a Lie group are one parameter subgroups.

[36]. Give an example of the following situation: } is a complex Lie algebra and ⊓ is a compact real form of }. Thus, $\} = ⊓ + i.⊓$ and the Lie subgroup U of ⊓ is a compact subgroup of the Lie group G of }.

hint:Consider $G = SL(2,)$ with Lie algebra $sl(2,\mathbb{C})$, $U = SU(2)$ so that $⊓ = su(2)$ is the real Lie algebra generated by the Pauli spin matrices.

[37]. Give an example of the following situation: } is a real Lie algebra and ⊓ is a compact Lie subalgebra of }. s is an automorphism of } such that the fixed points of s form precisely ⊓.

hint: Let $\} = sl(2,\mathbb{R})$, ie } is the Lie algebra of all real 2×2 real matrices having trace zero. Any $X \in \}$ can be uniquely decomposed as $X = Y + Z$ where X is real and symmetric and Z is real and skewsymmetric. Then take ⊓ to be the set of all real skew symmetric 2×2 matrices and let ⌐ denote the set of all real symmetric 2×2 matrices having trace zero. We have the direct sum decomposition $\} = ⊓ \oplus ⌐$. Define $s : \} \to \}$ so that $s(Y + Z) = Y - Z$ where $Y \in ⊓$ and $Z \in ⌐$. Then ⊓ is precisely the set of fixed points of s. This works even if $sl(2,\mathbb{R})$ is replaced by $sl(n,\mathbb{R})$ where n is any positive integer.

[38]. What do you understand by a diagonable linear operator T in a finite dimensional vector space V. Show that if the minimal polynomial p of T has the form $p(t) = \Pi_{i=1}^{r}(t - c_i)$, where $c_1, ..., c_r$ are distinct elements of the field, then T is diagonable.

hint: Define

$$p_i(t) = (\Pi_{j \neq i}(t - c_j))/\Pi_{j \neq i}(c_i - c_j), i = 1, 2, ..., r$$

Then show that $p_i(c_j) = \delta_{ij}$ and hence deduce that if f is a polynomial with $deg f \leq r - 1$, then

$$f(t) = \sum_i f(c_i)p_i(t)$$

In particular for $r \geq 2$, we have

$$1 = \sum_{i=1}^{r} p_i(t), t = \sum_{i=1}^{r} c_i p_i(t)$$

which implies that

$$I = \sum E_i, T = \sum c_i E_i, E_i = p_i(t)$$

and using the fact that p divides $p_i p_j$ if $i \neq j$, deduce that $E_i E_j = 0$ if $i \neq j$.

[39].Let E be a vector space and $S(E)$ the symmetric algebra over E corresponding to a given tensor product. Assume that a group G acts on the set E. This action induces an action of G on $S(E)$. For example, if $x_i \in E, i = 1, 2, ..., n$ then $x_1...x_n$ is the element

$$x_1...x_n = \frac{1}{n!} \sum_{\sigma \in S^n} x_{\sigma 1} \otimes ... \otimes x_{\sigma n}$$

For $g \in G$, the action of g on $x_1...x_n$ is given by $(gx_1)...(gx_n)$ which is the element

$$\frac{1}{n!} \sum_{\sigma \in S^n} (gx_{\sigma 1})...(gx_{\sigma n})$$

An invariant in $S(E)$ is an element $\xi \in S(E)$ for which $g.\xi = \xi$ for all $g \in G$.

[40].Transform the linear differential equation

$$y^{(n)}(t) + a_1 y^{(n-1)}(t) + ... + a_{n-1}y^{(1)}(t) + a_n y(t) = x(t)$$

into state variable form by means of the substitutions

$$\xi_k(t) = y^{(k-1)}(t), k = 1, 2, ..., n$$

[41].Compute $exp(tA)$ where A is a square matrix using (a) Cayley-Hamilton theorem, (b) Laplace transforms, (c) Jordan canonical form.

[42].Let $\mathfrak{g}$ be a Lie algebra and let G be a Lie group having $\mathfrak{g}$ as its Lie algebra. Assume $\mathfrak{g}$ is semisimple. A function $f : \mathfrak{g} \to \mathbb{C}$ is given. For $X, Y \in \mathfrak{g}$, consider the function

$$D_Y f(X) = \frac{d}{dt} f(Ad(exp(tY))(X))|_{t=0} = f(X; \partial([Y,X]))$$

Show that D_Y is derivation on the space of smooth functions. The function f is said to be globally invariant if $f(Ad(g)(X)) = f(X)$ for all $g \in G, X \in \mathfrak{g}$. Show that if f is globally invariant, then it is also locally invariant in the sense that $D_Y f = 0$ for all $Y \in \mathfrak{g}$.

[43]. Let $\mathfrak{g}$ be a Lie algebra and consider the operator

$$\sigma_X(Y) = L_{[X,Y]} + d_Y, X, Y \in \mathfrak{g}$$

on $S(\mathfrak{g})$, the symmetric algebra of $\mathfrak{g}$. Thus, if $p \in S(\mathfrak{g})$, then $\sigma_X(Y)p = [X,Y]p + d_X p$. Here, d_X is the extension of $ad(X)$ from $\mathfrak{g}$ to $S(\mathfrak{g})$. For example, if $Y_1, Y_2 \in \mathfrak{g}$, then $d_X(Y_1 Y_2) = [X, Y_1]Y_2 + Y_1[X, Y_2]$.

$$\sigma_X([Y, Z]) = L_{[X,[Y,Z]]} + d_{[Y,Z]}$$

so that

$$\sigma_X([Y, Z])p = [X, [Y, Z]]p + ad([Y, Z])p$$

on the other hand,

$$\sigma_X(Y)\sigma_X(Z)p - \sigma_X(Z)\sigma_X(Y)p$$

$$= ([X, Y] + ad(Y))([X, Z] + ad(Z))p - ([X, Z] + ad(Z))([X, Y] + ad(Y))p$$

$$= ad(Y)([X, Z]p) - ad(Z)([X, Y]p) + [X, Y]ad(Z)p - [X, Z]ad(Y)(p) + (ad(Y)ad(Z) - ad(Z)ad(Y))p$$

$$= [Y, [X, Z]]p + [X, Z]ad(Y)(p) - [Z, [X, Y]]p - [X, Y]ad(Z)(p) - [X, Z]ad(Y)(p) + [X, Y]ad(Z)p + ad([Y, Z])(p)$$

$$= [X, [Y, Z]]p + ad([Y, Z])p$$

Thus,

$$\sigma_X([Y, Z]) = \sigma_X(Y)\sigma_X(Z) - \sigma_X(Z)\sigma_X(Y)$$

and this proves that σ_X can be extended to a representation of the universal enveloping algebra of $\mathfrak{g}$.

Reference:HarishChandra, Collected papers.

Suppose we now define

$$\sigma_X(Y) = L_{[Y,X]} + d_Y$$

Then $Y \to \sigma_X(Y)$ extends to a representation of the universal enveloping algebra of $\mathfrak{g}$. This can be seen as follows.

$$\sigma_X(Y)\sigma_X(Z)p = (L_{[Y,X]} + d_Y)(L_{[Z,X]} + d_Z)p$$

$$= [Y, X][Z, X]p + [Y, X]ad(Z)(p) + ad(Y)([Z, X]p) + ad(Y)ad(Z)(p)$$

Interchanging Y and Z and subtracting, we get

$$\sigma_X(Y)\sigma_X(Z) - \sigma_X(Z)\sigma_X(Y) =$$

$$[Y,X]ad(Z)(p) - [Z,X]ad(Y)(p) + ad(Y)([Z,X]p) - ad(Z)([Y,X]p) + ad([Y,Z])(p)$$

$$= [Y,X]ad(Z)(p) - [Z,X]ad(Y)(p) + [Y,[Z,X]]p + [Z,X]ad(Y)(p) - [Z,[Y,X]]p - [Y,X]ad(Z)(p) + ad([Y,Z])(p)$$

$$= -[X,[Y,Z]]p + ad([Y,Z])(p) = (L_{[[Y,Z],X]} + ad([Y,Z]))p = \sigma_X([Y,Z])(p)$$

proving the claim.

[44].Project on the applications of group representation theory to image processing problems.

Determine the finite dimensional irreducible representations of $SL(2,\mathbb{C})$. Let π be one such representation. Let us denote by the same symbol π the corresponding representation of the proper, orthochronous Lorentz group G. We can regard G as acting on $\mathbb{R}^4$. Let $f : \mathbb{R}^4 \to \mathbb{R}$ be a time varying spatial signal. This signal corresponds to a definite pattern of variations in space-time, for example, it may correspond to the electromagnetic field produced by an antenna. When the pattern f undergoes spatial rotations and uniform motion, it gets changed to the pattern $f(g^{-1}x)$ where g is a Lorentz transformation. Fix a point $x_0 \in \mathbb{R}^4$ and define a function $f_1 : G \to \mathbb{R}$ so that $f_1(g) = f(g.x_0), g \in G$. Its Fourier transform at the representation π is denoted by $\hat{f}_1(\pi)$. Thus,

$$\hat{f}_1(\pi) = \int_G f_1(g)\pi^*(g)dg$$

Consider the Lorentz transformed pattern $h(x) = f(g_0^{-1}x)$. The corresponding function on G is given by $h_1(g) = f(g_0^{-1}gx_0) = f_1(g_0^{-1}g)$. Its Fourier transform at π is then given by

$$\hat{h}_1(\pi) = \int_G h_1(g)\pi^*(g)dg = \int_G f_1(g_0^{-1}g)\pi^*(g)dg = \int f_1(g)\pi^*(g_0g)dg = (\hat{f}_1(g)\pi^*(g)dg)\pi^*(g_0)$$

$$= \hat{f}_1(\pi)\pi^*(g_0)$$

Suppose π is unitary. Then it follows that

$$\hat{h}_1(\pi)\hat{h}_1(\pi)^* = \hat{f}_1(\pi)\hat{f}_1(\pi)^*$$

and hence we get an invariant for the spatio-temporal image. Suppose that we have an irreducible representation π of the Lie algebra of $SL(2,\mathbb{C})$, ie we choose the standard basis $\{H,X,Y\}$ of the Lie algebra $sl(2,\mathbb{C})$ satisfying the commutation relations $[H,X] = 2X, [H,Y] = -2Y, [X,Y] = H$ and specify the finite dimensional matrices $\pi(H), \pi(Y), \pi(Y)$ relative to a basis. A general element of $SL(2,\mathbb{C})$ then has the form

$$g = exp(t_1H + t_2X + t_3Y)$$

where t_1, t_2, t_3 are complex numbers. The representation π of $G = SL(2,\mathbb{C})$ corresponding to the representation π of $\} = sl(2,\mathbb{C})$ is then given by

$$\pi(g) = exp(t_1\pi(H) + t_2\pi(X) + t_3\pi(Y))$$

This need not be a unitary matrix. We have

$$H = \begin{pmatrix} 1 & 0 \\ 0 & -1 \end{pmatrix}, X = \begin{pmatrix} 0 & 1 \\ 0 & 0 \end{pmatrix},$$

$$Y = \begin{pmatrix} 0 & 0 \\ 1 & 0 \end{pmatrix}$$

Given a $g \in SL(2, \mathbb{C})$, we must determine complex numbers $t_i, i = 1, 2, 3$ such that $g = exp(t_1 H + t_2 X + t_3 Y)$ and then we can compute $\pi(g) = exp(t_1\pi(H) + t_2\pi(X) + t_3\pi(Y))$.

[45].Project on the applications of Schur orthogonality relations and the Peter-Weyl theorem to feature extraction in images.

Suppose G is a unimodular Lie group and π a representation of G in a vector space V. Let π_1 be a finite dimensional irreducible representation of G in $\mathbb{C}^d$. Consider the linear operator $P_\chi : V \to V$ defined by

$$P_\chi = \int_G \chi(g^{-1})\pi(g)dg$$

where dg is the left invariant Haar measure on G. For example if $V = L^2(G)$ and π is the left regular representation of G in V, then for any $f \in L^2(G)$, we have

$$P_\chi f(x) = \int \chi(g^{-1})f(g^{-1}x)dg$$

Let us decompose π so that the representation π_1 appears m times in this decomposition. Then relative to a basis for V, π has the form

$$\pi(g) = diag[\pi_1(g), ..., \pi_1(g), \pi_2(g), ..., ...]$$

where $\pi_i, i = 1, 2, ...$ are the inequivalent irreducible representations of G. Then by the Schur orthogonality relations, we have

$$\int_G \chi(g^{-1})\pi_k(g)dg = 0, k = 2, 3, ...$$

and

$$\int_G \chi(g^{-1})(\pi_1(g))_{ab}dg = \sum_c \int (\pi_1(g^{-1}))_{cc}(\pi_1(g))_{ab}dg = \sum_c d^{-1}\delta_{ca}\delta_{cb} = d_a^{-1}\delta_{ab}$$

It follows that

$$P_\chi = d\int_G \chi(g^{-1})\pi(g)dg = diag[I_d, ..., I_d, 0, ..., 0, ...]$$

From this, it immediately follows that P_χ is the orthogonal projection of V onto the space V_χ given by the direct sum of all the irreducible subspaces of V where the restriction of π is equivalent to π_1. We shall show that P_χ commutes with $\pi(h)$ for every $h \in G$. Indeed,

$$P_\chi\pi(h) = d.(\int \chi(g^{-1})\pi(g)dg)\pi(h) = d.\int_G \chi(g^{-1})\pi(gh)dg = d.\int_G \chi(hg^{-1})\pi(g)dg$$

by the right invariance of the Haar measure dg. Now,

$$\chi(hg^{-1}) = \chi(hg^{-1}hh^{-1}) = \chi(g^{-1}h)$$

since χ is a class function. It follows that

$$P_\chi \pi(h) = d. \int \chi(g^{-1}h)\pi(g)dg = d. \int_G \chi(g^{-1})\pi(hg)dg = \pi(h)P_\chi$$

proving the claim. It follows in particular that if π is the left regular representation of G, then for any $f \in L^2(G)$, we have

$$(P_\chi f)(g^{-1}x) = (\pi(g)P_\chi f)(x) = (P_\chi \pi(g)f)(x) = P_\chi f_1(x)$$

where

$$f_1(x) = f(g^{-1}x)$$

hence,

$$\int_G |P_\chi f_1(x)|^2 dx = \int_G |(P_\chi f)(g^{-1}x)|^2 dx = \int_G |P_\chi f(x)|^2 dx$$

In other words, the functional $I : L^2(G) \to \mathbb{R}$ defined by

$$I(f) = \int_G |P_\chi f(x)|^2 dx = \| P_\chi f \|^2$$

is invariant under left translation, ie, $I(\pi(g)f) = I(f)$ for all $g \in G$. This property can be used for feature extraction in images. Now suppose χ_1, χ_2 are two distinct characters of G. Then we have for $f \in L^2(G)$ with π as the left regular representation,

$$P_{\chi_i} \pi(g) = \pi(g)P_{\chi_i}, i = 1, 2,$$

and hence

$$I_{12}(f) = < P_{\chi_1} f, P_{\chi_2} f >$$

is invariant under left translation, ie,

$$I_{12}(\pi(g)f) = < \pi(g)P_{\chi_1} f, \pi(g)P_{\chi_2} f > = < P_{\chi_1} f, P_{\chi_2} f > = I_{12}(f), g \in G$$

[46]. Develop tools for computing $exp(tA)$ where A is an $n \times n$ matrix and t is a complex number. The tools include (a) diagonalization method if A is diagonable, (b) Cayley-Hamilton theorem (c) Jordan canonical form, (d) Laplace transform.

[47].Develop tools for computing the structure constants of a Lie algebra in different bases when the structure constants in one basis are known.

hint: Let $\{X_i, i = 1, 2, ..., n\}$ be a basis for a Lie algebra } and let $\{c^i_{jk}\}$ be the structure constants of } relative to this basis. Let $Y_i = \sum_j a_{ij}X_k, i = 1, 2, ..., n$ be another basis for }. Then, let (b_{ij}) be the inverse of the matrix (a_{ij}). It then follows that $X_i = \sum_j b_{ij}Y_j$ and so we get

$$[Y_i, Y_j] = \sum_{k,m} a_{ik}a_{jm}[X_k, X_m] = \sum_{k,m,p} a_{ik}a_{jm}c^p_{km}X_p = \sum_{k,m,p,q} a_{ik}a_{jm}c^p_{km}b_{pq}Y_q$$

$$= \sum_q f_{ij}^q Y_q$$

where

$$f_{ij}^q = \sum_{k,m,p} a_{ik} a_{jm} c_{km}^p b_{pq}$$

[48].Tabulate the finite dimensional irreducible representations of $SL(2, \mathbb{C})$ using MATLAB.

hint: We first construct the standard basis $\{H, X, Y\}$ of the Lie algebra $sl(2, \mathbb{C})$ of $SL(2, \mathbb{C})$. These matrices have trace zero and satisfy the commutation relations

$$[H, X] = 2X, [H, Y] = -2Y, [X, Y] = h$$

Let π be an irreducible representation of $sl(2, \mathbb{C})$. Then, we have

$$[\pi(H), \pi(X)] = 2\pi(X), [\pi(H), \pi(Y)] = -2Y, [\pi(X), \pi(Y)] = \pi(H)$$

Let V be the representation space of π. Then suppose $v \in V$ is an eigenvector of $\pi(H)$ corresponding to the eigenvalue λ. Then

$$\pi(H)\pi(X) - \pi(X)\pi(H) = 2\pi(X)$$

implies

$$\pi(H)\pi(X)v = \pi(X)(\pi(H) + 2)v = (\lambda + 2)\pi(X)v$$

so that $\pi(X)v$ is an eigenvector of $\pi(H)$ corresponding to the eigenvalue $\lambda + 2$. Likewise, $\pi(Y)v$ is an eigenvector of $\pi(H)$ corresponding to the eigenvalue $\lambda - 2$. Since V is finite dimensional and since eigenvectors of a linear operator in a finite dimensional vector space corresponding to distinct eigenvalues are linearly independent, it follows that there exists a nonzero vector $v_0 \in V$ such that $\pi(X)v_0 = 0$. v_0 is called a highest weight vector of the representation π. Likewise there exists a nonzero vector $v_1 \in V$ such that $\pi(Y)v_1 = 0$. From these facts and using the commutation relations of the Lie algebra, we can construct all the finite dimensional irreducible representations of $sl(2, \mathbb{C})$.

[49].Develop numerical methods for implementing differential operators on a Lie algebra. Suppose E is a vector space and E' is the dual of E. Let $S(E)$ denote the symmetric algebra of E and $P(E)$ the polynomial algebra of E. Then $S(E)$ can be identified with $P(E')$ and likewise $S(E')$ with $P(E)$. For example, suppose $X, Y \in E$. Then $XY = (X \otimes Y + Y \otimes X)/2$ is an element of $S(E)$. If $\lambda \in E'$, then $XY(\lambda) = \lambda(X)\lambda(Y)$ and more generally, if $X_1, ..., X_n \in E$, then

$$X_1...X_n = \frac{1}{n!} \sum_{\sigma \in S_n} X_{\sigma 1} \otimes ... \otimes X_{\sigma n}$$

so that for $\lambda \in E'$, we have

$$X_1...X_n(\lambda) = \Pi_{i=1}^n \lambda(X_i)$$

Suppose $X_1, ..., X_n, Y_1, ..., Y_n \in E$. We consider the implementation of the differential operator

$$D = \sum_{i=1}^n X_i \partial(Y_i)$$

Let f be an analytic function on E. Then

$$Df(X) = \sum_{i=1}^{n} B(X_i, X)\partial(Y_i)f(X) = \sum_{i=1}^{n} B(X_i, X)f(X; \partial(Y_i))$$

where $B : E \times E \to \mathbb{C}$ is a non-degenerate bilinear form on E. Using this bilinear form, we identify the element $Z \in E$ with the element $B(Z, .) \in E'$. Consider now linearly independent elements $X_1, ..., X_n \in E$ and positive integers $m_1, ..., m_n$. We consider the elements $p = X_1^{m_1} X_2^{m_2} ... X_n^{m_n} in S(E)$. We compute

$$\partial(p)(p) = m_1!...m_n!$$

We also note that if $q = X_1^{r_1}...X_n^{r_n}$, then $\partial(q)(p)(0) = 0$ if the n-tuples $(m_1, ..., m_n)$ and $(r_1, ..., r_n)$ are distinct. In other words,

$$p(0, \partial(q)) = m_1!...m_n!\delta_{m_1,r_1}...\delta_{m_n,r_n}$$

[50].Using the Baker-Campbell-Hausdorff formula to evaluate $exp(A + \epsilon B)$ where A, B are noncommuting matrices and hence deriving properties of perturbed dynamical systems of the form

$$\frac{d\mathbf{x}(t)}{dt} = (\mathbf{A} + \epsilon.\mathbf{B})\mathbf{x}(t)$$

[51]. Develop numerical techniques based on the Campbell-Baker-Hausdorff formula for computing $log(exp(A).exp(B))$ where A, B are non-commuting matrices.

  hint: Define

$$F(t) = exp(tA)exp(tB)$$

Then,

$$F'(t) = AF(t) + F(t)B = (L_A + R_B)F(t)$$

where L_A is left translation by A and R_B is right translation by B.

[52]. Develop MATLAB programs for tabulating the Fourier transform of a function defined on a compact group. If $f : G \to \mathbb{C}$ is a function and $\pi : G \to U(n, \mathbb{C})$ is an irreducible unitary representation of G, then the Fourier transform of f at π is defined by

$$\hat{f}(\pi) = \int_G f(g)\pi^*(g)dg$$

In order to compute this integral, we can use the Lie algebra coordinates and discretize the integral in that domain. Specifically, let $X_1, ..., X_n$ be a basis for the Lie algebra $\}$ of G. Then any $g \in G$ can be expressed as $g = exp(\sum_{i=1}^{n} t_i X_i)$. The Haar measure dg must be expressed as $dg = \psi(t_1, ..., t_n)dt_1...dt_n$. Then, if $\tilde{\pi}$ denotes the differential of π, we have

$$\hat{f}(\pi) = \int_{\mathbb{R}^n} f(exp(\sum_{i=1}^{n} t_i X_i))exp(t_1\tilde{\pi}(X_1)^* + ... + t_n\tilde{\pi}(X_n)^*)\psi(t_1, ..., t_n)dt_1...dt_n$$

[53]. Show how to compute the left invariant Haar measure on a Lie group using left invariant vector fields. Give examples of such computations by taking the group as $SO(3), SU(2), SL(n, \mathbb{R})$.

[54]. Suppose G is a Lie group and $\}$ is its Lie algebra. Let $f : \} \to \mathbb{C}$ be a smooth function. Consider the function $F : G \times \} \to \mathbb{C}$ defined by

$$F(g, X) = f(Ad(g)X)$$

In case G is a matrix group, we can write

$$F(g, X) = f(gXg^{-1})$$

Let $Y, Z \in \}$ and consider

$$F(g; Y, X; Z) = \frac{\partial^2}{\partial t \partial s} F(g.exp(tY), X + sZ)|_{t=s=0}$$

When G is a matrix group, this is the same as

$$\frac{\partial^2}{\partial t \partial s} f(g.exp(tY)(X + sZ).exp(-tY)g^{-1})|_{t=s=0}$$

We have

$$\frac{\partial}{\partial s} f(g.exp(tY)(X+sZ).exp(-tY)g^{-1})|_{s=0} = (\partial(g.exp(tY).Z.exp(-tY)g^{-1})f)(g.exp(tY).X.exp(-tY).g^{-1})$$

Hence,

$$\frac{\partial^2}{\partial t \partial s} f(g.exp(tY).(X + sZ).exp(-tY).g^{-1})|_{t=s=0}$$

$$= (\partial(g[Y, Z]g^{-1})f)(g.X.g^{-1}) + (\partial(g[Y, X]g^{-1})\partial(g.Z.g^{-1})f)(g.X.g^{-1})$$

This formula can be written as

$$F(g; Y, X; Z) = f^g(X; \partial([Y, Z] + [Y, X]Z))$$

where $f^g = foAd(g)$ and $[Y, X]Z \in S(\})$. We define $q \in S(\})$ so that

$$q = [Y, Z] + [Y, X]Z = (L_{[Y,X]} + ad(Y))Z = \sigma_X(Y)Z$$

Then,

$$F(g; Y, X; Z) = f^g(X; \partial(q)) = F(g, X; \partial(q))$$

[55]. Determination of the structure constants of a Lie algebra from a coordinate form of the group composition relations. The Lie group is parametrized by a vector parameter $\theta = \theta^a, a = 1, 2, ..., n$. Any

element of the group therefore has the form $T(\theta)$ where $T : \mathbb{R}^n \to GL_m(\mathbb{C})$ is a smooth mapping. We denote the generators of this group by t^a. Then,

$$T(\theta) = I + t_a \theta^a + \frac{1}{2} t_{ab} \theta^a \theta^b + \cdots$$

where $t_a, t_{ab}, \ldots$ are $n \times n$ matrices. Note that $T(\mathbf{0}) = \mathbf{I}$. Let $(\theta, \theta') \to f(\theta, \theta')$ be the group composition law. Then we have

$$f^a(\theta, 0) = \theta^a, f^a(0, \bar{\theta}) = \bar{\theta}^a$$

It follows that upto second order terms, we have the expansion

$$f^a(\theta, \bar{\theta}) = \theta^a + \bar{\theta}^a + f^a_{bc} \theta^b \bar{\theta}^c$$

Thus,

$$T(\theta)T(\bar{\theta}) = T(f(\theta, \bar{\theta}))$$

gives us

$$(I + t_a \theta^a + \frac{1}{2} t_{ab} \theta^a \theta^b + \cdots)(I + t_a \bar{\theta}^a + \frac{1}{2} t_{ab} \bar{\theta}^a \bar{\theta}^b + \cdots)$$

$$= I + t_a(\theta^a + \bar{\theta}^a) + t_a f^a_{bc} \theta^b \bar{\theta}^c + \frac{1}{2} t_{ab}(\theta^a + \bar{\theta}^a)(\theta^b + \bar{\theta}^b) + \cdots$$

It follows that

$$t_b t_c = t_a f^a_{bc} + \frac{1}{2}(t_{bc} + t_{cb})$$

and hence

$$t_b t_c - t_c t_b = t_a(f^a_{bc} - f^a_{cb})$$

This gives us the structure constants of the Lie algebra.

[56]. Develop MATLAB programs for computing the flow of a vector field on a differentiable manifold and on a Lie group. Suppose M is the manifold and $\chi(M)$ is the space of all vector fields on M. Let $f : M \to \mathbb{R}$ be a smooth map. Then for any vector field X on M, we denote the flow with initial condition x at $t = 0$ by $\Phi_X(t, x)$. In terms of components,

$$\frac{d\Phi^i_X(t, x)}{dt} = X^i(\Phi_X(t, x)), i = 1, 2, \ldots, n$$

We thus have

$$(Xf)(\Phi_X(t, x)) = \frac{d}{dt} f(\Phi_X(t, x))$$

It follows that

$$(Xf)(x) = \frac{d}{dt} f(\Phi_X(t, x))|_{t=0}$$

This equation can also be used as the defining relation for a vector field.

[57]. Let M denote the set of all diagonal 2×2 matrices with diagonal entries ± 1. Let A denote the set of all 2×2 diagonal matrices with positive diagonal entries and having determinant one. Let

N denote the set of all matrices of the form $\begin{pmatrix} 1 & s \\ 0 & 1 \end{pmatrix}$ where $s \neq 0$. Then what are the properties of matrices in MAN. Let K denote the set of all 2×2 real orthogonal matrices with determinant one. Then what are the properties of matrices in KAN.

[58]. Let X, Y, H be three matrices satisfying the commutation relations

$$[H, X] = 2X, [H, Y] = -2Y, [X, Y] = H$$

Give an example. Consider the real Lie algebra

$$\} = \mathbb{R}.H + \mathbb{R}.X + \mathbb{R}.Y$$

Show that $\}$ is semisimple and determine two non-conjugate Cartan subalgebras of $\}$.

hint: Use 2×2 matrices to realize these commutation relations and then show that $\dashv = \mathbb{R}.H$ and $\lfloor = \mathbb{R}.(X - Y)$ are two non-conjugate Cartan subalgebras of $\}$.

[59]. Let $\}$ be a Lie algebra and let π_i be a representation of $\}$ in the vector space V_i with $i = 1, 2$. For $X \in \}$, define

$$\pi(X)(v_1, v_2) = (\pi_1(X)v_1, \pi_2(X)v_2), v_i \in V_i, i = 1, 2$$

Show that π is a representation of $\}$ in the vector space $V_1 \times V_2$ (direct sum of V_1 and V_2). An alternate way of looking at this is as follows. Let V be a vector space and let $V = V_1 \oplus V_2$ where V_1, V_2 are subspaces of V. Let π_i be a representation of $\}$ in $V_i, i = 1, 2$. Then define

$$\pi(X)(v_1 + v_2) = \pi_1(X)v_1 + \pi_2(X)v_2, v_i \in V_i, i = 1, 2$$

Then π is a representation of $\}$ in V. π is called the direct sum of π_1 and π_2.

[60]. Let V_1, V_2 be vector spaces and let $\otimes$ be a tensor product of V_1 and V_2. Let $\}$ be a Lie algebra and let $\pi_i, i = 1, 2$ be representations of $\}$ in $V_i, i = 1, 2$ respectively. For $X \in \}$, define a linear operator $\pi(X)$ on $V_1 \otimes V_2$ so that

$$\pi(X)(v_1 \otimes v_2) = \pi_1(X)v_1 \otimes v_2 + v_2 \otimes \pi_2(X)v_2$$

ie,

$$\pi(X) = \pi_1(X) \otimes I + I \otimes \pi_2(X)$$

Then show that π is a representation of $\}$ in $V_1 \otimes V_2$. π is called the tensor product of the representations π_1 and π_2.

[61]. Let X, Y be two square matrices of the same size. Show that

$$exp(tX)exp(tY) = exp(t(X + Y) + t^2[X, Y]/2 + O(t^3)), t \to 0$$

hint:

$$log(exp(tX).exp(tY)) = log((I + tX + t^2X^2/2)(I + tY + t^2Y^2/2) + O(t^3))$$

$$= log(I + t(X+Y) + t^2(2XY + X^2 + Y^2)/2 + O(t^3)) = t(X+Y) + t^2(XY + X^2/2 + Y^2/2 - (X+Y)^2/2) + O(t^3)$$

$$= t(X+Y) + t^2((XY - YX))/2 + O(t^3) = t(X+Y) + t^2[X,Y]/2 + O(t^3)$$

This proves the claim.

[63]. Generalize the formula of the preceding problem to n matrices.

answer:

$$exp(tX_1).exp(tX_2)...exp(tX_n) = exp(t\sum_{i=1}^{n} X_i + \frac{t^2}{2} \sum_{1 \le i < j \le n} [X_i, X_j] + O(t^3))$$

for

$$exp(tX_1)...exp(tX_n) = (I + tX_1 + t^2X_1^2/2)(I + tX_2 + t^2X_2^2/2)...(I + tX_n + t^2X_n^2/2) + O(t^3)$$

$$= I + t\sum_{i=1}^{n} X_i + t^2(\sum_{i<j} X_iX_j + (1/2)\sum_i X_i^2) + O(t^3)$$

and on taking logarithms,

$$log((exp(tX_1)...exp(tX_n)) = t\sum_i X_i + t^2(\sum_{i<j} X_iX_j + (1/2)\sum_i X_i^2) - (\sum_i X_i)^2/2) + O(t^3)$$

$$= t.\sum_i X_i + (t^2/2)\sum_{i<j}[X_i, X_j] + O(t^3)$$

since

$$(\sum_i X_i)^2 = \sum_i X_i^2 + \sum_{i<j}(X_iX_j + X_jX_i)$$

[64]. Let G be a Lie group with Lie algebra $\}$. For $x \in G$ and $X \in \}$ and a smooth function $f : G \to \mathbb{C}$, we have

$$\frac{d^k}{dt^k} f(x.exp(tX))|_{t=0} = X^k f(x), k = 0, 1, 2, ...$$

and hence

$$f(x.exp(tX)) = \sum_{k=0}^{\infty} (t^k X^k f(x)/k!)$$

Generalize this formula to several vector fields obtain thus a multivariate version of the Taylor expansion formula on a Lie group.

[65]. Let $\}$ be a Lie algebra and ρ a nil representation of $\}$ in a finite dimensional vector space V. Then show that there exists a nonzero $v \in V$ and a $\lambda \in \}^*$ such that $\rho(X)v = \lambda(X)v$ for all $X \in \}$. Assume that the underlying field is algebraically closed.

[66]. Let $\}$ be a Lie algebra and let ρ be a representation of $\}$ in a vector space V. Let $\langle = Ker(\rho)$, ie, $\langle$ is the set of all $X \in \}$ for which $\rho(X) = 0$. Then consider the representation ρ_1 of the Lie algebra $\}/\langle$ in V induced by ρ. In other words, $\rho_1(X + \langle) = \rho(X)$ for all $X \in \}$. That ρ_1 is well defined follows from the fact that $X + \langle = Y + \langle$ for some $X, Y \in \}$ implies $Y - X \in \langle$ implies $\rho(Y) = \rho(X)$. Note that $\langle$ is an ideal in $\}$ since $Y \in \langle$ and $X \in \}$ imply that $\rho([X,Y]) = \rho(X)\rho(Y) - \rho(Y)\rho(X) = 0$ since $\rho(Y) = 0$ and hence $[X,Y] \in \langle$, ie, $[\},\langle] \subset \langle$ The fact that $\langle$ is an ideal in $\}$ implies that $\}/\langle$ is a Lie algebra. Indeed, for $X, Y \in \}$, we define $[X + \langle, Y + \langle] = [X,Y] + \langle$. This is well defined since $X_1 + \langle = X_2 + \langle$ and $Y_1 + \langle = Y_2 + \langle$ imply $X_2 - X_1, Y_2 - Y_1 \in \langle$ imply $[X_2, Y_2] - [X_1, Y_1] \in \langle$ since $\langle$ is an ideal in $\}$, ie, $[X_2, Y_2] + \langle = [X_1, Y_1] + \langle$.

[67]. Let G be a matrix Lie group and $\}$ its Lie algebra. Let $\pi : G \to GL_n(\mathbb{C})$ be a smooth representation of G and for $X \in \}$, define

$$\tilde{\pi}(X) = \frac{d}{dt}\pi(exp(tX))|_{t=0}$$

Then show that $\tilde{\pi}$ is a representation of the Lie algebra $\}$.

[68]. What do you understand by connectedness and arcwise connectedness of a topological space.

hint: If the topological space contains a proper nonempty subset that is both open and closed, then the space is said to be disconnected. Equivalently, a disconnected topological space can be expressed as a disjoint union of two non-empty open subsets. A topological space that is not disconnected is said to be connected.

[69]. The Campbell-Baker-Hausdorff formula. Derive a power series expansion for

$$F(t) = log(exp(tX).exp(tY))$$

where X, Y are $n \times n$ complex matrices.

[70]. Differential of the exponential map. Let X, Y be $n \times n$ matrices. Then derive a power series expansion for $\frac{\partial}{\partial s}exp(-tY)exp(tX + sY)|_{s=0}$.

[71]. Let G be a group and π a representation of G in a finite dimensional vector space V. Let V^* denote the dual of V and define for $g \in G$, the linear operator $\rho(g) = \pi(g^{-1})^T$ in V^*. Note that if $A \in L(V)$, then $A^T \in L(V^*)$ is defined by $A^T f(x) = f(Ax)$ for $f \in V^*, x \in V$. Show that ρ is a representation of G in V^*. The representation ρ is said to be contragredient to π.

[72]. When are two representations of a group G said to be equivalent ? What do you understand by an intertwining operator between two representations. Prove Schur's lemma for the intertwining operator between two irreducible representations.

[73]. Give two alternate definitions of a semisimple Lie algebra and prove their equivalence.

[74]. Let V be a finite dimensional vector space and V^* its dual. For $L \in L(V)$, define $L_{1,1} \in L(V \otimes V^*)$ so that

$$L_{1,1}(v \otimes v^*) = Lv \otimes v^* - v \otimes L^T v^*$$

Recall the definition of transpose: For $f \in V^*$ and $x \in V$, we have $(L^T f)(x) = f(Lx)$. Then show that if A is a nonsingular linear transformation on V and $\pi(A) \in L(V \otimes V^*)$ is defined so that

$$\pi(A)(v \otimes v^*) = Av \otimes A^{-T} v^*$$

ie,

$$\pi(A) = A \otimes A^{-T}$$

then

$$L_{1,1} = \frac{d}{dt}\pi(exp(tL))|_{t=0}, L \in L(V \otimes V^*)$$

In other words, $L \to L_{1,1}$ is therefore the differential of the representation π of $GL(V)$. This can be generalized to tensor products of arbitrary order. Specifically, let r, s be non-negative integers with $r + s \geq 1$. Let $V_{r,s} = V^r \otimes V^{*s}$ be the vector space obtained by taking the tensor product of r copies of V and s copies of V^*. For $L \in L(V)$, define $L_{r,s} \in L(V_{r,s})$ so that

$$L_{r,s}(v_1 \otimes ... \otimes v_r \otimes v_1^* \otimes ... \otimes v_s^*)$$

$$= Lv_1 \otimes v_2 \otimes ... \otimes v_r \otimes v_1^* \otimes ... \otimes v_s^*$$

$$+v_1 \otimes Lv_2 \otimes ... \otimes v_r \otimes v_1^* \otimes ... \otimes v_s^*$$

$$+... + v_1 \otimes v_2 \otimes ... \otimes Lv_r \otimes v_1^* \otimes ... \otimes v_s^*$$

$$-v_1 \otimes ... \otimes v_r \otimes L^T v_1^* \otimes v_2^* \otimes ... \otimes v_s^*$$

$$-v_1 \otimes ... \otimes v_s \otimes v_1^* \otimes L^T v_2^* \otimes ... \otimes v_s^*$$

$$-v_1 \otimes ... \otimes v_s \otimes v_1^* \otimes ... \otimes L^T v_s^*$$

Then if A is a nonsingular linear operator on V, define a linear operator $\pi_{r,s}(A)$ in $V_{r,s}$ so that

$$\pi_{r,s}(A)(v_1 \otimes ... \otimes v_r \otimes v_1^* \otimes ... \otimes v_s^*) = Av_1 \otimes ... \otimes Av_r \otimes A^{-T} v_1^* \otimes ... \otimes A^{-T} v_s^*$$

Then $\pi_{r,s}$ is a representation of the group $GL(V)$ and $L \to L_{r,s}$ is the differential of $\pi_{r,s}$.

[75] Define the following terms:(a) Arcwise connected topological space, (b) homotopy, (c) fundamental group.

hint:Let (X, τ) be a topological space. Given any two points $x, y \in X$, if there exists a continuous function $\gamma : [0,1] \to X$ such that $\gamma(0) = x, \gamma(1) = y$, then we say that X is an arcwise connected topological space. Let $x, y \in X$ and let $\gamma_i : [0,1] \to X, i = 1,2$ be two continuous functions such that $\gamma_i(0) = x, \gamma_i(1) = y, i = 1,2$. In other words, γ_i is a path connecting x and y. We say that the two paths $\gamma_i, i = 1,2$ are homotopic if there exists a continuous function $\Phi : [0,1] \times [0,1] \to X$ such that

$\Phi(s,0) = x, \Phi(s,1) = y$ and $\Phi(0,t) = \gamma_1(t), \Phi(1,t) = \gamma_2(t)$ for all $s,t \in [0,1]$. In other words, two paths are homotopic if any one of them can be continuously deformed into the other. A loop in X at x_0 is a continuous map $\gamma : [0,1] \to X$ such that $\gamma(0) = \gamma(1) = x_0$. Given two loops at the same point, we can concatenate them to get another loop. Two loops that are homotopic to each other are said to belong to the same equivalence class. The set of such equivalence classes of loops under concatenation is a group and this group is called the fundamental group of X. Note the method of defining concatenation: Let $\gamma_i : [0,1] \to X, i = 1,2$ such that $\gamma_i(0) = \gamma_i(1) = x_0$, we define $\gamma(t) = \gamma_1(2t)$ for $0 \le t \le 1/2$ and $\gamma(t) = \gamma_2(2t-1)$ for $1/2 \le t \le 1$. Then γ is denoted by $\gamma_1.\gamma_2$ and it is called the concatenation of γ_1 with γ_2.

[76] Prove the Jacobi identity for matrices:If A, B, C are $n \times n$ matrices, then

$$[A,[B,C]] + [B,[C,A]] + [C,[A,B]] = 0$$

where

$$[A,B] = AB - BA$$

[77] Prove that an infinitesimal rotation of $\mathbb{R}^3$ has the following representation upto second order differentials:

$$\delta r = \delta\phi \times r + \frac{1}{2}\delta\phi \times (\delta\phi \times r)$$

Use this formula to obtain the commutation relations between the generators of rotations.

[78] Let G be a Lie group with Lie algebra $\}$. For $g \in G$ and $X \in \}$, define $Ad(g)(X) \in \}$ so that $Ad(g)$ is the differential of the map $x \to gxg^{-1}$ from G into G. Prove that $g \to Ad(g)$ is a representation of G in $\}$.

[79] Suppose V and W are two vector spaces. Let G be a group and π, ρ two representations of G in V and W respectively. For $f : V \to W$ and $g \in G$, define $(\sigma(g)f)(x) = \rho(g)(f(\pi(g)^{-1}x)$. Then show that ρ is a representation of G in the space $F(V,W)$ of all functions from V into W.

[80] Let $G_i, i = 1,2$ be two groups and let $\pi_i, i = 1,2$ be representations of $G_i, i = 1,2$ respectively in the vector spaces $V_i, i = 1,2$. Let $\otimes$ be a tensor product of V_1 and V_2. For $g_i \in G_i, i = 1,2$, show that $(g_1,g_2) \to \pi_1(g_1) \otimes \pi_2(g_2)$ is a representation of $G_1 \times G_2$ in $V_1 \otimes V_2$.

[81] Let $\}_i, i = 1,2$ be two Lie algebras and let $\pi_i : \}_i \to L(V_i), i = 1,2$ be representations of $\}_i, i = 1,2$ in vector spaces $V_i, i = 1,2$ respectively. Let $\}_1 \times \}_2$ be the Lie algebra consisting of all ordered pairs (X_1, X_2) with $X_i \in \}_i, i = 1,2$ with the Lie bracket defined by

$$[(X_1,X_2),(Y_1,Y_2)] = ([X_1,Y_1],[X_2,Y_2]), X_i, Y_i \in \}_i, i = 1,2$$

Define

$$\pi((X_1,X_2)) = \pi_1(X_1) \otimes I + I \otimes \pi_2(X_2), X_i \in \}_i, i = 1,2$$

Then show that π is a representation of $\}_1 \times \}_2$ in $V_1 \otimes V_2$.

hint: Let $X_i, Y_i \in \}_i, i = 1, 2$. Then,

$$\pi((X_1, X_2))\pi((Y_1, Y_2)) = \pi_1(X_1)\pi_1(Y_1) \otimes I + I \otimes \pi_2(X_2)\pi_2(Y_2) + \pi_1(X_1) \otimes \pi_2(Y_2) + \pi_1(Y_1) \otimes \pi_2(X_2)$$

Interchanging (X_1, X_2) with (Y_1, Y_2) and subtracting gives us

$$\pi((X_1, X_2))\pi((Y_1, Y_2)) - \pi((Y_1, Y_2))\pi((X_1, X_2))$$

$$= (\pi_1(X_1)\pi_1(Y_1) - \pi_1(Y_1)\pi_1(X_1)) \otimes I + I \otimes (\pi_2(X_2)\pi_2(Y_2) - \pi_2(Y_2)\pi_2(X_2))$$

$$= \pi_1([X_1, Y_1]) \otimes I + I \otimes \pi_2([X_2, Y_2]) = \pi(([X_1, Y_1], [X_2, Y_2])) = \pi([(X_1, Y_1), (X_2, Y_2)])$$

[82] Define a right invariant vector field on a Lie group and give an example.

[83] Give an example in terms of permutation groups that illustrates Lagrange's theorem for finite groups, namely that if G is any finite group and H a subgroup of G, then $o(H)$ divides $o(G)$ where for any finite set E, $o(E)$ denotes the number of elements in E.

[84] Define the following terms and give examples. (a) range of a linear transformation, (b) nullspace of a linear transformation, (c) rank of a linear transformation, (d) nullity of a linear transformation.

[85] Prove the rank-nullity theorem. If V and W are finite dimensional vector spaces over the same field and $T : V \to W$ is a linear transformation, then $rank(T) + nullity(T) = dimV$.

[86] Describe geometrically the effect of the linear transformations T, U on $\mathbb{R}^2$ defined by $T(x_1, x_2) = (x_2, x_1)$ and $U(x_1, x_2) = (x_1, 0)$. Determine the actions of the linear transformations TU and UT on the element (x_1, x_2) of $\mathbb{R}^2$.

[87] Determine in the standard basis, the matrix of a reflection about the line $y = mx$ in $\mathbb{R}^2$. What are the eigenvalues of this matrix.

[88] What do you understand by the statement that $SU(2)$ is the covering group of $SO(3)$. Construct a homomorphism of $SU(2)$ onto $SO(3)$ and show that this homomorphism maps two distinct elements of $SU(2)$ onto the same element of $SO(3)$.

[89] Let $\}_i, i = 1, 2, ..., n$ be matrix Lie algebras over the same field. Construct the tensor product of these Lie algebras.

hint. If $n = 2$, then for $X_i \in \}_i, i = 1, 2$, we set $(X_1, X_2) = X_1 \otimes I + I \otimes X_2$ and observe that

$$(X_1, X_2).(Y_1, Y_2) = X_1Y_1 \otimes I + I \otimes X_2Y_2 + X_1 \otimes Y_2 + Y_1 \otimes X_2$$

implying that

$$(X_1, X_2)(Y_1, Y_2) - (Y_1, Y_2)(X_1, X_2) = (X_1 Y_1 - Y_1 X_1) \otimes I + I \otimes (X_2 Y_2 - Y_2 X_2)$$

[90] Here we define the exponential map on a Riemannian manifold. Let $\mathcal{M}$ be an n-dimensional Riemannian manifold. Introduce a local coordinate system on this manifold and identify $\mathcal{M}$ locally with $\mathbb{R}^n$. Let $p \in \mathcal{M}$ and identify $T\mathcal{M}_p$, the tangent space to $\mathcal{M}$ at p with $\mathbb{R}^n$. For $X \in T\mathcal{M}_p$, identify X with its coordinates $X = (X^i)_{i=1}^n$. Let $s \to \gamma_X(s)$ denote the geodesic on the Riemannian manifold $\mathcal{M}$ for which $\gamma_X(0) = p$ and $\gamma_X'(0) = X$. Define $Exp(X) \in \mathcal{M}$ by the equation $Exp(X) = \gamma_X(1)$. We are assuming that the parameter s measures the length of the geodesic, ie, the shortest distance on the manifold between the points p and $\gamma_X(s)$. If $Y \in \mathbb{R}^n$, then $(dExp)_X(Y)$ is defined to be $\frac{d}{dt} Exp(X + tY)|_{t=0}$ where the derivative is computed by using the local coordinate system. Note that $\frac{d}{dt} Exp(tX) = \gamma_X'(t)$ is the vector on the curve $\gamma_X(.)$ at the point $\gamma_X(t)$ obtained by displacing X at $\gamma_X(0) = p$ parallely along the curve $\gamma_X(.)$ to the point $\gamma_X(t)$. Given two points p, q on $\mathcal{M}$, we define the operator τ_{pq} so that it transforms a vector X in $T\mathcal{M}_p$ to the vector in $T\mathcal{M}_q$ obtained by parallely displacing X along the geodesic joining p to q. Then it is clear that $\frac{d}{dt} Exp(tX) = \tau_{pq} X$ where $q = Exp(tX)$. This is because $s \to Exp(sX), 0 \le s \le t$ is the geodesic joining p to $q = Exp(tX)$.

[91] $\mathcal{M}$ and $\mathcal{N}$ are two differentiable manifolds and $\Phi : \mathcal{M} \to \mathcal{N}$ is a diffeomorphism. Let $p \in \mathcal{M}$ and let $x \to \phi^i(x)$ be a local coordinate system around p in $\mathcal{M}$. We assume that the domain of the functions ϕ^i is an open set U in $\mathcal{M}$ containing p. Likewise, let V be an open set in $\mathcal{N}$ containing $\Phi(p)$ and let $\psi^i : V \to \mathbb{R}^N$ be local coordinates around $\Phi(p)$. Then locally, Φ has the coordinate representation $\psi o \Phi o \phi^{-1} : \phi(U) \to \mathbb{R}^N$ provided we assume that $\Phi(U) \subset V$. We denote this local coordinate representation of Φ by the same symbol Φ. Let X be a vector in $T\mathcal{M}_p$ with components X^i relative to a local coordinate system. Then $d\Phi(X) = Y$ is a vector in $T\mathcal{N}_{\Phi(p)}$ with components Y^i relative to a local coordinate system where

$$Y^i = \sum_j \frac{\partial \Phi^i(p)}{\partial x^j} X^j$$

This follows from the defintion of $d\Phi$ stating that if X and Y are regarded as local differential operators, then $d\Phi(X)(f) = X(f o \Phi)$ where f is a map $f : U \to V$ with U a neighbourhood of p and V a neighbourhood of $\Phi(p)$. Note that if (x^i) are the local coordinates of p and (y^i) the local coordinates of $\Phi(p)$, then X corresponds to the differential operator $X^i \frac{\partial}{\partial x^i}$ and Y corresponds to the local differential operator $Y^i \frac{\partial}{\partial y^i}$.

[92] Let $\mathcal{M}$ be a Riemannian manifold with metric $(X, Y) \to g(X, Y)$ where X, Y are vector fields and $g(X, Y)$ is a function on the manifold. Note that $g(X, Y)$ is bilinear and symmetric in (X, Y). Show that there exists a unique connection ∇ on the manifold such that the torsion of the connection vanishes and the covariant derivative of the metric also vanishes. Determine the connection components in terms of the components of the metric tensor in a local coordinate basis $\frac{\partial}{\partial x^i}$. Note that $\nabla_X g = 0$ iff

$$X(g(Y, Z)) = g(\nabla_X Y, Z) + g(Y, \nabla_X Z)$$

for all vector fields Y, Z and that the torsion T vanishes means that

$$\nabla_X Y - \nabla_Y X - [X, Y] = 0$$

for all vector fields X, Y. To solve this problem, take X, Y, Z as $\frac{\partial}{\partial x^i}, \frac{\partial}{\partial x^j}$ and $\frac{\partial}{\partial x^k}$.

[93] Suppose G is a Lie group with Lie algebra $\}$. Let H be a Lie subgroup of G with Lie algebra $\langle$, so that we can assume that $\langle \subset \}$. Suppose H is a normal subgroup of G. Then, choose $X \in \}$ and $Y \in \langle$. It follows that $exp(tX) \in G$ and $exp(sY) \in H$ for all $t, s \in \mathbb{R}$. Then

$$Ad(exp(tX))(exp(sY)) = exp(s.Ad(exp(tX)(Y)) \in H, s, t \in \mathbb{R}$$

so that

$$Ad(exp(tX))(Y) \in \langle$$

and hence differentiating with respect to t at $t = 0$ gives

$$ad(X)(Y) = [X, Y] \in \langle$$

or in other words,

$$[\}, \langle] \subset \langle$$

which means that $\langle$ is an ideal in $\}$.

[94] Let G be a Lie group with Lie algebra $\}$. Suppose D is a linear operator on $\}$ such that $exp(tD)$ is an automorphism of $\}$ for all $t \in \mathbb{R}$. Then for $X, Y \in \}$, we have

$$exp(tD)[X, Y] = [exp(tD)(X), exp(tD)(Y)]$$

Differentiate both sides with respect to t at $t = 0$ to get

$$D[X, Y] = [DX, Y] + [X, DY], X, Y \in \}$$

In other words, D is a derivation of $\}$.

Reference:S.Helgason, Differential geometry, Lie groups and Symmetric spaces.

[95] Suppose X, Y are two vector fields on a differentiable manifold $\mathcal{M}$. Suppose we view X and Y as linear first order differential operators on the space $C^\infty(\mathcal{M})$. Relative to a local coordinate chart,

$$X = X^i \frac{\partial}{\partial x^i}, Y = Y^i \frac{\partial}{\partial x^i}$$

with summation over the repeated index i implied. The Lie bracket of X and Y is the differential operator $[X, Y] = XY - YX$. Show that $[X, Y]$ is a vector field on $\mathcal{M}$, ie, $[X, Y]$ is a linear first order differential operator with components

$$[X, Y]^i = X^j Y^i_{,j} - Y^j X^i_{,j}$$

summation over the repeated index j being implied.

[96] Define the notion of a geodesic on a differentiable manifold with respect to a specified connection. For a Riemannian manifold endowed with the Riemannian connection, show that geodesics are actually extremals for the distance, ie, they are paths $s \to \gamma(s)$ that extremize the functional $\int_a^b g(\gamma'(s), \gamma'(s))^{1/2} ds$ when the endpoints $\gamma(a)$ and $\gamma(b)$ are kept fixed.

[97] Given a differentiable manifold $\mathcal{M}$ with a connection ∇, the Riemann-Christoffel curvature tensor is a mapping $(X, Y, Z) \to R(X,Y)Z$ where X, Y, Z and $R(X,Y)Z$ are vector fields and

$$R(X,Y)Z = \nabla_X \nabla_Y Z - \nabla_Y \nabla_X Z - \nabla_{[X,Y]} Z$$

Let $\{X_1(x), ..., X_n(x)\}$ be a basis for $T\mathcal{M}_x$ for each $x \in \mathcal{M}$ and let

$$[X_i, X_j](x) = c_{ij}^k(x) X_k(x)$$

Let $\{\omega^1(x), ...\omega^n(x)\}$ be the dual to $\{X_1(x), ..., X_n(x)\}$. Calculate

$$R_{kmj}^i(x) = \omega^i(R(X_k, X_m)X_j)(x)$$

In the special case when $X_i = \frac{\partial}{\partial x^i}$ with (x^i) a local coordinate basis, we have $c_{ij}^k = 0$ and then compute

$$R_{kmj}^i(x) = dx^i(R(\frac{\partial}{\partial x^k}, \frac{\partial}{\partial x^m})\frac{\partial}{\partial x^j})$$

[98] Let $\}$ be a three dimensional complex Lie algebra with a basis $\{H, X, Y\}$ satisfying the commutation relations $[H, X] = 2X, [H, Y] = -2Y, [X, Y] = H$. Show that this Lie algebra is isomorphic to $sl(2, \mathbb{C})$ which consists of all 2×2 complex matrices having zero trace.

[99] What do you understand by a Cartan matrix with reference to a given complex semisimple Lie algebra. Give an example.

[100] Describe Cartan's classification of the complex simple Lie algebras and write down the Cartan matrices associated with these.

[101] Let V be a real inner product space and let $\alpha \in V$ be nonzero. For $\beta \in V$, define

$$s_\alpha(\beta) = \beta - \frac{2 <\beta, \alpha>}{<\alpha, \alpha>}\alpha$$

Prove that s_α is a linear transformation on V and is in fact a reflection. Given a complex semisimple Lie algebra and a Cartan subalgebra $\langle$ of $\}$, show that if $dim\langle = l$, then there exists a set $\{\alpha_1, ..., \alpha_l\}$ of roots such that any root α is of the form $m_1\alpha_1 + ... + m_l\alpha_l$ where either all the $m_j's$ are non-negative integers or all are non-positive integers.

[102] If $[H, X] = \alpha(H)X, [H, Y] = \beta(H)Y$, then show that $[H, [X, Y]] = (\alpha(H)+\beta(H))[X, Y]$. Here, X, Y, H are three elements of a semisimple Lie algebra $\}$ and H is an element of a Cartan subalgebra $\langle$ of $\}$. α, β are linear functionals on $\langle$.

[103] Show that the space of root vectors corresponding to a given root of a complex semisimple Lie algebra is one dimensional.

[104] Show that the Cartan-Killing form $B(X, Y) = tr(ad(X).ad(Y))$ on a semisimple Lie algebra is non-degenerate. Show that the restriction of $B(.,.)$ to a Cartan-Subalgebra of the given semisimple Lie algebra is also non-degenerate.

[105] Let } be a semisimple Lie algebra, ie a Lie algebra whose Cartan-Killing form is non-singular. Show that it has a Cartan subalgebra, ie, a maximal Abelian subalgebra $\langle$ such that for all $H \in \langle$, adH is a semisimple linear operator on }.

[106] What do you understand by a regular element of a semisimple Lie algebra. Give an example.

[107] Prove the Schur orthogonality relations for the matrix elements of irreducible representations.

[108] Show that every finite dimensional representation of a compact group is equivalent to a unitary representation.

hint:From a given inner product on the vector space, construct an inner product on the same space that is invariant under the group using the method of group averaging.

[109] Let } be a Lie algebra and let $\{X_1, ..., X_n\}$ be a basis for } with corresponding structure constants as c^i_{jk} so that

$$[X_i, X_j] = \sum_k c^k_{ij} X_k$$

The Jacobi identity

$$[X, [Y, Z]] + [Y, [Z, X]] + [Z, [X, Y]] = 0$$

implies that the structure constants satisfy a certain set of identities. What are these identities.

[110] Assume that } is a three dimensional complex Lie algebra with basis elements H, X, Y satisfying the commutation relations

$$[H, X] = 2X, [H, Y] = -2Y, [X, Y] = H$$

Show that } is isomorphic to $sl(2, \mathbb{C})$. Determine a nonzero element in the centre of the universal enveloping algebra of }.

hint: Consider $\Omega = H^2 - 2H + 4XY$ in the universal enveloping algebra. Note that $\Omega = H^2 - 2H + 2XY + 2YX + 2[X, Y] = H^2 + 2XY + 2YX$. Then,

$$[H, \Omega] = [H, 4XY] = 4[H, X]Y + 4X[H, Y] = 8XY - 8XY = 0,$$

$$[X, \Omega] = [X, H^2 - 2H + 4XY] = [X, H]H + H[X, H] - 2[X, H] + 4X[X, Y]$$

$$= -2XH - 2HX + 4X + 4XH = 2(XH - HX) + 4X = 0$$

$$[Y, \Omega] = [Y, H^2 - 2H + 4XY] = H[Y, H] + [Y, H]H - 2[Y, H] + 4[Y, X]Y$$

$$= 2HY + 2YH - 2Y - 4HY = 2(YH - HY) - 2Y = 0$$

This proves that Ω is in the centre of }.

[111] Define the following terms in Lie algebra theory.

[1] Universal enveloping algebra of a Lie algebra.

hint:Let } be a Lie algebra and consider the associated tensor algebra

$$\tau = \mathbb{C} \oplus \bigoplus_{n=1}^{\infty} \}^{\otimes n}$$

Consider the two sided ideal $\mathcal{L}$ in τ generated by the elements $u(X, Y) = X \otimes Y - Y \otimes X - [X, Y]$. Then, $\mathcal{U} = \tau/\mathcal{L}$ is the universal enveloping algebra of }. It has got the following important property. Let $\mathcal{A}$ be any associative algebra and let $\xi : \} \to \mathcal{A}$ be a homomorphism, ie, ξ is linear and $\xi([X, Y]) = \xi(X)xi(Y) - \xi(Y)\xi(X)$ for all $X, Y \in \}$. Then there exists a unique homomorphism $\tilde{\xi} : \mathcal{U} \to \mathcal{A}$ such that $\tilde{\xi}(\gamma(X)) = \xi(X)$ for all $X \in \}$. Here, $\gamma : \} \to \mathcal{U}$ is the canonical map $\gamma(X) = X + \mathcal{L}$.

[2] Adjoint representation of a Lie group with the corresponding Lie algebra as the representation space.

hint: G is a Lie group and consider the map $\Phi_g : G \to G$ defined by $\Phi_g(h) = ghg^{-1}$. Clearly $\Phi_g(e) = e$ and hence $d\Phi_g : TG_e \to TG_e$ is well defined. Identifying TG_e with the Lie algebra } of G, we obtain the adjoint representation of G in } as $Ad(g) = (d\Phi_g)_e$. We have $Ad(g)$ is linear and $Ad(g_1 g_2) = Ad(g_1)Ad(g_2)$ for all $g_1, g_2 \in G$.

[3] Roots, root vectors.

[4] Direct sum of two Lie algebras, tensor product of two representations of a Lie algebra, direct sum of two representations of a Lie algebra.

[5] ideal of a Lie algebra, solvable Lie algebra, nilpotent Lie algebra.

[6] Simple Lie algebra, Semisimple Lie algebra, reductive Lie algebra.

[7] Exponential map from a Lie algebra into the associated Lie group.

[8] Centralizer of a subset of a Lie algebra, normalizer of a Lie subalgebra.

[9] Elements of the universal enveloping algebra of a Lie algebra as differential operators.

[10] Character of a representation of a Lie algebra.

[11] Cartan subalgebra.

[12] Cartan-Killing form on a Lie algebra.

[112] Let G be a matrix Lie group with Lie algebra $\}$. Let H be another matrix Lie group with Lie algebra $\langle$. Suppose $\Phi : G \to H$ is a homomorphism between the Lie groups G and H. For $X \in \}$, consider the map $t \to \Phi(exp(tX)) = h(t)$ from $\mathbb{R}$ into H. We have for $t, s \in \mathbb{R}$,

$$h(t).h(s) = \Phi(exp(tX)).\Phi(exp(sX)) = \Phi(exp(tX).exp(sX)) = \Phi(exp((t+s)X)) = h(t+s)$$

Since Phi is continuous, it follows that there exists a matrix Y such that

$$h(t) = exp(tY), t \in \mathbb{R}$$

Hence, $Y \in \langle$. We set $Y = \phi(X)$ and get

$$\Phi(exp(tX)) = exp(t\phi(X)), t \in \mathbb{R}, X \in \}$$

The mapping $\phi : \} \to \langle$ is a Lie algebra homomorphism, i.e, ϕ is linear and $\phi([X_1, X_2]) = [\phi(X_1), \phi(X_2)]$ for all $X_1, X_2 \in \}$. For let $X_1, X_2 \in \}$. Then

$$exp(t\phi(X_1 + X_2)) = \Phi(exp(t(X_1 + X_2))) = lim_{n \to \infty} \Phi((exp(tX_1/n).exp(tX_2/n))^n)$$

$$= lim_{n \to \infty}(\Phi(exp(tX_1/n).exp(tX_2/n)))^n = lim_{n \to \infty}(\Phi(exp(tX_1/n))/\Phi(exp(tX_2/n)))^n$$

$$= lim_{n \to \infty}(exp(t\phi(X_1)/n).exp(t\phi(X_2)/n))^n = exp(t(\phi(X_1) + \phi(X_2))$$

showing that $\phi(X_1 + X_2) = \phi(X_1) + \phi(X_2)$. Let $X \in \}$ and $c \in \mathbb{C}$. Then,

$$exp(t\phi(cX)) = \Phi(exp(tcX)) = exp(tc\phi(X))$$

proving that $\phi(cX) = c\phi(X)$. Thus, ϕ is linear. For $X_1, X_2 \in \}$, we have

$$exp(t\phi([X_1, X_2])) = \Phi(exp(t[X_1, X_2])) = lim_{n \to \infty} \Phi((exp(-tX_1/n)exp(-tX_2/n)exp(tX_1/n)exp(tX_2/n))^{2n^2})$$

$$= lim_{n \to \infty}(\Phi(exp(-tX_1/n)).\Phi(exp(-tX_2/n)).\Phi(exp(tX_1/n)).\Phi(exp(tX_2/n)))^{2n^2}$$

$$= lim_{n \to \infty}(exp(t\phi(X_1)/n).exp(-t\phi(X_2)/n).exp(t\phi(X_1)/n).exp(t\phi(X_2)/n))^{2n^2}$$

$$= exp(t[\phi(X_1), \phi(X_2)])$$

proving that

$$\phi([X_1, X_2]) = [\phi(X_1), \phi(X_2)], X_1, X_2 \in \}$$

[113] Determine the finite dimensional irreducible representations of $SU(3)$.

Note that $SU(3)$ has Lie algebra $su(3)$ which consists of all 3×3 skew-Hermitian matrices having trace zero. If $A \in SU(3)$ and $A = exp(X)$, then $A^*A = I$ implies $X^* = -X$ and $det A = 1$ implies $tr(X) = 0$. $su(3)$ is a real Lie algebra and its complexification is $sl(3, \mathbb{C})$ which consists of all 3×3 complex matrices having trace zero. $sl(3, \mathbb{C})$ is a complex Lie algebra and determining the irreducible representations of

$SU(3)$ is equivalent to determining the irreducible representations of $sl(3,\mathbb{C})$. A convenient basis for $sl(3,\mathbb{C})$ is given by

$$H_1 = \begin{pmatrix} 1 & 0 & 0 \\ 0 & -1 & 0 \\ 0 & 0 & 0 \end{pmatrix},$$

$$H_2 = \begin{pmatrix} 0 & 0 & 0 \\ 0 & 1 & 0 \\ 0 & 0 & -1 \end{pmatrix},$$

$$X_1 = \begin{pmatrix} 0 & 1 & 0 \\ 0 & 0 & 0 \\ 0 & 0 & 0 \end{pmatrix},$$

$$Y_1 = \begin{pmatrix} 0 & 0 & 0 \\ 1 & 0 & 0 \\ 0 & 0 & 0 \end{pmatrix},$$

$$X_2 = \begin{pmatrix} 0 & 0 & 0 \\ 0 & 0 & 1 \\ 0 & 0 & 0 \end{pmatrix},$$

$$Y_2 = \begin{pmatrix} 0 & 0 & 0 \\ 0 & 0 & 0 \\ 0 & 1 & 0 \end{pmatrix},$$

$$X_3 = \begin{pmatrix} 0 & 0 & 1 \\ 0 & 0 & 0 \\ 0 & 0 & 0 \end{pmatrix},$$

$$Y_3 = \begin{pmatrix} 0 & 0 & 0 \\ 0 & 0 & 0 \\ 1 & 0 & 0 \end{pmatrix}$$

Determine the commutation relations between these matrices.

[114] $\}_0$ is a real Lie algebra and $\}$ is its complexification. $\sqcap$ is a compact real form of $\}$ and $\tilde{\theta}$ denotes conjugation with respect to $\sqcap$. Thus, if $Z = X + iY$ with $X, Y \in \sqcap$, then $\tilde{\theta}(Z) = X - iY$. Let η denote the conjugation with respect to $\}_0$ and assume that $\eta(\sqcap) = \sqcap$. Let $\approx_0 = \sqcap \cap \}_0$ and $\rho = i\sqcap \cap \}_0$.

Then, $\}_0 = \approx_0 \oplus \rho$ Indeed, let $X \in \}_0$. Then $X = Y + iZ$ for some $Y, Z \in \sqcap$. Applying η to both sides gives

$$X = \eta(Y) - i\eta(Z)$$

and hence,

$$X = (Y + \eta(Y))/2 - (Z - \eta(Z)/2i$$

It is clear that $(Y + \eta(Y))/2, (Z - \eta(Z))/2i \in \}_0$ and since $\eta(Y), \eta(Z) \in \sqcap$, it follows that

$$(Y + \eta(Y))/2 \in \}_0 \cap \sqcap = \approx_0, (Z - \eta(Z))/2i \in \}_0 \cap i\sqcap = \rho$$

and the claim is now obvious. Let $\approx$ and $\sqrt{\ }$ denote respectively the complexifications of $\approx_0$ and $\sqrt{\ }_0$, ie, $\approx = \approx_0 + i\approx_0$ and $\sqrt{\ } = \sqrt{\ }_0 + i.\sqrt{\ }_0$. Let $\theta = \eta o \tilde{\theta}$. It is clear that $\} = \approx \oplus \sqrt{\ }$. Let $Z \in \}$. Then $Z = X + iY$ where $X, Y \in \}_0$. Set

$$X = X_1 + X_2, Y = Y_1 + Y_2, X_1, Y_1 \in \approx_0, X_2, Y_2 \in \sqrt{\ }_0$$

Then,

$$Z = X_1 + X_2 + i(Y_1 + Y_2) = (X_1 + iY_1) + (X_2 + iY_2)$$

and

$$X_i + iY_1 \in \approx, X_2 + iY_2 \in \sqrt{\ }$$

This shows that $\} = \approx + \sqrt{\ }$. That the sum is direct follows immediately from $\approx_0 \cap \sqrt{\ }_0 = \{0\}$.

Suppose $Z \in \approx$. Then $Z = X + iY$ with $X, Y \in \approx_0$. Then,

$$\tilde{\theta}(Z) == \tilde{\theta}(X) - i\tilde{\theta}(Y) = X - iY$$

since $\approx_0 \subset \sqcap$. Then,

$$\theta(Z) = \eta o\tilde{\theta}(Z) = \eta(X) + i\eta(Y) = X + iY = Z$$

since $\approx_0 \subset \}_0$. Likewise suppose $Z \in \sqrt{\ }$. Then, $Z = X + iY$ with $X, Y \in \sqrt{\ }_0$. Then since $\sqrt{\ }_0 \subset i\sqcap$, it follows that

$$\tilde{\theta}(Z) = \tilde{\theta}(X) - i\tilde{\theta}(Y) = -X + iY$$

$$\theta(Z) = \eta o\tilde{\theta}(Z) = -\eta(X) - i\eta(Y) = -X - iY = -Z$$

since $\approx_0 \subset \}_0$.

[115] Let π be an irreducible unitary representation of a compact group G in $\mathbb{C}^n$. Consider the $n \times n$ matrix

$$\mathbf{P} = \int_G \pi(g)^* \mathbf{X} \pi(g) dg$$

where $\mathbf{X}$ is any $n \times n$ complex matrix and dg is the Haar measure on G. Then for any $h \in G$, we have

$$\mathbf{P}\pi(h) = \int_G \pi(g)^* \mathbf{X} \pi(gh) dg = \int_G \pi(gh^{-1})^* \mathbf{X} \pi(g) dg$$

$$= \pi(h)\mathbf{P}$$

In other words, $\mathbf{P}$ is an intertwining operator for the representation π. From Schur's lemma, it follows that $\mathbf{P}$ is a constant multiple of the identity matrix. Hence derive the orthogonality relations for the matrix elements of π.

[116] Calculate the volume of the n dimensional real unit sphere, ie,

$$V_n = \int_{\sum_{i=1}^n x_i^2 \leq 1} dx_1...dx_n$$

[117] Give an example of a real semisimple Lie algebra that has two non-conjugate Cartan subalgebras.

hint: $sl(2, \mathbb{R})$, $\langle_1 = so(2, \mathbb{R})$, $\langle_2 = \mathbb{R}.diag[1, -1]$. Note that

$$so(2, \mathbb{R}) = \mathbb{R}. \begin{pmatrix} 0 & 1 \\ -1 & 0 \end{pmatrix}$$

Can a complex semisimple Lie algebra have two or more non-conjugate Cartan subalgebras.

[118] Let $\}$ be a semisimple Lie algebra and $\langle$ a Cartan subalgebra of $\}$. Using the root space decomposition show that

$$\} = \langle \oplus \backslash_+ \oplus \backslash_-$$

where $\backslash_+$ and $\backslash_-$ are nilpotent subalgebras of $\}$ and both are invariant under $ad(H), H \in \langle$.

[119] Let $\}_0$ be a real semisimple Lie algebra and $\langle_0$ a Cartan subalgebra of $\}_0$. Let $\}$ be the complexification of $\}_0$ and $\langle$ the complexification of $\langle_0$. Then, $\langle$ is a Cartan subalgebra of $\}$. Let $\sqcap$ be a compact real form of $\}$ and let $\tilde{\theta}$ be the conjugation with respect to $\sqcap$ and let η be the conjugation with respect to $\}_0$. Assume that $\sqcap$ is invariant under η. Then set $\theta = \tilde{\theta}o\eta$. Define $\approx_0 = \sqcap \cap \}_0$ and $\sqrt{}_0 = i\sqcap \cap mathcalg_0$. Let $\approx$ and $\sqrt{}$ be respectively the complexifications of $\approx_0$ and $\sqrt{}_0$. Let $\langle_{t_0} = \langle_0 \cap \approx_0$ and $\langle_{p_0} = \langle_0 \cap \sqrt{}_0$. Assume that $\langle_0$ is invariant under θ. Then it is clear that

$$\langle_0 = \langle_{t_0} \oplus \langle_{p_0}, \langle = \langle_t \oplus \langle_p$$

where $\langle_t$ and $\langle_p$ are respectively the complexifications of $\langle_{t_0}$ and $\langle_{p_0}$. Note that $\theta : \} \to \}$ is a Lie algebra automorphism and satisfies $\theta(X + Y) = X - Y$ for $X \in \approx, Y \in \sqrt{}$. It follows that $\theta^2 = I$ where I is the identity operator on $\}$. Let α be a root of $(\}, \langle)$. α assumes real values on $\langle_{p_0}$ and pure imaginary values on $\langle_{t_0}$. Suppose α vanishes identically on $\langle_{p_0}$. Then consider the equation

$$[H_1 + H_2, X_\alpha] = (\alpha(H_1) + \alpha(H_2))X_\alpha = \alpha(H_1)X_\alpha, H_1 \in \langle_{t_0}, H_2 \in \langle_{p_0}$$

Applying the automorphism θ to both sides gives us

$$[H_1 - H_2, \theta(X_\alpha)] = \alpha(H_1)\theta(X_\alpha)$$

for all $H_1 \in \langle_{t_0}$ and $H_2 \in \langle_{p_0}$. Changing H_2 to $-H_2$ gives

$$[H_1 + H_2, \theta(X_\alpha)] = \alpha(H_1)\theta(X_\alpha)$$

It follows that since the root space corresponding to each root is one dimensional, $\theta(X_\alpha) = c_\alpha.X_\alpha$ for some non-zero constant c_α. Another application of θ gives

$$X_\alpha = \theta^2(X_\alpha) = c_\alpha\theta(X_\alpha) = c_\alpha^2 X_\alpha$$

and hence $c_\alpha^2 = 1$ or $c_\alpha = \pm 1$. If $c_\alpha = 1$, it is clear that $X_\alpha \in \approx$ and if $c_\alpha = -1$, then $X_\alpha \in \sqrt{}$. In the former case, we say that α is a compact root and in the latter case, we say that α is a non-compact root.

Reference:Collected papers of HarishChandra.

[120] Let $(V, < .,. >)$ be a finite dimensional inner product space, let G be a compact Lie group and let $\pi : G \to GL(V)$ be a representation of G in V. Define another product $< .,. >_1$ on V so that

$$< \xi, \eta >= \int_G < \pi(g)\xi, \pi(g)\eta > dg, \xi, \eta \in V$$

where dg is the Haar measure on G. Then show that $< .,. >_1$ is an invariant inner product on V in the sense that $< \pi(g)\xi, \pi(g)\eta >_1=< \xi, \eta >_1$.

[121] Show by means of an example that if a Lie group is non-compact, then a representation of the group may not have an invariant inner product.

[122] Let G_0 be a real semisimple Lie group and let $\}_0$ be its Lie algebra. Let $\}$ be the complexification of $\}_0$ and let η denote the conjugation of $\}$ with respect to $\}_0$. Let $\sqcap$ be a compact real form of $\}$ that is invariant under η. Let $\tilde{\theta}$ denote the conjugation of $\}$ with respect to $\sqcap$. Set $\theta = \tilde{\theta}o\eta$. Then θ is an involutive automorphism of $\}$. Let $\approx_0 = \sqcap \cap \}_0$ and $\sqrt{}\rho = (i\sqcap) \cap \}_0$. Let $\approx$ and $\sqrt{}$ denote respectively the complexifications of $\approx_0$ and $\sqrt{}\rho$. Let $B(X, Y) = Tr(ad(X).ad(Y)), X, Y \in \}$. We have for $g \in G_c$ (G_c is the complex analytic Lie group corresponding to $\}$)

$$B(Ad(g)(X), Ad(g)(Y)) = B(X, Y), g \in G, X, Y \in \}$$

We have used the fact that

$$ad(Ad(g)(X))(Y) = [Ad(g)(X), Y] = Ad(g)([X, Ad(g)^{-1}(Y)]) = Ad(g)oad(X)oAd(g)^{-1}$$

Let $< .,. >$ be any inner product on $\}$ that is invariant with respect to U, the Lie group corresponding to $\sqcap$. ie,

$$< Ad(g)(X), Ad(g)(Y) >=< X, Y >, g \in \mathcal{U}, X, Y \in \}$$

Such an inner product exists since U is a compact group. Then it follows that if $X \in \sqcap$, $ad(X)$ is skew-Hermitian with respect to the inner product $< .,. >$. This follows from the identity

$$\frac{d}{dt} < Ad(exp(tX))(Y), Ad(exp(tX))(Z) > |_{t=0} =< ad(X)(Y), Z > + < Y, ad(X)(Z) >$$

Hence, all the eigenvalues of $ad(X)$ are pure imaginary if $X \in \sqcap$. Thus for $X \in \sqcap$, we have $B(X, X) = Tr(ad(X)^2) \le 0$, and hence $-B$ is positive semidefinite on $\sqcap$. Since $\sqcap$ is a semisimple Lie algebra, it follows that B is non-degenerate on $\sqcap \times \sqcap$. Hence $-B$ is positive definite on $\sqcap$. Since $\sqcap = \approx_0 \oplus i \sqrt{}_0$, it follows that for $X \in \approx_0$, $B(X,X) \leq 0$, for $Y \in \sqrt{}_0$, $B(Y,Y) \geq 0$. Let $X \in \approx_0, Y \in \sqrt{}_0$. Then $ad(X)$ is skew-Hermitian with respect to the inner product $< .,. >$ while $ad(Y)$ is Hermitian with respect to the same inner product. Then, with $*$ denoting adjoint with respect to the inner product $< .,. >$, we have

$$\bar{B}(X,Y) = Tr((ad(X).ad(Y))^*) = Tr(ad(Y)^*.ad(X)^*) = -Tr(ad(Y).ad(X)) = -B(X,Y)$$

and hence $B(X,Y)$ is pure imaginary. But, $X, Y \in \}_0$ and hence $B(X,Y) = Tr(ad(X).ad(Y))$ is real since $\}_0$ is a real Lie algebra. Thus, $B(X,Y) = 0$.

[123] Let $\}$ be a Lie algebra and let G be the corresponding Lie group. Let $\pi_1, \pi_2, ..., \pi_r$ be representations of G in the vector spaces $V_1, V_2, ..., V_r$ respectively and let $V_1 \otimes V_2 \otimes ... \otimes V_r$ be a tensor product of $V_1, V_2, ..., V_r$. Let $X \in \}$. Compute

$$\tilde{\pi}(X) = \frac{d}{dt}\pi_1(exp(tX)) \otimes ... \otimes \pi_r(exp(tX))|_{t=0}$$

and show that $\tilde{\pi}$ is a representation of $\}$ in $V_1 \otimes ... \otimes V_r$.

[124] What do you understand by the following terms. (a) Maximal compact subgroup of a semisimple Lie group (b) Cartan decomposition of a Lie algebra. Give an example of each.

[125] Explain how a representation of a Lie group leads to a representation of the Lie algebra of the Lie group.

[126] Compute the Haar measure for $SO(3)$ in terms of the Euler angles.

[127] Suppose G is a Lie group and $\mathfrak{g}$ is its Lie algebra. Then what is the Lie algebra of the Lie group $G \times G$.

[128] Suppose $G_1, ..., G_n$ are Lie groups with Lie algebras $\mathfrak{g}_1, ..., \mathfrak{g}_n$ respectively. Then what is the Lie algebra of the Lie group $G = G_1 \times ... \times G_n$.

hint:Work out the case when $G'_i s$ are matrix Lie groups. A typical one parameter subgroup of G will have the form

$$g(t) = (exp(tX_1), ..., exp(tX_n)), X_i \in \mathfrak{g}_i, i = 1, 2, ..., n$$

It follows that a typical element of the Lie algebra $\mathfrak{g}$ of G will have the form

$$g'(0) = (X_1, ..., X_n)$$

The composition law for G is given by

$$(g_1, ..., g_n).(h_1, ..., h_n) = (g_1 h_1, ..., g_n h_n), g_i, h_i \in G_i, i = 1, 2, ..., n$$

from which it is easily deduced that the Lie bracket in $\mathfrak{g}$ is given by

$$[(X_1, ..., X_n), (Y_1, ..., Y_n)] = ([X_1, Y_1], ..., [X_n, Y_n]), X_i, Y_i \in \mathfrak{g}_i, i = 1, 2, ..., n$$

[129] Let V be any finite dimensional vector space over $\mathbb{C}$ and let $T : V \to V$ be a linear operator. Show that $T = D + N$ where D is diagonable, N is nilpotent and $DN - ND = 0$. Show that this decomposition is unique, ie, if $T = D_1 + N_1$ where D_1 is diagonable, N_1 is nilpotent and $D_1 N_1 - N_1 D_1 = 0$, then $D = D_1, N = N_1$.

[130] Let $\mathfrak{g}$ be a Lie algebra and let α be an automorphism on $\mathfrak{g}$. Then for $X, Y \in \mathfrak{g}$, we have

$$\alpha([X, Y]) = [\alpha(X), \alpha(Y)]$$

or

$$(\alpha o ad(X))(Y) = ((ad\alpha(X)) o \alpha)(Y)$$

Since this holds for every X, Y, it follows that

$$\alpha o ad(X) o \alpha^{-1} = ad(\alpha(X))$$

[131] Let $\mathfrak{g}$ be a Lie algebra and $D : \mathfrak{g} \to \mathfrak{g}$ a derivation. Then $exp(tD)$ is an automorphism of $\mathfrak{g}$ where t takes values in the underlying field $\mathbb{F}$. To prove this, let

$$F(t) = exp(tD)([X, Y]), G(t) = [exp(tD)(X), exp(tD)(Y)]$$

Then,

$$F(0) = [X, Y] = G(0)$$

and

$$F'(t) = DF(t),$$

$$G'(t) = [D.exp(tD)X, exp(tD)Y] + [exp(tD)X, D.exp(tD)Y] = D([exp(tD)X, exp(tD)Y]) = DG(t)$$

These equations imply $F(t) = G(t)$ for all t. We are assuming that $\mathbb{F}$ is either $\mathbb{R}$ or $\mathbb{C}$.

[132] Let G_1 and G_2 be two Lie groups and $T : G_1 \to G_2$ a homomorphism. Show that T induces a homomorphism $dT : \mathfrak{g}_1 \to \mathfrak{g}_2$ between the corresponding Lie algebras. Note that T preserves the group composition operation while dT preserves the Lie bracket operation.

[133] Let V, W be finite dimensional vector spaces over the same field $\mathbb{F}$ and let $T : V \to W$ be a linear transformation. Show that if $dimV > dimW$, T can be onto but not one-one. Show that if $dimW > dimV$, then T can be one-one but not onto. Show that if $dimV = dimW$, then T is one-one iff T is onto.

Chapter 8

Statistical Mechanics

Introduction: This chapter introduces partition function computation and the Boltzmann kinetic transport equation in non-equilibrium thermodynamics.

[1].Project on numerical simulation of the Boltzmann kinetic transport equation with applications to semiconductor physics using MATLAB.

Let $f(t, \mathbf{x}, \mathbf{v})$ be the particle density function in phase space at time t. This means that the number of particles that fall in the spatial volume V and have velocities in the volume V_1 at time t is given by $\int_{V \times V_1} f(t, \mathbf{x}, \mathbf{v}) d^3 x d^3 v$. Assuming no collisions, f satisfies the Boltzmann equation which corresponds to the conservation of the number of particles:

$$f_{,t} + (\mathbf{v}, \nabla_x f) + (\mathbf{F}/m, \nabla_v f) = 0$$

This equation can be numerically simulated by discretization of the spatial and velocity space. Here, $\mathbf{F} = \mathbf{F}(t, \mathbf{x}, \mathbf{v})$ is the force acting on a particle having mass m when it is located at the position $\mathbf{x}$ and has a velocity $\mathbf{v}$. For example, if motion is one dimensional,

$$f_{,t}(t, x, v) + v.f_{,x}(t, x, v)) + (F(t, x, v)/m).f_{,v}(t, x, v) = 0$$

Discretization of this equation in time, position and velocity with respective time steps of τ, Δ_x, Δ_v leads to the partial difference equation

$$(f[n+1, k, m] - f[n, k, m])/\tau + k(\Delta_v/\Delta_x)(f[n, k+1, m] - f[n, k, m])$$

$$+(F[n, k, m]/m)(f[n, k, m+1] - f[n, k, m])/\Delta_v = 0$$

where

$$f[n, k, m] = f(n\tau, k\Delta_x, m\Delta_v)$$

To implement this using MATLAB, for each time index n, we store the numbers $f[n, k, m], k = 0, 1, ..., N-1, m = 0, 1, ..., M-1$ in an array (vector) of size NM using Lexicographic ordering. We call this array of numbers as $\mathbf{f}[n]$. Then, the above difference equation can be expressed as

$$\mathbf{f}[n+1] = \mathbf{A}[n]\mathbf{f}[n]$$

where $\mathbf{A}[n]$ is a time dependent matrix. The solution to this vector difference equation is given by

$$\mathbf{f}[n] = \mathbf{A}[n-1]....\mathbf{A}[0]\mathbf{f}[0]$$

This product may be calculated iteratively using MATLAB.

[2] (a) Computation of the collision integral in the Boltzmann equation for small perturbations of the Maxwellian equilibrium distribution.

(b) Derivation of the equation of mass and momentum conservation for a fluid from the Boltzmann equation.

(c) Computation of the thermal conductivity and viscosity from the Boltzmann kinetic transport equation.

[3] Computation of the statistical partition function for various kinds of Hamiltonians. The Hamiltonian is a Hermitian operator H in a Hilbert space $\mathcal{H}$. The partition function is defined for $\beta > 0$ by

$$Z(\beta) = Tr(exp(-\beta H))$$

$T = 1/k\beta$ is the temperature where k is Boltzmann's constant. Let $\{e_i : i = 1, 2, ...\}$ be an orthonormal eigenbasis for H with corresponding eigenvalues $\{E_i : i = 1, 2, ...\}$. Then,

$$Z(\beta) = \sum_i exp(-\beta E_i)$$

The average energy of the system at temperature $T = 1/k\beta$ is given by

$$U = Z(\beta)^{-1} \sum_i E_i.exp(-\beta E_i) = -Z'(\beta)/Z(\beta) = -\frac{d}{d\beta}log(Z(\beta))$$

The quantum mechanical partition function defined above will in general differ from the classical partition function which is defined for a classical Hamiltonian $H(\mathbf{q}, \mathbf{p})$ by

$$Z_{cl}(\beta) = \int exp(-\beta H(\mathbf{q}, \mathbf{p})d^{3N}q \, d^{3N}p$$

[4] Compute the classical an quantum mechanical partition function for a one dimensional harmonic oscillator.

hint: The Hamiltonian of the oscillator is given by

$$H(q, p) = \frac{p^2}{2m} + \frac{m\omega^2 q^2}{2}$$

The quantized energy levels for this oscillator are

$$E_n = (n+1/2)h\omega/2\pi, n = 0, 1, 2, ...$$

and hence the quantum mechanical partition function is given by

$$Z(\beta) = \sum_{n=0}^{\infty} exp(-\beta(n + 1/2)h\omega/2\pi)$$

On the other hand, the classical partition function is given by

$$Z_{cl}(\beta) = \int_{-\infty}^{\infty} \int_{-\infty}^{\infty} exp(-\beta(p^2/2m + m\omega^2 q^2/2))dqdp$$

This is a standard Gaussian integral and can be computed in closed form.

[5] Suppose that the energy of a particle in a semiconductor as a function of position and momentum is $\epsilon(\mathbf{x}, \mathbf{p})$, then write down the Boltzmann equation assuming an equilibrium probability density $f_{eq}(\mathbf{x}, \mathbf{p})$.

    hint:

$$\frac{\partial f}{\partial t} + (\nabla_{\mathbf{p}}\epsilon, \nabla_{\mathbf{x}}f) - (\nabla_{\mathbf{x}}\epsilon, \nabla_{\mathbf{p}}f) = (f - f_{eq})/\tau$$

where $f = f(t, \mathbf{x}, \mathbf{p})$ is the particle distribution function in phase space as a function of time. Note that we have made use of the Hamilton equations of motion of the particle

$$d\mathbf{x}/dt = \nabla_{\mathbf{p}}\epsilon(\mathbf{x}, \mathbf{p}), dp/dt = -\nabla_{\mathbf{x}}\epsilon(\mathbf{x}, \mathbf{p})$$

τ is the relaxation time and in writing the above, we have approximated the non-linear collision term $(\frac{\partial f}{\partial t})_{coll}$ in the Boltzmann equation by $(f - f_{eq})/\tau$.

[6] Derive the equations of fluid dynamics (the mass conservation equation and the Navier-Stokes equation) starting from the Boltzmann kinetic transport equation.

    hint:Let $\chi(v)$ be a conserved quantity, ie, if v_1, v_2 are the velocities of the particle before the collision and v_1', v_2' the corresponding velocities after the collision, then $\chi(v_1) + \chi(v_2) = \chi(v_1') + \chi(v_2')$. For such a quantity, show that $\int (\frac{\partial f}{\partial t})_{coll}\chi d^3v = 0$. Multiply the Boltzmann equation by $\chi(v)$ and integrate over v using this identity and thus derive a partial differential equation satisfied by $\int f\chi d^3v$. Now successively take χ to be (a) the particle mass m, (b) the particle momentum mv and (c) the internal kinetic energy of the particle, ie, $\frac{1}{2}|v- <v>|^2$ and derive the mass conservation equation, the Navier-Stokes equation and the energy equation for the fluid.

[7] Simulate the Boltzmann transport equation for the distribution function in six dimensional phase space when the external force is derived from a constant electromagnetic field. Assume that the particles of the fluid are electrons for which the equilibrium distribution is the Fermi-Dirac statistics.

    hint:

$$f_{,t} + v_x f_{,x} + v_y f_{,y} + v_z f_{,z} - \frac{e}{m}(E_x + v_y B_z - v_z B_y)f_{,v_x} - \frac{e}{m}(E_y + v_z B_x - v_x B_z)f_{,v_y}$$

$$-\frac{e}{m}(E_z + v_x B_y - v_y B_x)f_{,v_z} = \frac{f_{eq} - f}{\tau}$$

where

$$f_{eq}(x, y, z, v_x, v_y, v_z) = \frac{1}{exp((E - E_F)/kT) + 1}$$

with

$$E = \frac{m}{2}(v_x^2 + v_y^2 + v_z^2) + e(xE_x + yE_y + zE_z)$$

Chapter 9

Functional Analysis

Introduction: The concepts discussed in this chapter include metric spaces, Riemannian manifolds, operators in Banach and Hilbert spaces and basic spectral theory in a Hilbert space.

[1].Define a metric space and give three examples.

[2].How does one convert a Riemannian manifold into a metric space.

hint:The distance between two points on a Riemannian manifold is defined as the length of the geodesic joining the two points.

[3]. Give two equivalent ways of defining a geodesic on a Riemannian manifold.

hint:The first is the curve of least length joining the two points and the second is the curve for which the tangent at any point on it is obtained by parallel displacement of the tangent at a given fixed point.

[4].Suppose V is the space of all $n \times n$ complex matrices. For $A, B \in V$, define $< A, B >= Tr(AB^*)$ where star denotes conjugate transpose. Then show that $< ., . >$ is an inner product on V.

hint:

$$Tr(AB^*) = \sum_{i,j} a_{ij}\bar{b}_{ij}, Tr(AA^*) = \sum_{i,j} |a_{ij}|^2$$

and hence

$$|Tr(AB^*)|^2 \leq Tr(AA^*).Tr(BB^*)$$

[5].State and prove the singular value decomposition.

int: Let A be a rectangular matrix of size $n \times m$. Then $A = UDV^*$ where U, V are respectively unitary matrices of sizes n and m and D is a diagonal matrix. Consider the cases $n \geq m$ and $n < m$ separately.

[6].Prove the polar decomposition :If A is a complex square matrix, then $A = UH$ where U is a unitary matrix and H is a positive semidefinite matrix.

hint:Consider the spectral decomposition of the positive semidefinite matrix A^*A.

[7].Define a normed linear space and a Banach space. Show that a Hilbert space is a special case of a Banach space. Give an example of a Banach space that is not a Hilbert space.

[8].Prove the following version of the triangle inequality:Suppose $f_i(t), i = 1, 2$ are in $L^2([0, T])$. Then,

$$\left(\int_0^T |f_1(t) + f_2(t)|^2 dt\right)^{1/2} \leq \left(\int_0^T |f_1(t)|^2 dt\right)^{1/2} + \left(\int_0^T |f_2(t)|^2 dt\right)^{1/2}$$

[9].Prove the following version of Holder's inequality:Suppose $f_i(t), i = 1, 2$ are measurable functions over $[0, T]$ and $p, q > 1$ are such that $1/p + 1/q = 1$. Then,

$$\left|\int_0^T f_1(t)\bar{f}_2(t)dt\right| \leq \left(\int_0^T |f_1(t)|^p dt\right)^{1/p}\left(\int_0^T |f_2(t)|^q dt\right)^{1/q}$$

[10].Prove using Holder's inequality the following version of the triangle inequality:Suppose $f_i(t), i = 1, 2$ are measurable functions over $[0, T]$ and $p \geq 1$. Then,

$$\left(\int_0^T |f_1(t) + f_2(t)|^p dt\right)^{1/p} \leq \left(\int |f_1(t)|^p dt\right)^{1/p} + \left(\int_0^T |f_2(t)|^p dt\right)^{1/p}$$

[11] Define a contraction mapping on a metric space. State and prove the Banach fixed point theorem for such such a mapping on a complete metric space.

hint: Let (X, d) be a complete metric space and let $T : X \to X$ be a contraction mapping so that there exists an $\alpha \in (0, 1)$ such that $d(Tx, Ty) \leq \alpha d(x, y)$ for all $x, y \in X$. Let $x_0 \in X$ be arbitrary. Prove that $T^n x_0, n = 0, 1, 2, \dots$ is a Cauchy sequence in X and show that its limit is independent of the choice of x_0.

[12] Explain how using Banach's fixed point theorem that an integral equation of the form

$$y(t) = x(t) + \int_0^t K(t, s)y(s)ds, t \in [a, b]$$

has a unique solution for y under certain conditions on the functions x, K.

[13] Let (X, d) be a complete metric space and $T : X \to X$ a continuous mapping such that for some positive integer N, $T^N : X \to X$ is a contraction mapping. Then by the Banach fixed point theorem, $lim_{m\to\infty} T^{Nm} x_0$ converges to a unique fixed point y of T^N for any $x_0 \in X$. Then,

$$Ty = T.lim_m T^{Nm} x_0 == lim_m T^{Nm+1} x_0 = lim_m T^{Nm} Tx_0 = y$$

proving that y is a fixed point of T.

[14] Define the spectrum of a linear operator in a normed linear space.

[15] Give an example of a linear operator in a vector space that has no eigenvalues in the complex field. Can this be true for a finite dimensional vector space.

[16] Define the norm of a linear functional on a normed linear space. State and prove the principle of uniform boundedness.

[17] Prove using the Hahn-Banach theorem that if X is a normed linear space and $x \in X$, then there exists a bounded linear functional g on X such that $\| g \| = 1$ and $g(x) = \| x \|$.

hint: Let $W = \mathbb{C}.x = span\{x\}$ and define a linear functional g on W so that $g(cx) = c \| x \|$ for all $c \in \mathbb{C}$. Then, $|g(\xi)| = \| \xi \|$ for all $\xi \in W$. It follows from the Hahn-Banach theorem on extension of linear functionals that g can be extended to a linear functional f on X so that $|f(\xi)| \leq \| \xi \|$ for all $\xi \in X$. Since $g = f$ on W, it follows that $\| g \| = 1$ and $g(x) = \| x \|$

[18] Suppose T is a bounded linear operator in a Banach space X. Show that

$$(I - aT)^{-1} = \sum_{n=0}^{\infty} a^n T^n$$

where a is a complex number such that $|a| \leq \| T \|^{-1}$. The convergence of the above series is in the operator norm.

[19] Let T be a linear invertible operator in a Banach space X and let A be a linear operator in X. Consider the operator $S = T + A$. We have

$$S^{-1} = (I + T^{-1}A))^{-1}T^{-1}$$

so that

$$\| S^{-1} \| \leq \| T^{-1} \| \sum_{n=0}^{\infty} \| T^{-1}A \|^n$$

$$\leq \frac{\| T^{-1} \|}{1 - \| T^{-1}A \|}$$

provided that

$$\| T^{-1}A \| < 1$$

[20] Show that if $\| . \|$ is the operator norm in a Banach space X, then if T is the product of the operators A_i appearing n_i times with $i = 1, 2, ..., r$ taken in any order, then

$$\| T \| \leq \Pi_{i=1}^{r} \| A_i \|^{n_i}$$

[21] Let X be a Banach space and T a linear operator in X. We say that T is closed if whenever u_n is a sequence in X such that $u_n \to u$ and $Tu_n \to v$, then $v = Tu$. Show that T is closed iff $Gr(T)$ is closed where

$$Gr(T) = \{(u, Tu) : u \in X\}$$

[22] Consider an Hermitian operator H in a Hilbert space $\mathcal{H}$ with spectral representation

$$H = \int_a^b x.dE(x)$$

where $\{E(x)\}$ is the spectral family associated with H. Show that if f is a well bounded function on $\mathbb{R}$, then

$$f(H) = \int_a^b f(x).dE(x)$$

Show that the spectral norm of $f(H)$ is given by

$$\| f(H) \| = sup\{|f(x)| : x \in [a, b]\}$$

Show that for $y \in \mathbb{R}$,

$$|H - y| = \int_a^b |x - y| dE(x)$$

and hence

$$U(y) = |H - y|^{-1}(H - y) = \int_a^b sgn(x - y)dE(x) = \int_{-\infty}^{\infty} sgn(x - y)dE(x) = I - E(y) - E(y - 0)$$

Show that

$$U(y)^2 + U(y) = (I - E(y) - E(y - 0))^2 + (I - E(y) - E(y - 0))$$

$$= I + E(y) + E(y - 0) - 2E(y) - 2E(y - 0) + 2E(y - 0) - I - E(y) - E(y - 0) = I - 2E(y)$$

and hence

$$E(y) = I - (U(y) + U(y)^2)/2$$

This formula tells us how to compute the spectral family associated with a Hermitian operator H in a Hilbert space from the polar decomposition of the operators $H - yI, y \in \mathbb{R}$.

[23] Let T be a closed linear operator in a Banach space X. Suppose $x \in P(T)$ where $P(T)$ is the resolvent set of T. Then, $R(T, x) = (T - xI)^{-1}$ is in $\mathcal{B}(X)$, where $\mathcal{B}(X)$ is the algebra of bounded linear operators in X. Suppose y is a real number. Then show that

$$R(T, y) = (T - yI)^{-1} = (T - xI + (x - y)I)^{-1} = \sum_{n \geq 0} (y - x)^n (xR(T, x)^{n+1}$$

with convergence in the operator norm provided that $|y - x| \leq \parallel R(T,x) \parallel^{-1}$. Deduce that the resolvent set of an operator is an open set and hence the spectrum of an operator (the complement of the resolvent) is a closed set.

[24] Let X be a Banach space and let D be a dense linear manifold of X. Suppose $T : D \to X$ is bounded. Then show that T admits a unique extension $\tilde{T} : X \to X$.

hint:For $x \in X$ choose a sequence $\{x_n\} \subset D$ such that $\parallel x_n - x \parallel \to 0$. Show that $\{Tx_n\}$ is a Cauchy sequence in X and hence there exists a $y \in X$ such that $\parallel Tx_n - y \parallel \to 0$. Define $\tilde{T}x = y$. Show that $\tilde{T}$ is a well defined bounded linear operator in X.

[25] Project on functional analysis: This project is based on extending the notions of linear algebra to infinite dimensional vector spaces, applying it to study differential and partial differential equations in mathematical physics especially problems involving the perturbation of the spectrum of a linear operator when the operator is perturbed which is important in quantum mechanics.

[26] Define the notion of adjoint of a linear operator in a Hilbert space.

Suppose X is a Hilbert space and let Y be another Hilbert space. Let T be a linear operator from $D(T) \subset X$ into Y. Assume that $D(T)$ is dense in X. The graph of T is given by

$$G(T) = \{(x, Tx) : x \in D(T)\} \subset X \times Y$$

For $E \subset X \times Y$, define $E' \subset Y \times X$ so that

$$E' = \{(y, x) : (x, y) \in E\}$$

$$Let(\xi, \eta) \in \{G(T)'\}^{\perp}$$

Then $(\xi, \eta) \perp (Tx, x)$ for all $x \in D(T)$. It follows that for all $x \in D(T)$, we have $< Tx, \xi > + < x, \eta >= 0$. In particular, the map $x \to < Tx, \xi >$ from $D(T)$ into $\mathbb{C}$ is a bounded linear functional and since $D(T)$ is dense in X, it follows that this map can be extended uniquely to a bounded linear functional f on X. It follows from the Riesz representation theorem that $f(x) = < x, T^*(\xi) >$ for all $x \in X$ where $T^*(\xi)$ is an element in X uniquely determined by T, ξ. It follows therefore that $< x, T^*(\xi) >= - < x, \eta >$ for all $x \in D(T)$ and hence $\eta = -T^*(\xi)$. It is easily seen that the map T^* is linear. We have from the above that $< Tx, \xi >=< x, T^*\xi >$ for all $x \in D(T)$ and all $\xi \in D(T^*)$ where $D(T^*)$ is the set of all $\xi \in Y$ for which $(\xi, eta) \in \{G(T)'\}^{\perp}$ for some $\eta \in Y$. Further, from the above discussion, it follows that

$$G(T^*) = \{(\xi, T^*\xi) : \xi \in D(T^*)\} = \{(\xi, -\eta) : (\xi, \eta) \in \{G(T)'\}^{\perp}\} = \{G(-T)'\}^{\perp}$$

and hence $G(T^*)$ is closed in $Y \times X$.

[27] Study the notions of weak and strong convergence in a Banach space. Let X be a Banach space and X^* the space of all bounded linear functionals on X. A sequence $\{x_n\}$ in X is said to converge weakly to $x \in X$ if for all $f \in X^*$, $f(x_n) \to f(x)$. The same sequence is said to converge strongly to x if $\parallel x_n - x \parallel \to 0$ as $n \to \infty$. Strong convergence implies weak convergence since

$$|f(x_n) - f(x)| = |f(x_n - x)| \leq \parallel f \parallel \cdot \parallel x_n - x \parallel$$

for all $f \in X^*$.

Reference:T.Kato, perturbation theory for linear operators, Springer.

[28] Suppose $\mathcal{H}$ is a separable Hilbert space. Then show that $\mathcal{H}$ has a countable orthonormal basis.
hint:Make use of the Gram-Schmidt orthonormalization process.

[29] Solve the differential equation

$$-u''(x) - \zeta u(x) = y(x), 0 \leq x \leq 1, u(0) = 0, u'(1) + \chi u(1) = 0$$

Note that if $T(\chi)$ is the differential operator $T(\chi)u = -u''$ with the above boundary condition, then the solution to u can be written as $(T(\chi) - \zeta I)^{-1}y$, ie $R(\zeta, \chi)y$, where $R(\zeta, \chi)$ is the resolvent of $T(\chi)$.

hint: Taking unilateral Laplace transforms, we have

$$-s^2\hat{u}(s) + su(0) + u'(0) - \zeta\hat{u}(s) = \hat{y}(s)$$

so that

$$\hat{u}(s) = -\frac{\hat{y}(s)}{s^2 + \zeta} + \hat{f}(s)$$

where $f(t)$ is the general solution to the associated homogeneous equation. Thus,

$$u(x) = A.cos(\sqrt{\zeta}x) + B.sin(\sqrt{\zeta}x) - \frac{1}{\sqrt{\zeta}} \int_0^x sin(\sqrt{\zeta}(x - x'))y(x')dx'$$

The condition $u(0) = 0$ implies $A = 0$ and B is then determined from the condition $u'(1) + \chi u(1) = 0$.

[30] Let $\mathcal{H}$ be a Hilbert space and let $T : D(T) \rightarrow \mathcal{H}$ be a linear operator in $\mathcal{H}$, where $D(T)$ is a dense linear manifold in $\mathcal{H}$. Let $\mathcal{S}$ denote the set of all $y \in \mathcal{H}$ for which the mapping $x \rightarrow < Tx, y >$ from $D(T)$ into $\mathbb{C}$ is continuous. Let $y \in \mathcal{S}$ and then extend the continuous linear functional $x \rightarrow < Tx, y >$ from $D(T)$ to the whole of H. Denote this linear functional by f_y. f_y is a continuous linear functional on $\mathcal{H}$. By the Riesz representation theorem, there exists a unique $T^*(y) \in \mathcal{H}$ such that $f_y(x) = < x, T^*(y) >$ for all $x \in \mathcal{H}$. In particular, $< Tx, y >=< x, T^*(y) >$ for all $x \in D(T)$ and $y \in \mathcal{S}$. It is easy to see that T^* is a linear operator in $\tilde{S}$ so that $\tilde{S} = D(T^*)$. Note that uniqueness of T^* follows from the fact that if $< x, y_1 >=< x, y_2 >$ for all $x \in D(T)$, then since $\bar{D}(T) = H$, we have that $< x, y_1 >=< x, y_2 >$ for all $x \in \mathcal{H}$ which implies that $y_1 = y_2$. T^* is called the adjoint of the operator T.

[31] Definition of a closed operator. Let X, Y be two Banach spaces and $T : X \rightarrow Y$ be a linear operator. We define $Gr(T)$ to be the Linear manifold in $X \times Y$ consisting of all elements of the form $(x, T(x)), x \in X$. $Gr(T)$ is called the graph of the operator T. Suppose $x_n \in X$ is a sequence such that $x_n \rightarrow x$ and $T(x_n - x_m) \rightarrow 0$ as $n, m \rightarrow \infty$. This is equivalent to assuming that both x_n and $T(x_n)$ are convergent. If this implies that $T(x_n - x) \rightarrow 0$ no matter what the sequence x_n in X is, then we say that T is a closed operator. It is easy to see that T is a closed operator iff $Gr(T)$ is a closed subset of $X \times Y$.

A linear operator $T : X \to Y$ is said to be closable if $\bar{G}r(T)$ is the graph of a linear operator $\tilde{T} : X \to Y$. For T to be closable, it is necessary that $x_n \to 0$ and $T(x_n) \to y$ should imply $y = 0$.

[32] What do you understand by an equivalence relation in a set. What do you understand by the equivalence classes generated by an equivalence relation. Prove that the equivalence classes are pairwise disjoint.

[33] Suppose T is a linear operator in a Banach space X. Let $\| T \|$ denote the operator norm of T, ie, $\| T \| = sup \| Tx \| / \| x \|$, the supremum being taken over all nonzero $x \in X$. Show that if $\| T \| < 1$, then the series

$$\sum_{n=0}^{\infty} T^n$$

converges in operator norm to $(I - T)^{-1}$. Deduce that if T is a linear operator in X and α is a complex number such that $\| I - \alpha T \| < 1$, then

$$T^{-1} = \alpha \sum_{n=0}^{\infty} (I - \alpha.T)^n$$

the convergence being in the operator norm.

[34] Let $\mathcal{H}$ be a Hilbert space and H a self-adjoint operator in $\mathcal{H}$. Show that if the spectral representation of H is given by

$$H = \int_{-\infty}^{\infty} \lambda.dE(\lambda)$$

and if ϕ is a bounded measurable function on $\mathbb{R}$, then the operator

$$\phi(H) = \int \phi(\lambda).dE(\lambda)$$

is defined and satisfies

$$< \phi(H)x, y > = \int_{-\infty}^{\infty} \phi(\lambda)d < E(\lambda)x, y >$$

and if ϕ_1, ϕ_2 are two bounded measurable functions on $\mathbb{R}$, then

$$< \phi_1(H)x, \phi_2(H)y > = \int_{-\infty}^{\infty} \phi_1(\lambda)\bar{\phi}_2(\lambda)d < E(\lambda)x, y >$$

[35] Let T be a bounded linear operator in a Banach space X. Let $\alpha = limsup_{n \to \infty} \| T^n \|^{1/n}$. Show that if $|z| < \alpha^{-1}$, then the operator $\sum_{n=0}^{\infty} z^n T^n$ is defined with convergence being in the operator norm.

[36] Let X be a Banach space and M a closed subspace of X, for $x \in X$, define

$$d(x, M) = inf\{\| x - y \| : y \in M\}$$

Show that (a) $d(x, M) = 0$ iff $x \in M$, (b) $|d(x, M) - d(y, M)| \leq \| x - y \|$ for all $x, y \in M$.

hint: Let $z \in M$. Then

$$\| x - z \| \leq \| x - y \| + \| y - z \|$$

Take infimum on both sides over all $z \in M$ to get the second part. For the first part, choose a sequence $x_n \in M$ such that $\| x - x_n \| < 1/n$. Then $\{x_n\}$ converges to x and since M is closed, it follows that $x \in M$.

[37] Using Zorn's lemma, prove the Hahn-Banach theorem, namely that if X is a Banach space, M is a linear manifold of X and f is a linear functional on M such that $|f(x)| \leq K \| x \|$ for all $x \in M$ and some finite K, then there exists a linear functional g on X such that $g(x) = f(x)$ for all $x \in M$ and $|g(x)| \leq K \| x \|$ for all $x \in X$.

Reference:Kreyszig, Functional Analysis.

[38] Lagrange interpolation:Given n distinct points $t_1, ..., t_n$ in $\mathbb{C}$, construct n polynomials $p_i(t), i = 1, 2, ..., n$ having degree at most $n - 1$ such that $p_i(t_j) = \delta_{ij}$ for $1 \leq i, j \leq n$. Show that given any polynomial g of degree at most $n - 1$, we have

$$g(t) = \sum_{i=1}^{n} g(t_i)p_i(t)$$

[39] Consider the differential operator L such that $Lf = (p.f')' + qf$ where p, q are smooth functions. Assume that this operator acts on smooth functions f defined on $[a, b]$. Calculate L^* assuming appropriate boundary conditions.

hint:

$$< g, Lf >= \int_a^b g((pf')' + qf)dx = g(b)p(b)f'(b) - g(a)p(a)f'(a) + \int_a^b (qf - pf'g')dx$$

$$= g(b)p(b)f'(b) - g(a)p(a)f'(a) - p(b)g'(b)f(b) + p(a)g'(a)f(a) + \int_a^b f(q + (pg')')dx$$

[40] If $\mathcal{H}$ is a Hilbert space and T is a densely defined operator in $\mathcal{H}$ such that $T \subset T^*$, then T is said to be a symmetric operator. Show that a densely defined operator T is symmetric iff

$$< Tx, y >=< x, Ty >, x, y \in D(T)$$

[41] Let X be a Banach space and suppose an operator T in X is invertible and closed. Then T^{-1} is also closed.

hint: Let $(x_n, T^{-1}x_n) \to (x, y)$. Then define $y_n = T^{-1}x_n$. We have $y_n \to y$ and $Ty_n = x_n \to x$ and since T is closed, it follows that $y \in D(T)$ and $x = Ty$, ie, $y = T^{-1}x$.

[42] Let $\mathcal{H}$ be a Hilbert space and let T be a linear operator in $\mathcal{H}$. Let $x_n \in D(T^*), x_n \to x$ and $T^*x_n \to y$. Then for any $z \in D(T)$, we have $< T^*x_n, z > \to < y, z >$ or $< x_n, Tz > \to < y, z >$ or $< x, Tz >=< y, z >$. Assuming that $D(T)$ is dense in $\mathcal{H}$, it follows that $x \in D(T^*)$ and $T^*x = y$. Note the uniqueness of y. For suppose $< x, Tz >=< y', z >$ also for all $z \in D(T)$. Then $y - y' \perp D(T)$ and since $D(T)$ is dense in $\mathcal{H}$, it would follow that $y = y'$. The argument above shows that T^* is a closed operator in $\mathcal{H}$.

[43] Suppose T is an operator in a Hilbert space $\mathcal{H}$ satisfying $Re < Tu, u > \geq 0$ for all $u \in \mathcal{H}$. Then for $c \in \mathbb{C}$,

$$\| (T+c)u \|^2 = \| Tu \|^2 + |c|^2 \| u \|^2 + 2Re(\bar{c} < Tu, u >)$$

Let $c = c_1 + ic_2$ with $c_1, c_2 \in \mathbb{R}$ and $c_1 > 0$. Then from the above,

$$\| (T+c)u \|^2 = \| Tu \|^2 + c_1^2 \| u \|^2 + c_2^2 \| u \|^2 + 2(c_1 Re < Tu, u > + c_2 Im < Tu, u >)$$

$$\geq c_1^2 \| u \|^2 + \| Tu \|^2 + c_2^2 \| u \|^2 - 2|c_2| \| Tu \| . \| u \|$$

since

$$|c_2 Im < Tu, u > | \leq |c_2| \| Tu \| . \| u \|$$

It follows that

$$\| (T+c)u \|^2 \geq (\| Tu \| - c_2 \| u \|)^2 + c_1^2 \| u \|^2 \geq c_1^2 \| u \|^2$$

It follows that $T + c$ is one-one. Let v be any vector in $R(T + c)$. Then, $(T + c)^{-1}v$ is a well defined vector in $\mathcal{H}$. We get from the above formula,

$$\| v \|^2 = \| (T+c)(T+c)^{-1}v \|^2 \geq c_1^2 \| (T+c)^{-1}v \|^2$$

or

$$\| (T+c)^{-1}v \| \leq \| v \| / Re(c)$$

Reference:T.Kato, Perturbation theory for linear operators.

[44] Define the following notions in general topology.

[1] Topology on a set.

[2] Open and closed sets.

[3] Continuous function from one topological space to another.

[4] Base for a topological space.

[5] Standard topology on a metric space.

[6] Product topology.

[7] Compact set in a topological space.

[8] Connected set.

[9] Path connected set.

[10] Homotopy.

[45] Define the following terms.

(a) Compact subset of a topological space, (b) locally compact topological space, (c) countable base for a topological space, (d) Hausdorff topological space, (e) Normal topological space.

(a) Let (X, τ) be a topological space. A set $A \subset X$ is said to be compact if given any open covering $\{O_\alpha : \alpha \in \Gamma\}$ of E (ie, $O_\alpha \in \tau$ and $E \subset \bigcup_{\alpha \in \Gamma} O_\alpha$, there is a finite subcovering, (ie a finite integer n and $\alpha_1, ..., \alpha_n \Gamma$ such that $E \subset \bigcup_{i=1}^{n} O_{\alpha_i}$). (b) A topological space (X, τ) is said to be locally compact if given any $x \in X$, there exists a $U \in \tau$ such that $x \in U$ and $\bar{U}$ is compact where $\bar{U}$ denotes closure of U. Note that if E is a subset of X, then $\bar{E}$ is the set of all points $x \in X$ for which given any open set O containing x, we have $O \cap E \neq \phi$. (c) A base for a the topological space (X, τ) is a collection of open sets $\{B_\alpha : \alpha \in \Gamma\}$ such that given any $x \in X$ and an open subset U of X containing x, there exists an $\alpha \in \Gamma$ such that $x \in B_\alpha \subset U$. It follows from this immediately, that given any base $\{B_\alpha : \alpha \in \Gamma\}$ for (X, τ) and any open set U (ie, $U \in \tau$), there exists a set $I \subset \Gamma$ such that $O = \bigcup_{\alpha \in \Gamma} B_\alpha$. A base $\{B_\alpha : \alpha \in \Gamma\}$ for which Γ is a countable set is called a countable base for the topological space. A topological space (X, τ) is called a Hausdorff space if given any $x, y \in X$ such that $x \neq y$, there exist $U, V \in \tau$ such that $x \in U, y \in V$ and $U \cap V = \phi$. (e) The topological space (X, τ) is said to be normal if given any two disjoint closed subsets A, B of X, there exist disjoint open subsets U, V of X, such that $A \subset U$ and $B \subset V$.

Suppose X is a locally compact topological space and has a countable base $\{B_n : n = 1, 2, ...\}$. Let $x \in X$ be arbitrary. Then there exists an open set U_x such that $x \in U_x$ and $\bar{U}_x$ is compact. Further there exists a positive integer $n(x)$ such that $x \in B_{n(x)} \subset U_x$. Hence, $\bar{B}_{n(x)}$ is compact. Note that $I = \{n(x) : x \in X\}$ is a subset of the positive integers and is hence countable. Let now U be any open set and $x \in U$. Then there exists an integer k such that $x \in B_k \subset U$. It follows that $x \in B_{n(x)} \cap B_k \subset U$ and since the closure of $B_{n(x)} \cap B_k$ is contained in the closure of $B_{n(x)}$, it follows that the closure of $B_{n(x)} \cap B_k$ is compact. It follows that $\{B_m \cap B_k : m \in I, n = 1, 2, ...\}$ is a countable base, each member of which has a compact closure. We have thus shown that given any locally compact space having a countable base, there exists a countable base for the same space, each member of which has compact closure.

[46] Let X, Y be Banach spaces and let T be a linear operator from $D(T) \subset X$ into Y. Here, $D(T)$ is a linear manifold. $D(T)$ is called the domain of T. The graph $Gr(T)$ of T is defined to be the set $\{(x, Tx) : x \in D(T)\}$. It is easy to see that $Gr(T)$ is a linear manifold, with $Gr(T) \subset X \times Y$. We say that T is a closed operator if $Gr(T)$ is closed in $X \times Y$. Let $x_n \in D(T), n = 1, 2, ...$ be such that $x_n \to x \in X$ and $Tx_n \to y \in Y$. If $Gr(T)$ is closed, then $(x, y) \in Gr(T)$ and hence $y = Tx$. Conversely, if for every sequence $x_n, n = 1, 2, ...$ in $D(T)$ with $x_n \to x$ and $Ty_n \to y$, if $y = Tx$, then $Gr(T)$ is closed. Thus, T is closed iff $x_n \in D(T), x_n \to x \in X$ and $Ty_n \to y \in Y$ imply $y = Tx$. It is clear that every bounded linear operator is closed.

Problem:Give an example of a closed linear operator that is not bounded.

[47] Let X be a Banach space and M a proper closed subspace of X. Let $x \in X - M$ so that $\| x \| = 1$. Let f be a linear functional on $span\{x, M\}$ such that $f(x) = 1, f(y) = 0, y \in M$. Then, by the Hahn-Banach theorem, f can be extended to the whole of X so that $\| f \| = 1$. Let $dist(x, M) = inf\{\| x - y \|: y \in M\} = \epsilon$. Since M is closed, $\epsilon > 0$. Let $y_n \in M$ be a sequence such that $\| x - y_n \| < \epsilon + 1/n, n = 1, 2, ...$Then, let $z_n = (x - y_n)/\epsilon$. It follows that

$$1 \leq \| z_n \| < 1 + 1/n\epsilon,$$

$$dist(z_n, M) = dist(x - y_n)/\epsilon, M) = dist(x, M)/\epsilon = 1$$

Define $u_n = z_n/ \| z_n \|$. Then,
$$\| u_n \| = 1,$$

$$dist(u_n, M) = dist(z_n, M)/ \| z_n \| \in (1/(1 + 1/n\epsilon), 1]$$

In particular, $\| u_n \| = 1$ for all n and $lim_n dist(u_n, M) = 1$. Note further that

$$f(z_n) = 1/\epsilon, f(u_n) = 1/ \| z_n \|$$

and hence

$$lim f(u_n) = 1$$

[5] Let X be a Banach space and $x \in X$ such that $\| x \| = 1$. Let f be any bounded linear functional on X. Then, $|f(x)| \leq \| f \|$ and hence if $x \in X$ is regarded as a linear functional l_x on X^*, the dual to X, so that $l_x(f) = f(x), f \in X^*$, then $\| l_x \| \leq 1$. Now define $f \in X^*$ so that $f(x) = 1$ and extend f to the whole of X so that $\| f \| = 1$, then $l_x(f) = f(x) = 1$ which implies that $\| l_x \| = 1$.

Chapter 10

Optimal Control of Dynamical Systems

Introduction: This chapter discusses optimal control of dynamical systems described by ordinary and partial differential equations.

[1]. Implementing the optimal control equations for putting a satellite into space. Let $\mathbf{y}(t)$ denote the position and velocity components of the satellite and $\mathbf{u}(t)$ the fuel input to the satellite. These satisfy the differential equations

$$d\mathbf{x}(t)/dt = \mathbf{A}(t)\mathbf{x}(t) + \mathbf{G}(t)\mathbf{u}(t), \mathbf{y}(t) = \mathbf{C}(t)\mathbf{x}(t), 0 \leq t \leq T$$

The final condition is that the satellite's final phase variables $\mathbf{y}(T)$ should satisfy a constraint of the form $\mathbf{D}\mathbf{y}(T) = \mathbf{a}$ where $\mathbf{D}$ is an appropriate matrix and $\mathbf{a}$ an appropriate vector. Based on this data, we must find an input function $\mathbf{u}(t), 0 \leq t \leq T$ so that the cost

$$\int_0^T J_1(\mathbf{u}(t), \mathbf{y}(t), t)dt$$

is minimized subject to the above constraints, ie the equations of motion and the terminal condition on the satellite's phase variables. Typically, the function J has the form

$$J(\mathbf{u}(t), \mathbf{y}(t), t) = \| \mathbf{y}(t) - \mathbf{d}(t) \|^2 + \mathbf{u}(t)^T \mathbf{Q}\mathbf{u}(t)$$

The first term corresponds to the discrepancy between the actual trajectory and the desired trajectory while the second term is the fuel cost. The required optimization can be carried out using variational calculus.

[2] Synopsis of a thesis on optimal control applied to electromagnetic and quantum mechanical problems.

Optimal control theory deals with designing the input process to a dynamical system such that the output tracks a desired output for a certain duration of time and at the end of the duration, the output is close in distance to a given value or set of values. The dynamical system is usually specified by a system of first order linear/nonlinear differential equations of the form

$$\frac{dx_i(t)}{dt} = a_i(x_1(t), ..., x_n(t), u_1(t), ..., u_p(t), t), i = 1, 2, ..., n$$

which in vector notation can be expressed as

$$\frac{d\mathbf{x}(t)}{dt} = \mathbf{a}(\mathbf{x}(t), \mathbf{u}(t), t)$$

$\mathbf{u}(t) = (u_1(t), ..., u_p(t))$ is the input process and $\mathbf{x}(t) = (x_1(t), ..., x_n(t))$ is the state process. The problem of designing the input $\mathbf{u}(t)$ over the duration $[0, T]$ so that $\mathbf{x}(t)$ tracks a given process and simultaneously the input energy conserved is a minimum can be generalized to the problem of minimizing a cost functional

$$\int_0^T C(\mathbf{x}(t), \mathbf{u}(t), t)dt$$

subject to the dynamical equations. If in addition, a constraint is imposed on the state $\mathbf{x}(T)$ at the end point in the form $\psi(\mathbf{x}(T), T) = \mathbf{0}$, then the functional to be minimized can be expressed using a Lagrange multiplier as

$$\int_0^T C(\mathbf{x}(t), \mathbf{u}(t), t)dt + \int_0^T \lambda(t)^T(\mathbf{x}'(t) - \mathbf{a}(\mathbf{x}(t), \mathbf{u}(t), t))dt + \mu^T \psi(\mathbf{x}(T), T)$$

The minimization of this functional of the state process $\mathbf{x}(t)$, the input process $\mathbf{u}(t)$ and the Lagrange multipliers $\mu, \lambda(.)$ leads to a system of first order differential equations which are in a certain sense, a generalization of the Euler-Lagrange differential equations of variational calculus. The aim of this thesis to develop analogous equations for electromagnetism and quantum mechanics. Maxwell's equations inside a finite region of space-time constitute a system of partial differential equations for the electromagnetic field with the inputs being the charge and current density. We wish to design the charge and current density fields subject to the charge conservation equation such that the electromagnetic field inside the box tracks a given pattern. Let $\mathbf{E}_d(t, \mathbf{x})$ and $\mathbf{H}_d(t, \mathbf{x})$ denote the desired electric and magnetic fields inside the box and $\mathbf{E}(t, \mathbf{x})$ and $\mathbf{H}(t, \mathbf{x})$ are the actual electric and magnetic fields. $\rho(t, \mathbf{x})$ and $\mathbf{J}(t, \mathbf{x})$ are the charge density and current density fields which are controllable by us subject to the charge conservation equation. Incorporating Lagrange multiplier fields, $\lambda_i(t, \mathbf{x}), i = 1, 2, 3, 4$, the functional to be minimized can be expressed as

$$\int_{[0,T] \times V} \| \mathbf{E}(t, \mathbf{x}) - \mathbf{E}_d(t, \mathbf{x}) \|^2 dt d^3 x$$

$$+ \int_{[0,T] \times V} \| \mathbf{H}(t, \mathbf{x}) - \mathbf{H}_d(t, \mathbf{x}) \|^2 dt d^3 x$$

$$+ \int_{[0,T] \times V} \lambda_1(t, \mathbf{x})(\nabla.\mathbf{E}(t, \mathbf{x}) - \rho(t, \mathbf{x})/\epsilon)dt d^3 x$$

$$+ \int_{[0,T] \times V} \lambda_2(t, \mathbf{x}).(\nabla \times \mathbf{E}(t, \mathbf{x}) - \mu \mathbf{H}_{,t}(t, \mathbf{x}))dt d^3 x$$

$$+ \int_{[0,T]\times V} \lambda_3(t,\mathbf{x})\nabla.\mathbf{H}(t,\mathbf{x})dtd^3x$$

$$+ \int_{[0,T]\times V} \lambda_4(t,\mathbf{x}).(\nabla \times \mathbf{H}(t,\mathbf{x}) - \mathbf{J}(t,\mathbf{x}) - \epsilon.\mathbf{E}_{,t}(t,\mathbf{x}))dtd^3x$$

Here λ_1 and λ_3 are scalar fields while λ_2 and λ_4 are vector fields, each having three components. The problem of tracking the electromagnetic field may be simplified considerably by considering the electromagnetic four potential and the equation it obeys under the Lorentz gauge condition. The equations are

$$\nabla^2 \mathbf{A} - \frac{1}{c^2}\mathbf{A}_{,tt} = -\mu_0\mathbf{J}$$

$$\nabla^2 \Phi - \frac{1}{c^2}\Phi_{,tt} = -\rho/\epsilon_0$$

The problem of field tracking can be reduced to the problem of tracking a four potential. The Lorentz gauge condition $\Phi_{,t} + c^2\nabla.\mathbf{A} = 0$ in conjunction with the above two equations implies charge conservation $\rho_{,t} + \nabla.\mathbf{J} = 0$. Hence the potential tracking problem can be cast as a variational problem of minimizing the functional

$$\int_{[0,T]\times V} \| \Phi(t,\mathbf{x}) - \Phi_d(t,\mathbf{x}) \|^2 \, dtd^3x$$

$$+ \int_{[0,T]\times V} \| \mathbf{A}(t,\mathbf{x}) - \mathbf{A}_d(t,\mathbf{x}) \|^2 \, dtd^3x$$

$$+ \int_{[0,T]\times V} \lambda_1(t,\mathbf{x})^T(\nabla^2\mathbf{A}(t,\mathbf{x}) - \frac{1}{c^2}\mathbf{A}_{,tt}(t,\mathbf{x}) + \mu_0\mathbf{J}(t,\mathbf{x}))dtd^3x$$

$$+ \int_{[0,T]\times V} \lambda_2(t,\mathbf{x})^T(\nabla^2\Phi(t,\mathbf{x}) - \frac{1}{c^2}\Phi_{,tt}(t,\mathbf{x}) + \rho(t,\mathbf{x})/\epsilon_0)dtd^3x$$

For the purpose of obtaining dynamical equations for the optimal potentials, it is more convenient to cast this optimal control problem in a way that only first order derivatives in time appear. This can be done by introducing additional fields $\mathbf{U} = \mathbf{A}_{,t}$ and $\psi = \Phi_{,t}$. Then the equations for the potentials become

$$\mathbf{A}_{,t} = \mathbf{U}, \mathbf{U}_{,t} = c^2\nabla^2\mathbf{A} + \mathbf{J}/\epsilon_0$$

and

$$\Phi_{,t} = \psi, \psi_{,t} = c^2\nabla^2\Phi + \rho/\epsilon_0$$

[3] Optimal control in partial differential equations. Let L be a linear partial differential operator acting on functions defined on $\mathbb{R}^n$. If f is the input and g the output, then $L(g) = f$. The cost functional has the form

$$C(f,g) = \int K(f(x),g(x))d^nx$$

We wish to choose the input f such that C is a minimum subject to the constraint $L(g) = f$. Assuming for the moment that L is invertible, this amounts to choosing the input f such that $C(f, L^{-1}(f)) =$

$\int K(f(x), L^{-1}f(x))d^n x$ is a minimum. Using Lagrange multiplier functions $\lambda(x)$, we must extremize the functional

$$S(f, g, \lambda) = \int K(f(x), g(x))dx + \int \lambda(x)^T (Lg(x) - f(x))d^n x$$

We are here assuming that

$$f(x) \in \mathbb{R}^p, g(x) \in \mathbb{R}^q, \lambda(x) \in \mathbb{R}^p$$

f, g, λ are obtained from the variational principle

$$\delta S(f, g, \lambda) = 0$$

We have

$$\delta S(f, g, \lambda) = \int (\delta f^T \nabla_f K(f, g) + \delta g^T \nabla_g K(f, g))d^n x + \int \delta \lambda^T (Lg - f)d^n x$$

$$+ \int \delta g^T L^* \lambda d^n x - \int \lambda^T \delta f d^n x$$

Where L^* is the adjoint of L defined by

$$\int (u^T Lv)d^n x = \int (v^T L^* u)d^n x$$

Setting the coefficients of $\delta f, \delta g, \delta \lambda$ to zero gives us

$$\nabla_f K(f, g) - \lambda = 0, \nabla_g K(f, g) + L^* \lambda = 0, Lg - f = 0$$

From these, we derive

$$\nabla_f K(Lg, g) - \lambda = 0, \nabla_g K(Lg, g) + L^* \lambda = 0$$

and hence,

$$\nabla_g K(Lg, g) + L^* \nabla_f K(Lg, g) = 0$$

[4] Optimal control of the heat equation with source. $u(t, x)$ is the temperature and $f(t, x)$ the heat input. D is the diffusion coefficient and u satisfies the partial differential equation

$$u_{,t}(t, x) - D.u_{,xx}(t, x) = f(t, x), t \geq 0, x \in \mathbb{R}$$

The cost functional is assumed to have the form

$$C(f, g) = \int K(u, u_{,x}, f)dtdx$$

The problem is to find the input f so that the cost is mimimized. Introducing the Lagrange multiplier function $\lambda(t, x)$ to take into account the constraints imposed by the heat equation with source, the problem amounts to extremizing the functional

$$S(f, g, \lambda) = \int K(u, u_{,x}, f)dtdx + \int \lambda.(u_{,t} - D.u_{,xx} - f)dtdx$$

We have using integration by parts,

$$\delta S = \int (K_{,u}\delta u + K_{,u_x}\delta u_{,x} + K_{,f}\delta f)dtdx$$

$$+ \int \delta\lambda.(u_{,t} - D.u_{,xx} - f)dtdx + \int \lambda.(\delta u_{,t} - D.\delta u_{,xx} - \delta f)dtdx$$

$$= \int \delta u(K_{,u} - (K_{,u_x})_{,x} - \lambda_{,t} - D\lambda_{,xx})dtdx$$

$$+ \int \delta\lambda(u_{,t} - D.u_{,xx} - f)dtdx + \int \delta f.(K_{,f} - \lambda)dtdx$$

Setting the coefficients of $\delta u, \delta\lambda, \delta f$ to zero gives us

$$K_{,u} - (K_{,u_x})_{,x} - \lambda_{,t} - D.\lambda_{,xx} = 0$$

$$u_{,t} - D.u_{,xx} - f = 0,$$

$$K_{,f} - \lambda = 0$$

References

[1] S.Helgason, Differential geometry, Lie groups and Symmetric Spaces.

[2] Collected Papers of HarishChandra, Edited by V.S.Varadarajan, Springer.

[3] K.R.Parthasarathy, An introduction to Quantum Stochastic Calculus.

[4] T.Kato, Perturbation theory for linear operators, Springer.

[5] D.W.Stroock and S.R.S.Varadhan, Multidimensional Diffusion Processes, Springer.

[6] Steven Weinberg, The Quantum Theory of Fields, Vols.I and II, Cambridge University Press.

[7] Vireshwar Sharma, M.Tech thesis, NSIT.

[8] Sudipta Mazumdar, Ph.D thesis, NSIT.

[9] Arthi Vaish, Ph.D thesis, NSIT.

Index